The Handbook of Infrared and Raman Characteristic Frequencies of Organic Molecules

The Handbook of Infrared and Raman Characteristic Frequencies of Organic Molecules

Daimay Lin-Vien
Shell Development Company
Houston, Texas

Norman B. Colthup
Stamford, Connecticut

William G. Fateley
Kansas State University
Manhattan, Kansas

Jeanette G. Grasselli
Ohio University
Athens, Ohio

ACADEMIC PRESS, INC.
Harcourt Brace Jovanovich, Publishers
Boston San Diego New York
London Sydney Tokyo Toronto

This book is printed on acid-free paper. ♾

ACADEMIC PRESS, INC.
1250 Sixth Avenue, San Diego, CA 92101-4311

United Kingdom Edition published by
ACADEMIC PRESS LIMITED
24–28 Oval Road, London NWI 7DX

Library of Congress Cataloging-in-Publication Data

The Handbook of infrared and Raman characteristic frequencies of organic molecules / Daimay Lin-Vien . . . (et al.).
p. cm.
Includes bibliographical references and index.
ISBN 0-12-451160 -0 (alk. Paper)
1. Organic compounds — Spectra. 2. Molecular spectroscopy. 3. Infrared spectroscopy. 4. Raman spectroscopy. I. Lin-Vien, Daimay.
QC462.5.H343 1991
547.3'08583 — dc20 91-8611
CIP

Printed in the United States of America
93 94 95 96 EB 9 8 7 6 5 4 3 2

Dedicated to

Lionel J. Bellamy

His monumental works on summarizing and explaining group frequencies in vibrational spectroscopy stand alone. His wit and his wisdom were legendary.

and to

Tomas B. Hirschfeld

His brilliant contributions to the theory and practice of spectroscopy enlightened us all, and were characterized by an unbounded enthusiasm and delight in the field.

"If a poet at the same time be a physicist, he might convey to others the pleasure, the satisfaction, almost the reverence, which the subject inspires.—Especially is its fascination felt in the branch that deals with light."

Albert A. Michelson
Light Waves and Their Uses

PUDDLE

Our Cover

The lithograph by the artist, M. C. Escher, on our cover is entitled, "Puddle." I purchased the original from Mr. Escher at his home in Baarn, Holland, in 1968. I believe it gives us a preview of what is to come in our interpretation of spectra. In Puddle we find two different bicycle and human tracks (are they different?) and two different auto and truck tracks. Likewise, the frequency of absorption in our spectra are the tracks for the identification of molecules. We know these frequencies provide important data about the molecules, chemical groups, and ions. It is the organization of this information, to be found in this Handbook, which elucidates the molecular structure; so let the "tracks" be your guide to molecular structure.

Bill Fateley
1991

Contents

Preface

Infrared and Raman spectroscopy are two of the most widely used techniques for the determination of molecular structure and for the identification of compounds. One needs only to glance at the thousands of references in the *Fundamental Reviews on Infrared and Raman Spectroscopy* published by Analytical Chemistry biennially to verify this statement. Sample handling is very simple and usually non-destructive. Instrumentation has improved dramatically in the last 20 years. Information can be obtained in seconds rather than hours, and energy-limited samples (low concentration or small size) that would have presented difficulty in previous days are now readily examined. The "FT" revolution has involved both IR and Raman techniques, and the computer capabilities of all analytical instrumentation today provide formidable routines for data manipulation.

The information content of the infrared and Raman spectra are completely complementary, and both are necessary for the vibrational analysis of a molecule. The spectroscopist, organic chemist, or physical chemist using these techniques depends on theory, on group frequency information on molecules that has been developed empirically over the years, and on evaluated collections of reference spectra. The seminal texts of L. J. Bellamy, *The Infrared Spectra of Complex Molecules*, *Vols. I & II*, first published in 1954 and 1968, are still among the most widely cited books in the world. Later books by Colthup, Daly and Wiberley, and A. Lee Smith continued to explore and expand on our understanding of group frequencies and how they can be used to analyze materials. The revitalization of Raman spectroscopy occurred in the 1970s with the advent of the laser as a source. In 1974, the first book on *Characteristic Raman Frequencies of Organic Compounds* was published by Dollish, Fateley and Bentley. Although the Colthup *et al.* book is in a third edition, the Bellamy and Dollish books are out of print. It was this fact, coupled with the continuing rapid growth in the field of vibrational analysis, that led to this book.

We have attempted to integrate the discussion of the infrared and Raman group frequencies for 16 classes of organic molecules as chapters in the text. Literature through 1990 was searched for information on group frequencies, and results of this search are evaluated and organized in the discussion. In rigorous interpretation of spectra, it is important to understand the origin of group frequencies and to recognize and assign them in the spectrum. It is equally important to understand why group frequencies shift. Our text should provide assistance, as well as references to more thorough discussions in the literature, in the use of group frequencies. We have also included a set of representative pairs of IR and Raman spectra in Chapter 18 and in the Appendices, which complement the discussion in the text by illustrating the spectral features. We gratefully acknowledge the Aldrich Chemical Co., VCH publishers of the B. Schrader Raman/Infrared Atlas, and the Sadtler Research Laboratories, Division of Bio-Rad Laboratories, Inc., for permission to use selections from their infrared and Raman collections in order to present the pairs of spectra for each compound. A summary of characteristic Raman frequencies from Dollish *et al.*, and the Colthup correlation charts for infrared frequencies are presented in Appendix Three.

The purpose of this book is to provide a reference to aid in the interpretation of the infrared and Raman spectra of organic compounds. We hope it accomplishes that. Although excellent instrumentation and software are now readily available to chemists everywhere, it is perhaps more important than ever that the interpretations of the data are based on sound spectroscopic principles. There is still no better way to do this than to use and understand the group frequencies of organic molecules.

D. Lin-Vien
N. B. Colthup
W. G. Fateley
J.G. Grasselli

Acknowledgment

The authors wish to thank Mr. Dick Nyquist for the wonderful help and guidance he has provided us in the preparation of this book. For many years he has published outstanding papers on the relationship between vibrational group frequencies and structure in molecules. It is work such as this that made this book possible

Many thanks to you, Dick!

All of us

CHAPTER 1

Introduction

Infrared (IR) and Raman spectroscopies provide information on molecular vibrations. These methods cause molecules to undergo changes in vibrational energy state by subjecting them to excitation radiation in selected spectral regions. IR and Raman spectroscopy differ in the means by which photon energy is transferred to the molecule and in the instrumentation used. Thus, the information extracted exhibits different characteristics. Infrared and Raman spectroscopy are complementary rather than competing techniques. The molecular vibrational frequencies observed by both techniques are nearly the same, but the vibrational band intensities differ (sometimes markedly so) because of the different excitation mechanisms and therefore different selection rules.

1-1. PRINCIPLES AND SELECTION RULES

The theory and selection rules of IR and Raman transitions have been discussed by Herzberg [1] and Steinfeld [2] in detail. Fateley *et al.* [19] have discussed the infrared and Raman selection rules for molecular and lattice vibrations by the correlation method. Considerations on the selection rules from the view of molecular symmetry are given by Cotton [3] and Drago [4]. Only qualitative descriptions of the principles and selection rules of IR Raman spectroscopies are given here.

1-1.1. Infrared Spectroscopy

In IR spectroscopy, the vibrational excitation is achieved by radiating the sample with a broad-band source of radiation in the infrared region, which is generally 4000–200 cm^{-1} (2.5–50 μm). The wavenumber, $\bar{\nu}$ in cm^{-1}, is the number of waves per centimeter. It is equal to the reciprocal of the wavelength λ in cm, and is equal to the frequency ν divided by c, the velocity of light (cm sec^{-1}). In the IR region, the wavelength λ is given in micrometers μm or 10^{-6} m. As illustrated in Fig. 1-1, the molecule is excited to a higher vibrational state by directly absorbing the infrared radiation. The transmittance at a given wavenumber can be calculated according to Eq. 1. The transmission spectrum is then obtained by plotting the transmittance versus the IR wavenumber. Similarly, the absorbance at a given wavenumber can be obtained by using the Beer–Lambert equation (Eq. 2). A plot of absorbance versus wavenumber yields an absorption spectrum.

$$T = \frac{I}{I_0}, \tag{1}$$

$$\begin{aligned} A &= \log\left(\frac{I_0}{I}\right) \\ &= abc, \end{aligned} \tag{2}$$

where T is transmittance, A is absorbance, I_0 is the intensity of the entering radiation (before sample absorption), I is the intensity of the transmitted light (after sample absorption), a is absorptivity, b is cell thickness, and c is concentration.

1-1.2. Raman Spectroscopy

The origin of Raman spectroscopy is an inelastic scattering effect. In this case, the excitation radiation source is monochromatic and is much more energetic than infrared radiation. Elastic and inelastic scattering of radiation by the sample is observed in a Raman experiment. In elastic scattering (Rayleigh scattering), the molecule is excited to a virtual state, and then relaxes to the original vibrational state by re-emitting a photon at the same frequency as the incident light. The molecule "absorbs" no energy from the incident radiation in this case. See Fig. 1-1.

Only a very small fraction of molecules undergo inelastic scattering (Raman scattering). When Raman scattering occurs, the excited molecule relaxes to a different vibrational level, rather than to the original state. The energy carried by an inelastically scattered photon is different from that of

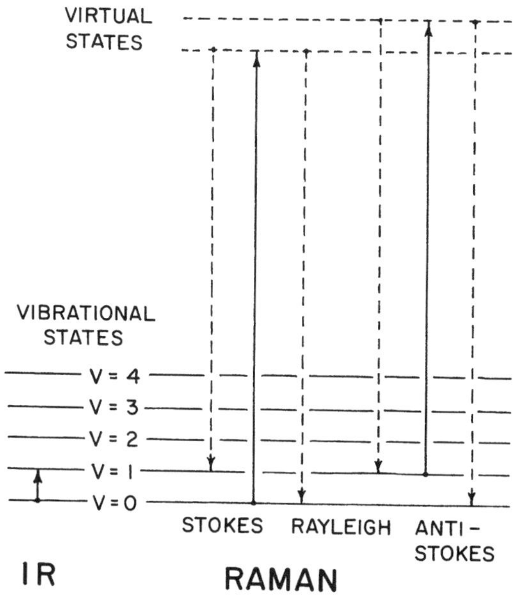

Fig. 1-1. Energy states involved in IR and Raman spectroscopies.

the incident light. In a Raman spectrum, the energy difference between the incident and scattered lights appears as a frequency shift between the scattered light ν' and the excitation frequency ν. These two frequencies, ν and ν', are related to the vibrational energy by the following equation (see also Fig. 1-1):

$$h\nu = h\nu' + \Delta E_{\text{vibration}} \tag{3}$$

It should be noted that in Raman experiments, the final vibrational state of the molecule can be either higher or lower in energy than the original state. In the case where the final vibrational state is lower in energy than the original one, the scattered photon will exhibit a higher frequency than the incident radiation. Thus, a blue shift from the excitation frequency is observed. Raman bands of this type are called anti-Stokes lines. Similarly, a red shift from the incident radiation is observed when the final state exhibits higher energy than the original state; these Raman bands are referred to as Stokes lines. Since most of the molecules are at ground vibrational state ($v = 0$) at room temperature, the Stokes lines exhibit higher Raman intensity than the corresponding anti-Stokes lines, which originate from an elevated vibrational state ($v = 1$). Therefore, the Stokes lines are more commonly used for molecular characterization.

If the direction of observation of the radiation scattered by the sample is perpendicular to the direction of the incident polarized laser beam source, and also to its electron vector, polarized Raman spectra can be obtained. A polarizer is placed between the sample and the spectrometer. It can be oriented parallel and perpendicular to the electric vector of the incident radiation. The Raman band intensity ratio, namely the perpendicular polarization intensity divided by the parallel polarization intensity, is called the depolarization ratio ρ. With a laser source, a totally symmetric vibration will generate a polarized Raman band where ρ has a value between 0 and $\frac{3}{4}$. A vibration that is not totally symmetric will generate a depolarized Raman band where $\rho = \frac{3}{4}$.

1-1.3. Selection Rules

In brief, a vibrational mode is IR-active when there is a change in the molecular dipole moment during the vibration; a vibrational mode is Raman-active when there is a change in polarizability during the vibration. Qualitatively, antisymmetric vibrational modes and vibrations due to polar groups are more likely to exhibit prominent IR absorption, while a symmetric vibrational mode is often strong in Raman. Molecules with a center of symmetry have no fundamental lines in common in the IR and Raman spectra.

1-2. CHARACTERISTIC FREQUENCIES

The terms "characteristic frequencies" and "group frequencies" are widely used in this text and deserve a clarification. According to Dollish *et al.* [5], the term "characteristic frequencies" is used to describe the Raman bands that are common for molecules containing the same functional groups. In infrared spectra, these common bands are more often referred to as "group frequencies." Based on empirical correlation, Colthup *et al.* [6] defined the term "group frequency" as the vibrational band consistently produced by a certain submolecular group of atoms in a characteristic spectral region. Smith [7] further categorized the group frequencies into three types: (i) those that have a rather constant frequency regardless of their environment; (ii) those that move around according to their molecular environment in a somewhat predictable manner; and (iii) those that have unpredictable positions or intensities.

Table 1-1.

Approximate X—H Stretching Wavenumbers (cm^{-1})

BH 2500	CH 3000	NH 3400	OH 3600	FH 3960
AlH 1820	SiH 2150	PH 2350	SH 2570	ClH 2890
	GeH 2070	AsH 2150	SeH 2300	BrH 2560
	SnH 1850	SbH 1890		IH 2230

An example of a type of group frequency is the band that arises from the stretching of the X—H bond. For a given atom X, the band tends to appear in a characteristic wavenumber region of the spectrum. Some examples are listed in Table 1-1, where the X atoms are arranged according to the periodic table, showing that the X—H stretching wavenumber tends to increase as the electronegativity of X increases [6], upward or to the right in the table.

1-3. THE NEED TO RE-EXAMINE THE IR AND RAMAN CHARACTERISTIC FREQUENCIES

Extensive studies were done on IR group frequencies during the period from the 1950s to the early 1970s. These early studies are reviewed by Bellamy [8] and Colthup *et al.* [6]. Recently, an extensive study was reported by Nyquist on the correlation of gas-phase IR group frequencies [18]. Raman characteristic frequencies, on the other hand, have received considerably less attention in the past. Surveys for the correlations of Raman characteristic frequencies before 1974 were given by Kohlrausch [9] and Dollish *et al.* [5].

The recent interest in vibrational spectroscopy has been focused on instrumentation, method development, and vibrational analysis. The rapid development of Fourier transform infrared spectrometry (FT-IR) has revolutionized the applications of infrared spectroscopy [10]. Sampling methods of FT-IR such as diffuse reflectance (DRIFT), photoacoustic (PAS), IR microscope, and attenuated internal reflectance (ATR), etc., have shortened the time required for obtaining high-quality IR spectra for various types of samples. In addition, complex systems are analyzed with added features such as spectral subtraction, spectral smoothing, and spectral deconvolution.

In Raman spectroscopy, the discoveries of resonance and surface-enhanced Raman scattering effects have provided not only ways to increase the Raman scattering efficiency, but also methods for reducing fluorescent interferences [11, 12]. Therefore, the detection limit of the Raman technique has been greatly improved. Furthermore, the recently developed FT-Raman [13] and Hadamard Transform HT-Raman [14] instruments suggest the possibility of obtaining both IR and Raman spectra from the same spectrometer. Hence, all the data processing techniques enjoyed by FT-IR spectrometry will become available for the Raman method as well. As a result of all these advancements, it is no surprise to observe that IR and Raman spectroscopic methods have progressed from mere methods of compound identification to a comprehensive system of material characterization. Numerous applications of IR and Raman methods include identification of unknown structures and specific chemical groups, surface analysis, information about chemical bonding and electronic structures, molecular geometry, and kinetic and equilibrium studies.

In comparison with the vast number of studies on instrumentation, method development, and vibrational analysis of individual compounds, the number of publications on the correlation of IR and Raman characteristic frequencies seems to be diminishing [7]. As a matter of fact, many vibrational spectra are poorly interpreted in the area of IR and Raman applications. As Smith [7] once stated, "good instrumentation cannot compensate for poor spectroscopy." A good understanding of group frequencies is necessary in order to make full use of the modern instruments. It is the large amount of literature on vibrational analysis since 1975, the increasing applications of IR and Raman spectroscopies, and the lack of studies on group frequencies in the last decade that collectively prompted us to undertake this investigation.

In this work, an extensive literature search was carried out to verify the existing correlations and determine ways in which correlations between vibrational frequencies and molecular structure could be enhanced. Particular attention has been paid to the correlation between Raman characteristic frequencies and molecular structure. The spectra published in *The Aldrich Library of FT-IR Spectra* [15], *The Sigma Library of FT-IR Spectra* [16], and *The Sadtler Standard Raman Spectra* [17] are carefully examined. The characteristic IR and Raman frequencies of compounds in the following categories were investigated: alkanes, halocompounds, alcohols and phenols, ethers and peroxides, alkenes, acetylenes, $-C{\equiv}N$ and $-\overset{+}{N}{\equiv}\overset{-}{C}$ groups, carbonyls, compounds containing $-NH_2$, $-NHR$, and $-NR_2$ groups, nitrocompounds, double bonds containing nitrogen atoms, cumulative double bonds, sulfur compounds, silicon compounds, phosphorus compounds, and aromatic and heteroaromatic rings.

REFERENCES

1. G. Herzberg, "Molecular Spectra and Molecular Structure," 2nd Ed., van Nostrand Reinhold Company, New York, 1950.
2. J. I. Steinfeld, "Molecules and Radiation, An Introduction to Molecular Spectroscopy," The MIT Press, Cambridge, Massachusetts, 1985.
3. F. A. Cotton, "Chemical Applications of Group Theory," Wiley-Interscience, New York, 1971.
4. R. S. Drago, "Physical Methods in Chemistry," Saunders, Philadelphia, 1977.
5. F. R. Dollish, W. G. Fateley, and F. F. Bentley, "Characteristic Raman Frequencies of Organic Compounds," John Wiley & Sons, New York, 1974.
6. N. B. Colthup, L. H. Daly, and S. E. Wiberley, "Introduction to Infrared and Raman Spectroscopy," 3rd Ed., Academic Press, San Diego, 1990.
7. A. L. Smith, *Appl. Spectrosc.* **41**, 1101, 1987.
8. L. J. Bellamy, "The Infrared Spectra of Complex Molecules," 3rd Ed., 1975, "Advances in Infrared Group Frequencies," 2nd Ed., 1980, John Wiley & Sons, New York.
9. K. W. F. Kohlrausch, "Ramanspektren," Heyden, New York, 1972.
10. P. R. Griffiths and J. A. de Haseth, "Fourier Transform Infrared Spectrometry," John Wiley & Sons, New York, 1986.
11. "Advances in Infrared and Raman Spectroscopy" (R. J. H. Clark and R. E. Hester, eds.), Vol. 7, Chapters 4 and 5, Wiley Heyden Ltd., 1984.
12. R. K. Chang and T. E. Furtak, "Surface Enhanced Raman Scattering," Plenum Press, New York, 1982.
13. T. Hirschfeld and B. Chase, *Appl. Spectrosc.* **40,** 133, 1986.
14. D. C. Tilotta, R. D. Freeman, and W. G. Fateley, *Appl. Spectrosc.* **41,** 1280, 1987.
15. C. J. Pouchert, "The Aldrich Library of FT-IR Spectra," Aldrich Chemical Company, Inc., 1985.
16. R. J. Keller, "The Sigma Library of FT-IR Spectra," Sigma Chemical Company, Inc., 1986.
17. "The Sadtler Standard Raman Spectra," Sadtler Research Laboratories, 1976.
18. R. A. Nyquist, "The Interpretation of Vapor-Phase Infrared Spectra, Group Frequency Data," Sadtler Research Laboratories, 1984.
19. W. G. Fateley, F. R. Dollish, and N. T. McDevitt, "Infrared and Raman Selection Rules for Molecular and Lattice Vibrations: The Correlation Method," John Wiley & Sons, Inc., 1972.

CHAPTER 2

Alkanes

2-1.

ACYCLIC ALKANES

The characteristic vibrations of alkanes can be approximately categorized as CH motions and skeletal (i.e., C—C skeleton) vibrations. Since many organic compounds contain alkyl groups, these vibrational bands are common in the IR and Raman spectra of organic compounds. A very good theoretical understanding of these vibrations is provided by Schachtschneider and Snyder, who evaluated the normal coordinates of the *n*-paraffins from C_2H_6 to $C_{14}H_{30}$ and polyethylene by a perturbation method [1]. Their results on the vibrational frequencies for this homologous series are listed in Table 2-1. The influence of rotational isomerism on the vibrations of CH_2 and CH_3 moieties has been extensively studied [1, 3, 4, 42–47, 55–58, 95, 99]. The experimental correlations of Raman and IR group frequencies are discussed by several authors [2, 5, 39, 59–61]. We have re-examined the literature spectra and summarized the characteristic IR and Raman group frequencies of alkanes in Table 2-2.

The vibrations involving CH_2 and CH_3 stretchings (3000–2840 cm^{-1}) give rise to bands that are strong and characteristic in both IR and Raman spectra, whereas bending modes (1470–1340 cm^{-1}) give rise to strong IR bands but variable Raman bands. See Fig. 2-1. On the other hand, the CH_2 and CH_3 rocking, wagging, and twisting modes are very sensitive to the immediate environments, and the frequencies and intensities vary considerably. Two useful long-chain vibrations are the $-(CH_2)_n-$ in-phase

Table 2-1.

Vibrational Analysis of n-Alkanes, C_nH_{2n+2}

Vibration	Frequency Range (cm^{-1})	
Methyl antisymmetric C—H stretching	For $n = 3$	2969–2965
	For $n > 3$	2967
Methyl symmetric C—H stretching	For $n = 3$	2884 and 2883
	For $n > 3$	2884
Methylene antisymmetric C—H stretching		2929–2912
Methylene symmetric C—H		2861–2849
Methyl out-of-plane HCH deformation	For $n = 3$	1466 and 1465
	For $n > 3$	1465
Methylene and methyl in-plane HCH deformation		1473–1466
Methyl symmetric HCH deformation	For $n = 3, 4$	1385–1368
	For $n > 4$	1376
Methylene wagging		1411–1174
C—C stretching		1132–885
Methyl terminal rocking	For $n < 10$	975–835
	For $n > 10$	895
CCC deformation		535–0
Methylene twisting-rocking		1310–1175
Methylene rocking-twisting		1060–719
CH_3—CH_2, CH_2—CH_2 torsion		~220, 153–0

rocking mode, which appears in the IR at 730–720 cm^{-1} with medium intensity, and the —$(CH_2)_n$— in-phase twisting mode, which appears in the Raman at 1305–1295 cm^{-1} with medium-strong intensity. The C—C stretchings (1132–885 cm^{-1}) and C—C—C bendings (below 540 cm^{-1}) usually demonstrate weak or no spectral features in IR spectra. Raman spectroscopy, on the other hand, reveals more information on the skeletal vibrations. In solid states, the longitudinal acoustical (below 900 cm^{-1}) and transverse acoustical (below 200 cm^{-1}) modes were reported to be skeletal and lattice vibrations [3, 7, 8]. These vibrational modes are dependent on the chain length and/or crystal structure.

2-1.1. CH_3 Stretchings

The CH_3 antisymmetrical and symmetrical stretchings occur near 2962 and 2872 cm^{-1}, respectively. The intensities of these vibrations are strong, and their frequencies are characteristic. For example, no more than ±10 cm^{-1}

variation is expected for the CH_3 stretching modes if the CH_3 group is attached to another saturated carbon. The term "antisymmetric" used in group frequency descriptions has the implied meaning "non-totally symmetric," where the vibrationally distorted group has less symmetry than the group at equilibrium [60].

Table 2-2.

IR and Raman Characteristic Frequencies (cm^{-1}) of Alkanes

Vibrations		Frequency	IR	R
CH_3 Vibrations:	antisym. str.	2972–2952	vs	vs
	sym. str.	2882–2862	vs	vs
	antisym. bend.	1470–1440	ms	ms
	sym. bend.:			
	$-C(CH_3)$	1380–1370	m	w-0
	$-C(CH_3)_2$	1385–1380	m	w-0
		1373–1365	m	w-0
	$-C(CH_3)_3$	1395–1385	m	w-0
		1373–1365	ms	w-0
CH_2 Vibrations:	antisym. str.	2936–2916	vs	vs
	sym. str.	2863–2843	vs	vs
	sci.	1475–1445	ms	ms
Skeletal Vibrations:				
n-alkanes				
$-(CH_2)_n-$ in-phase twist		1305–1295	—	s
$-(CH_2)_n-$ in-phase rock, $n > 3$		726–720	m	—
C—C skel. str.:		1180–1120	(doublet)	m
		1100–1040	—	s-m
		900–800	—	s-m
Branched alkanes				
$-CH(CH_3)_2$, C—C skel str.:		1175–1165	m	vw
		1170–1140	m	vw
		1060–1040		m
		950–900		m
	sym. C—C—C str.:	830–800		s, p
	C—C skel. def.:	500–460		s-m
		320–250		s-m
$-CHR_2$	C—C skel. str.:	1060–1040		m
$R \neq CH_3$		950–900		m
		850–800		s, p
	sym. C—C—C str.	800–750		s, p
	C—C skel. def.:	450–410		m
		350–250		m
$-CR_3$	C—C skel. str.:	1255–1245	m	m
		1250–1200	m	m
	sym. CC_4 str.	750–650		s, p
	C—C skel. def.	480–250		

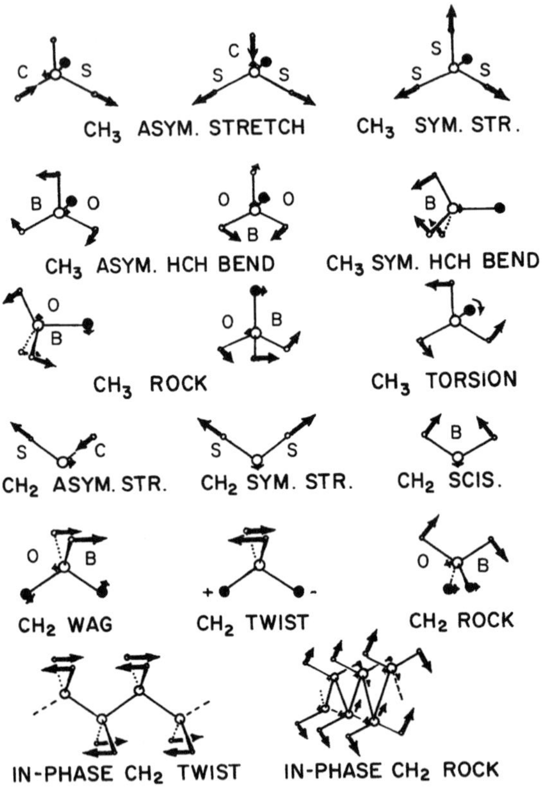

Fig. 2-1. Vibrations of CH_3 and CH_2 groups. C and S stand for contract and stretch, and B and O stand for bend and open.

If the CH_3 group is attached to an aromatic ring, two bands are regularly seen in the IR and Raman near 2925 and 2865 cm^{-1}, both assigned to the CH_3 symmetrical stretch vibration in Fermi resonance with the overtone of the CH_3 bending vibration band near 1450 cm^{-1} [96]. Larger shifts of the CH_3 stretching bands will occur when the methyl group is attached to a heteroatom, as in $O-CH_3$ and $N-CH_3$ (to be discussed later).

2-1.2. CH_2 Stretchings

The antisymmetric and symmetric CH_2 stretchings appear strongly at about 2926 and 2853 cm^{-1} in IR and Raman. Again, the spectral positions vary no more than $\pm 10\ cm^{-1}$ for acyclic $-CH_2-$ groups. However, the stretching frequencies will increase to higher values when the CH_2 group is included in a strained ring, as will be discussed in Section 2-2. Large shifts of the CH_2

stretch frequencies are also observed if there is an adjacent oxygen atom ($-CH_2-O-$) or nitrogen atom ($>N-CH_2$), as discussed in later chapters.

2-1.3. CH Stretching

The R_3C-H stretching mode absorbs near 2890 cm^{-1}. This band is found in a broader region (3000–2850 cm^{-1}) [100] when the CH is adjacent to a non-carbon atom. This band is weak, and its presence is often obscured by the much stronger CH_3 and CH_2 stretching bands in both IR and Raman spectra.

2-1.4. CH_3 Bendings

The antisymmetric and symmetric CH_3 bendings occur at 1470–1430 and 1395–1365 cm^{-1}, respectively. The CH_3 antisymmetric bend band near 1460 cm^{-1} has medium intensity in both IR and Raman spectra, while the CH_3 symmetric bend near 1380 cm^{-1} is medium in the IR but is very weak in the Raman except when a C=C, C≡C, C=O, or aromatic ring is directly attached to the CH_3 group (hyperconjugation). In these cases, the CH_3 symmetric bend band is medium to strong in the Raman near 1380 cm^{-1}. The 1380 cm^{-1} band does not overlap with other bands of alkanes and is usually a good indicator for the presence of CH_3 on a carbon atom.

When the CH_3 group is attached to a non-carbon atom, the $X-CH_3$ symmetrical deformation wavenumbers vary from 1475 cm^{-1} (FCH_3) to 1165 cm^{-1} ($PbCH_3$) [97]. These wavenumbers correlate linearly with the negative of the electron density on the CH_3 carbon [98], calculated from molecular orbital theory. The electron density on the carbon decreases as the electronegativity of the attached atom X increases [97]. Approximate wavenumbers for examples of $X-CH_3$ symmetrical deformation are given in Table 2-3 for various X atoms, arranged according to the periodic table [97].

Table 2-3.

Approximate Wavenumbers for $X-CH_3$ Symmetrical Deformation for Various X Atoms

B 1310	C 1380	N 1410	O 1445	F 1475
	Si 1265	P 1295	S 1310	Cl 1355
	Ge 1235	As 1250	Se 1252	Br 1305
	Sn 1190	Sb 1200		I 1252
	Pb 1165			

2-1.5. CH_2 Bendings

The four bending vibrations of CH_2 groups are scissoring, rocking, wagging, and twisting. The scissoring mode of the CH_2 group gives rise to a characteristic band near 1465 cm^{-1} in IR and Raman spectra. This band often overlaps with CH_3 antisymmetrical bending in the 1470–1430 cm^{-1} region. The rocking, wagging, and twisting appear in the region of 1422–719 cm^{-1}. These vibrations are affected by interaction with neighboring CH_2 groups and give rise to weak IR and medium Raman bands for alkanes.

2-1.6. CH Bending

The tertiary C—H bending occurs near 1360–1350 cm^{-1} as weak IR and Raman bands. Therefore, the use of this vibrational mode as a group frequency is limited unless there is no spectral interference in this region.

2-1.7. The Skeletal Vibrations below 1300 cm^{-1}

The skeletal vibrations are those involving the carbon-chain stretchings and deformations. IR absorption of these vibrational modes is very weak for *n*-alkanes. Although more IR details are observed for branched alkanes in the skeletal vibration region, determination of skeletal structures by IR spectroscopy is always challenging. On the other hand, Raman spectra of alkanes are rich in skeletal information in the region below 1300 cm^{-1}. These characteristic Raman frequencies greatly aid in the identification of molecular skeletons.

2-1.8. *n*-Alkanes

The only prominent IR band between 1300 and 200 cm^{-1} is located near 720 cm^{-1}. This IR band is assigned to the $-(CH_2)_n-$ in-phase rocking mode. Typically, the following characteristic bands are expected in Raman spectra of *n*-alkanes: a strong band in the 1305–1295 cm^{-1} region, a doublet in the 1180–1120 cm^{-1} region, two groups of medium to strong bands in the 1100–1040 and 900–800 cm^{-1} regions, and a band located below 400 cm^{-1}. For the doublet in the 1180–1120 cm^{-1} region, the peak at the higher-frequency side is normally weak, and the peak at the lower frequency is medium in intensity.

Table 2-4.

CH_2 Rocking Frequencies (cm^{-1}) of $-(CH_2)_n-$, $n > 3$

n	$CH_3(CH_2)_nCH_3$	$CH_3(CH_2)_nOH$	$CH_3(CH_2)_nCl$	$CH_3(CH_2)_nBr$
4	727	729	731	729
5	725	725	727	726
6	723	724	726	724
7	721 ± 1	723	724	723
8	721 ± 1	721 ± 1	723	721 ± 1
>9	721 ± 1	721 ± 1	721 ± 1	721 ± 1

The $-(CH_2)_n-$ in-phase rocking mode appears in the IR as a medium-weak band in the 726–720 cm^{-1} region when n is greater than 3. For $n = 3$, 2, or 1, the regions are 729–726, 743–734, and 785–770 cm^{-1}, respectively [59]. In the crystalline solid state for long-chain CH_2 groups, this band splits into a doublet at 730 and 720 cm^{-1}. In the Raman spectrum, this band is too weak to be useful. See Table 2-4.

The $-(CH_2)_n-$ in-phase twisting mode occurs in Raman in the 1305–1295 cm^{-1} region with medium to strong intensity. Its relative Raman intensity increases with the n value of the $(CH_2)_n$ chain. In IR spectra of alkanes it is usually not observable.

Raman spectroscopy is especially valuable in studying the C—C stretching modes in the regions of 1100–1040 and 900–800 cm^{-1} [13, 40, 53, 54]. Varsanyi has correlated the spectral ranges of 1110–1080 and 910–855 cm^{-1} to antisymmetrical CCC stretching and symmetrical CCC stretching modes, respectively, and those of 1120–1090, 1005–930, and 910–880 cm^{-1} to various types of CCCC stretchings [40]. Therefore, multibands are expected in these spectral ranges, since various combinations of carbon atoms can be involved in the vibrations.

Another vibrational mode that gives rise to a Raman band in this region is the longitudinal acoustical vibration in solid state. Mizushima and Simanouti [3] and Schaufele and Shimanouchi [7] have reported two series of longitudinal acoustical vibrations as the skeletal vibrations of n-paraffins in solid phase. The first series appears in the region of 900–800 cm^{-1} (Table 2-5) and is attributed to a combination of in-plane methyl

Table 2-5.

The Longitudinal Acoustical Vibrational Frequencies (cm^{-1}) of n-Paraffins, C_nH_{2n+2}, in the Range of 900–800 cm^{-1}

n	4	5	6	7	8	9	10	12	16
cm^{-1}	837	869	898	905	899	888	886	892	888

rocking, C—C stretching, and CCC deformation. The frequency is closer to the lower limit of the correlated range when the number of carbon atoms (i.e., n) is small. For example, the vibration occurs at 837 cm^{-1} for n-C_4H_{10} and at 869 cm^{-1} for n-C_5H_{12}; it then stays at a fairly constant spectral region of 888–905 cm^{-1} when n is greater than five.

The second series of longitudinal acoustic vibrations is a chain expansion–contraction motion where all C—C—C angles change in-phase. This gives rise to a highly polarized Raman band. If all the carbons are in a linear progression of trans zigzag C—C—C—C units, the wavenumber $\bar{\nu}$ of this band is predicted to be inversely proportional to the number of carbons in the n-paraffin chain by the equation

$$\bar{\nu} = (a'/N),$$

where N is the carbon number in the paraffin chain, and a' is a calculated value involving Young's modulus and the density. Data are shown in Table 2-6, where the wavenumber $\bar{\nu}$ decreases gradually from 373 cm^{-1} for C_6H_{14} to 26 cm^{-1} for $C_{94}H_{190}$ in the solid state. The basic linearity between the wavenumber $\bar{\nu}$ and $1/N$ in the solid state can be demonstrated by a plot of $\bar{\nu}$ vs. $1/N$. The data reported in Table 2-6 are from different references [7, 8], and the alkanes are of several different crystal structures, which does not appear to affect the results. The vibration is also observed in the Raman spectra of long-chain lipids [54].

The Raman spectra of liquid n-alkanes exhibit a broad polarized Raman band at 307 cm^{-1} for C_7H_{16}, which shifts gradually to about 220 cm^{-1} when the carbon chain length increases to 18 carbons. In the liquid state there can be some gauche C—C—C—C units, so the linear progression of trans zigzag C—C—C—C units may be shorter than the full chain length.

Crystallized n-paraffins show Raman scattering below 200 cm^{-1} because of the lattice vibrations. Olf and Fanconi reported several transverse

Table 2-6.

The Chain Expansion–Contraction Frequencies (cm^{-1}) of n-Paraffins (C_nH_{2n+2}) in the Solid State

n	6	7	8	9	10	11	12	13	14	15
cm^{-1}	373	311	282	249	230	206	195	183	169	158
n	16	17	18	19	20	21	22	23	24	25
cm^{-1}	150	141	131	126	120	115	113	106	104	97
n	26	28	32	35	36	44	94			
cm^{-1}	96	86	76	70	67	55	26			

acoustical vibrations as lattice vibrations from n-C_5H_{12} to n-$C_{26}H_{54}$ [8]. These vibrations were found to be influenced by the crystal structures and size of the Bravias cell (i.e., the number of chains in the cell).

2-1.9. Tertiary Carbon Atom (C_3CH)

The isopropyl group $-CH(CH_3)_2$ is the simplest structure unit that contains a tertiary carbon. The coupling between the symmetric bending modes of the two neighboring methyl groups [98] leads to a characteristic IR doublet at 1390–1380 and 1372–1365 cm^{-1} [59]. The components of the doublet are approximately equal in intensity (medium) in the IR spectra, but they are weak in Raman. There are two more IR bands located in the 1175–1165 cm^{-1} and 1170–1140 cm^{-1} regions due to vibrations involving $C-C(CH_3)_2$ stretchings [9]. These bands are of weak to medium IR and Raman intensities. Furthermore, their frequencies may shift to a higher values if another branched carbon is adjacent to the isopropyl group.

In addition to the above IR absorptions, the Raman spectra of the $R-CH(CH_3)_2$-containing compounds present characteristic bands in the following regions: 1360–1350 (m-w, tertiary C—H deformation), 1060–1040 (m), 950–900 (m), 830–800 (s, polarized), 500–460 (s-m, polarized) and 320–250 (s-m, polarized). According to Dollish *et al.*, the 830–800 cm^{-1} band is due to the symmetrical C—C stretching mode of the $C-C(CH_3)_2$ moiety [5], and its Raman intensity is among the strongest bands observed below 1,000 cm^{-1}. The assignment is supported by the vibrational studies of *t*-butylhalides $X-C(CH_3)_2$ (X=Cl, Br). In the Raman spectra of these compounds, a polarized strong band at the 810–805 cm^{-1} region is attributed to a C—C stretching mode [62, 63]. The assignment of this vibrational mode, however, is sometimes interfered with by the CH_2 rocking mode. Crowder and Lin have assigned the 819 cm^{-1} frequency, which is the only strong Raman band in the 950–800 cm^{-1} region, to a mixing of CH_2 rocking and C—C stretching for 1-chloro-2-methylpropane ($(CH_3)_2CHCH_2Cl$) [64]. In addition, the symmetrical C—C stretching vibration will shift to a higher frequency when the tertiary carbon atom is unsaturated. For example, the symmetrical C—C—C stretching mode in $C{=}C(CH_3)_2$ is observed at 870–840 cm^{-1} as a strong polarized Raman band [40]. The $C{=}C(CH_3)_2$ symmetric CH_3 bending band near 1375 cm^{-1} is more intense in the Raman compared with alkanes, but is not distinctly split into a doublet in the Raman or the IR spectrum.

The identification of internal tertiary carbon atoms (R_3CH) by the IR technique is difficult, since the doublet of CH_3 symmetric bendings is no longer available and the CC_3 stretchings in the 1175–1165 and 1170–1140 cm^{-1} regions are not consistent in IR frequencies. Fortunately, the Raman spectra provide more information on the skeletal vibrations of these branched alkanes. Two strong polarized bands are observed in the 850–800 cm^{-1} and 800–750 cm^{-1} regions due to the symmetrical CC_3 stretching modes [41]. For instance, these bands are found at 806 and 790 cm^{-1} in Raman spectra of 2-(chloromethyl)-2-methyl-1,3-dichloropropane ($(ClCH_2)_2(CH_3)CCl$) [65].

The CC_3 deformations are observed in the 450–410 and 350–250 cm^{-1} regions. When there is another branched carbon atom (tertiary or quaternary) next to the tertiary carbon atom under study, another Raman band in the 540–510 cm^{-1} region will be expected.

2-1.10. Quaternary Carbon Atom (CC_4)

The *t*-butyl group is best recognized in IR by an unequal-intensity doublet at 1400–1393 cm^{-1} (m) and 1374–1366 cm^{-1} (s) [59]. These are due to three interacting CH_3 groups, each performing the CH_3 symmetric bend vibration. There are two components in the band near 1368 cm^{-1}, which makes it about twice as strong as the band near 1395 cm^{-1} [60]. The *t*-butyl group in alkanes also gives rise to medium-weak IR bands near 1245, 1200, and 930 cm^{-1} [59].

The gem-dimethyl group in alkanes, $R_2C(CH_3)_2$, has CH_3 IR bands nearly the same as those in the isopropyl group. The CH_3 symmetric bending doublet is seen in the IR at 1391–1381 and 1368–1366 cm^{-1}, where the intensity ratio for the higher to the lower frequency band is about 4 to 5 [59]. In the IR, additional bands are usually seen at 1221–1206 cm^{-1} (w) and 1191–1185 cm^{-1} (m) [59].

The Raman characteristic frequencies of the *t*-butyl group and other internal quaternary carbon atoms are quite similar. The most distinctive feature in the Raman spectrum of a quaternary carbon atom is due to the symmetrical CC_4 stretching mode, which occurs in the 750–650 cm^{-1} range. This band is polarized, and very often it is the most intense Raman band below 1000 cm^{-1} [9, 64]. Recently, Klaboe *et al.* examined the vibrational spectra of compounds with the *t*-butyl groups attached to a triple bond. They observed elevated symmetrical CC_4 stretching frequencies for di-*tert*-butylethyne ($Me_3C{-}C{\equiv}C{-}CMe_3$) at 846 cm^{-1} (IR : O, R : vs, p) [68]. On the other hand, Kim *et al.* reported lower CC_4 stretching frequencies (627 cm^{-1}, IR : O, R : s; 652 cm^{-1}, IR : vvw, R : mw) for isobornyl acetates, in which the two quaternary carbon atoms are included in a bicyclic skeleton [91]. Kim and coworkers further indicated that the quaternary carbon atom with higher local symmetry will give rise to a stronger Raman band than the one with lower local symmetry. Interference on assigning the symmetrical CC_4 stretching mode will be encountered when the molecule contains $O{-}CC_3$ and $N{-}CC_3$ types of structure. These two skeletons also exhibit polarized strong bands in Raman near this region due to the symmetric $O{-}CC_3$ and $N{-}CC_3$ stretching modes. Nevertheless, they can be differentiated from the symmetrical CC_4 stretching mode by their strong IR absorption in the majority of cases. Examples can be found in the Raman spectra of tertiary alcohols [48b], *t*-butyl formate [66], and *t*-butyl peroxide ([67], see discussion in Chapter 4). Other characteristic Raman bands for quaternary carbon are located near 1250, near 1200, and in the 480–250 cm^{-1} region.

2-2. CYCLOALKANES

The CH_2 stretchings of small cycloalkanes (C_nH_{2n}, $n < 6$) occur at higher frequencies than those of acyclic alkanes, because of the ring strain. The effect of ring strain on CH_2 bending, however, is to lower the frequency. Since cyclopropanes exhibit greater ring strain than the other larger cycloalkanes, the frequency shifts are more prominent in the vibrational spectra of cyclopropanes. The CH_2 stretching and bending frequencies will gradually approach those of acyclic alkanes as the ring size increases. Eventually, the CH_2 stretching and bending frequencies are located in the same regions as acyclic alkanes when the ring size large enough to release the ring strain, as in the case of cyclohexanes.

Another characteristic vibration of cycloalkanes is the ring breathing mode. An excellent discussion on this vibration is given by Colthup *et al.* based on a simple mathematical model [60]. They indicated that the ring breathing frequency of cycloalkanes is proportional to the $\cos(\alpha/2)$ value, where α is the CCC angle of the ring. Therefore, the frequency is higher for cyclopropanes (near 1190 cm^{-1}), but gradually decreases to about 800 cm^{-1} for cyclohexanes.

2-2.1. Cyclopropane and Its Derivatives

The vibrational assignments of cyclopropane were reported by Baker and Lord in 1955 [10], and re-investigated by Duncan and McKean in 1968 [11]. Table 2-7 lists the vibrational frequencies of cyclopropane. The CH_2

Table 2-7.

Characteristic Frequencies (cm^{-1}) of Cyclopropane

	Vibration		cm^{-1}	IR	R
CH_2	stretching	antisym.	3102	vs	—
			3082	vs	s
		sym.	3038	vs	s
			3024	vs	s
	scissoring		1438	m	m
	wagging		1028	vs	—
Ring	breathing		1188	w	s
	deformation		868	vs	s

stretchings of this molecule occur at frequencies above 3000 cm^{-1}, which is higher than the corresponding frequencies of acyclic CH_2 stretchings. This high frequency shift is caused by the ring strain due to the deviation of valence angles from the tetrahedral values in cyclopropanes. The ring breathing mode is a symmetric motion. Thus, the ring breathing mode displays a polarized strong Raman band at 1188 cm^{-1}, but exhibits weak IR absorption.

The spectroscopic data and information on cyclopropane derivatives have been published for the alkylated cyclopropanes [72, 80–83], halogenated cyclopropanes [14–20, 69–74], cyclopropyl cyanides [21, 75–79], cyclopropylborane compounds [18, 22, 23], the halomethyl cyclopropanes [24–26], and the compounds with a cyclopropyl group attaching to a C=O or a C≡C moiety [83, 84]. A review of the vibrational spectra of cyclopropane derivatives is given by Wurrey and Nease [12]. Table 2-8 presents the correlations between the characteristic IR and Raman frequencies for cyclopropane derivatives. As with cyclopropane, the CH_2 stretching frequencies of these compounds are significantly shifted by the ring strain toward higher regions. Therefore, the $\nu_{as}(CH_2)$ of substituted propanes occurs at 3105–3085 cm^{-1} (IR : s, R : m-w) and the $\nu_s(CH_2)$ at 3040–3020 cm^{-1} (R : s). In the cases where more than one possible conformational isomer is possible, the CH_2 stretching will split into two bands in Raman spectra (3100–3080 cm^{-1} and 3085–3075 cm^{-1}). With the exception of 1,1,2,2-tetrafluorocyclopropane, 1,1,2,2-tetrahalogenation lowers the $\nu(CH_2)$ stretching frequencies in comparison to other cyclopropane derivatives (Tables 2-8) and 2-9). The IR absorption of CH_2 wagging (1030–900 cm^{-1}), rocking (805–785 cm^{-1}), and α-CH bending (1365–1295 cm^{-1}) is intense and characteristic.

Table 2-8.

Characteristic Frequencies (cm^{-1}) of Cyclopropane Derivatives

	Vibration		cm^{-1}	IR	R
CH_2	stretching	antisym.	3105–3085	s	m-w
		sym.	3040–3020	—	m
			3020–3000	s	s
	scissoring		1420–1400	s	w
	twisting		1170–1070	w	w
	wagging		1070–1050	m	w
			1030–990	s	w
	rocking		805–785	m	w
α-CH	stretching		3060–3020	var.	m-s
	bending		1365–1295	s	w
Ring	breathing		1200–1180	s-m	var.
	deformation		960–900	s	var.
			880–800	var.	var.

Table 2-9.

CH_2 Stretching and Rocking Frequencies (cm^{-1}) for 1,1,2,2-Tetrahalogenated Cyclopropanes

Antisymmetric Stretch			Symmetric Stretch			Rocking		
(cm^{-1})	IR	R	(cm^{-1})	IR	R	(cm^{-1})	IR	R
3115	w	m	3022	w	m	871	m	w
3074	ms	m	2992	m	s	990	vs	w
3077	ms	m	2996	m	s	987	vs	vw
3084	m	m	2998	s	s	1000	s	vw

Since the molecular symmetry is reduced by the presence of substituents, the ring breathing mode (1200–1180 cm^{-1}) becomes active in both IR and Raman spectra. In addition, the ring deformations are observed to be at 960–900 cm^{-1} and 880–800 cm^{-1} regions with variable intensities.

2-2.2. Cyclobutane and Its Derivatives

Raman and IR measurements of cyclobutane are reported by Miller *et al.* [27], and the vibrational assignments are listed in Table 2-10. In addition, Durig *et al.* [28–31] and Rothschild [32] have examined the vibrational frequencies of halocyclobutane and cyclobutanol. The vibrational spectra

Table 2-10.

Characteristic Frequencies (cm^{-1}) of Cyclobutane

	Vibration		cm^{-1}	IR	R
CH_2	stretching	antisym.	2974	vs	—
			2965	vs	s
		sym.	2945	s	m
	scissoring		1443	m	w
	wagging		1260	s	vw
	twisting		1224	m	w
	rocking		626	m	w
Ring	breathing		1005	—	vs
			926	—	s
			901	s	w
	puckering		199	vw	vw

of cyclobutane carboxyl halides and cyclobutane carboxylic acid have been investigated by Hanai and Katon [33], Katon *et al.* [34], and Bardet *et al.* [85], respectively. There are very few spectroscopic data for disubstituted cyclobutanes. Assignments for 1,2-cyclobutane dicarbonitrile and 1,1- and 1,3-cyclobutane dicarboxylic acids are given by Compton *et al.* [35] and Fabrégue *et al.* [36, 86]. Carter and coworkers have examined the IR absorption of a total of 18 cyclobutane derivatives in the region below 1100 cm^{-1} [37].

The characteristic Raman and IR frequencies of substituted cyclobutanes are summarized in Table 2-11. The $\nu_{as}(CH_2)$ and $\nu_s(CH_2)$ frequencies of cyclobutanes are located at 3000–2975 cm^{-1} and 2956–2945 cm^{-1}, respectively, which are lower frequencies than those of cyclopropanes but still higher than those of acyclic alkanes. Similarly, the CH_2 scissoring of cyclobutanes is located in a region (1450–1440 cm^{-1}) between that of cyclopropanes (1420–1400 cm^{-1}) and acyclic alkanes (1485–1445 cm^{-1}). For monosubstituted cyclobutanes, the α-CH bending (1360–1250 cm^{-1}, IR : s) is another characteristic frequency in the IR spectra.

Because of the molecular symmetry, the ring breathing mode of cyclobutane at 1005 cm^{-1} (R : s, p) is Raman-active but IR-forbidden. However, as the molecular symmetry is altered by the substituents, this vibration becomes observable in Raman as well as IR spectra of many cyclobutane derivatives. The characteristic spectral region for the ring breathing mode of substituted cyclobutanes is between 960 and 950 cm^{-1} (IR : s-m, R : m). Several ring deformations are also observed in the range of 1080–700 cm^{-1} (Table 2-11). The ring puckering occurs in the 180–140 cm^{-1} range and is usually of very weak intensity.

Table 2-11.

Characteristic Frequencies (cm^{-1}) of Monosubstituted Cyclobutanes

	Vibration		cm^{-1}	IR	R
CH_2	stretching	antisym.	3000–2985	s	—
			2990–2970	m	var.
			2956–2945	s	s
		sym.	2985–2975	—	s
			2970–2955	s	m
	scissoring		1450–1440	m	m
	wagging		1245–1220	s-m	m-w
	twisting		1250–1050	m-w	w
	rocking		~800	s	w
α-CH	stretching		~2970	w	w
	bending		1360–1250	s	w
Ring	breathing		960–950	s-ms	m
	deformation		1080–1050	s-m	m
			900–880	m	w
			780–700	s	m
	puckering		180–140	w	w

2-2.3. Cyclopentane and Its Alkylated Derivatives

The vibrational assignments of cyclopentane and monosubstituted cyclopentanes have been reported [49–52, 87]. The Raman frequencies of cyclopentane and more than 60 of its alkylated derivatives are tabulated by Sterin and coworkers [38]. The cyclic CH_2 stretching modes are found in the regions near 2965–2960 cm^{-1} (ν_{as}, IR : s, R : s) and at 2880–2870 cm^{-1} (ν_s, IR : s, R : s). The CH_2 scissoring is located at 1490–1430 cm^{-1}, and its intensity is usually very strong in IR spectra and medium to strong in Raman spectra.

The ring breathing mode of cyclopentane at 889 cm^{-1} is very strong in Raman, but is IR-forbidden. When the ring is substituted by other groups, the intensity of the ring breathing mode in Raman is weakened by the reduction in molecular symmetry, whereas several ring deformations involving C—C stretchings (1100–800 cm^{-1}) become strong in Raman. Therefore, the determination of the ring breathing frequency is more difficult for substituted cyclopentanes. Nevertheless, the strongest Raman band of monoalkyl cyclopentanes in the region below 1000 cm^{-1} is located at the 910–880 cm^{-1} range, as illustrated by Table 2-12. It is interesting to note that the characteristic frequencies for *trans*- and *cis*-1,2-dialkyl

Table 2-12.

The Ring Breathing Frequencies (cm^{-1}) of Cyclopentane and Its Derivatives

Compound	cm^{-1} (Raman)
Cyclopentane	888 vs
Alkylated cyclopentanes	910–813 var. (m-w)
Methylene cyclopentane	897 s
α-alkyl methylene cyclopentane (α-alkyl = CH_3, C_2H_5, C_3H_7)	898–894 ms
α,α′-dimethyl methylene cyclopentane	905 m

cyclopentanes in the 894–887 cm^{-1} and 910–890 cm^{-1} regions, respectively, can be used to distinguish between the two isomers. The characteristic Raman frequency will shift to 880–840 cm^{-1} and 770–760 cm^{-1} regions for 1,3-dialkyl and 1,2,3- or 1,2,4-trialkyl cyclopentanes, respectively.

2-2.4. Cyclohexane and Its Alkylated Derivatives

The CH_2 antisymmetrical and symmetrical vibrational modes in the 2933–2915 cm^{-1} and 2897–2852 cm^{-1} regions are very strong in Raman and IR spectra [38, 2, 5]. The CH_2 scissoring is located at 1450 cm^{-1} for cyclohexane and in the region of 1460–1440 cm^{-1} (IR : s, R : m) for alkylated cyclohexanes [2]. It should be pointed out here that the ring strain is largely released as the ring size reaches six; thus, the CH_2 stretching and bending frequencies occur in the same regions as those found for open-chain alkanes. In some cases, the vibrational bands are split by the coexistence of different ring conformations [92–94].

The skeletal vibrations are prominent in the region below 1100 cm^{-1}. For cyclohexane, the strong IR band at 903 cm^{-1} can be assigned to an antisymmetric ring stretching mode [5, 6]. This band is observed in the range between 910 and 880 cm^{-1} in the IR spectra of monoalkylated cyclohexanes [6].

The ring breathing mode at 802 cm^{-1} was observed to be very strong in the Raman spectrum of chair-form cyclohexane [88–90]. Upon alkylation, however, the Raman intensity of this vibrational mode is diminished by the reduction in molecular symmetry. Nevertheless, the strongest Raman band in the region below 1000 cm^{-1} is usually located between 800 and 730 cm^{-1} with medium to weak intensity [38]. Table 2-13 illustrates the characteristic Raman bands below 1000 cm^{-1}. As with cyclopentane derivatives, the cis and trans isomers of dialkyl cyclohexanes can be distinguished by their characteristic Raman frequencies below 1000 cm^{-1}.

Table 2-13.

The Characteristic Raman Frequencies (cm^{-1}) of Alkylated Cyclohexanes in the 900–500 cm^{-1} Region

Compounds		cm^{-1} (Raman)
Monoalkyl cyclohexane		865–840 m-vw, 800–770 var.
1-methyl,2-alkyl cyclohexane	trans:	850–840 w-vw, 830–820 var., 770–750 m-vw.
	cis:	850–840 w-vw, 760–730 var.
1-methyl,4-alkyl cyclohexane	trans:	780–735 s-m.
	cis:	775–750 s-m, 637–623 w-vw.
1-methyl,3-alkyl cyclohexane	trans:	760–750 w-vw, 637–620 vw.
	cis:	785–770 w.
1-methyl,1-alkyl cyclohexane		835–820, 715–695, 565–555.

REFERENCES

1. J. H. Schachtschneider and R. G. Snyder, *Spectrochim. Acta* **19,** 117, 1963.
2. L. J. Bellamy, "The Infrared Spectra of Complex Molecules," Vol. 1, 3rd Ed., John Wiley & Sons, Inc., New York, 1975, pp. 13-36.
3. S. Mizushima and T. Simanouti, *J. Am. Chem. Soc.* **71,** 1320, 1949.
4. R. G. Snyder and J. H. Schachtschneider, *Spectrochim. Acta* **19,** 85, 1963.
5. F. R. Dollish, W. G. Fateley, and F. F. Bentley, "Characteristic Raman Frequencies of Organic Compounds," John Wiley & Sons, Inc., New York, 1974, pp. 1-11.
6. C. J. Pouchert, "The Aldrich Library of FT-IR Spectra," 1st Ed., Aldrich Chemical Company, Inc., 1985.
7. R. F. Schaufele and T. Shimanouchi, *J. Chem. Phys.* **47,** 3605, 1967.
8. H. G. Olf and B. Fanconi, *J. Chem. Phys.* **59,** 534, 1973.
9. G. B. B. Sutherland and D. M. Simpson, *J. Chem. Phys.* **15,** 153, 1947.
10. A. W. Baker and R. C. Lord, *J. Chem. Phys.* **23,** 1636, 1955.
11. J. L. Duncan and D. C. McKean, *J. Mol. Spectrosc.* **27,** 117, 1968.
12. C. J. Wurrey and A. B. Nease, "Vibrational Spectra and Structure," Vol. 7 (J. R. Durig, Ed.), Elsevier Scientific Publishing Company, New York, 1978, Chapter 1 (and references therein).
13. J. L. Lippert and W. L. Peticolas, *Proc. Nat. Acad. Sci.*, U.S.A., 1572, 1971.
14. G. Schrumpf, *Spectrochim. Acta* **39A,** 487, 1983.
15. C. J. Wurrey, R. B. Blatt, and A. B. Nease, *J. Phys. Chem.* **81,** 2279, 1977.
16. C. J. Wurrey, R. J. Berry, Y. Y. Yeh, T. S. Little, and V. F. Kalasinsky, *J. Raman Spectrosc.* **14,** 87, 1983.
17. Y. Y. Yeh and C. J. Wurrey, *J. Raman Spectrosc.* **15,** 55, 1984.
18. J. D. Odom, Z. Szafran, S. A. Johnston, Y. S. Li, and J. R. Durig, *J. Am. Chem. Soc.* **102,** 7173, 1980.
19. G. Schrumpf, *Spectrochim. Acta* **39A,** 505, 1983.

20. C. J. Wurrey, A. B. Nease, R. B. Blatt, and D. W. Solas, *Spectrochim. Acta* **35A,** 151, 1979.
21. G. Schrumpf, *Spectrochim. Acta* **39A,** 511, 1983.
22. J. R. Durig, P. L. Trowell, Z. Szafran, S. A. Johnston, and J. D. Odom, *J. Mol. Struct.* **74,** 85, 1981.
23. J. D. Odom, S. V. Saari, A. B. Nease, Z. Szafran, and J. R. Durig, *J. Raman Spectrosc.* **12,** 111, 1982.
24. C. J. Wurrey, Y. Y. Yeh, M. D. Weakley, and V. F. Kalasinsky, *J. Raman Spectrosc.* **15,** 179, 1984.
25. C. J. Wurrey, R. Krishnamoorthi, S. Pechsiri, and V. F. Kalasinsky, *J. Raman Spectrosc.* **12,** 95, 1982.
26. V. F. Kalasinsky and C. J. Wurrey, *J. Raman Spectrosc.* **9,** 315, 1980.
27. F. A. Miller, R. J. Capwell, R. C. Lord, and D. G. Rea, *Spectrochim. Acta* **28A,** 603, 1972.
28. J. R. Durig and A. C. Morrissey, *J. Chem. Phys.* **46,** 4854, 1967.
29. J. R. Durig, J. N. Willis, Jr., and W. H. Green, *J. Chem. Phys.* **54,** 1547, 1971.
30. J. R. Durig and W. H. Green, *J. Chem. Phys.* **47,** 673, 1967.
31. J. R. Durig and W. H. Green, *Spectrochim. Acta* **25A,** 849, 1969.
32. W. G. Rothschild, *J. Chem. Phys.* **45,** 3599, 1966.
33. K. Hanai and J. E. Katon, *J. Mol. Struct.* **70,** 127, 1981.
34. J. E. Katon, R. O. Carter, and W. Yellin, *J. Mol. Struct.* **11,** 347, 1972.
35. D. A. C. Compton, J. L. Green, M. A. S. Hazle, and J. G. Grasselli, *Spectrochim. Acta* **39A,** 541, 1983.
36. E. Fabrégue, G. Cassanas, L. Bardet, R. Escale, and J. Chapat, *J. Raman Spectrosc.* **14,** 206, 1983.
37. R. O. Carter, J. E. Katon, and F. F. Bentley, *Appl. Spectrosc.* **25,** 378, 1972.
38. K. E. Sterin, V. T. Aleksanyan, and G. N. Zhizhin, "Raman Spectra of Hydrocarbon Compounds," Hayka, Moscow, 1976 (in Russian, title only translated).
39. E. J. Rosenbaum, A. V. Grosse, and H. F. Jacobson, *J. Am. Chem. Soc.* **61,** 689, 1939.
40. G. Varsanyi, "Assignments for Vibrational Spectra of Seven Hundred Benzene Derivatives," John Wiley & Sons, New York, 1974, Vol. 1, Chap. 1.
41. R. M. P. Jaiswal and G. A. Crowder, *Can. J. Spectrosc.* **28,** 160, 1983.
42. G. A. Crowder and R. M. P. Jaiswal, *J. Mol. Struct.* **102,** 145, 1983.
43. G. A. Crowder and L. Gross, *J. Mol. Struct.* **102,** 257, 1983.
44. G. A. Crowder and L. Gross, *J. Mol. Struct.* **118,** 135, 1984.
45. G. A. Crowder and L. Gross, *Chem. Scr.* **25,** 150, 1985.
46. G. A. Crowder and D. Hill, *J. Mol. Struct.* **145,** 69, 1986.
47. G. A. Crowder, *J. Mol. Struct.* **145,** 77, 1986.
48. "The Sadtler Standard Raman Spectra," Sadtler Research Laboratories, 1976. (a) See subject index: alkanes. (b) Spectra 213R, 215R, 216R, 605R, 647R, 649R.
49. F. A. Miller and R. G. Inskeep, *J. Chem. Phys.* **18,** 1519, 1950.
50. F. H. Kruse and D. W. Scott, *J. Mol. Spectrosc.* **20,** 276, 1966.
51. I. O. C. Ekejiuba and H. E. Hallam, *Spectrochim. Acta* **26A,** 59, 1970.
52. J. R. Durig, J. M. Karriker, and W. C. Harris, *Spectrochim. Acta* **27A,** 1955, 1971.
53. K. Kalyanasundaram and J. K. Thomas, *J. Phys. Chem.* **80,** 1462, 1976.
54. A. M. A. da Costa, C. F. G. C. Geraldes, and J. J. C. Teixeira-Dias, *J. Raman Spectrosc.* **13,** 56, 1982.
55. I. Harada, H. Takeuchi, M. Sakakibara, H. Matsuura, and T. Shimanouchi, *Bull. Chem. Soc. Jpn.* **50,** 102, 1977.
56. R. G. Snyder, *J. Chem. Phys.* **47,** 1316, 1967.

57. M. Sakakibara, H. Matsuura, I. Harada, and T. Shimanouchi, *Bull. Chem. Soc. Jpn.* **50,** 111, 1977.
58. M. Ohta, Y. Ogawa, H. Matsuura, I. Harada, and T. Shimanouchi, *Bull. Chem. Soc. Jpn.* **50,** 380, 1977.
59. H. L. McMurry and V. Thornton, *Anal. Chem.* **24,** 318, 1952.
60. N. B. Colthup, L. H. Daly, and S. E. Wiberley, "Introduction to Infrared and Raman Spectroscopy," 3rd Ed., Academic Press, New York, 1990.
61. K. Nakanishi and P. H. Solomon, "Infrared Absorption Spectroscopy," 2nd Ed., Holden-Day, Inc., San Francisco, 1977.
62. J. E. Bertie and S. Sunder, *Can. J. Chem.* **51,** 3344, 1973.
63. S. Sunder, *Can. J. Chem.* **57,** 846, 1979.
64. G. A. Crowder and M.-Y. Lin, *J. Mol. Struct.* **64,** 193, 1980.
65. K. Martinsen, D. L. Powell, C. J. Nielsen, and P. Klaeboe, *J. Raman Spectrosc.* **17,** 437, 1986.
66. Y. Omura, J. Corset, and R. M. Moravie, *J. Mol. Struct.* **52,** 175, 1979.
67. K. O. Christe, *Spectrochim. Acta.* **27A,** 463, 1971.
68. P. Klaboe, D. Bougeard, B. Schrader, P. Paetzold, and C. von Plotho, *Spectrochim. Acta* **41A,** 53, 1985.
69. C. J. Wurrey and L. E. Firment, *Spectrochim. Acta* **30A,** 1115, 1974.
70. G. Schrumpf and H. Dunker, *J. Raman Spectrosc.* **17,** 433, 1986.
71. G. Schrumpf, *J. Raman Spectrosc.* **17,** 183, 1986.
72. G. Schrumpf, *Spectrochim. Acta* **39A,** 497, 1983.
73. W. G. Rothschild, *J. Chem. Phys.* **44,** 3875, 1966.
74. F. A. Miller and K. O. Hartman, *Spectrochim. Acta* **23A,** 1609, 1967.
75. J. M. Mallols, V. Tabacik, and S. Sportouch, *J. Raman Spectrosc.* **11,** 312, 1981.
76. P. M. Green, C. J. Wurrey, R. Krishnamoorthi, and Y. Y. Yeh, *J. Raman Spectrosc.* **17,** 355, 1986.
77. G. Schrumpf and H. Dunker, *Spectrochim. Acta* **41A,** 841, 1985.
78. R. J. Mureinik and C. W. F. Pistorius, *Spectrochim. Acta* **23A,** 1603, 1967.
79. R. W. Mitchell and J. Nakovisch, Jr., *Spectrochim. Acta* **29A,** 1153, 1973.
80. V. T. Aleksanyan, M. R. Aliev, M. Yu. Lukina, O. A. Nesmeyanova, and G. A. Khotimskaya, *Izv. Akad. Nauk. USSR, Ser. Kim.* 807, 1968; *Chem. Abstract* **69,** 31640n, 1968.
81. G. Schrumpf and T. Alshuth, *Spectrochim. Acta* **41A,** 1335, 1985.
82. J. D. Bartleson, R. E. Burk, and H. P. Lankelma, *J. Am. Chem. Soc.* **68,** 2513, 1946.
83. G. Schrumpf and T. Alshuth, *J. Mol. Struct.* **101,** 47, 1983.
84. J. E. Katon, W. R. Féairheller, Jr., and J. T. Miller, Jr., *J. Chem. Phys.* **49,** 823, 1968.
85. L. Bardet, G. Cassanas-Fabre, and E. Bourret, *J. Mol. Struct.* **28,** 45, 1975.
86. E. Fabrégue, G. Cassanas, and L. Bardet, *J. Raman Spectrosc.* **8,** 218, 1979.
87. V. F. Kalasinsky and T. S. Little, *J. Raman Spectrosc.* **9,** 224, 1980.
88. G. B. Carpenter and R. S. Halford, *J. Chem. Phys.* **15,** 99, 1947.
89. H. Takahashi, T. Shimanouchi, K. Fukushina, and T. Miyazawa, *J. Mol. Spectrosc.* **13,** 43, 1964.
90. C. W. Beckett, K. S. Pitzer, and R. Spitzer, *J. Am. Chem. Soc.* **69,** 2488, 1967.
91. S. B. Kim, R. M. Hammaker, and W. G. Fateley, *Spectrochim. Acta* **41A,** 167, 1985.
92. T. Woldbaek, C. J. Nielsen, and P. Klaeboe, *J. Mol. Struct.* **66,** 31, 1980.
93. T. Woldbaek, O. H. Ellestad, J. E. Gustavsen, and P. Klaeboe, *J. Mol. Struct.* **62,** 9, 1980.
94. T. Woldbaek, C. J. Nielsen, and P. Klaeboe, *Spectrochim. Acta* **41A,** 43, 1985.

95. J. J. C. Teixeira-Dias, L. A. E. Carvalho, A. M. M. de Costa, I. M. S. Lampreia, and E. F. G. Barbosa, *Spectrochim. Acta* **42A,** 589, 1986.
96. A. B. Dempster, D. B. Powell, and N. Sheppard, *Spectrochim. Acta* **28A,** 373, 1972.
97. N. Sheppard, *Trans. Faraday Soc.* **51,** 1465, 1955.
98. N. B. Colthup, *Appl. Spectrosc.* **34,** 1, 1980.
99. J. H. Lii and N. L. Allinger, *J. Am. Chem. Soc.* **111,** 8566, 1989.
100. P. L. Polavarapu and H. E. Smith, *J. Phys. Chem.* **92,** 1774, 1988.

CHAPTER 3

Halocompounds

3-1. MONOHALOALKANES

Most haloalkanes show strong bands in the low frequency region of their infrared and Raman spectra because of the C—X (X = F, Cl, Br, I) stretching vibrations. The frequencies of these C—X vibrations are affected by the mass of the halogen atom, X, and the force constant of the C—X bond. Progressive displacement to lower frequency regions for fluorine (1400–1000 cm^{-1}), chlorine (760–550 cm^{-1}), bromine (700–510 cm^{-1}), and iodine (600–480 cm^{-1}) are observed. In addition, the type of the halogen substitution (i.e., primary, secondary, and tertiary) as well as the molecular structure in the vicinity of the C—X bond (i.e., rotational isomers) demonstrate effects on the C—X stretching vibration. The C—X stretching frequencies, with the exception being the C—F mode, are fairly characteristic in their spectral positions. These characteristic frequencies are shifted progressively to lower values for primary, secondary, and tertiary halides. The correlation between C—X (X = Cl, Br, I) characteristic frequencies and the rotational conformations for the halides was recognized in the 1960s by the IR technique (1–3). A more recent vibrational analysis of bromides was carried out by Crowder and Jallian [4, 5]. The Raman C—X stretching frequencies of 55 halides are catalogued in a book by Dollish *et al.* [6]. In addition, vibrational analysis for *n*-alkylchlorides [7], *n*-alkylbromides [5], and *n*-alkyliodides [8–10] have been reported.

Since these C—X characteristic frequencies in haloalkanes are sensitive to rotational conformation within the molecule, the isomeric conformations of haloalkanes will be discussed along with their associated characteristic vibrational frequencies. Alkylfluorides do not possess the same correlation and group frequency as the other alkylhalides (C—Cl, C—Br, and C—I); hence, they will be discussed later in Section 3-6.

According to the notation system suggested by Mizushima *et al.* [11] and Krimm *et al.* [1, 12, 13], P, S, and T are used to specify the chemical type of the halides: primary, secondary, and tertiary, respectively. The subscript C specifies that a carbon atom is trans to the halogen atom, and the subscript H specifies that a hydrogen atom is trans to the halogen atom (Fig. 3-1). A summary of the rotational isomers of monohaloalkanes and their associated characteristic C—X stretching frequencies is given in Table 3-1.

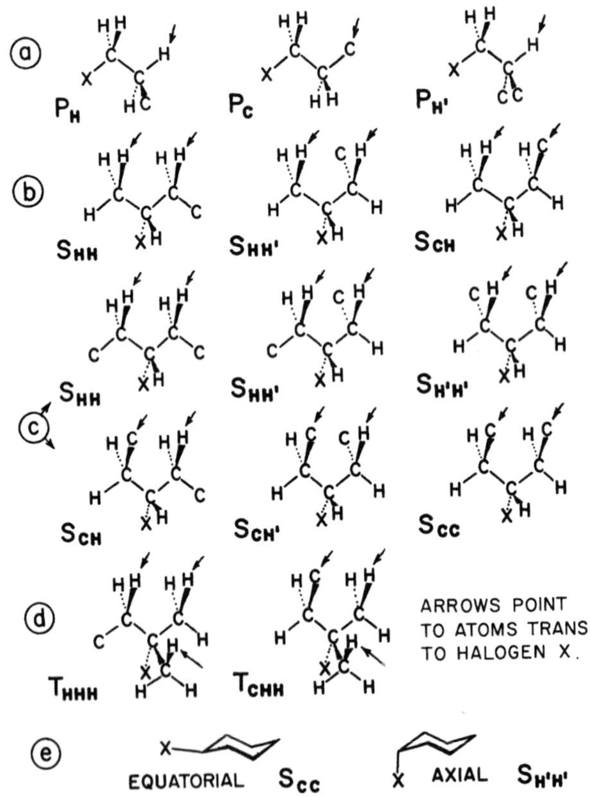

Fig. 3-1. Rotational isomers in haloalkanes, primary (P) (CH_2—X), secondary (S) (CH—X) and tertiary (T) (C—X). Atoms trans to halogen are indicated as subscripts (P_H) and with arrows on formulas.

Table 3-1.

Characteristic C—X Stretching Frequencies (cm^{-1}) of Monohaloalkanes[a]

	Cl			Br			I		
	Frequency	R	IR	Frequency	R	IR	Frequency	R	IR
A. Primary									
CH_3X	709			609			523		s
Other R—X:									
P_H	660–650	s	s	565–560	vs	s	510–500	vs	s
$P_{H'}$	690–680	s	s	625–615	s	s	590–580	s	s
P_C	730–720	s	s	650–640	vs	s	600–590	vs	s
B. Secondary									
S_{HH}	615–605	vs	s	540–535	s	s	495–485	s	s
$S_{HH'}$	635–630	ms	mw	590–575	m	m-w	590–575	m	w
$S_{H'H'}$	690–680	—	—	670–650	—	—	—		
S_{CH}	675–655	ms	m	620–605	m	s	585–575	m	s
$S_{CH'}$	670	s	—	—			—		
S_{CC}	760–740	var.	m	700–680	—	—	—		
C. Tertiary									
T_{HHH}	570–560	s	vs	520–510	vs	var.	495–485	s	s
$T_{HHH'}$	540		ms	—			—		
T_{CHH}	620–590	s-m	ms	590–580	m	s	580–570	s	s
$T_{CHH'}$	590		ms	—			—		

[a] The spectral ranges and Raman intensities are from Ref. 6. The infrared intensities are from Ref. 2.

3-1.1. Primary Haloalkanes (R—CH_2—X)

Figure 3-1 illustrates the possible rotational isomers. The C—X stretching vibrations give rise to strong IR and Raman bands. Ethylhalides are the simplest primary alkylhalides. There are three equivalent rotational isomers (P_H) for such compounds. Consequently, only one C—X vibrational band is possible. The frequencies for ethylchloride, bromide, and iodide are at 657 cm^{-1}, 559 cm^{-1}, and 499 cm^{-1}, respectively. Two vibrational bands are observed for other *n*-alkyl halides because of P_H and P_C (Fig. 3-1a) conformations [100]. Neu and Gwinn [14] showed that the P_C form is the dominant species in the liquid phase and becomes the only conformation in the solid phase. These two bands are Raman- and IR-active and occur at 730–720 cm^{-1} (P_C) and 660–650 cm^{-1} (P_H) for *n*-alkylchlorides, 650–640 cm^{-1} (P_C) and 560–565 cm^{-1} (P_H) for *n*-alkylbromides, and 600–590 cm^{-1} (P_C) and 510–500 cm^{-1} (P_H) for *n*-alkyliodides.

In the trans zigzag X—C—C—C (P_C) or X—C—C—H (P_H) group, the α carbon moves towards and away from the halogen during the X—C stretch vibration. Colthup *et al.* [86, 103] have shown that during this vibration in the P_H case, the C—C—H angle does not change appreciably, but that in the P_C case, the C—C—C angle is sharply bent in the X—C—C—C group. This adds an extra restoring force that increases the X—C stretching frequency in P_C related to P_H types.

The $P_{H'}$ and P_C forms are the possible conformations when branching occurs at the β-atom of a primary halide (Fig. 3-1a) [4, 6, 41–44]. The C—X vibrational frequency of the P_C form remains in the same region as the *n*-alkylhalides [4, 6, 41–45], but the frequencies of $P_{H'}$ are in a higher frequency region than those for the P_H form. $P_{H'}$ forms have bands in the regions of 690–680 cm^{-1}, 625–615 cm^{-1}, and 590–580 cm^{-1} for beta-branched alkylchlorides, alkylbromides, and alkyliodides, respectively. Only the band due to the P_C form will be observed when there are no hydrogens in the β position(s). The C—X characteristic frequencies will be in the same regions as *n*-alkylhalides if a primary halide branches at the third (γ) or higher carbon number in the skeletal chain.

The group CH_2—Cl has a strong IR band (or bands) at 1300–1240 cm^{-1} caused by the CH_2 wag vibration. Comparable IR bands are seen for CH_2-Br near 1230 cm^{-1} and CH_2—I near 1170 cm^{-1} [86].

3-1.2. Secondary Haloalkanes (R_2CH—X)

With the secondary halides, the situation is somewhat more complex than with the primary halides, since it is now necessary to consider and include the rotations about two C—C bonds. For example, the simplest secondary halide, 2-halopropane, has three equivalent rotational isomers of the S_{HH} form. Consequently, only one band is expected. This band is located at 617 cm^{-1}, 537 cm^{-1}, and 490 cm^{-1} for 2-chloro-, 2-bromo-, and 2-iodo-propanes, respectively, in liquid phase [45]. The 2-halobutanes and higher 2-halo homologs can exist in three isomeric forms: S_{HH}, S_{SC}, and $S_{HH'}$ (Fig. 3-1b) [46]. In the S_{HH} conformation, the first four carbon atoms are planar, but these carbons are nonplanar in the $S_{HH'}$ conformation. It must be noted that the H′ here is used to specify a rotation of the carbon skeleton away from the zigzag carbon plane, at the β position where the trans hydrogen is, rather than a branching on the β carbon atom as is found in the case of primary halides. See Figs. 3-1a and 3-1b. Published spectra [1–3, 6] show that S_{CH}, S_{HH}, and $S_{HH'}$, forms are all observable from the vibrational spectra of small secondary haloalkanes. For example, 2-chlorobutane

(Et—CHCl—Me) has three C—Cl bands at 670 cm^{-1} (S_{CH}, IR:m, R:m), 628 cm^{-1} ($S_{HH'}$, IR:m, R:m), and 609 cm^{-1} (S_{HH}, IR:s, R:s) in vibrational spectra [47]. However, the C—X band of $S_{HH'}$ isomers decreases in intensity progressively as the alkyl chain length increases. Eventually, this C—X ($S_{HH'}$) disappears for halides with alkyl groups larger than the butyl group. In this case, the two vibrational bands observed in the regions of 674–670 cm^{-1} and 615–611 cm^{-1} are attributed to S_{CH} and S_{HH} conformers, respectively. Similarly, infrared data show three bands for 2-bromoalkanes (Me—CHBr—R) with eight or fewer carbon atoms. These bands are in the ranges of 615–610 cm^{-1} (medium, S_{CH}), 578–579 cm^{-1} (medium, $S_{HH'}$), and 535–536 cm^{-1} (strong, S_{HH}) [6]. As the number of carbon atoms is equal to or greater than eight, only the C—Br (S_{CH}) and C—Br (S_{HH}) bands are observed at room temperature. The same trend is found for 2-iodoalkanes. For instance, 2-iodobutane (Me—CHI—Et) has three IR bands at 575 cm^{-1} (S_{CH}), 550 cm^{-1} ($S_{HH'}$), and 483 (S_{HH}) cm^{-1}, whereas only two IR bands at 581 (S_{CH}) and 490 (S_{HH}) cm^{-1} are found for 2-iodooctane (C_6H_{13}—CHI—Me).

Six rotational conformations are theoretically possible if the halogen atom is bound to a carbon atom located at the third or higher position from the end of an alkyl chain (Fig. 3-1c). Nevertheless, the concentrations of S_{CC} and $S_{H'H'}$ isomers are negligible in the majority of noncyclic cases because of the steric hindrance, and the more stable isomers such as S_{CH} and S_{HH} forms are predominant (see Fig. 3-1c). Thus, one will not necessarily see all of the six C—X stretching bands in the vibrational spectra for secondary halides of this type. The 3-chloroalkanes absorb infrared radiation in the ranges of 749–733 cm^{-1} (S_{CC}, IR:m), 667–661 cm^{-1} (S_{CH}, IR:m), 635–628 cm^{-1} ($S_{HH'}$, IR:w), and near 609 cm^{-1} (S_{HH}, IR:s).

3-1.3. Tertiary Haloalkanes (R_3C—X)

Although rotations about the three C—C bonds predict many rotational isomers in tertiary haloalkanes, the number of stable conformations is reduced by the steric hindrance effect. Therefore, the T_{HHH} and T_{CHH} isomers (Fig. 3-1d) are the predominant species for most tertiary halides; in some cases the $T_{HHH'}$ form is observed. The T_{HHH} conformation is the only rotational isomer of 2-halo-2-methylpropanes (Me_2C—X), which are the simplest tertiary halides. The vibrational bands at 570 cm^{-1}, 520 cm^{-1}, and 492 cm^{-1} were assigned to the C—X stretching frequencies of 2-chloro-, 2-bromo-, and 2-iodo-2-methylpropanes, respectively [6, 46]. These bands are quite strong in both Raman and infrared spectra. As the tertiary alkyl group increases its size, a second conformation, T_{CHH}, becomes observable in the vibrational spectra. For example, 2-halo-2-methylalkanes ((R$-$CMe$_2$$-$Cl), R > CH_3) usually have two vibrational bands in the regions of 570–560 cm^{-1} (R : s) and 620–590 cm^{-1} (R : m-s) due to C$-$Cl stretching modes of T_{HHH} and T_{CHH} forms, respectively. In addition to the commonly observed T_{HHH} and T_{CHH} isomers, a small amount of $T_{HHH'}$ isomers is found for 3-chloro-3-methylalkanes. Therefore, three bands ascribable to the C$-$Cl modes of T_{HHH}, T_{CHH}, and $T_{HHH'}$ conformations are present in the vibrational spectra.

3-2. EFFECT OF OTHER SUBSTITUENTS

3-2.1. α-Substituted Halocompounds

The substitution of a C≡N group to the same carbon atom that holds the halogen atom results in a relatively pronounced frequency shift on the C$-$X stretching mode. This effect is exemplified by the C$-$X stretching of Cl$-$CH$_2$$-$CN (746 cm^{-1}), Br$-CH_2$$-$CN (647 cm^{-1}), and I$-CH_2$$-$CN (597 cm^{-1}) [38, 39], as compared with the values for CH$_3$$-$Cl (709 cm^{-1}), CH$_3$$-$Br (609 cm^{-1}), and CH$_3$$-$I (523 cm^{-1}) [6]. In the case of allylic halides (=CH$-$CH$_2$X), the C$-$X bond can assume three different positions, namely cis, gauche, and trans, with respect to the double bond because of rotation about the single bond of the =C$-$C moiety. However, Diallo reported only two isomers by IR spectra. He found that the two C$-$Br stretching bands are at 650 cm^{-1} and 580 cm^{-1} for liquid (CH$_3$)$_2$C=CCH$_2$Br [35]. There is only one C$-$X stretching band when the $-$CH$_2$X group is placed on a C≡C group. The C$-$X band due to this ≡C$-$CH$_2$X grouping seems to occur in a region very close to the corresponding C$-$X frequencies of methylhalides (CH$_3$X). For example, the C$-$Cl stretching of H$-$C≡CCH$_2$Cl was assigned to an IR band at 710 cm^{-1} in the solution IR spectrum, while the corresponding Raman band at 707 cm^{-1} (liquid) was reported to be strong and polarized [40]. Similarly, the vibrational bands at 689 cm^{-1} and 609 cm^{-1} are assigned to the C$-$X stretching mode for CH$_3$$-$C≡C$-CH_2$Cl and CH$_3$$-$C≡CH$_2$Br, respectively [36]. Other examples are Cl$-$C≡C$-$CH$_2$Cl and Br$-$C≡C$-$CH$_2$Br, for which the CH$_2$$-$X stretching mode occurs at 709 cm^{-1} (solution) and 613 cm^{-1} (solution), respectively [37].

3-2.2. β-Substituted Halocompounds

The existence of other functional groups at the *β*-position with respect to the halogen atom can shift the C—X stretching frequencies out of the correlation region for monohaloalkanes. For instance, the C—X stretching frequencies of liquid $CH_2Cl—CH_2CN$ occurs at 755 cm^{-1} (P_C) and 663 cm^{-1} (P_H), and those of liquid $CH_2Br—CH_2CN$ are observed at 679 cm^{-1} (P_C) and 579 cm^{-1} (P_H) [34]. The coexistence of trans and gauche isomers for $Ar—O—CH_2CH_2—X$ (X=Cl, Br) is reported by Furer *et al.* [31]. For $Ar—O—CH_2CH_2—Cl$, the trans isomer (P_C) absorbs in a region near 750 cm^{-1} and the gauche isomer (P_H) at 674 cm^{-1}. The trans isomer of the bromo analogue exhibits a C—Br stretching band at 677 cm^{-1}, and the gauche isomer, at 564 cm^{-1}. Although two rotational isomers are expected for bromoethylcyclopropane (cyclo-$C_3H_5—CH_2CH_2Br$) and epibromohydrin (cyclo-$C_2H_3O—CH_2CH_2Br$), the C—Br stretching band is not useful in differentiating the isomers. Only one C—Br band at 622 cm^{-1} (liquid) is reported for the bromoethylcyclopropane compound, and the C—Br band of the epibromohydrin in liquid phase is observed at 640 cm^{-1} [48].

3-3. DIHALOGENATED AND POLYHALOGENATED ALKANES

3-3.1. Dihalocompounds

The C—X frequencies of dihalogenated *n*-alkanes, $X(CH_2)_nX$, which contain one halogen atom at each end of the straight carbon chain, can be approached by the correlations developed for primary monohaloalkanes. Application of such correlations was demonstrated by Sakakibara and co-workers [32, 33], who determined the conformational structures of $X(CH_2)_nX$ (X = Cl, Br; n = 5, 6; or X = I; n = 4–6). According to their results, these compounds possess the following C—X stretching frequencies: $Cl(CH_2)_nCl$ (P_C: 730–710 cm^{-1}; P_H: near 650 cm^{-1}), $Br(CH_2)_nBr$ (P_C: 660–626 cm^{-1}), and $I(CH_2)_nI$ (P_C: 615–576 cm^{-1}; P_H: near 497 cm^{-1}). It is apparent from the above results that the spectral ranges of the $X(CH_2)_nX$ type of compounds fall into wider regions than those of primary monohaloalkanes.

Vibrational spectra of dihalobutanes, in which the two halogen atoms are attached to different carbon atoms, have been extensively studied [9, 49–51]. The vibrational analysis of 2,4-dichloropentane and 2,4,6-trichloroheptane have been reported [47]. In general, the C—X frequencies of these compounds fall in a region near that of the corresponding monohalogenated alkanes (see Table 3-1). However, the spectra are usually more complicated, since the rotational isomers involving two different C—X bonds need to be considered.

The attachment of two halogen atoms to the same carbon atom, $-CX_2-$, results in symmetric and antisymmetric CX_2 stretching modes. Vibrational studies for compounds containing $-CCl_2-$ groups have been carried out [6, 52–56]. The CCl_2 antisymmetric stretching frequencies fall in a region between 705 and 650 cm^{-1}, and the symmetric CCl_2 stretching mode is located in the 641–547 cm^{-1} region. As a result of the increase in atomic mass, the CBr_2 stretching modes occur at lower frequency ranges than those of the CCl_2 grouping. Vibrational analysis on $CH_3CBr_2CH_3$ by Klaeboe [52] reveals that the antisymmetric and symmetric CBr_2 stretchings are situated at 597 cm^{-1} and 486 cm^{-1}, respectively, with strong IR and Raman intensities. Diallo reported the IR spectra of $C_2H_5CBr_2CH_3$ [57] and found three IR bands, possibly involving C—Br stretching of the various conformers, at 645 cm^{-1}, 580 cm^{-1}, and 503 cm^{-1}. According to the potential-energy distribution calculations, the CCl_2 and CBr_2 antisymmetric stretching modes are less coupled by other vibrational modes than the symmetric mode [55, 57]. Hence, the antisymmetric stretching mode exhibits more localized frequencies.

3-3.2. Trichloromethyl ($-CCl_3$) and Tribromomethyl ($-CBr_3$) Compounds

Nyquist reported the IR group frequency for the $-CCl_3$ moiety in vapor phase as 840–740 cm^{-1} (ν_{asy} CCl_3) and 730–660 cm^{-1} (ν_{sym} CCl_3) [99]. However, large frequency shifts may occur upon substitution, such as those found in CCl_3OH, which has the antisymmetric CCl_3 stretching band at 740 cm^{-1} and the symmetric mode at 398 cm^{-1} [60]. The CBr_3 antisymmetric stretching mode was predicted to occur near 665 cm^{-1} [99], whereas the $\nu_{asy}(CBr_3)$ of CBr_3CN is shifted to 712 cm^{-1} upon substitution of a $-CN$ group [61]. The $\nu_{sy}(CBr_3)$ frequency of this particular compound is located at 449 cm^{-1} [61]. CCl_3-CHO and CBr_3-CHO have been studied [59].

3-4. CYCLOALKYL HALIDES

3-4.1. Monohalocycloalkanes

The C—X stretching frequencies for some cycloalkyl monohalides are given in Table 3-2. Two C—X stretching frequencies are possible when the ring size is larger than three carbon atoms. The two C—X stretchings arise from the axial and equatorial conformations that are generated by the ring puckering or pseudorotation (Fig. 3-1e) [15, 16]. In general, the equatorial C—X is at higher stretching frequencies than the corresponding axial conformers [15, 17, 19, 27–30, 63, 64]. The intensities of these C—X stretching modes are usually medium to strong in both Raman and infrared spectra.

Alkylation on the cyclic group appears to increase the C—Br stretching frequency of bromocyclopropanes. Schrumpf [20] reported the C—Br stretching frequencies of *cis*- and *trans*-1-bromo-2,3-dimethylcyclopropanes to be located at above 600 cm^{-1}, which is about 50 cm^{-1} higher than the C—Br stretching frequency of bromocyclopropane (547 cm^{-1}). Similarly, elevated C—Br stretching frequencies are observed for other alkylated bromocyclopropanes such as *trans*-1,2-dimethyl-3-bromocyclopropane (656 cm^{-1}) and *cis*-1,2-dimethyl-3-bromocyclopropane (634 cm^{-1}) [20]. The same effects are observed for *cis*- and *trans*-1,2-dibromocyclopropanes, which exhibit C—Br stretching bands at 603 cm^{-1} and 551 cm^{-1}, as well as 604 cm^{-1} and 594 cm^{-1}, respectively [62].

In addition to the C—X stretching bands, vibrational bands involving the C—X bending vibrations are reported in the region below 500 cm^{-1} for cycloalkyl monohalides. Very strong Raman bands at 296 cm^{-1}, 298 cm^{-1}, and 365 cm^{-1} were assigned to the C—X bending mode of cyclopropyl bromide [17], cyclobutyl bromide [18], and cyclobutyl chloride [5], respectively. As in the case of the C—Br stretching mode, alkylation on the cyclic moiety shifts the C—Br bending mode to above 300 cm^{-1} in the vibrational spectra of for *cis*- and *trans*-1-bromo-2,3-dimethylcyclopropanes [20].

3-4.2. Dihalocycloalkanes

The vibrational spectra of cyclopropanes with more than one chloro or bromo substituent have been subjected to extensive studies [21–25, 62). Yeh and Wurrey suggested the tabulated characteristic IR and Raman vibrational

Table 3-2.

C—X Stretching Frequencies (cm^{-1}) of Cycloalkyl Monohalides

	F				Cl				Br				I			
	Frequency	R	IR	Ref.	Frequency	R	IR	Ref.	Frequency	R	IR	Ref.	Frequency	R	IR	Ref.
Cyclopropyl halide					623			15	547	s	s	17, 26				
Cyclobutyl halide	1098		vs	15	E[a]: 618 A[a]: 528	vw m	m m	19, 27	E: 534 A: 481	w s	w m	18, 19, 28				
Cyclopentyl halide	975 968		s s		E: 624 A: 588	vw vw	m m	15, 28 29	516	s	s	15	487	s		29
Cyclohexyl halide	1062, 1020 or 1129, 1053			15	E: 733 A: 688	s m		30, 17	E: 689 A: 660	s m		30	654			15

[a] E: equatorial, A: axial.

frequencies for cyclopropanyl chloride [23].

		IR and Raman
$-CClH-$	str.	900–500 cm^{-1}
	bend	300–200 cm^{-1}
$-CCl_2-$	antisym. str.	900–700 cm^{-1}
	sym. str.	600–500 cm^{-1}
	bend	below 500 cm^{-1}

In the case of gem-dibromo cyclopropanes, the characteristic CBr_2 bands are the antisymmetric CBr_2 stretching at 730–640 cm^{-1} and the symmetric mode at 520–355 cm^{-1} [22, 24, 25]. Woldbeak *et al.* [63, 64] examined the vibrational spectra of eight 1,4-dihalocyclohexanes. They reported prominent IR and Raman bands for these compounds in the spectral regions near the C—X stretching frequencies of monohalocyclohexanes.

3-5.

HALOGENATED ALKENES AND ETHYNES

Vibrational assignments of *cis*- and *trans*-1-chloropropene ($HClC{=}CH(CH_3)$) have been studied by Durig and Guirgis [65] who assigned the ethylenic C—Cl stretch mode of the cis isomer to 758 cm^{-1} (vap.; IR : s; R : vs, p) and that of the trans isomer to 805 cm^{-1} (vap.; IR : vs; R : m). In the case of $CH_2{=}C(CH_2X)Y$ (X, Y: halogen atoms), the =C—Cl stretching frequency is sensitive to the rotational structure. Anti and gauche isomers are two possible conformational isomers due to the rotation around the C—C single bond in the $=C-CH_2X$ structure. According to the normal coordinate analysis given by Klaeboe *et al.* [66], the C—Y stretching mode couples with other vibrations differently in anti and gauche forms. As a result, the C—Y stretching frequency of the anti isomer is about 50 cm^{-1} higher than that of the corresponding gauche isomer. For example, the ethylenic C—Cl stretching frequency of the anti conformer is located at 695 cm^{-1}, and that of the gauche isomer is at 633 cm^{-1} for $CH_2{=}CCl-CH_2Cl$ [66]. Similarly, the ethylenic C—Br stretching vibrations of anti and gauche $CH_2{=}CBr-CH_2Br$ were found at 594 cm^{-1} and 546 cm^{-1}, respectively.

Antisymmetric and symmetric $=CCl_2$ stretchings are characteristic vibrations for compounds with two chlorine atoms substituted on the same ethylenic carbon atom. In comparison with the vibrational frequencies of the saturated $-CCl_2-$ moiety, the antisymmetric mode of the $=CCl_2$ grouping can be found at a higher frequency region, whereas the symmetric mode is somewhat lower. The IR spectra of $CH_2{=}CCl_2$ is reported by Winther and Hummel [68] and the Raman spectra by Joyner and Glockler [69]. The bands at 785 cm^{-1} and 598 cm^{-1} were assigned to the antisymmetric and symmetric CCl_2 stretching modes, respectively. For $(CF_2H)ClC{=}CCl_2$, the antisymmetric and symmetric modes are located at 928 cm^{-1} (liq.; IR : vs; R : w, p) and 407 cm^{-1} (liq.; IR : m; R : vs, p) [66]. Miller *et al.* [67] suggested that the antisymmetric $=CCl_2$ stretching mode should occur between 1000 cm^{-1} and 800 cm^{-1}. These authors also assigned the strong IR band at 609 cm^{-1} to symmetric CCl_2 stretching of $CCl_2{=}CH{-}CH{=}CCl_2$.

The vibrational spectra of compounds containing the $\equiv C{-}X$ moiety (X: halogen atoms) have been subjected to extensive studies [70–77]. The C—Cl stretching mode of $HC{\equiv}C{-}Cl$ is found at 756 cm^{-1} (vap; IR) [71]. However, substitution of other functional groups, such as $-CH_3$, $-C{\equiv}CH$, $-CN$, and $-CHO$, lowers the stretching frequency to the region of 579–473 cm^{-1} [70, 73–76]. A similar effect on substitution is found for bromide analogues. For $HC{\equiv}C{-}Br$, the C—Br stretching mode absorbs at 618 cm^{-1} (vap; IR) [71], but is shifted to the region of 474–395 cm^{-1} for $CX{\equiv}C{-}Br$ (X = $-CH_3$, $-C{\equiv}CH$, $-CN$, $-CHO$) [70, 72–76]. Compounds containing $-C{\equiv}C{-}I$ grouping exhibit C—I stretching bands between 405 cm^{-1} and 360 cm^{-1} [70, 72–76].

3-6. FLUOROCOMPOUNDS

3-6.1. Monofluoroalkanes

A series of normal coordinate analyses for *n*-alkylfluorides has been carried out by Crowder and co-workers [79–82]. Unlike other alkylhalides, the C—F stretching mode couples strongly with other vibrational modes, especially the C—C stretchings. However, the presence of the C—F moiety may be recognized by the abnormally strong multiple IR bands in the 1400–1000 cm^{-1} regions. There is no correlation between the C—F vibrational frequencies and the conformational structures. Crowder and Lightfoot [82] observed that the IR bands involving C—F stretching are in a region from 1100 to 900 cm^{-1} for liquid *n*-alkylfluorides. These authors also found that the C—F bands appear to be broad and poorly resolved in the liquid state spectra, possibly because of the coexistence of trans and gauche isomers. However, the solid-state spectra consist of strong, sharp bands due to the predominant gauche (P_H) isomers. The C—F stretching band in the IR spectrum of fluorocyclobutane is assigned by Durig *et al.* [83] to a very intense IR absorption at 1098 cm^{-1} in gas phase. However, this band has not been observed in the Raman spectrum of the molecule. In the case of 1,4-difluoro-2-butyne ($CH_2F-C{\equiv}C-CH_2F$), the C—F stretching mode is observed to be at 986 cm^{-1} with very strong IR and Raman intensity [84]. Baldacci *et al.* [85] proposed that the strong IR absorption of gas-phase $CHFCl_2$ at 1079 cm^{-1} is due to a vibration involving the C—F stretching mode.

3-6.2. Difluoro ($-CF_2-$) and Trifluoromethyl ($-CF_3$) Compounds

Colthup *et al.* [86] suggested that the $-CF_2-$ moiety absorbs IR strongly in the 1280–1120 cm^{-1} region. This correlation is confirmed by vibrational analysis on chlorodifluoromethane ($CHClF_2$) [87] and 2,2-difluoropropane ($CH_3-CF_2-CH_3$) [88]. The antisymmetric and symmetric CF_2 stretching vibrations of $CHClF_2$ are assigned to the bands at 1127 cm^{-1} and 1107 cm^{-1}, respectively. For $CH_3-CF_2-CH_3$, the $\nu_{as}(CF_2)$ and $\nu_s(CF_2)$ vibrations are observed at 1246 cm^{-1} and 1200 cm^{-1} as very intense IR bands. The CF_2 stretching frequencies may shift beyond the characteristic region of 1280–1120 cm^{-1} when the moiety is included in a small ring structure. For example, the CF_2 vibrations of hexafluorocyclopropane (cyclo-C_4F_6) were found at 1369 cm^{-1}, 1280 cm^{-1}, 1278 cm^{-1}, and 737 cm^{-1} [90]. It was suggested the four- and five-membered cyclic CF_2 compounds absorb in the 1350–1140 cm^{-1} region [89].

Compounds containing $-CF_3$ grouping have been subjected to extensive vibrational analysis [91–98]. The strong IR absorptions in the 1350–1120 cm^{-1} region are attributable to vibrations involving the $-CF_3$ group [86]. However, trifluoromethanol exhibits CF_3 frequencies outside this spectral region. The antisymmetric and symmetric CF_3 stretching frequencies of CF_3OH are located at 1401 cm^{-1} and 905 cm^{-1}, respectively [97]. Furthermore, it is difficult to distinguish CF_3 from the CF_2 moiety simply by vibrational spectra, since they absorb in very similar regions.

3-7. ARYL HALIDES

When a halogen is directly attached to a benzene ring, the C—halogen stretch vibration tends to interact with ring vibrations. One substituent sensitive ring vibration that involves some stretching of the C—X bond is found in the IR at about 1270–1100 cm^{-1} for aryl-F compounds. Aryl-Cl IR bands are seen at 1057–1034 cm^{-1} for ortho, 1078–1074 cm^{-1} for meta, and 1096–1089 cm^{-1} for para. Aryl-Br IR bands are seen at 1042–1028 cm^{-1} for ortho, 1073–1065 cm^{-1} for meta, and 1073–1068 cm^{-1} for para [101, 102].

REFERENCES

1. J. J. Shipman, V. L. Folt, and S. Krimm, *Spectrochim. Acta* **18,** 1603, 1962.
2. F. F. Bentley, N. T. McDevitt, and A. L. Rozek, *Spectrochim. Acta* **20,** 105, 1964.
3. P. N. Gates, E. F. Mooney, and H. A. Willis, *Spectrochim. Acta* **23A,** 2043, 1967.
4. G. A. Crowder and M.-R. Jallian, *Spectrochim. Acta* **34A,** 707, 1978.
5. G. A. Crowder and M.-R. Jallian, *Can. J. Spectrosc.* **22,** 1, 1977.
6. F. R. Dollish, W. G. Fateley, and F. F. Bentley, "Characteristic Raman Frequencies of Organic Compounds," John Wiley & Sons, Inc., 1974.
7. R. G. Snyder and H. H. Schachtschneider, *J. Mol. Spectrosc.* **30,** 290, 1969.
8. G. A. Crowder, *J. Mol. Spectrosc.* **18,** 467, 1973.
9. G. A. Crowder and S. Ali, *J. Mol. Struct.* **27,** 43, 1975.
10. G. A. Crowder and M. Jallian, *J. Mol. Struct.* **33,** 127, 1976.
11. S. Mizushima, T. Shimanouchi, K. Nakamura, M. Hayushi, and T. Tsuchiya, *J. Chem. Phys.* **26,** 970, 1957.
12. C. G. Opaskar and S. Krimm, *Spectrochim. Acta* **23A,** 2261, 1967.
13. J. J. Shipman, V. L. Folt, and S. Krimm, *Spectrochim. Acta* **24A,** 437, 1968.
14. J. Neu and W. Gwinn, *J. Chem. Phys.* **18,** 1642, 1950.
15. W. G. Rothschild, *J. Chem. Phys.* **45,** 1214, 1966.
16. I. O. C. Ekejiuba and H. E. Hallam, *J. Mol. Struct.* **6,** 341, 1970.
17. C. J. Wurrey, R. J. Berry, Y. Y. Yeh, T. S. Little, and V. F. Kalasinsky, *J. Raman Spectrosc.* **14,** 87, 1983.
18. J. R. Durig and W. H. Green, *J. Chem. Phys.* **47,** 673, 1967.
19. J. R. Durig, W. H. Green, and N. C. Hammond, *J. Phys. Chem.* **70,** 1989, 1966.
20. G. Schrumpf, *Spectrochim. Acta* **39A,** 505, 1983.
21. C. J. Wurrey, R. B. Blatt, and A. B. Nease, *J. Phys. Chem.* **81,** 24, 1977.
22. G. Schrumpf, *Spectrochim. Acta* **39A,** 487, 1983.
23. Y. Y. Yeh and C. J. Wurrey, *J. Raman Spectrosc.* **15,** 55, 1984.
24. C. J. Wurrey and L. E. Firment, *Spectrochim. Acta* **30A,** 1115, 1974.
25. C. J. Wurrey, A. B. Nease, R. B. Blatt, and D. W. Solas, *Spectrochim. Acta* **35A,** 151, 1979.
26. W. G. Rothschild, *J. Chem. Phys.* **44,** 3875, 1966.
27. J. R. Durig and A. C. Morrissey, *J. Chem. Phys.* **48,** 4854, 1966.
28. J. R. Durig, J. M. Karriker, and D. W. Wertz, *J. Mol. Spectrosc.* **31,** 237, 1969.
29. I. O. C. Ekejiuba and H. E. Hallam, *Spectrochim. Acta* **26A,** 59, 1969.
30. M. Rey-Lafon, C. Rouffi, M. Camiade, and M. Forel, *J. Chim. Phys.* **67,** 2030, 1970.
31. V. L. Furer, V. E. Kataev, A. M. Salikhova, S. N. Valitova, and A. N. Vereshchagin, *Bull. Acad. Sci. USSR, Div. Chem.* **35** (part I), 1372, 1986.
32. M. Sakakibara, H. Matsuura, and H. Murata, *J. Mol. Struct.* **55,** 21, 1979.
33. M. Sakakibara, Y. Yonemura, Z. Tanaka, H. Matsuura, and H. Murata, *J. Mol. Struct.* **70,** 169, 1981.
34. P. Klaeboe and J. Grundnes, *Spectrochim. Acta* **24A,** 1905, 1968.

35. A. O. Diallo, *Spectrochim. Acta* **39A,** 613, 1983.
36. R. D. McLachlan, *Spectrochim. Acta* **23A,** 1793, 1967.
37. R. A. Nyquist, A. L. Johnson, and Y.-S. Lo, *Spectrochim. Acta* **21,** 77, 1965.
38. R. G. Jones and W. J. Orville-Thomas, *J. Chem. Soc. London*, 4632, 1965.
39. J. R. Durig and D. W. Wertz, *Spectrochim. Acta* **24A,** 21, 1968.
40. R. A. Nyquist, T. L. Reder, G. R. Ward, and G. J. Kallos, *Spectrochim. Acta* **27A,** 541, 1971.
41. G. A. Crowder, C. Harper, and M. R. Jalilian, *J. Mol. Struct.* **64,** 193, 1980.
42. G. A. Crowder, C. Harper, and M. R. Jalilian, *J. Mol. Struct.* **64,** 193, 1980.
43. P. Klaeboe, C. J. Nielsen, and D. L. Powell, *Spectrochim. Acta* **41A,** 1315, 1985.
44. K. Martinsen, D. L. Powell, C. J. Nielsen, and P. Klaeboe, *J. Raman Spectrosc.* **17,** 437, 1986.
45. P. Klaeboe, *Spectrochim. Acta* **26A,** 87, 1970.
46. S. Sunder, *Can. J. Chem.* **57,** 846, 1979.
47. W. O. George, J. E. Goodfield, and W. F. Maddams, *Spectrochim. Acta* **41A,** 1243, 1985.
48. C. J. Wurrey, R. Krishnamoorthi, S. Pechsiri, and V. F. Kalasinsky, *J. Raman Spectrosc.* **12,** 95, 1982.
49. G. A. Crowder and C. Smith, *J. Mol. Struct.* **51,** 157, 1979.
50. G. A. Crowder and S. Ali, *J. Mol. Struct.* **34,** 47, 1976.
51. R. G. Snyder, *J. Mol. Spectrosc.* **28,** 273, 1968.
52. P. Klaeboe, *Spectrochim. Acta* **26A,** 977, 1970.
53. G. A. Crowder and W.-Y. Lin, *J. Mol. Struct.* **62,** 1, 1980.
54. G. A. Crowder, *J. Mol. Struct.* **70,** 193, 1981.
55. R. J. A. Riberiro-Claro, A. M. D'A Rocha Gonsalves, and J. J. C. Teixeira-Dias, *Spectrochim. Acta* **41A,** 1055, 1985.
56. A. Hacura, T. W. Zerda, and M. Kaczmarski, *J. Raman Spectrosc.* **11,** 437, 1981.
57. A. O. Diallo, *Spectrochim. Acta* **38A,** 687, 1982.
58. J. R. Durig and W. J. Natter, *J. Raman Spectrosc.* **11,** 32, 1981.
59. G. Lucazeau and A. Novak, *Spectrochim. Acta* **25A,** 1615, 1969.
60. H. F. Shurvell and D. W. James, *J. Raman Spectrosc.* **14,** 36, 1983.
61. J. F. Arenas, J. I. Maros, and M. T. Suero, *J. Raman Spectrosc.* **15,** 132, 1984.
62. G. Schrumpf and H. Dunker, *J. Raman Spectrosc.* **17,** 433, 1986.
63. T. Woldbeak, O. H. Ellestad, J. E. Gustavsen, and P. Klaeboe, *J. Mol. Struct.* **62.** 9, 1980.
64. T. Woldbeak, C. J. Nielsen, and P. Klaeboe, *J. Mol. Struct.* **66,** 31, 1980.
65. J. R. Durig and G. A. Guirgis, *J. Raman Spectrosc.* **13,** 160, 1982.
66. P. Klaeboe, G. Neerland, and S. H. Schei, *Spectrochim. Acta* **38A,** 1025, 1982.
67. F. A. Miller, W. F. Elbert, and W. Pingitore, *J. Mol. Struct.* **41,** 19, 1977.
68. F. Winther and D. O. Hummel, *Spectrochim. Acta* **23A,** 1839, 1967.
69. P. Joyner and G. Glockler, *J. Chem. Phys.* **20,** 302, 1952.
70. A. G. Meister, *J. Chem. Phys.* **16,** 950, 1948.
71. G. R. Hunt and M. K. Wilson, *J. Chem. Phys.* **34,** 1301, 1961.
72. R. A. Nyquist, *Spectrochim. Acta* **21,** 1245, 1965.
73. P. Klaeboe, E. Kloster-Jensen, and S. J. Cyvin, *Spectrochim. Acta* **23A,** 2733, 1967.
74. P. Klaeboe and E. Kloster-Jensen, *Spectrochim. Acta* **23A,** 1981, 1967.
75. D. H. Christensen, I. Johnsen, P. Klaeboe, and E. Kloster-Jensen, *Spectrochim. Acta* **25A,** 1569, 1969.
76. E. Lagset, P. Klaeboe, E. Kloster-Jensen, S. J. Cyvin, and F. M. Nicolaisen, *Spectrochim. Acta* **29A,** 17, 1973.

77. E. Augdahl, E. Kloster-Jensen, V. Devarajan, and S. J. Cyvin, *Spectrochim. Acta* **29A,** 1329, 1973.
78. G. A. Crowder and S. Ali, *J. Mol. Struct.* **25,** 377, 1975.
79. G. A. Crowder and H. K. Mao, *J. Mol. Struct.* **18,** 33, 1973.
80. G. A. Crowder and H. K. Mao, *J. Mol. Struct.* **16,** 165, 1973.
81. G. A. Crowder and H.-S. Mao, *J. Mol. Struct.* **23,** 162, 1974.
82. G. A. Crowder and J. M. Lightfoot, *J. Mol. Struct.* **99,** 77, 1983.
83. J. R. Durig, J. M. Willis, Jr., and W. H. Green, *J. Chem. Phys.* **54,** 1547, 1971.
84. A. Karlsson, P. Klaeboe, K.-M. Marstokk, H. Mollendal, and C. J. Nielsen, *Acta Chem. Scand.* **40A,** 374, 1986.
85. A. Baldacci, A. Pesserini, and S. Ghersetti, *Spectrochim. Acta* **40A,** 165, 1984.
86. N. B. Colthup, L. H. Daly, and S. E. Wiberley, "Introduction to Infrared and Raman Spectroscopy," 3rd Ed., Academic Press, New York, 1990.
87. J. V. Magill, K. M. Gough, and W. F. Murphy, *Spectrochim. Acta* **42A,** 705, 1986.
88. G. A. Crowder and D. Jackson, *Spectrochim. Acta* **27A,** 2505, 1971.
89. D. G. Weiblen, in "Fluorine Chemistry," Vol. 2 (J. H. Simons, ed.), Academic Press Inc., New York, 1954, Chapter 7.
90. F. A. Miller and K. O. Hartman, *Spectrochim. Acta* **23A,** 1609, 1967.
91. C. V. Berney, L. R. Cousins, and F. A. Miller, *Spectrochim. Acta* **19,** 2019, 1963.
92. C. V. Berney, *Spectrochim. Acta* **20,** 1437, 1964.
93. G. Hagen, *Acta Chem. Scand.* **25,** 813, 1971.
94. C. V. Berney, *Spectrochim. Acta* **27A,** 663, 1971.
95. H. Wolff, D. Horn, and H.-G. Rollar, *Spectrochim. Acta* **29A,** 1835, 1973.
96. M. Perttila, *Spectrochim. Acta* **35A,** 585, 1979.
97. J. S. Francisco, *Spectrochim. Acta* **40A,** 923, 1984.
98. R. A. Yadav and I. S. Singh, *Spectrochim. Acta* **41A,** 191, 1985.
99. R. A. Nyquist, "The Interpretation of Vapor-Phase Infrared Spectra," Group Frequency, Sadtler Research Laboratories, 1984, pp. 523–608.
100. J. K. Brown and N. Sheppard, *Proc. Roy. Soc.* **231A,** 555, 1955.
101. A. Stojiljkovic and D. Whiffen, *Spectrochim. Acta* **12,** 47, 1958.
102. A. R. Katritzky and J. M. Lagowski, *J. Chem. Soc. London*, 2421, 1960.
103. N. B. Colthup, *Spectrochim. Acta* **20,** 1843, 1964.

CHAPTER 4

Alcohols and Phenols

4-1. INTRODUCTION

The characteristic vibrations involving the O—H grouping are the stretching and bending modes of the O—H and C—O moieties. A list of the group frequencies for alcohols and phenols is given in Table 4-1.

Excellent discussions of the group frequencies of alcohols and phenols in IR and Raman spectra have been given by Bellamy [1] and by Dollish *et al.* [2]. The normal coordinate analysis has been reported for methanol [3], *n*-propanol, *n*-butanol [5], and *n*-$C_{12}H_{25}OH$ through *n*-$C_{37}H_{72}OH$ [6]. The IR and Raman spectra of ethanol, isopropanol and their deuterated analogs were studied by Tanaka [7] and Perchard and Josien [8]. Durig and coworkers have reported the vibrational spectra of cyclobutanol [33] and cyclopentanol [34] in vapor and liquid phases. The vibrational assignments of β-cyclopropylidenic alcohols were proposed by Guiliano *et al.* [51]. The Raman spectra of CF_3CH_2OH and CCl_3CH_2OH were reported by Perttilä [52]. The vibrational spectra of phenol [10–14, 18–22] and substituted phenols [23–32] have been subjected to extensive studies.

4-2. O—H STRETCHING

Since the O—H stretching shows very weak scattering in the Raman spectrum, only the characteristic IR absorption will be emphasized here for

Table 4-1.

Characteristic Frequencies (cm^{-1}) of Alcohols and Phenols

		IR	R
Free OH Stretch (CCl_4 solution):			
primary alcohols	3644–3635	m ($\Delta\nu$ = 20–26)	w
secondary alcohols	3637–3626	m ($\Delta\nu$ = 20–33)	w
tertiary alcohols	3625–3614	m ($\Delta\nu$ = 14–23)	w
tertiary bicyclic alcohols	3612–3606	m ($\Delta\nu$ = 14–23)	w
phenols	3612–3593	m	w
Hydrogen-Bonded OH Stretch:			
Intermolecular:			
dimeric	3600–3450	m	
polymeric	3400–3200	vs, br	w
Intramolecular:			
single bridge	3600–3500	m	
chelation	3200–2500	var. br.	
Out-of-phase C—C—O Stretch:			
primary alcohols	1075–1000	s	m-s
secondary alcohols	1150–1075	s	m-s
tertiary alcohols	1210–1100	s	m-s
phenols	1260–1180	s	
In-phase C—C—O Stretch:			
primary alcohols	900–800	m	vs-m
secondary alcohols	900–800 (mostly ca. 820)	m	vs-m
tertiary alcohols	800–750	m	vs, p
	~1000	m-s	w-m
Involving C—O—H Bend:			
primary alcohols	1430–1200	w	m-w
secondary	1430–1200	w	m-w
tertiary	1410–1310	w	m-w
phenols	1410–1310	m	
C—C—O Bend:			
primary alcohols	460–430		m-w
secondary alcohols	~500		m-w
tertiary alcohols	~360		m-w

the O—H stretching mode. The O—H stretching band in the IR spectrum is by far the most characteristic IR band of alcohols and phenols. The free O—H vibration occurs as a sharp IR band at above 3600 cm^{-1}, while for the majority of condensed-phase (liquid or solid) samples, this OH band is broadened and shifted to 3500–2500 cm^{-1} due to various types of hydrogen bond formation. Characteristically, the presence of this sharp or broad band suggests a good possibility of the presence of the OH group. However, it should be noted here that the NH stretching frequencies occur in this same spectral range; hence, the NH stretchings can be an interference in the identification of the OH group, and vice versa. The OH bands of water also come in this region.

When the concentrations are high enough to allow self-association, hydrogen bonding is commonly observed for compounds containing hydroxyl groups. The presence of hydrogen bonds can be verified not only by the lowering of the O—H stretching band frequency, but also by the increase in the band width and intensity. The O—H stretching frequencies were found by Piaggio *et al.* [9] to be correlated with the lengths of hydrogen bonds in hydroxyl chains. In some cases, there exists more than one type of hydrogen bonding in a given solution. Consequently, more than one O—H band will be observed. For example, a multiple-band feature in the O—H stretch region is reported for β-cyclopropylidenic alcohols, measured as dilute solution, due to the O—H stretch vibrations of free OH groups and various types of hydrogen-bonded OH moieties [51].

4-2.1. Free O—H Stretching in Aliphatic Alcohols

VAPOR PHASE

Nyquist has correlated the sharp IR bands in the regions of 3680–3670, 3670–3658, and 3648–3640 cm^{-1} with the O—H stretching mode of primary, secondary, and tertiary alcohols in the vapor phase. The progressive lowering of O—H stretching frequencies from primary alcohols through secondary to tertiary alcohols is due to the inductive effect of the alkyl groups [32]. Differentiation of the three types of alcohols by their O—H stretching frequencies is possible since their spectral regions of ν(O—H) are fairly well separated. Ring strain appears to have little effect on the ν(O—H) frequency because both cyclobutanol [33] and cyclopentanol [34] exhibit a sharp IR band within the region expected for open-chain secondary alcohols (3670–3658 cm^{-1}). The O—H vibration is very sensitive to the molecular configuration; hence, a splitting of the O—H stretching bands in vapor phase is possible when there is coexistence of nonequivalent rotational conformers [33, 58].

DILUTE SOLUTION (IN NONPOLAR SOLVENTS)

When measured as monomers in a dilute solution with a nonpolar solvent (e.g., CCl_4 and CS_2), the sharp IR band is shifted toward lower frequency regions as compared with those measured in vapor phase. For example,

Bellamy [1] has quoted the OH absorption ranges of primary, secondary, and tertiary alcohols as 3643–3630 cm^{-1}, 3635–3620 cm^{-1}, and 3620–3600 cm^{-1}, respectively, for IR frequencies obtained from dilute solution. Recently, van der Maas and Lutz [15] carried out IR measurements on 70 monohydric saturated alcohols in dilute CCl_4 solution (approximately 0.01 M). According to these authors, the IR absorptions of O—H stretching frequencies fall in the 3644–3635 cm^{-1}, 3637–3626 cm^{-1}, and 3625–3614 cm^{-1} regions for primary, secondary, and tertiary alcohols, respectively. Their results agreed remarkably well with the spectral regions quoted by Nakanishi and Solomon [16]. The decrease in the O—H stretch frequency in CCl_4 solution, as compared with the vapor-phase data, can be attributed to the very weak hydrogen-bond interaction between the O—H group and the chlorine atoms of the surrounding solvent molecules (CCl_4). Such interaction between the O—H group and the solvent molecules is further supported by the data measured in dilute CS_2 solution, where another 10–20 cm^{-1} downward shift from the frequency measured in CCl_4 solution is expected [27, 35, 53–57].

Regardless of the interaction between the OH moiety and the CCl_4 (or CS_2) molecules, the OH bands measured in these solvents as dilute solutions are quite characteristic and relatively sharp. The band width ($\Delta\nu$) is generally measured at half the maximum band intensity. In this case, $\Delta\nu$ = 15–36 cm^{-1}. Many authors consider the O—H bands measured in dilute CCl_4 solutions to be free O—H bands. Useful structural information, such as rotational conformation, fixation effect of the β-CH_2 moieties, the shielding effects of the neighboring alkyl groups, and other intramolecular nonbonding interactions, etc., can be derived from the O—H stretching frequencies, the band width, and the band shape measured in the nonpolar solvents. As in the vapor-phase IR spectra of alcohols, the presence of different rotational isomers causes antisymmetric band shape or a doublet character of the O—H stretching band [59–66]. Comprehensive studies on the relationship between the molecular structure and the O—H stretching frequency, the band width, and the band shape of bands in dilute CCl_4 and CS_2 solutions have been recently carried out by Lutz, Visser, and van der Maas [53–56].

Some typical free OH stretching frequencies measured in dilute CCl_4 solution are summarized in Table 4-2. As illustrated by this table, the secondary monocyclic alcohols (Table 4-2d) absorb infrared in a similar region as open-chain secondary alcohols in CCl_4 dilute solution (Table 4-2b). Tertiary monocyclic alcohols (Table 4-2e) absorb in the same region as that found for open-chain tertiary alcohols (Table 4-2c). These results demonstrate again that ring strain has a negligible effect on the O—H stretching frequency. However, the free O—H stretching frequencies of

Table 4-2.

Free O—H Stretching Frequencies (cm^{-1}) in CCl_4 Solution (after Ref. 15)

(a) Primary Alcohols		(b) Secondary Alcohols	
CH_3OH	3643.7	$CH_3-CH(OH)-CH_3$	3626.1
C_2H_5OH	3634.8	$CH_3-CH(OH)-C_2H_5$	3627.0
n-C_3H_7OH	3638.5	$CH_3CH(OH)$-n-C_3H_7	3626.3
n-C_4H_9OH	3638.8	$CH_3-CH(OH)$-n-C_5H_{11}	3626.3
n-$C_9H_{19}OH$	3637.6	$CH_3-CH(OH)$-n-C_6H_{13}	3627.4
n-$C_{12}H_{25}OH$	3638.2	$CH_3-CH(OH)$-i-C_4H_9	3628.5
$(CH_3)_2CHCH_2OH$	3640.2	$(C_2H_5)_2CHOH$	3631.1
$(CH_3)_3C(CH_2)_2OH$	3636.8	(n-$C_4H_9)_2CHOH$	3632.9
(c) Tertiary Alcohols		(d) Cyclic Secondary Alcohols	
$(CH_3)_3COH$	3617.0	cyclobutanol	3626.7
$(CH_3)_2C(OH)-C_2H_5$	3616.9	cyclopentanol	3625.5
$(CH_3)_2C(OH)$-n-C_3H_7	3616.9	cyclohexanol	3623.9
$(CH_3)_2C(OH)$-n-C_4H_9	3617.0	cycloheptanol	3621.7
$(C_2H_5)_2C(OH)-CH_3$	3618.1	cyclooctanol	3621.4
(n-$C_3H_7)_2C(OH)-CH_3$	3617.0	cyclododecanol	3625.7
(e) Cyclic Tertiary Alcohols		(f) Secondary Alcohols β-Branched	
1-ethyl-cyclopentanol	3617.6	(i-$C_3H_7)_2CHOH$	3641.5, 3614.0
1-methyl-cyclohexanol	3612.7	(s-$C_4H_9)_2CHOH$	3640.4, 3615.0
1-ethyl-cyclohexanol	3615.0	(t-$C_4H_9)_2CHOH$	3651.7, 3624.6
		(*cyclo*-$C_6H_{11})_2CHOH$	3641.5, 3615.0
(g) Bicyclic Tertiary Alcohols		t-$C_4H_9-CH(OH)-C_2H_5$	3641.7, 3616.3
adamantanols	3612–3605	t-$C_4H_9-CH(OH)$-i-C_3H_7	3642.8, 3624.0

some tertiary bicyclic alcohols, such as adamantanols (Table 4-2g), demonstrate a characteristically lower frequency region, 3612–3605 cm^{-1}. This frequency lowering was attributed to the fixation of β-CH_2 groups of these bicyclic alcohols, according to Lutz and van der Maas. The rotation of β-CH_2 moieties is limited by the ring structure, so the bulky cyclic group does not sterically interfere with the OH group. This effect shifts the OH band to a lower frequency value because of the enhanced weak hydrogen bond to the CCl_4 chlorine. On the other hand, if the ring structure of the molecule allows the steric hindrance effect on the OH group (shielding effect), then the OH stretching frequency can be expected to have a higher frequency value [15, 67]. Band splitting, possibly arising from rotational isomerism, will be observed when β-carbon atoms are highly branched (see Table 4-2f). In this type of alcohol, containing a highly shielded O—H group, the previously mentioned group frequency regions for the ν(O—H) band of primary, secondary, and tertiary alcohols will not be applicable.

4-2.2. Free O—H Stretching in Phenols

VAPOR PHASE

The O—H stretching frequencies of para- and meta-substituted phenols in vapor phase appear as a single band in the region of 3660–3642 cm^{-1} [32]. The situation in ortho-substituted phenols, however, is more complicated and warrants some discussion. For the majority of 2-alkylated phenols, the O—H stretch bands fall in the same region as that for para- and meta-substituted phenols (i.e., 3660–3642 cm^{-1}). Nevertheless, substitution of a bulky alkyl group at the ortho position may result in two O—H stretching bands due to rotational isomers. For example, the vapor-phase IR spectrum of 2-*tert*-butyl phenol exhibits two O—H bands at 3670 and 3642 cm^{-1}, with the higher-frequency component being the cis form and the lower frequency, the trans form* [32]. The introduction of a second *tert*-butyl group to the other ortho position, as in the case of 2,6-di-*tert*-butyl-phenol, limits the number of stable rotational isomers to one, and only one single ν(O—H) band is observed at 3670 cm^{-1} [32, 68]. In this particular case, the abnormally high ν(O—H) frequency is a result of the repulsion between the planar O—H group and protons on the *t*-butyl groups [68]. Similar ν(O—H) bands are observed for other substituted phenols with bulky group(s) on the ortho position.

When a heteroatom (such as O, S, N, F, Cl, Br, and I) or an unsaturated moiety (such as C=O, C=C, C≡C, and C≡N) is present in the ortho position, a lower frequency will be observed due to the intramolecular hydrogen bonding effect. This type of intramolecular hydrogen bonding will be discussed further later.

DILUTE SOLUTION IN NONPOLAR SOLVENTS

The ν(O—H) band of phenol occurs at 3611 cm^{-1} in dilute CCl_4 solution. The characteristic IR absorptions of meta- and para-substituted phenols fall in the region of 3613–3592 cm^{-1} in dilute CCl_4 solution. An approximately 20 cm^{-1} lowering in the O—H stretching frequencies will be observed when the IR spectra of meta- and para-substituted phenols are measured in dilute CS_2 solution, and a region of 3595–3578 cm^{-1} should be expected [27]. In the case of *ortho-tert*-butyl phenols, two IR bands at the region of

* The *cis* and *trans* isomers arise from rotation of the O—H bond about the C—O axis. For the *cis* isomer, the O—H group is pointing toward the neighboring *tert*-butyl group on the ortho position, while in the *trans* isomer, the O—H bond points away from the *tert*-butyl group.

3650–3640 and 3615–3607 cm^{-1} are attributed to cis and trans rotational isomers, respectively. Again, the possibility of intramolecular hydrogen bonding should be considered for ortho-substituted phenols. In general, less frequency lowering in the ν(O—H) mode will be observed for intramolecular hydrogen bonded O—H groups as they change from vapor phase to dilute solution.

4-2.3. Hydrogen Bonded O—H Stretching

Hydrogen bonding arises from the interaction between the OH proton and the lone-pair or π electrons of a nearby basic site. Examples for the basic sites include nitrogen, oxygen, sulfur, and halogen atoms and functional groups that contain π electrons. The interaction can be either intermolecular or intramolecular. Infrared spectroscopy has been an excellent technique to study hydrogen bonding. The effect of hydrogen bond formation is to increase the IR intensity and lower the O—H stretching frequency. Generally, the stronger the hydrogen bond, the more the OH stretching frequency is shifted to a lower frequency.

The strength of intermolecular hydrogen bonding depends on the local molecular structure [54, 70], concentration, solvent, and temperature [69, 71, 72]. The hydrogen bond strength increases as the basicity of the lone electron-pair hydrogen-bond acceptor site increases, and as the acidity of the hydrogen-bond donor hydrogen increases. The spectral position and intensity of the O—H stretching mode are influenced by these factors. The extent of intermolecular hydrogen bond interaction decreases with solute concentrations, and the hydrogen bond strength is minimized in the vapor phase or in dilute CCl_4 solution.

Intramolecular hydrogen bonding should be expected for those compounds simultaneously containing an OH group and an accessible basic site. An intramolecular hydrogen bond is usually formed via a five- or six-membered ring structure. However, other ring sizes were also reported for intramolecular hydrogen bonds [32]. Unlike intermolecular hydrogen bonding, the intramolecular hydrogen bonding is relatively unaffected by the solute-solvent interaction. There are two types of intramolecular hydrogen bonding—namely, single-bridge and chelation. In "single bridge," the "bonding" is due to the mutual attraction between the OH proton and the lone-pair or π electrons on the basic site. The O—H stretching of this kind of intramolecular hydrogen bonding is in a region very similar to that observed for the intermolecularly hydrogen-bonded O—H moiety. In the case of chelation, the OH group is involved in molecular resonance structure while the OH group is on the β carbon of a C=C conjugated with a C=O or NO_2 group.

This kind of hydrogen bonding results in a lower O—H stretching frequency (3200–2500 cm^{-1}), and the IR band is sometimes extremely broad.

As a result of interference from the very strong and broad O—H stretching band, the spectral features of unsaturated C—H stretching are often obscured in the important region above 3000 cm^{-1}. This obstruction can lead to uncertainty in the assignment or identification of some unsaturated C—H groups in IR. Fortunately, the weak Raman scattering from the O—H stretch mode provides a good window for observing the unsaturated CH moieties.

4-2.4. Intermolecular Hydrogen Bonding in Concentrated Solution or Neat Sample

Self-association of the aliphatic alcohol molecules through intermolecular hydrogen bonding is very common when the concentration is high. Depending on the solute concentration, the intermolecular hydrogen bonding results in association of solute molecules to form either cyclic or linear dimers, trimers, tetramers and "polymers" $(\cdot\cdot\text{O—H}\cdots\text{O—H}\cdots\text{O—H}\cdot\cdot)_n$. The dimeric OH stretch gives rise to a rather sharp band ($\Delta\nu$ = 30–90 cm^{-1}) of medium intensity at 3550–3450 cm^{-1} [1, 17]. Although the dimeric OH stretching band is often encountered in the IR spectra of dilute solutions, for most alcohols measured in condensed phase (liquid or solid), little dimer is present, and its band is masked by the very strong and broad polymeric OH band. The exceptions are those alcohols and phenols containing sterically hindered OH groups. In the IR spectra of these compounds, both dimeric and polymeric OH (or even free OH) can be observed. For example, tricyclohexyl methanol has a free OH band at 3625 cm^{-1} and a dimeric OH band at 3472 cm^{-1} [17].

The polymeric OH stretching frequency occurs between 3400 and 3200 cm^{-1} [1, 17]. This band is very strong in IR spectra and the band width at half maximum can be as great as 450 cm^{-1} in many instances [17]. Among the 295 IR spectra of aliphatic alcohols reported in "The Aldrich Library of FT-IR Spectra" [17], more than 250 compounds have a band maximum in the 3400–3300 cm^{-1} region when measured as neat sample of Nujol mull. Examples of some H-bonded OH stretch bands of some alcohols can be found in Table 4-3. It is interesting to note that tertiary

Table 4-3.

Intermolecular Hydrogen-Bonded O—H Stretching Frequencies (cm^{-1}) of Some Alcohols in the Condensed State [17]

Primary Alcohols		Secondary Alcohols	
CH_3OH	3342	$CH_3-CH(OH)-CH_3$	3342
C_2H_5OH	3332	$CH_3-CH(OH)-C_2H_5$	3343
n-C_3H_7OH	3325	$CH_3-CH(OH)$-n-C_3H_7	3351
n-C_4H_9OH	3324	$CH_3-CH(OH)$-n-C_4H_9	3355
Tertiary Alcohols		Cyclic Secondary Alcohols	
$(CH_3)_3COH$	3362	cyclobutanol	3304
$(CH_3)_2C(OH)-C_2H_5$	3386	cyclopentanol	3335
$(CH_3)_2C(OH)$-i-C_3H_7	3388	cyclohexanol	3332
$(C_2H_5)_2C(OH)-CH_3$	3382	cycloheptanol	3343
$(C_2H_5)_3COH$	3404		

alcohols usually exhibit the H-bonded ν(O—H) band at the higher portion of the characteristic frequency region (i.e., closer to 3400 cm^{-1}), primary alcohols at the lower portion of the spectral range (i.e., closer to 3300 cm^{-1}), and the secondary alcohols in between. This observation shows an opposite trend in the O—H stretching frequency for the three types of alcohols as compared to what is observed in vapor phase and in dilute CCl_4 solution, where the tertiary alcohols exhibit the lowest ν(O—H) frequencies among the three types, and the primary alcohols exhibit the highest frequencies. The shielding effect can be used to explain this result. Depending on the structure, the alkyl groups around the O—H group may partially shield the O—H from intermolecular hydrogen bonding. Primary alcohols are less well shielded than secondary or tertiary alcohols and can make more effective hydrogen bonds, whereas tertiary alcohols may be somewhat shielded and may make somewhat less effective hydrogen bonds. When several bulky groups are present on the α-carbon, the hydrogen bonded O—H bands may be above 3400 cm^{-1}. The compound 2,3-dimethyl-3-pentanol in the liquid state has an OH band at 3428 cm^{-1} [17]. In some bicyclic alcohols such as adamantanols, the rotation of the β-CH_2 groups is fixed by the ring structure so that the bulky cyclic group does not shield the O—H group. The more effective hydrogen bonding results in an O—H band at 3285–3223 cm^{-1}. On the other hand, if the structure allows steric hindrance of the O—H group, then the O—H stretching frequency can be higher [15, 67].

As listed in Table 4-3, cyclic alcohols also exhibit polymeric O—H bands within this characteristic 3400–3300 cm^{-1} range. This again supports the observation that the ring strain has little effect on the O—H stretching

Table 4-4.

The O—H Stretching Frequencies (cm^{-1}) of Substituted Phenols (Nujol Mull)

unsubstituted phenol	3373	4-methyl	3364
2-methyl	3448	4-ethyl	3350
2-ethyl	3438	4-*t*-butyl	3301
2-*n*-propyl	3451	4-chloro	3383
2-*t*-butyl	3535	4-bromo	3340
		4-methoxy	3346
3-methyl	3333	2-*t*-butyl,6-methyl	3573[a]
3-ethyl	3343	2,6-di-*t*-butyl	3645[b]
3-*t*-butyl	3361	2,6-di-*i*-propyl	3570[a]
3-chloro	3360		
3-bromo	3353		
3-methoxy	3395		

[a] Dimeric.
[b] Free.

frequency (Section 4-2.1). However, there are exceptions for some polycyclic alcohols absorbed below 3300 cm^{-1} and for some above 3400 cm^{-1}. An explanation of these exceptions can be made by utilizing Lutz and van der Maas's discussion on the fixation and shielding effects.

The ν(O—H) band of phenol is located at 3373 cm^{-1} when measured as Nujol mull. For most of the para- and meta-substituted phenols, this vibrational mode falls in the region of 3400–3180 cm^{-1} (Nujol mull). A higher ν(O—H) range, 3535–3419 cm^{-1}, is found for ortho-alkylated phenols due to the shielding effect from the neighboring alkyl group(s). Some examples of the substituted phenols measured as Nujol mulls are given in Table 4-4. It is not surprising to observe a splitting on the ν(O—H) band in the IR spectra of ortho-substituted phenols due to the rotational isomerism of the dimeric ν(O—H) mode. When there is intramolecular hydrogen bonding, the O—H stretching frequency may shift out of the correlated region.

4-2.5. Intramolecular Hydrogen Bonding

The single bridge is observed for alcohols or phenols containing another polar functional group (e.g., —OR, —OH, —X, X = F, Cl, Br, I) or π bond [1, 32, 35–50, 55, 56, 73, 74]. The most common geometry for single bridge formation is via five- or six-membered ring formation. Nevertheless, single bridges through a larger ring size have been observed in vapor phase [32].

Table 4-5.

The O—H Stretching Frequencies (cm^{-1}) of Intramolecular Hydrogen Bonded *o*-Substituted Phenols in the Vapor State

Ortho Substituent	O—H Stretch (cm^{-1})	Ortho Substituent	O—H Stretch (cm^{-1})
—F	3635–3630	>C=O	3278–3170 (—OH···O=C<)
—Cl	3592–3555	—OR	3500–3470 (—OH···O<)
—Br	3560–3544	$-NO_2$	3275–3238 (broad)
—I	3538–3525	—C≡N	3585[b]
—OR	3595–3569	>C=C<	3599–3585[b]
—SMe	3445	$-CH_2-CH=CH_2$	3585[b]
$-NH_2$	3662[a]		

[a] This IR frequency is due to the free ν(O—H) of 2-aminophenol. Nyquist has suggested that the —N—H···O type of hydrogen bond interaction occurs predominently within the 2-aminophenol molecule. Therefore, no bonded O—H stretching band is observed in this case.
[b] Hydrogen bonding to the π electrons.

The single-bridged O—H stretching frequency in vapor-phase aliphatic alcohols and ortho-substituted phenols has been discussed in detail by Nyquist [32], and the phenols are summarized in Table 4-5. In general, the O—H stretching frequency decreases as the ring size increases. Furthermore, the hydrogen bond strength increases (i.e., the O—H stretching frequency decreases) as the atomic radius of the basic site increases. For instance, progressive lowering of the O—H stretching frequency of ortho-halogenated phenols was found as the halogen atom varies from —F (3635—3630 cm^{-1}) through —Cl (3592–3555 cm^{-1}) and —Br (3560–3544 cm^{-1}) to —I (ca. 3538 cm^{-1}) [32, 75]. A similar trend in frequency lowering is observed for single bridges involving oxygen and sulfur atoms.

In dilute solution, this type of hydrogen bonding gives rise to a sharp IR band in the 3570–3540 cm^{-1} region with medium intensity. Since single-bridge hydrogen bonding is not a strong interaction, it is not surprising to find the coexistence of other types of O—H grouping [17, 55, 56, 73, 74]. Therefore, the single-bridged O—H band is usually observed in conjunction with polymeric hydrogen bonding absorption. That is to say, a sharp band is contained in a broad band corresponding to a single bridge and polymeric-bonded OH stretchings, respectively, in IR spectra. The 1,2-diols and *o*-halophenols represent typical spectral examples for this type of hydrogen bonding (Table 4-5). In dilute CCl_4 solutions, α, β or β, γ unsaturated alcohols can show OH-to-π-type intramolecular hydrogen bonding at 3623–3581 cm^{-1}. For example, primary, secondary, and tertiary alcohols with α, β unsaturation have absorption at 3623–3601 cm^{-1}. Primary and secondary alcohols with β, γ unsaturation have absorption at 3596–3581 cm^{-1}, and the tertiary alcohols of this type are seen near 3611 cm^{-1} [81]. The free OH bands for all these are seen at 3638–3608 cm^{-1}, higher frequencies than these for OH-to-π H-bonded bands [81].

The so-called "chelation" refers to the interaction between the OH group and a neighboring conjugated NO_2 or C=O group, where the OH group is on the β carbon of the conjugated C=C bond. Some examples are seen here for this type of O—H···O stretch [17].

~2900 cm^{-1}

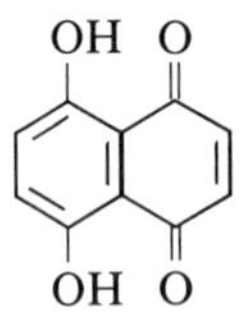

~3200 cm^{-1}

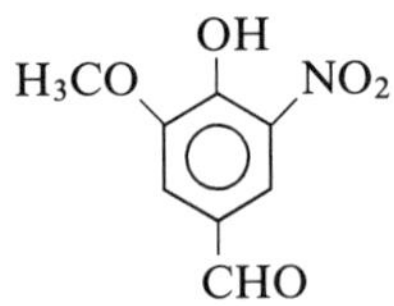

~2900 cm^{-1}

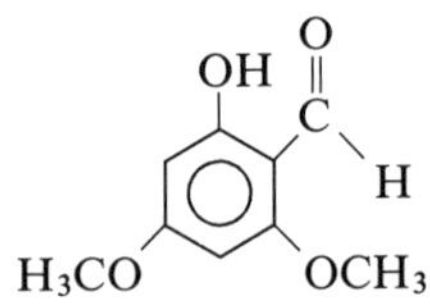

The chelation structure results in very strong resonance-enhanced hydrogen bonding, and hence only the O—H stretching band corresponding to this kind of hydrogen bonding will be observed in the IR spectrum. The bonded O—H stretching is shifted to the very broad 3200–2500 cm^{-1} region. It is observed that the stronger the chelated hydrogen bond, the lower the intensity recorded in this region. Large variations in intensity in the IR spectra are observed for these types of alcohols and phenols. Sometimes the bonded OH band may be overlooked because of broad and weak features. Schreiber has studied the hydrogen bonding in *o*-nitrophenol [82].

Nyquist [80] has also examined some γ(OH) or out-of-plane wag bands in ortho-substituted phenols that are seen in the IR at 860–300 cm^{-1} in solution, while the ν(OH) or OH stretch bands are seen for these at 3610–3050 cm^{-1}. The two types of bands shift in opposite directions in these ortho-substituted phenols. Stronger hydrogen bonding decreases the ν(OH····X) stretch wavenumber and increases the out-of-plane γ(OH····X) wag wavenumber [80].

4-3.

O—H BENDING

The in-plane bending or deformation is coupled to the adjacent CH_2 or CH wagging and gives rise to some broad, weak IR bands in the 1440–1220 cm^{-1} region [1, 77–79]. The Raman intensities are medium to weak.

In the IR spectra, primary alcohols (CH_2-OH) in the bonded state have two sets of weak broad bands about 1420 and 1330 cm^{-1} (± 20 cm^{-1}). In dilute solution, these bands shift to about 1385 and 1300–1200 cm^{-1}. Secondary alcohols ($CH-OH$) have similar bands. Tertiary alcohols have no α-CH to interact with the OH bend. They absorb near 1410 cm^{-1} (bonded) and near 1320 cm^{-1} (in solution). While these bands are not prominent, one should be aware of them because of their nearness to the 1375 cm^{-1} CH_3 bands.

4-4. SKELETAL VIBRATIONS

The skeletal vibrational frequencies involving C—C and C—O stretch of *n*-alkyl alcohols show little shift from those of *n*-alkanes. The C—O bond force constant is nearly the same as that of the C—C bond, and the masses of O and C do not differ greatly. This means that C—O and C—C bonds interact strongly. While the skeletal vibrational band of primary, secondary, and tertiary alcohols have different characteristics in the Raman spectra, these are similar to the features in the Raman spectra of similarly branched alkanes. However, because of the polar nature of the C—O bond, skeletal C—C—O stretching modes can give rise to strong IR bands that are considerably more intense than comparable C—C—C IR bands in alkanes. (See Table 4-1.)

4-4.1. CCO Out-of-phase Stretching

The intense IR band in the 1260–1000 cm^{-1} region, which is referred to as C—O stretching in some works, is assigned to the out-of-phase C—C—O stretching mode by the normal coordinate analysis method [2, 4, 5]. This vibration also gives rise to a medium to strong intensity in Raman spectra. Primary alcohols absorb strongly in the IR at 1075–1000 cm^{-1}, secondary alcohols at 1150–1075 cm^{-1}, tertiary alcohols at 1210–1100 cm^{-1}, and phenols at 1260–1180 cm^{-1} [79]. These may be single bands or band clusters. All these bands shift somewhat when there are changes in hydrogen bonding, usually decreasing in frequency in dilute solution. The study of factors influencing the C—O stretching frequencies has been reported in detail by Nyquist [32].

4-4.2. CCO In-phase Stretching

The CCO in-phase stretching frequencies occur at a lower spectral region than the corresponding out-of-phase vibrational mode. For example, in primary and secondary alcohols, Raman bands appear in the 900–800 cm^{-1} region. Infrared bands here are weaker and less useful. In the Raman spectra of small tertiary alcohols, a strong polarized band is found at the 760–730 cm^{-1} region. Rather than assigning this band to the in-phase CCO stretching mode, it seems more reasonable to attribute this band to the in-phase $C_3C{-}O$ stretching mode. This assignment is based on the fact that a similar band is observed for the in-phase CC_4 stretching mode in the same region in alkanes. See Section 2-1.10. The IR intensity of this in-phase $C_3C{-}O$ stretching band is medium to weak, while it can be the strongest Raman band below 1500 cm^{-1} for small alcohols; but the relative Raman intensity of this stretching mode in the whole spectrum decreases gradually as the carbon number in the skeletal chain increases. Eventually this band is buried by the many spectral features arising from other skeletal vibrational modes and is no longer easy to identify. For example, the Raman spectra of *n*-alkyl alcohols and corresponding *n*-alkanes show great similarities when the carbon number in the chain is greater than eight [76].

4-4.3. C—O Stretching of Rotational Isomers

In *n*-alkyl alcohols, many skeletal stretch vibrations involving the C—C—C—O—H bonds are conformation dependent, which can complicate the spectrum. For example, trans and gauche isomers have skeletal stretch bands at 971 and 917 cm^{-1}, respectively, for *n*-propanol, and at 998 and 953 cm^{-1}, respectively, for *n*-butanol.

4-4.4. Skeletal Bendings

A medium to weak Raman band in the 460–350 cm^{-1} region, which is likely due to some bending mode, was reported to be characteristic for the specific types of alcohol [2]. For example, primary alcohols absorb at 460–430 cm^{-1}, secondary alcohols around 495 cm^{-1}, and tertiary alcohols at 350–350 cm^{-1}. However, there usually exist several medium to weak bands in this region arising from other vibrations and different tones of the molecule; thus, it is difficult to select one frequency to distinguish the type of alcohol. Nevetheless, useful information may be obtained when several different characteristic frequencies are identified for elucidating the molecular structure.

REFERENCES

1. L. J. Bellamy, "The Infrared Spectra of Complex Molecules," John Wiley and Sons, Inc., New York, 1975, pp. 107–128.
2. F. R. Dollish, W. G. Fateley, and F. F. Bentley, "Characteristic Raman Frequencies of Organic Compounds," John Wiley and Sons, Inc., New York, 1974, pp. 27–34.
3. C. Tanaka, K. Kuratani, and S. Mushima, *Spectrochim. Acta* **9**, 265, 1957.
4. K. Fukushima and B. J. Zwolinski, *J. Mol. Spectrosc.* **26**, 368, 1968.
5. G. A. Crowder and M. J. Townsend, *J. Mol. Struct.* **42**, 27, 1977.
6. M. Tasumi, T. Shimanouchi, A. Watanabe, and R. Goto, *Spectrochim. Acta* **20**, 629, 1964.
7. C. Tanaka, *Nippon Kagaku Zasshi* **83**, 521, 1962 (CA58 : 10882d); **83**, 655, 1962 (CA58 : 10882e); **83**, 661, 1962 (CA58 : 10882b); **83**, 398, 1962 (CA58 : 940b).
8. (a) J. P. Perchard and M. L. Josien, *J. Chim. Phys. Physiochim. Biol.* **65**, 1834, 1968. (b) *Ibid.*, **65**, 1856, 1968.
9. P. Piaggio, R. Tubino, and G. Dellepiane, *J. Mol. Struct.* **96**, 277, 1983.
10. J. C. Evans, *Spectrochim. Acta* **16**, 1382, 1960.
11. J. H. S. Greene, *J. Chem. Soc.* 2236, 1961.
12. H. D. Bist, J. C. D. Brand, and D. R. Williams, *J. Mol. Spectrosc.* **24**, 402, 1967.
13. H. D. Bist, J. C. D. Brand, and D. R. Williams, *J. Mol. Spectrosc.* **24**, 413, 1967.
14. H. D. Bist, J. C. D. Brand, and D. R. Williams, *J. Mol. Spectrosc.* **21**, 76, 1966.
15. J. H. van der Maas and E. T. G. Lutz, *Spectrochim. Acta.* **30A**, 2005, 1974.
16. K. Nakanishi and P. H. Solomon, "Infrared Absorption Spectroscopy," Holden-Day, Inc., San Francisco, 1977.
17. C. J. Pouchert, "The Aldrich Library of FT-IR Spectra," Aldrich Chemical Company, Inc., 1985, Vol. I, pp. 109–202 and pp. 1069–1172.
18. J. Christoffersen, J. M. Hollas, and G. H. Kirby, *Proc. R. Soc. Ser. A* **307**, 97, 1968.
19. V. N. Sarin, M. M. Rai, H. D. Bist, and D. P. Khandelwal, *Chem. Phys. Lett.* **6**, 473, 1970.
20. H. W. Wilson, *Anal. Chem.* **46**, 962, 1974.
21. E. Mathier, D. Weilti, A. Bauder, and Hs. H. Guthard, *J. Mol. Spectrosc.* **27**, 63, 1967.
22. H. W. Wilson, R. W. MacNamee, and J. R. Durig, *J. Raman Spectrosc.* **11**, 252, 1981.
23. R. J. Jakobsen and E. J. Brewer, *Appl. Spectrosc.* **16**, 32, 1962.
24. V. N. Verma and D. K. Rai, *Appl. Spectrosc.* **24**, 445, 1970.
25. G. Varsanyi, "Assignments for Vibrational Spectra of Seven Hundred Benzene Derivatives," Vols. I & II, John Wiley & Sons, New York, 1974.
26. N. S. Sundar, *Spectrochim. Acta* **41A**, 1449, 1985.
27. E. T. G. Lutz and J. H. van der Mass, *Spectrochim. Acta* **39A**, 1007, 1983.
28. P. V. Khadikar, B. Pol, and S. M. Ali, *Spectrochim. Acta* **42a**, 755, 1986.
29. J. A. Faniran, *Spectrochim. Acta* **35A**, 1257, 1979.
30. J. A. Faniran and H. F. Shurvell, *J. Raman Spectrosc.* **9**, 73, 1980.
31. J. A. Faniran and H. F. Shurvell, *Spectrochim. Acta* **38A**, 115, 1982.
32. R. A. Nyquist, "The Interpretation of Vapor-Phase Infrared Spectra, Group Frequency Data," The Sadtler Research Laboratories, 1985, pp. 109–172.
33. J. R. Durig and W. H. Green, *Spectrochim. Acta* **25A**, 849, 1969.
34. J. R. Durig, J. M. Karriker, and W. C. Harris, *Spectrochim. Acta* **27A**, 1955, 1971.
35. T. Visser and J. H. van der Maas, *Spectrochim. Acta* **42A**, 599, 1986.
36. M. Oki and H. Iwamura, *Bull. Chem. Soc. Jpn,* **32**, 955, 1959.
37. M. Oki and H. Iwamura, *Bull. Chem. Soc. Jpn.* **35**,, 1552, 1962.

38. I. D. Campbell, G. Eglington, and R. A. Rapheal, *J. Chem. Soc. B*, 338, 1968.
39. H. Iwamura and K. Hanaya, *Bull. Chem. Soc. Jpn.* **43,** 3901, 1970.
40. N. Mori, M. Yoshifuji, Y. Asabe, and Y. Tsuzuki, *Bull. Chem. Soc. Jpn.* **44,** 1137, 1971.
41. F. H. Hon, Matsumara, H. Tanida, and T. T. Tidwel, *J. Org. Chem* **37,** 1778, 1972.
42. Z. Yamashita, K. Hara, S. Aizama, and M. Hirota, *Bull. Chem. Soc. Jpn.* **47,** 2508, 1974.
43. C. Laurence and M. Berthelot, *Spectrochim. Acta* **34A,** 1127, 1978.
44. N. Mori, S. Kasuya, H. Miyazaki, and T. Takezawa, *Bull. Chem. Soc. Jpn.* **52,** 1423, 1979.
45. M. Takasuka and H. Tanida, *J. Chem. Soc. Perkin II*, 486, 1980.
46. M. Ito and M. Hirota, *Bull. Chem. Soc. Jpn.* **54,** 2093, 1981.
47. K. Hanaya, S. Onodera, Y. Ikegami, H. Kudo, and K. Shimaya, *J. Chem. Soc. Perkin II*, 944, 1981.
48. J. M. Bakke and G. B. Lorentzen, *Acta Chim. Scand. B* **28,** 650, 1974.
49. R. J. Abraham and J. M. Bakke, *Acta Chim. Scand. B* **37,** 865, 1983.
50. R. J. Abraham, J. M. Bakke, and T. Skjetne, *Acta Chim. Scand. B* **38,** 547, 1984.
51. M. Guiliano, G. Mille, J. Chouteau, G. Lēandri, and A. Mèou, *J. Raman Spectrosc.* **13,** 223, 1983.
52. M. Perttilä, *Spectrochim. Acta* **35A,** 585, 1979.
53. E. T. G. Lutz and J. H. van der Maas, *Spectrochim. Acta* **36A,** 177, 1980.
54. E. T. G. Lutz and J. H. van der Maas, *Spectrochim. Acta* **36A,** 805, 1980.
55. T. Visser and J. H. van der Maas, *Spectrochim. Acta* **39A,** 241, 1983.
56. T. Visser and J. H. van der Maas, *Spectrochim. Acta* **41A,** 757, 1985.
57. E. T. G. Lutz and J. H. van der Maas, *Spectrochim. Acta* **41A,** 943, 1985.
58. J. R. Durig, Ş. F. Bush, and W. C. Harris, *J. Chem. Phys.* **50,** 2851, 1969.
59. R. Piccolini and S. Winstein, *Tetrahedron Lett.* **13,** 4, 1959.
60. E. L. Saier, L. R. Cousins and M. R. Basila, *J. Chem. Phys.* **41,** 40, 1964.
61. O. Oki and H. Iwamura, *Bull. Chem. Soc. Jpn.* **32,** 950, 1959.
62. L. Joris, P. von Rague Schleyer, and E. Osawa, *Tetrahedron* **24,** 4759, 1968.
63. J. Weinman and S. Weinman, *Bull. Soc. Chim. France* **3,** 925, 1969.
64. H. S. Aaron and C. P. Radar, *J. Am. Chem. Soc.* **85,** 3046, 1963.
65. H. S. Aaron and C. P. Ferguson, and C. P. Rader, *J. Am. Chem. Soc.* **89,** 1431, 1967.
66. F. Dalton, G. D. Meakins, J. H. Robinson, and W. Zaharia, *J. Chem. Soc.*, 1566, 1962.
67. F. G. Dijkman, *Mol. Phys.* **36,** 705, 1978.
68. K. U. Ingold, *Can. J. Chem.* **38,** 1029, 1960.
69. A. W. Baker, H. O. Kerlinger, and A. T. Shulgin, *Spectrochim. Acta* **20,** 1477, 1964.
70. E. T. G. Lutz and T. H. van der Maas, *Spectrochim. Acta* **38A,** 743, 1982.
71. T. Visser and J. H. van der Maas, *Spectrochim. Acta* **38A,** 293, 1982.
72. K. U. Ingold and R. D. Taylor, *Can. J. Chem.* **39,** 481, 1961.
73. T. Visser and J. H. van der Maas, *Spectrochim. Acta* **39A,** 921, 1983.
74. T. Visser and J. H. van der Mass, *Spectrochim. Acta* **40A,** 959, 1984.
75. T. S. Lein and E. Fishman, *Spectrochim. Acta* **23A,** 491, 1967.
76. "The Sadtler Standard Raman Spectra," Sadtler Research Laboratories, 1976.
77. A. V. Stuart and G. B. B. M. Sutherland, *J. Chem. Phys.* **24,** 559, 1956.
78. S. Krimm, C. Y. Liang, and G. B. B. M. Sutherland, *J. Chem. Phys.* **25,** 778, 1956.
79. N. B. Colthup, L. H. Daly, and S. E. Wiberley, "Introduction to Infrared and Raman Spectroscopy," 3rd Ed., Academic Press, New York, 1990.
80. R. A. Nyquist, *Spectrochim. Acta* **19,** 1655, 1963.
81. J. F. Bacon and J. H. van der Maas, *Can. J. Chem.* **67,** 250, 1989.
82. V. M. Schreiber, *J. Mol. Struct.* **197,** 73, 1989.

CHAPTER 5

Ethers and Peroxides

5-1. INTRODUCTION

Characteristic frequencies of ethers and peroxides are due to vibrations involving C—O linkage and O—O bonds, respectively. Furthermore, the C—H vibrations for $-OCH_3$, $-OCH_2-$, and $-OCH-$ ethers occur at somewhat different frequencies than those observed in alkanes. A summary of the characteristic frequencies of ethers is given in Table 5-1.

The characteristic COC vibrations in ethers are approximated as antisymmetric COC stretch, symmetric COC stretch, and COC deformation. However, these vibrations are far more complex than simply COC vibrations because of strong interaction between the C—O and all the C—C bonds in the molecule. Such interactions arise from the fact that the force constants of C—O and C—C bonds are of similar value. Furthermore, the mass of the oxygen atom is very close to that of the CH_2 group. As a result of the vibrational coupling, the COC vibrations are sensitive to the skeletal structure.

The antisymmetric or out-of-phase COC stretching mode of aliphatic ethers usually gives rise to a strong broad IR band in the 1150–1060 cm^{-1} region, while the characteristic IR band of vinyl ethers is located in a somewhat higher region, at 1225–1200 cm^{-1}. Aryl—O—CH_2 ethers absorb at 1310–1210 and 1050–1010 cm^{-1} in the IR. The symmetric or in-phase C—O—C stretching modes are more easily identified by Raman spectroscopy. For aliphatic ethers, the in-phase COC stretch occurs at 890–820 cm^{-1}.

Table 5-1.

Characteristic Frequencies of Ethers

		cm^{-1}	IR	R
Aliphatic ethers				
antisym. str. COC		1150–1060	vs	w
sym. str. COC		890–820	w	vs, p
def. COC		500–400		vs, p
Acetals				
antisym. str. COCOC	$ROCH_2OR$	1140–1115	s	—
	$ROCH_2OR$	1050–1040	s	var.
	$(RO)_2CHCH_3$	1140–1130	s	var.
sym. str. COCOC		870–850	s	m
		1115–1080	s	m, p
def. COCOC		660–600	vs	m
		540–450	s	m-s
		400–320	s	m-s, p
Vinyl ethers				
out-of-phase stretch COC		1225–1200	vs	w
in-phase stretch COC		850–840	w	s
Aromatic ethers				
aryl—O stretch		1310–1210	vs	w
O—CH_2		1050–1010	s	w
Epoxy compounds				
CH_2 antisym. str.		3075–3030	m	s
CH_2 sym. str.		3020–2990	m	s
ring sym. str.		1280–1230	m	s
ring antisym. def.		950–815	s	mw
ring sym. def.		880–750	s	m

In the Raman spectra of aromatic ethers, the in-phase COC stretching falls in a wider range of 1050–800 cm^{-1}. The above COC stretching bands can be complicated by the rotational isomers in some cases.

In the vibrational spectra of peroxides, the characteristic frequency is due to the O—O stretching mode. Raman spectroscopy is a far better technique to investigate the presence of the peroxy group, since the O—O stretching gives rise to strongly polarized Raman bands about 900–700 cm^{-1}, but only to weak IR absorption. The O—O stretch vibration interacts with other skeletal bond C—O and C—C vibrations. Moreover, the symmetric stretching of tertiary and quaternary carbon atoms have Raman bands in the same spectral region [16, 17].

5-2. ALIPHATIC ETHERS

The IR absorption in the region 1140–1085 cm^{-1} and the Raman band in the 890–820 cm^{-1} range were correlated with COC vibrations experimentally [1–4]. Complete assignments on the vibrational frequencies of ethers were limited to simple molecules such as dimethyl ether [5], diethyl ether [6], and methyl ethyl ether [7]. The COC stretching frequencies for selected simple ethers are summarized in Table 5-2. In general, the existence of different rotational isomers [6, 7] complicates the vibrational spectrum in the frequency range of COC stretching modes.

5-2.1. C—O—C Antisymmetric Stretching

According to the *Aldrich Library of FT-IR Spectra* [8], most of the di-*n*-alkyl ethers absorb near 1123 cm^{-1} if the alkyl group is small. The IR band shifts to a slightly lower frequency as the number of carbon atoms in the

Table 5-2.

Characteristic Frequencies (cm^{-1}) of Aliphatic Ethers [2]

		ν_{as}(COC)	ν_s(COC)	δ(COC)
A. Symmetric substitution				
CH_3-O-CH_3		1102	924	428
$CH_3CH_2-O-CH_2CH_3$	trans-trans:	1120	845	440
	trans-gauche:	1140	835	499
$CH_3(CH_2)_2-O-(CH_2)_2CH_3$		1120	890	480
$CH_3(CH_2)_3-O-(CH_2)_3CH_3$		1115	840	460
$CH_3(CH_2)_4-O-(CH_2)_4CH_3$		1120	850	498
$(CH_3)_2CHCH_2-O-CH_2CH(CH_3)_2$		1125	840	420
$(CH_3)_2CHCH_2CH_2-O-CH_2CH_2CH(CH_3)_2$		1140	870	420
$(CH_3)_2CH-O-CH(CH_3)_2$		1125	798[a]	492
$(CH_3)_3C-O-C(CH_3)_3$		1205	700[a]	425
B. Antisymmetric substitution				
$CH_3-O-CH_2CH_2$	trans:	1120	853	475
	gauche:	1068	843	—
$CH_3CH_2-O-CH(CH_3)_2$		1125	830	492
$CH_3CH_2-O-(CH_2)_3CH_3$		1125	860	465
$CH_3CH_2-O-CH_2CH(CH_3)_2$		1140	845	480
$CH_3CH_2-O-C(CH_3)_3$		1115	890[a]	490

[a] The assignments to these bands are only tentative since a strong polarized Raman band due to tertiary (or quaternary) carbon atoms is also expected in the same spectral region (see Chapter 2).

n-alkyl group increases. For example, octyl ether, $(n\text{-}C_8H_{17})_2{-}O$, shows a strong IR band at 1114 cm^{-1} instead of 1123 cm^{-1}. For those ethers with the structural unit $R(OCH_2CH_2)_xOR'$, the absorption occurs at 1105–1115 cm^{-1}.

In comparison with n-alkyl ethers, the effect of branching on the COC stretching bands is to reduce the vibrational frequency. Often, multiple-band structure is observed for the COC stretching of branched ethers. For example, three very strong bands at 1169, 1112, and 1015 cm^{-1} are observed for isopropyl ether, and at 1201, 1117, and 1076 cm^{-1} for t-butyl ether [8]. A possible explanation for such spectral complication is the interaction of COC stretching modes with the tertiary (or quaternary) skeletal vibrations.

There is less effect when the branched atom is located beyond the α-position. Taking isoamyl ether as an example, the antisymmetric COC stretching found in this γ-branched ether is lowered to 1113 cm^{-1} [8], as compared with the 1123 cm^{-1} value of di-n-alkyl ethers. Regardless of these minor frequency shifts, the COC asymmetric stretching frequencies are correlated to a characteristic region of 1150–1060 cm^{-1}.

Unsaturation seems to have little effect on the COC antisymmetrical stretching as long as the oxygen atom is not attached directly to the double bond. This observation is demonstrated by the IR absorption of allyl ethyl ether ($H_2C{=}CH{-}CH_2{-}O{-}CH_2CH_3$) at 1108 cm^{-1} [8], which is within the normal 1150–1060 cm^{-1} range for aliphatic ethers.

Rotational isomerism sometimes causes a complication in the COC stretching region. Examples on $CH_3OCH_2CH_2CH_3$, $CH_3CH_2OCH_2CH_2CH_3$, $CH_3OCH_2CH_2CH_2CH_3$, and $CH_3OCH_2CH_2OCH_3$ are reported by Shimanouchi and co-workers [20, 21].

5-2.2. Symmetric C—O—C Stretching

The symmetric COC stretch and deformation of aliphatic ethers are responsible for intense Raman bands in the 890–830 cm^{-1} and 500–400 cm^{-1} regions, respectively. Again, the existence of rotational isomers, as well as the coupling of COC vibrations with skeletal vibrations, complicates the Raman spectral feature.

5-2.3. Deformation Vibrations

According to Dollish *et al.* [2], compounds containing more than one ether linkage, such as $CH_3OCH_2CH_2OCH_2CH_2OCH_3$ and $CH_3(OCH_2CH_2)_2OCH_3$, usually exhibit three broad Raman bands at 380, 440, and 530 cm^{-1} in most cases. For $CH_3OCH_2CH_2OCH_3$, COC bending modes appear at 390 and 365 cm^{-1} [21].

5-3.

ACETALS

The IR and Raman spectra of dialkoxymethanes ($ROCH_2OR'$) and ethylidene dialkyl ethers ($ROCHCH_3OR'$) were investigated by Nukada [9–11]. Results of the characteristic COCOC vibration are summarized in Table 5-3. According to Nukada, ethylidene dialkyl ethers exhibit more than one rotational isomer at room temperature. Hence, more than one of each of the COCOC antisymmetric and symmetric stretching bands are expected. Antisymmetric COCOC stretches for dialkoxymethanes occur in the 1140–1115 cm^{-1} and 1050–1040 cm^{-1} regions, while those for ethylidene dialkyl ethers appear in 1140–1130 cm^{-1} and 870–850 cm^{-1} regions. Symmetric COCOC stretching frequencies for these two types of compound fall in characteristic regions of 1115–1080 cm^{-1} and 870–800 cm^{-1}. In the low frequency region, there are three very characteristic Raman bands in the 550–600 cm^{-1}, 540–450 cm^{-1}, and 400–320 cm^{-1} ranges, where these frequencies are assigned to COCOC deformations.

5-4.

VINYL ETHERS

As shown in Table 5-4, the intense out-of-phase COC stretching IR bands of vinyl ethers ($CH_2{=}CH{-}OR$) are at 1225–1200 cm^{-1} [3, 22]. Also seen in Table 5-4 is 2-methoxypropene, which is a 1,1-disubstituted ethylene type ($CH_2{=}CMe{-}OMe$). The out-of-phase COC stretch band is at 1286 cm^{-1}. The band is at 1229 cm^{-1} for *cis*-1-bromo-2-ethoxyethylene (a 1,2-disubstituted ethylene type). Although very strong in-phase COC stretching Raman bands for vinyl ether (850 cm^{-1}) and vinyl *n*-butyl ether (840 cm^{-1}) are well characterized [12], additional work is necessary to provide a better understanding of the Raman spectra of vinyl ethers.

5-5.

AROMATIC ETHERS

Aromatic ethers of the type aryl-O$-CH_2$ usually have strong IR bands at 1310–1210 cm^{-1} (aryl-O) and 1050–1010 cm^{-1} ($O{-}CH_2$ or $O{-}CH_3$)

Table 5-3.

Characteristic Frequencies (cm^{-1}) of Dialkoxymethanes and Ethylidene Dialkyl Ethers [9–11]

		Antisymmetric COCOC Stretch			Symmetric COCOC Stretch		COCOC Deformation		
Dialkoxy methanes									
$MeOCH_2OMe$	IR	1138 (s)	1043 (s)		1110 (s)	867 (s)	602 (s)	455 (s)	325 (s)
	R	—	1045 (m, dp)		1113 (m, dp)	858 (s, p)	602 (m, dp)	452 (m, dp)	327 (s, p)
$EtOCH_2OEt$	IR	1117 (s)	1048 (s)		1095 (s)	850 (m)	629 (s)	468 (s)	344 (s)
	R	—	1041 (w, dp)		1103 (s, p)	841 (s, p)	634 (m, dp)	471 (m, dp)	349 (m, p)
$MeOCH_2OEt$	IR	1139 (s)	1045 (s)		1096 (s)	846 (m)			
	R	—	1045 (w, dp)		1100 (s, p)	845 (s, p)	618 (m, dp)	457 (m, dp)	355 (m, p)
Ethylidene dialkyl ethers									
MeO–CH(Me)–OMe	IR	1135 (s)		867 (s)	1084 (s)	810 (w)			
	R	1140 (s, dp)		869 (m, dp)	1087 (s)	808 (s, p)	631 (m, dp)	537 (s, p)	396 (s, p)
EtO–CH(Me)–OEt	IR	1133 (s)	1060 (s)	850 (m)	1099 6s)	809 (m)			
	R	1135 (w, dp)	1061 (w, dp)	850 (s, dp)	1100 (s, p)	806 (s, p)	654 (m, dp)	526 (s, p)	351 (s, p)
MeO–CH(Me)–OEt	IR	1134 (s)		857 (s)	1099 (s)	810 (w)			
	R	1136 (m, dp)		855 (m, dp)	1098 (m, p)	805 (s, p)	646 (m, dp)	530 (s, p)	354 (m, p)

Table 5-4.

Characteristic Frequencies (cm^{-1}) of Vinyl Alkyl Ethers [8, 12]

	cm^{-1}	IR	R	cm^{-1}	IR	R
$CH_2{=}C(H)OCH_2CH_3$	1208	vs	w	~850	w	s
$CH_2{=}C(H)OCH_2CH_2Cl$	1209	vs				
$CH_2{=}C(H)O(CH_2)_3CH_3$	1204	vs	w	~840	w	s
$(Br)(H)C{=}C(OCH_2CH_3)(H)$	1229	vs				
$(H)(H)C{=}C(CH_3)(OCH_3)$	1286	vs				

[1, 3, 23–25]. The 1310–1210 cm^{-1} band is in about the same region as the aryl-O band in phenols. The 1050–1010 cm^{-1} band is in the same region as the C—O stretch band in methanol, or the highest C—C—O IR frequency in primary alcohols. Diphenyl ether has only the aryl-O band at 1240 cm^{-1} and has no $O{-}CH_2$ band in the 1050–1010 cm^{-1} region. The "aryl-O" band, of course, is not pure aryl-O stretch, but involves out-of-phase C—O—C stretching and also mixes with ring vibrations [3, 13]. This means that the "aryl-O" band may be affected by interaction with other substituents on the ring, which can complicate the interpretation. Also, the "$O{-}CH_2$" band is a mixed vibration as in the alcohols and involves all the attached C—C bonds interacting with the C—O bonds. Phenetol (phenyl ethyl ether) has IR bands involving mixed aryl-O, O—C, and C—C stretch [3] at 1240, 1040, and 920 (weak) cm^{-1} in the IR.

5-6. EPOXY COMPOUNDS

Group frequencies for epoxy compounds have been summarized in the IR by Bellamy [1] and in the Raman by Dollish *et al.* [2]. A recent summary of IR group frequencies for 1,2-epoxyalkanes (monoalkyl ethylene oxides) by Nyquist [27] is given as follows: ring in-phase stretch or ring breathing, 1268–1248 cm^{-1}; symmetric ring deformation, 877–830 cm^{-1}; antisymmetric ring deformation, 932–883 cm^{-1}; antisymmetric oxirane CH_2 stretch, 3065–3038 cm^{-1}; and symmetric oxirane CH_2 stretch, 3001–2990 cm^{-1}. Oxirane CH_2 bend, twist, and wag are assigned to the regions 1501–1478, 1412–1404, and 1130–1125 cm^{-1}, respectively.

General IR and Raman regions for variously substituted ethylene oxides include 1280–1230 cm^{-1} (ring breathing, strong in Raman), 950–815 cm^{-1} (antisymmetric ring deformation), and 880–750 cm^{-1} (symmetric ring deformation) [1–3, 27–29). The last band tends to shift to lower frequencies with increasing substitutions, it is usually seen at 880–805 cm^{-1} for monosubstituted oxirane rings, 850–775 cm^{-1} for 1,1- or 1,2-disubstituted rings, and 770–750 cm^{-1} for trisubstituted rings. The CH_2 in the ring gives rise to bands at 3075–3030 cm^{-1} and 3020–2990 cm^{-1}.

The characteristic frequencies shift in larger rings. In the cyclic ethers $O(CH_2)_n$ where $n = 2$-5, the antisymmetric C—O—C stretch (usually strong in the IR) increases in frequency as the ring gets larger. The wavenumbers for three-, four-, five-, and six-membered rings are 892, 983, 1071, and 1098 cm^{-1}, respectively. The symmetric C—O—C stretch (ring breathing, which is strong in the Raman) decreases in frequency as the ring gets larger. The wavenumbers for three-, four-, five-, and six-membered rings are 1270, 1028, 913, and 813 cm^{-1}, respectively [3].

5-7. PEROXIDES

In general, IR spectroscopy is a poor technique for detecting the peroxy O—O vibration. In contrast, the change in polarizability induced by the O—O stretching vibration gives rise to an intense Raman band for most peroxy compounds (Table 5-5) In the normal coordinate analysis of CH_3OOCH_3, Bell and Laane [16] indicated that there is an extensive coupling between the ν(O—O), ν(C—O), and δ(C—O—O) modes. Nevertheless, a characteristic region of 900–800 cm^{-1} is suggested for the O—O stretching frequency [1]. After examining the Raman spectra of 29 alkyl

Table 5-5.

Characteristic Raman Frequencies (cm^{-1}) of Peroxides

Compound	cm^{-1}	Compound	cm^{-1}
$H-OO-H$	880	$C_6H_5\overset{\overset{O}{\|}}{C}-OO-\overset{\overset{O}{\|}}{C}C_6H_5$	883
CF_3-OO-H	860	$Me\overset{\overset{O}{\|}}{C}-OO-\overset{\overset{O}{\|}}{C}C_6H_5$	894, 855
CF_3-OO-D	863	$MeCH_2CH_2\overset{\overset{O}{\|}}{C}-OO-\overset{\overset{O}{\|}}{C}C_6H_5$	898, 870
CF_3-OO-F	875	$Me\overset{\overset{O}{\|}}{C}-OO-\overset{\overset{O}{\|}}{C}(p\text{-}C_6H_4-Me)$	895, 857
$Me-OO-Cl$	830	$Me\overset{\overset{O}{\|}}{C}-OO-\overset{\overset{O}{\|}}{C}(p\text{-}C_6H_4-OMe)$	896, 858
$Me-OO-Me$	779	$Me\overset{\overset{O}{\|}}{C}-OO-\overset{\overset{O}{\|}}{C}(p\text{-}C_6H_4-Cl)$	894, 856
$CD_3-OO-CD_3$	713	$Me\overset{\overset{O}{\|}}{C}-OO-\overset{\overset{O}{\|}}{C}(p\text{-}C_6H_4-Br)$	895, 857
$Et-OO-Et$	882	$Me\overset{\overset{O}{\|}}{C}-OO-\overset{\overset{O}{\|}}{C}(p\text{-}C_6H_4-I)$	894, 840
$Me_3C-OO-CMe_3$	771[a] (or 850)	$Me\overset{\overset{O}{\|}}{C}-OO-\overset{\overset{O}{\|}}{C}(m\text{-}C_5H_4-NO)$	910, 892, 844
$Me(CF_3)_2C-OO-C(CF_3)_2Me$	774[a]	$Me\overset{\overset{O}{\|}}{C}-OO-\overset{\overset{O}{\|}}{C}(m\text{-}C_6H_4-F)$	910, 895, 848
$(CF_3)_3C-OO-C(CF_3)_3$	781[a] (or ~850)	$Me\overset{\overset{O}{\|}}{C}-OO-\overset{\overset{O}{\|}}{C}(m\text{-}C_6H_4-Cl)$	899, 873
$Me(OH)CH-OO-H$	881		
$HOCH_2CH_2-OO-CH_2CH_2OH$	883		

[a] The assignments are only tentative. This vibrational band can also be attributed to the symmetric OCC_3 stretching mode (see Chapter 4).

peroxides, Budinger *et al.* [17] concluded, without reporting Raman spectra and O—O stretching frequencies for each individual compound, that O—O stretching frequencies of peroxides fall in the 900–700 cm^{-1} region.

The characteristic vibration of the peroxy group is assigned by some authors by simply attributing the strongest Raman band to the O—O stretching mode. Examples can be found in the assignments of $(CH_3)_3COOC(CH_3)_3$ [14, 15], $(CH_3)(CF_3)_2COOCC(CF_3)_2(CH_3)$, and $(CF_3)_3COOC(CF_3)_3$ [14] (see Table 5-5). These assignments are, however, still open to discussion, since a medium Raman band near 860 cm^{-1} is also observed in Raman spectra of both $(CH_3)_3COOCC(CH_3)_3$ and $(CF_3)_3COOC(CF_3)_3$. Based on a recent report by DesMarteau and Hammaker [19], who studied the Raman spectra of SF_5OOH, SF_5OOF, and SF_5OOCl, the O—O stretch does not necessarily correspond to the strongest Raman band for these compounds. The above authors assigned the strongest Raman bands for the SF_5-containing peroxides to the symmetrical SF_4 stretching modes [19]. A similar argument seems to apply to peroxides with quaternary carbon atoms. Thus, the strongest Raman bands at 771 and 781 cm^{-1} for $(CH_3)_3COOC(CH_3)_3$ and $(CF_3)_3COOF(CF_3)_3$, respectively, could be assigned to the skeletal CC_4 stretching mode. Further support of this assignment is that a polarized strong Raman band in this range is also observed for tertiary alcohols and ethers [2]. By the same reasoning, one can assume that the strongest band in the Raman spectrum of $CH_3(CF_3)_2COOC(CF_3)_2CH_3$ at 774 cm^{-1} is also due to quaternary carbon atoms.

Table 5-5 shows some characteristic Raman bands for selected alkanoyl benzoyl peroxides ($RCOOC_6H_5$). For these compounds, the bands in the regions of 900–890 cm^{-1} and 856–545 cm^{-1} are assigned to the O—O stretch and a dihedral angle vibration, respectively, of the peroxide chain [18].

In summary, the O—O stretch may be correlated to a strong polarized Raman band at 900–800 cm^{-1}. However, caution must be exercised because this mode is subjected to interference from carbon skeletal vibrations. In addition, the O—O stretch may be highly coupled to the C—O stretching and the C—O—O bending modes. Such coupling will certainly influence the frequency of the O—O stretch.

Table 5-6.

The CH Vibrational Frequencies (cm^{-1}) of $O{-}CH_3$ and $O{-}CH_2$ Groups

	$\nu_{as}(CH)$	$\nu_s(CH)$	Deformation	Wag	Rock
$O{-}CH_3$	3000–2980 2970–2920	2850–2815	1470–1435	—	1200–1185
$O{-}CH_2$	2955–2922	2878–2835	1475–1445	1400–1360	—

5-8. CH VIBRATIONS IN $O-CH_3$ AND $O-CH_2$ GROUPS

McKean *et al.* [26, 30–32) have shown that the CH bonds in $O-CH_3$ groups do not have identical force constants. The CH bonds that are trans to the major lobe of either lone pair of the oxygen have longer bond lengths and smaller force constants than the CH bond that has a trans orientation to the $C-O$ bond in $C-O-CH_3$ groups. These have been studied by substituting CD_2H for CH_3 groups and observing the spectra, and also by molecular orbital calculations. This causes the $O-CH_3$ stretching bands to be spread over a larger region than that for the $C-CH_3$ group. The bands are further complicated by Fermi resonance with $O-CH_3$ bending overtones. See Table 5-6.

Study of the *Aldrich Library of FT-IR Spectra* [8] shows that the $O-CH_3$ group usually has IR bands in the regions 3000–2980, 2970–2920, and 2850–2815 cm^{-1}. These bands are seen in methyl ethers including anisoles, and also in methyl esters.

The $O-CH_3$ symmetrical deformation band wavenumber is higher than the 1375 cm^{-1} $C-CH_3$ band and is seen near 1440 cm^{-1} [3]. The symmetrical and antisymmetrical $O-CH_3$ deformation bands overlap in the 1470–1435 cm^{-1} region [3].

Our examination of the *Aldrich Library* [8] shows that most of the ethers with O-CH_3 groups have a characteristic IR band at 1200–1185 cm^{-1}. In CH_3-O-CH_3, a strong IR band in this region at 1197 cm^{-1} is assigned to CH_3 rock [5].

The CH_2 stretching bands in $O-CH_2$ groups are seen in the IR at 2955–2922 and 2878–2835 cm^{-1} [3], and $O-CH_2$ deformation bands are seen at 1475–1445 cm^{-1} [3]. The $O-CH_2$ wag vibration absorbs in the IR at 1400–1360 cm^{-1} in the C-CH_3 deformation region, and in $O-CH_2-CH_3$ groups, interaction can occur, usually giving two bands in the 1400–1360 cm^{-1} region [3] that resemble the IR *t*-butyl bands here.

The $O-CH-O$ group has a $C-H$ wag vibration that gives rise to a band in the IR at 1350–1325 cm^{-1} [3].

REFERENCES

1. L. J. Bellamy, "The Infrared Spectra of Complex Molecules," John Wiley & Sons, Inc., New York, 1975, and references therein.
2. F. R. Dollish, W. G. Fateley, and F. F. Bentley, "Characteristic Raman Frequencies of Organic Compounds," John Wiley & Sons, Inc., New York, 1974.

3. N. B. Colthup, L. H. Daly, and S. E. Wiberley, "Introduction to Infrared and Raman Spectroscopy," 3rd Ed., Academic Press, New York, 1990.
4. K. Nakanishi and P. H. Solomon, "Infrared Absorption Spectroscopy," Holden-Day, Inc., San Francisco, 1977.
5. Y. Kanazawa and K. Nukada, *Bull. Chem. Soc.* (*Jap.*) **35,** 612, 1962.
6. H. Wieser, W. G. Laidlaw, P. J. Krueger, and H. Fuhrer, *Spectrochim. Acta* **24A,** 1055, 1968.
7. J. P. Perchard, *Spectrochim. Acta* **26A,** 707, 1970.
8. C. J. Pouchert, "The Aldrich Library of FT-IR Spectra," Aldrich Chemical Company Inc., 1985.
9. K. Nukada, *Bull. Chem. Soc.* (*Jap.*) **35,** 3, 1962.
10. K. Nukada, *Bull. Chem. Soc.* (*Jap.*) **34,** 1615, 1961.
11. K. Nukada, *Spectrochim. Acta* **18,** 745, 1962.
12. "Sadtler Standard Raman Spectra," Sadtler Research Laboratories, 1976.
13. S. Varsanyi, "Assignments for Vibrational Spectra of Seven Hundred Benzene Derivatives," John Wiley and Sons, Inc., New York, 1974.
14. A. J. Melveger, L. R. Anderson, C. T. Ratcliffe, and W. B. Fox, *Appl. Spectrosc.* **26,** 381, 1972.
15. K. O. Christie, *Spectrochim. Acta* **27A,** 463, 1971.
16. M. E. Bell and J. Laane, *Spectrochim. Acta* **28A,** 2239, 1972.
17. P. A. Budinger, J. R. Mooney, J. G. Grasselli, P. S. Fay, and A. T. Guttman, *Anal. Chem.* **53,** 884, 1981.
18. I. P. Zyat'kov and D. I. Sagaidak, *J. Appl. Spectrosc.* (*U.S.S.R.*) **38,** 94, 1984 (translated from *Zhur. Spectrosk.* **38,** 110, 1983).
19. D. D. DesMarteau and R. M. Hammaker, *Israel J. Chem.* **17**, 103, 1978.
20. T. Shimanouchi, Y. Ogawa, M. Ohta, H. Matsuura, and I. Harada, *Bull. Chem. Soc.* (*Jap.*) **49,** 2999, 1976.
21. Y. Ogawa, M. Ohta, M. Sakakibara, H. Matsuura, I. Harada, and T. Shimanouchi, *Bull. Chem. Soc.* (*Jap.*) **50,** 650, 1977.
22. T. Mikawa, *Bull. Chem. Soc.* (*Jap.*) **29,** 110, 1956.
23. A. R. Katritzky and N. A. Coats, *J. Chem. Soc., London*, 2062, 1959.
24. L. H. Briggs, L. D. Colebrook, H. M. Fales, and W. C. Wildman, *Anal. Chem.* **29,** 904, 1957.
25. K. J. Sax, W. S. Saari, C. L. Mahoney, and J. M. Gordon, *J. Org. Chem.* **25,** 1590, 1960.
26. D. C. McKean, J. L. Duncan, and L. Batt, *Spectrochim. Acta* **29A,** 1037, 1973.
27. R. A. Nyquist, *Appl. Spectrosc.* **40,** 275, 1986.
28. R. A. Nyquist, C. L. Putzig, and N. B. Skelly, *Appl. Spectrosc.* **40,** 821, 1986.
29. J. Bomstein, *Anal. Chem.* **30,** 544, 1958.
30. D. C. McKean, *Chem. Soc. Rev.* **7,** 399, 1978.
31. D. C. McKean, J. E. Boggs, and L. Schäfer, *J. Mol. Struct.* **116,** 313, 1984.
32. D. C. McKean, *Croat. Chem. Acta* **61,** 447, 1988.

CHAPTER 6

Alkenes

6-1. ALKYL-SUBSTITUTED ETHYLENES

Vibrational spectra of olefinic hydrocarbons were studied extensively between 1930 and 1950 [1–15, 21]. Some useful reviews have been given by Bellamy [16], Colthup *et al.* [17], Dollish *et al.* [18], Nyquist [19], Piaux [20], Sheppard and Simpson [15], Gruzder [22], Sverdlov [23], and Rea [24]. Vibrational assignments of ethylene [25, 28], propene [36], 1-butene [13], 1-hexene [37], 3,3-dimethyl-1-butene [38], isobutene [36, 39, 40), *cis*- and *trans*-2-butenes [26], *cis*-3-hexene [37], and tetramethyl ethylene [41] have been reported. In addition, Raman frequencies of 75 hydrocarbon olefins were reported by Sterin *et al.* [27] in 1976. Spectral positions as well as the relative IR and Raman intensities of various alkyl-substituted ethylenes are summarized in Table 6-1.

6-2. C=C STRETCHING

The C=C stretching frequency of alkyl-substituted ethylenes occurs in the neighborhood of 1650 cm^{-1} and is very characteristic for this type of substitution. The strong, highly polarized nature of this vibrational mode has made it a very useful characteristic Raman band. Freeman [115] reports that

Table 6-1.

Characteristic Frequencies (cm^{-1}) of Alkylated Ethylenes

Group	cm^{-1}	IR	R	Assignment
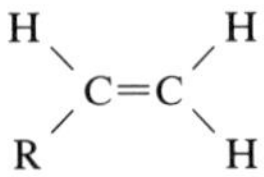 Monoalkyl (vinyl)	3090–3075	m	m	CH_2 antisym. str.
	3020–2995	w	s, p	CH str.
	3000–2980	m	w	CH_2 sym. str.
	1840–1805	w	—	$2 \times 910\ cm^{-1}$
	1650–1638	ms	vs, p	C=C str.
	1420–1412	mw	s, p	CH_2 scis. def.
	1309–1288	w	s, p	CH rock
	995–985	vs	w	trans CH in-phase wag
	910–905	vs	w	CH_2 wag
	688–611	w	w	cis CH in-phase wag
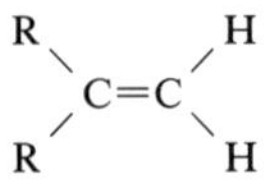 1,1-Dialkyl (vinylidine)	3090–3075	m	w	CH_2 antisym. str.
	3000–3075	m	s, p	CH_2 sym. str.
	1810–1770	w	—	$2 \times 890\ cm^{-1}$
	1660–1640	ms	vs, p	C=C str.
	1420–1400	w	m, p	CH_2 scis. def.
	900–885	vs	w	CH_2 wag
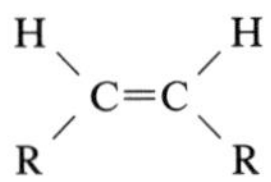 *cis*-Dialkyl	3020–2995	m	m	CH str.
	1662–1631	mw	vs, p	C=C str.
	1429–1397	m	—	cis CH antisym. rock
	1270–1250	w	s	cis CH sym. rock
	730–650	ms	vw	cis CH wag
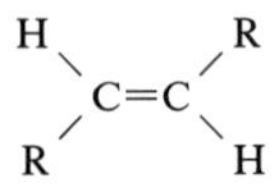 *trans*-Dialkyl	3010–2995	m	m	CH str.
	1676–1665	vw	vs, p	C=C str.
	1325–1300	—	s	trans CH sym. rock
	ca. 1295	mw	—	trans CH antisym. rock
	980–965	vs	vw	trans CH wag
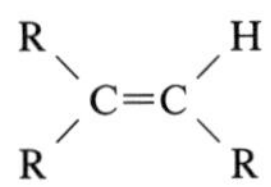 Trialkyl	3040–3020	w	m	CH str.
	1680–1664	w	vs, p	C=C str.
	1360–1322	—	w	CH rock
	840–790	m	vw	CH wag
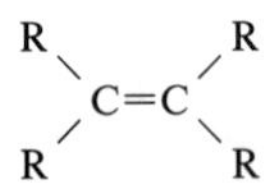 Tetraalkyl	1680–1665	—	vs, p	C=C str.

for acyclic alkenes, the depolarization ratio ρ for vinyls and vinylidines is about 0.04; for cis, ρ is about 0.05; for trans, ρ is about 0.08; and for tri- and tetrasubstitution, ρ is about 0.1. The IR intensity is variable and is very dependent on the local symmetry of the C=C group. As a general rule, monosubstituted and 1,1-disubstituted alkenes give rise to a medium-strong IR band. The IR band for 1,2-*cis*-dialkyl ethylenes is medium-weak, but still prominent, and serves as a good characteristic frequency. The C=C stretching of *trans*-dialkyl, trialkyl, and tetraalkyl ethylenes exhibits very weak IR absorption.

The C=C stretching frequencies for ethylene and 1-butene were reported to be at 1623 cm^{-1} and 1645 cm^{-1}, respectively. For other 1-olefins with carbon numbers higher than four, the C=C stretching frequency falls in a narrow region, 1640–1642 cm^{-1} [24, 29]. As the number of substituents on the double bond increases, the C=C stretching frequency occurs at a higher frequency (Table 6-1). It should be noted that the presence of a methyl group attached to one of the ethylenic carbon double bonds will increase the characteristic spectral region shown in Table 6-1 by approximately 5 cm^{-1} relative to other alkane groups [24]. The presence of the ethylenic group with a CH_3 substitution can be further confirmed by the enhanced CH_3 symmetric deformation in the Raman spectra about 1375 cm^{-1}.

The effect of the α-branched alkyl group on the C=C stretching vibration is demonstrated by Table 6-2. According to this table, the C=C vibration moves 5–14 cm^{-1} toward lower wavenumbers when there is steric hindrance between two alkane groups on one C=C carbon. This may decrease the C=C—C angle a little. See the discussion in Section 6-3. This occurs for 1,1-disubstituted, trisubstituted, and tetrasubstituted ethylenes. However, the vibrational frequency is not lowered for monosubstituted and trans or cis disubstituted ethylene where there is little or no steric hindrance [24].

The IR intensity of the C=C stretching band is usually only medium-strong at best, and in some cases such as *trans*-di-, tri-, or tetraalkyl substituted ethylenes, it can be very weak or absent. In some conjugated systems, the C=C band may even fall near aromatic bands below 1620 cm^{-1}. Also, C=O bands are very strong in the IR and some can fall in the C=C region. In general, the C=C stretch band is much more distinctive in the Raman spectrum because of its very strong intensity and its highly polarized character. The C=O stretch band, by contrast, has a much weaker Raman intensity. In unsymmetrically substituted ethylenes such as XCH=CHY, where X is an electron donor and Y is an electron withdrawer, the C=C stretch band can be strong in the IR.

Table 6-2.

The Effect of Branchings on the C=C Stretching Frequencies (cm^{-1}) of Alkylated Ethylenes

Unbranched	cm^{-1}	α-Branched	cm^{-1}
$CH_2{=}CH$-*n*-C_3H_7	1641	$CH_2{=}CH{-}CH(CH_3){-}C_2H_5$	1640
		$CH_2{=}CH{-}C(CH_3)_2{-}C_2H_5$	1640
		$CH_2{=}CH{-}CH(CH_3){-}CH(CH_3)_2$	1640
		$CH_2{=}CH{-}CH(C_2H_5)_2$	1640
		$CH_2{=}CH{-}CH(CH_3){-}C_3H_7$	1642
trans-$CH_3{-}CH{=}CH{-}C_2H_5$	1673	*trans*-$CH_3{-}CH{=}CH{-}CH(CH_3)_2$	1670
		trans-$CH_3{-}CH{=}CH{-}C(CH_3)_3$	1673
trans-$CH_3{-}CH{=}CH$-*n*-C_3H_7	1672	*trans*-$CH_3{-}CH{=}CH{-}CH(CH_3){-}C_2H_5$	1673
		trans-$C_2H_5{-}CH{=}CH{-}CH(CH_3)_2$	1668
		trans-$C_2H_5{-}CH{=}CH{-}C(CH_3)_3$	1665
		trans-$(CH_3)_2CH{-}CH{=}CH{-}CH(CH_3)_2$	1666
cis-$CH_3{-}CH{=}CH{-}C_2H_5$	1658	*cis*-$CH_3{-}CH{=}CH{-}CH(CH_3)_2$	1660
		cis-$CH_3{-}CH{=}CH{-}C(CH_3)_3$	1652
$CH_2{=}C(CH_3){-}C_2H_5$	1651	$CH_2{=}C(CH_3){-}CH(CH_3)_2$	1646
		$CH_2{=}C(CH_3){-}C(CH_3)_3$	1640
$CH_2{=}C(CH_3)$-*n*-C_3H_7	1652	$CH_2{=}C(CH_3){-}CH(CH_3){-}C_2H_5$	1646
		$CH_2{=}C(CH_3){-}CH(C_2H_5)_2$	1644
$CH_3{-}CH{=}C(CH_3){-}C_2H_5$	1673	$CH_3{-}CH{=}C(CH_3){-}CH(CH_3)_2$	1668
		$CH_3{-}CH{=}C(CH_3){-}C(CH_3)_3$	1659
$(CH_3)_2C{=}C(CH_3){-}C_2H_5$	1670	$(CH_3)_2C{=}C(CH_3){-}CH(CH_3)_2$	1659, 1666

6-3.

CYCLIC C=C STRETCHING

In cyclic C=C compounds, shifts can occur in the C=C stretching frequency in five-, four-, and three-membered rings, relative to those of six-membered rings or noncyclic olefins. The carbon atom substituents on the C=C groups stay nearly stationary during the C=C stretch vibration and as a result, all the attached C—C bonds are changed in length somewhat, adding to the C=C restoring force. When one or both C=C carbon atoms are part of a ring, this C—C vibrational interaction varies if the C=C—C angle varies, causing some C=C stretch frequency shifts. The C=C force constant is relatively unaffected. This effect was summarized by Colthup *et al.* [17], and some examples are listed in Table 6-3. This interaction is at a minimum when all the C=C—C angles are 90° (cyclobutene) and increases as each C=C—C angle gets larger or smaller than 90°.

Table 6-3.

Cyclic C=C Stretching Frequencies (cm^{-1}) [17]

Both C=C Carbons in Ring	Substitution on C=C		
	CH=CH	CH=CR	CR=CR
Cyclohexene	1655–1640[a]	1682–1668	1685–1677
Cyclopentene	1617–1611[a]	1660–1640	1686–1671
Cyclobutene	1566	1641	1675
Cyclopropene	1656	1788	1900–1865

One C=C Carbon in Ring	Substitution on C=C	
	$C{=}CH_2$	$C{=}CR_2$
Methylene cyclohexane	1651	1668
Methylene cyclopentane	1657	1687
Methylene cyclobutane	1678	—
Methylene cyclopropane	1781	—

[a] Substituents may be present on single-bonded ring carbons.

6-4. OLEFINIC C—H STRETCHINGS

Olefinic C—H stretchings can be roughly divided into three types: namely, the antisymmetric $=CH_2$ stretching (3090–3075 cm^{-1}), the symmetric $=CH_2$ stretching that appears about 100 cm^{-1} lower (3000–2880 cm^{-1}), and the =CH stretching of a =CHR group (3040–2995 cm^{-1}). Since C—H stretchings are fairly localized, and the spectral pattern is only partially affected by substituents [29] on the C=C unit, the presence or absence of C—H stretching IR and Raman bands (in cm^{-1}) listed in Table 6-4 can be used to establish the possible presence of $=CH_2$ and =CHR groups.

The $=CH_2$ antisymmetric bands in monoalkyl and 1,1-dialkyl ethylenes (3090–3075 cm^{-1}) are easily identified because their frequencies are considerably higher than those of saturated CH stretching modes, which fall below 3000 cm^{-1}. The $=CH_2$ symmetric stretch band in 1.1-dialkyl ethylenes (3000–2980 cm^{-1}) is easily missed in compounds with CH_3 groups (2972–2952 cm^{-1}). Monoalkyl ethylenes have a doublet at 3020–2980 cm^{-1} caused by =CH(R) and $=CH_2$ symmetric stretch. In the IR, the component below 3000 cm^{-1} is the stronger ($=CH_2$), and in the Raman, the component above 3000 cm^{-1} is the stronger (=CH).

Table 6-4.

Wavenumbers for $=CH_2$ and $=CH$ Stretch

	$=CH_2$ Antisymmetric Stretch	$=CH_2$ Symmetric Stretch	$=CH$ Stretch
cis and trans $RCH=CHR$, $R_2C=CHR$	—	—	3040–2995
$R_2C=CH_2$	3080–3075	3000–2980	—
$RCH=CH_2$	3090–3075	3000–2980	3020–2995
cis and trans $ClCH=CHCl$, $Cl_2C=CHCl$	—	—	3086–3079
$Cl_2C=CH_2$	3125	3030	—
$ClCH=CH_2$	3121	3030	3086

Olefinic C—H stretching bands of 1,2-disubstituted and trisubstituted ethylenes appear as sharp bands of medium intensity in a higher frequency region of 3040–2995 cm^{-1}. Sometimes, the C—H stretching band of a trisubstituted double bond is of variable intensity and is overlooked in IR and Raman spectra.

In non-hydrocarbons, electronegative substituents on the C=C shift the olefinic C—H stretching bands to somewhat higher wavenumbers, as seen in Table 6-5 [29, 62–68].

6-5.

OLEFINIC C—H OUT-OF-PLANE DEFORMATIONS

By far the most useful IR bands to elucidate the type of alkylated ethylenes are the CH out-of-plane wagging of the ethylenic group [15, 16, 30–33, 116]. According to Colthup *et al.* [17], the two most characteristic out-of-plane wag modes for monoalkyl ethylenes $RCH=CH_2$ are described as $=CH_2$ wag (near 910 cm^{-1}, IR : s), and trans CH=CH in-phase wagging (near 990 cm^{-1}, IR : s). A less useful weak IR band near 630 cm^{-1} involves cis CH=CH in-phase wagging. Examples of such modes are given in Fig. 6-1. Generally, the Raman intensities of the wagging modes are weak and thus, of little practical use in elucidating the molecular structure. The CH_2 wagging band is accompanied by a medium to weak overtone in the IR that occurs in the neighborhood of 1825 cm^{-1}. A unique feature of this overtone

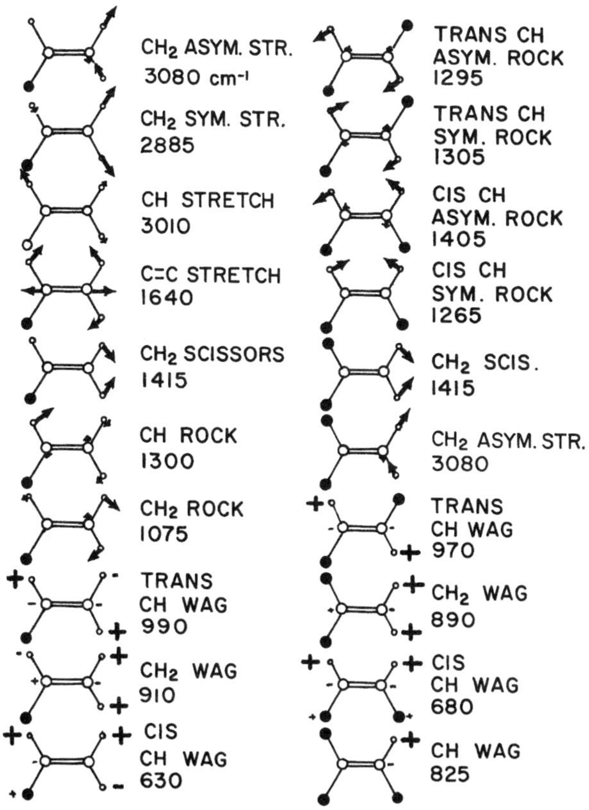

Fig. 6-1. Vibrations of vinyl groups or monosubstituted ethylenes in the left column and selected vibrations of di- and trisubstituted ethylenes in the right column. Approximate wavenumbers given are for ethylenes with alkane substituents.

band is its negative anharmonicity, which makes the overtone frequency of the CH_2 wagging mode 5–15 cm^{-1} higher than twice that of the fundamental wavenumber. The $=CH_2$ wag and trans CH=CH in-phase wag for 1-alkenes are found in rather narrow spectral regions in IR spectra. However, the cis CH=CH in-phase wag is more sensitive to mechanical interaction effects of the substituents and appears mainly in a broader region (683–611 cm^{-1}). The study by Nyquist [30] suggested that vibrational frequencies of the cis in-phase CH wagging mode of 1-alkenes are correlated to the conformational structure as well as the α-branching situation. The correlation is tabulated (next page).

For the dialkyl-substituted ethylenes, the trans dialkylated ethylenes are characterized by very strong trans CH wag IR bands between 980 and 965 cm^{-1}, and the vinylidene hydrocarbons are characterized by very

	Alkene	cm^{-1}
	$CH_3-CH=CH_2$	582
R−C−C=C, cis	$R-CH_2-CH=CH_2$	545–568
R−C−C=C, gauche	$R-CH_2-CH=CH_2$	611–635
	$(R-)_2CH-CH=CH_2$	661–672
	$(R-)_3C-CH=CH_2$	683–638

intense $=CH_2$ wagging bands in the 895–885 cm^{-1} region. The cis CH wagging bands for cis dialkyl ethylenes are located in the 720–680 cm^{-1} region. This band is less dependable for structural elucidation, especially when the substituent attached to the C=C double bond is non-hydrocarbon. The out-of-plane deformation of trialkyl ethylenes are observed in the 830–800 cm^{-1} region.

6-5.1. XYC=CH_2 Wag with Non-alkane Substituents

Potts and Nyquist [116] have shown that for vinyls and 1,1-disubstituted ethylenes, with non-alkane substituents, the $=CH_2$ wag IR frequencies extend over a wide but predictable range. The $=CH_2$ wag band increases or decreases in wavenumber when the substituents mesomerically withdraw or donate electrons to the $=CH_2$ carbon [121]:

$$N\equiv C-CH=CH_2 \leftrightarrow \bar{N}=C=CH-\overset{+}{C}H_2 \quad (960\ cm^{-1}),$$

$$R-CH=CH_2 \quad (910\ cm^{-1}),$$

$$R-O-CH=CH_2 \leftrightarrow R-\overset{+}{O}=CH-\bar{C}H_2 \quad (813\ cm^{-1}).$$

The effect is additive in 1,1-disubstituted ethylenes:

$$(N\equiv C)_2C=CH_2 \quad (985\ cm^{-1}),$$

$$R_2C=CH_2 \quad (890\ cm^{-1}),$$

$$(R-O)_2C=CH_2 \quad (711\ cm^{-1}).$$

Alkane substituents are weak electron donors, which accounts for the fact that $R_2C=CH_2$ wag bands are about 20 cm^{-1} lower than $RCH=CH_2$ wag bands.

The $(XY)C=CH_2$ wag wavenumber can be approximated by the use of Δ cm^{-1} shift values for the X and Y substituents in vinyls and vinylidines [17]:

$$\bar{\nu}[(XY)C=CH_2\ \text{wag}] = 930\ cm^{-1} + \Delta_X + \Delta_Y \pm 10\ cm^{-1}.$$

Δ values in cm^{-1} are −CO−O−OR, +30; −C≡N, +25; −H, O; $-CH_2-Cl$, O; R, −20; $-C_6H_5$, −20; −Cl, −30; $-O-CO-CH_3$, −50; −F, −65; and −O−R, −115.

6-5.2. In-Phase *trans*-XCH=CHY Wag

In vinyls and 1,2-*trans*-disubstituted ethylenes, the "trans CH wag" wavenumber shows sensitivity to the inductive effect of the substituents. Electronegative substituents tend to lower the wavenumber relative to alkane substituents [116]:

$$R{-}CH{=}CH_2 \quad 990\ cm^{-1},$$

$$Cl{-}CH{=}CH_2 \quad 938\ cm^{-1}.$$

The effect is additive in 1,2-*trans*-disubstituted ethylenes:

$$R{-}CH{=}CH{-}R\ (trans) \quad 964\ cm^{-1},$$

$$R{-}CH{=}CH{-}Cl\ (trans) \quad 926\ cm^{-1},$$

$$Cl{-}CH{=}CH{-}Cl\ (trans) \quad 892\ cm^{-1}.$$

6-5.3. Out-of-Phase CH=CH Wag

In cis, trans, and vinylidine olefins with alkane substituents, there are CH wag vibrations where the hydrogens wag out-of-phase. These are inactive or very weak in the IR. In the Raman spectra, cis isomers have a medium band at 970–952 cm^{-1}, and trans isomers have a very weak band at 776–745 cm^{-1} [18] involving out-of-phase CH wag vibrations [120]. Also, vinylidine isomers have a weak Raman band at 711–684 cm^{-1} [18] for $=CH_2$ twist [120].

6-5.4. Conjugated CH=CH—CH=CH Wag

As seen previously, the out-of-phase cis CH wag and the in-phase trans CH wag both come near 965 cm^{-1}. In conjugated polyenes with cis- and trans-substituted double bonds, interaction can occur [119]. Approximate IR wavenumbers and intensities for conjugated double bonds are as follows: isolated trans, 965 ms; trans-trans, 986 s; cis-trans, 986 m, and 944 m; trans-trans-trans, 994 vs, and 936 w; cis-trans-trans, 994 s, and 965 m; and cis-trans-cis, 994 w, and 936 w [119]. In *trans*-1,3-pentadiene, the conjugated trans and vinyl groups interact to give IR CH wag bands at 1001, 948, and 895 cm^{-1} [116]. Vibrational analyses have been done on conformers of 1,3,5-hexatriene that also show interacting trans and vinyl bands. The trans-trans-trans conformer has IR bands at 1009 s, 941 m, and 900 s [122].

6-6.

OLEFINIC C—H IN-PLANE DEFORMATIONS

For the vinyl group, the two in-plane deformations that are characteristic are $=CH_2$ scissors deformation and C—H rock (see Fig. 6-1). The $=CH_2$ scissors give rise to prominent IR and Raman bands in the 1420–1415 cm^{-1} region. The =CH rocking is usually seen best in the Raman and is located in the 1309–1288 cm^{-1} region. In Raman spectra, a possible spectral interference on assigning the =CH rocking mode is due to the in-phase $-(CH_2)_n-$ twisting vibration at around 1300 cm^{-1}. Nevetheless, it is often possible to differentiate the $-(CH_2)_n-$ in-phase twisting from the =CH rocking mode because, in general, the =CH rocking mode is polarized while the $-(CH_2)_n-$ in-phase twisting is nonpolarized.

In 1,1-disubstituted alkenes ($H_2C{=}C\langle$), the Raman band at 1420–1405 cm^{-1} is due to the $=CH_2$ scissors deformation vibration. The IR intensity is weak.

In cis dialkyl ethylenes, the cis CH=CH antisymmetric rocking mode appears near 1404 cm^{-1} in the IR, and the cis CH=CH symmetric rock is located at 1270–1251 cm^{-1} in Raman spectra.

The trans CH=CH symmetric rock for trans dialkyl ethylenes gives rise to a strong Raman band at 1325–1300 cm^{-1}, and the trans CH=CH antisymmetric rock appears near 1295 cm^{-1} in the IR spectra. However, the IR intensity is variable, and IR absorptions from other vibrations occur in this region in substituted ethylenes [29].

6-7.

SKELETAL VIBRATIONS

In the Raman spectra of tetraalkylethylenes, the symmetric C—C stretching band appears as a strong band in the 690–678 cm^{-1} region [18], and the in-plane skeletal deformation occurs in the 510–485 cm^{-1} region. According to Varsanyi [35], the symmetric C_3 stretching mode of $C{=}CC_2$ is expected to be in the region of 840–810 cm^{-1} [35]; a shift in frequency to near 870 cm^{-1} in this vibrational mode is reported for $=C(CH_3)_2$ [35]. Investigations in the Raman spectra of $(CH_3)_2C{=}CH(CH_2CH_3)$ [34a] and $(C_3H_7)(CH_3)C{=}CH_2$ [34b] show bands for these two compounds near 840 cm^{-1} and 850 cm^{-1}, respectively.

6-8. THE SYMMETRIC CH_3 DEFORMATION OF $C{=}C{-}CH_3$ GROUPS

In compounds (including alkanes) with $C{-}CH_3$ groups, the symmetric CH_3 deformation band near 1375 cm^{-1} has medium intensity in the IR. In Raman spectra, this band is quite weak when the CH_3 group is in an alkane, but the intensity is enhanced when the CH_3 group is attached to an unsaturated carbon atom such as C=C, C≡C, C=O, or an aromatic ring [34] (hyperconjugation).

6-9. ALLYL COMPOUNDS

Dollish *et al.* [18] correlated the C=C stretching of allyl compounds ($H_2C{=}CH{-}CH_2X$) with the strong Raman band between 1649 and 1625 cm^{-1} on the basis of 24 allyl compounds. This region is virtually the same as that of monoalkylethylenes (1650–1638 cm^{-1}). Vibrations involving ethylenic C—H stretchings and bendings are in regions near that of monoalkylethylenes.

Recently, the vibrational analysis of several allyl compounds has been carried out with an interest in their rotational conformations. The rotation around the C—C single bond of the $={C}{-}CH_2X$ moiety results in two conformations, cis and gauche, for simple allyl compounds. These isomers may exhibit different C=C and ethylenic CH vibrational frequencies. Therefore, the spectral complication due to the coexistence of conformational isomers becomes a problem for group frequency study, since there is no simple rule to determine which conformer absorbs at a higher frequency. For example, the C=C stretchings frequencies for gauche allyl fluoride [42], allyl chloride [42, 45], and ally iodide [42] are lower than those for the corresponding cis conformers. But the gauche isomer of allyl amine [46–48] absorbs at a higher frequency than the cis form. In some cases the C=C stretching bands of cis and gauche rotatomers are overlapped, such as in the vibrational spectra of allyl cyanide [43], isocyanate [44], and isothiocyanate [44]. Therefore, the C=C band of allyl compounds can appear as a single band, two bands, or a shoulder adjacent to a stronger band in the IR and Raman spectra.

In spite of the spectral complication just stated, the vibrational assignments of many allyl compounds in different rotational conformations are available. It is therefore possible to choose an adequate allyl molecule as a model compound for a specific application. The molecules that are found by vibrational analysis to exist as cis isomers in the solid state are allyl fluoride ($H_2C{=}CH{-}CH_2F$) [42], allyl cyanide ($H_2C{=}CH{-}CH_2CN$) [43], allyl isocyanate ($H_2C{=}CH{-}CH_2NCO$), and allyl isothiocyanate ($H_2C{=}CH{-}CH_2NCS$) [44]. Examples for the allyl compounds, which exist predominantly as the gauche isomer in the crystalline state, are allyl chloride ($H_2C{=}CH{-}CH_2Cl$) [42, 45], allyl bromide ($H_2C{=}CH{-}CH_2Br$) [42, 45], allyl iodide ($H_2C{=}CH{-}CH_2I$) [42], and allyl amine ($H_2C{=}CH{-}CH_2NH_2$) [46–48]. At room temperature, small amounts of the less stable rotational isomers are found to coexist with the more stable conformer. In addition to the preceding examples, the following compounds are reported to exist as cis/gauche mixtures at room temperature: 2,3-dihalo-1-propene ($H_2C{=}CYCH_2X$, X, Y: halogen atoms) [49–57], 1,1,2-trichloro-3,3-difluoro-1-propene ($Cl_2C{=}CCl{-}CHF_2$) [58], 1-bromo-3-methyl-2-butene ($(CH_3)_2C{=}CH{-}CH_2Br$) [59], 3-methyl-3-butenenitrile ($H_2C{=}C(CH_3){-}CH_2CN$) [60], and 2-methyl-3-butenenitrile ($H_2C{=}CH{-}CH(CH_3)CN$) [61].

6-10.

OTHER SUBSTITUTED ETHYLENES

Group frequencies of selected vinyl compounds are listed in Table 6-5. With the exception of $H_2C{=}CH{-}(SiX_3)$, X = H or —R, the C=C stretching mode of these compounds absorbs IR strongly. Furthermore, the ν(C=C) Raman band is polarized and intense. In comparison with the C=C stretching frequency of ethylene (R: 1620 cm^{-1}), the presence of a fluorine atom on the C=C moiety increases the frequency by 34 cm^{-1}. The substitution of the other halogen atoms on the C=C group, however, induces an opposite shift on the C=C stretching frequency. The progressive decrease of the C=C stretching frequency on passing from vinyl chloride via vinyl bromide to vinyl iodide is demonstrated in Table 6-5. It is noteworthy that multiple-band structures arising from rotational isomerism may appear in the C=C stretching region for $H_2C{=}CH{-}OR$ because of =C—O—R rotational isomerism. The gauche isomer C=C band is at 1660–1635 cm^{-1}, and the cis isomer band is at 1620–1610 cm^{-1} [116, 117]. The conjugated ketone, $H_2C{=}CH{-}C({=}O){-}CH_3$ has an s-trans isomer C=C band at

Table 6-5.
Group Frequencies (cm^{-1}) for Some Vinyl Compounds

IR	Raman	Assignment	IR	Raman	Assignment
$H_2C{=}CH{-}F$ vap. [62, 63]			$CH_2{=}CH{-}Br$ vap. [6, 29, 64, 66, 67]		
3150	—	CH_2 asym. str.	3112 m	3112 m	CH_2 asym. str.
3115	—	CH str.	3087 m	3087 m	CH_2 str.
3080	—	CH_2 sym. str.	3027 s	3027 m	CH_2 sym. str.
1645 vs	—	C=C str.	1602 vs	1601 vs	C=C str.
929 vs	—	trans CH wag	1373 s	1373 s	CH_2 def.
863 s	—	CH_2 wag	1258 vs	1258 s	CH rock
711 m	—	cis CH wag	941 vs	—	trans CH wag
			901 vs	—	CH_2 wag
			582 vs	—	cis CH wag
$H_2C{=}CH{-}Cl$ vap. [64–66]			$CH_2{=}CH{-}I$ [68]		
3121 s	3122 m	CH_2 asym. str.	3115	3092	CH_2 asym. str.
3086 s	3086 s	CH str.	3050	3062	CH str.
3030 s	3033 s	CH_2 sym. str.	2995	3003	CH_2 sym. str.
1608 vs	1607 vs	C=C str.	1593	1581	C=C str.
1369 m	1368 s	CH_2 def.	1376	1369	CH_2 def.
1279 m	1279 vs	CH rock	1242	1229	CH rock
941 vs	—	trans CH rock	946	—	trans CH wag
896 vs	—	CH_2 wag	909	—	CH_2 wag
620 vs	—	cis CH wag			
$CH_2{=}CH{-}OR$ [18, 72, 117, 118]			$CH_2{=}CH{-}O{-}C({=}O){-}CH_3$ [69]		
3125–00 w		CH_2 asym. str.	—	3125	CH_2 asym. str.
3078–60 w		CH str.	—	3050	CH str.
3050–00 w	3024–18	CH_2 sym. str.	1644 m	1647 vs	C=C str.
1660–35 s	1654–51	C=C gauche	—	1353 w	CH rock
1643–37 s	1643–37	C=C	948 m	—	trans CH wag
1620–10 s	1612–10	C=C cis	874 m	—	CH_2 wag
	1419–10	CH_2 def.	712 m	—	cis CH wag
1323–20 s	1310–04	CH rock			
970–60 s		trans CH wag			
820–10 s		CH_2 wag			
$CH_2{=}CH{-}SiH_3$ vap. [70, 71]			$CH_2{=}CH{-}Si(CH_3)_3$ [73]		
3067 w	3066 w	CH_2 asym. str.	3050 w	3050 m	CH_2 asym. str.
2994 w	2994 s	CH_2 sym. str.	3010 w	3009 w	CH str.
1599 vw	1599 s	C=C str.	—	2965 vs	CH_2 sym. str.
1406 w	1408 w	CH_2 def.	1660 vw	1595 s	C=C str.
—	1269 m	CH rock	1010 m	—	trans CH wag
961 m		trans CH wag	952 m	—	CH_2 wag
			515 w	—	cis CH wag
$CH_2{=}CH{-}SCH_3$ [14, 74, 77]					
3095 m	3089 m	CH_2 asym. str.			
3033 s	—	CH str.			
3014 m	3010 s	CH_2 sym. str.			
1584 s	1587–85 vs	C=C str.			
1390 s	1389 s	CH_2 def.			
1277 w	1275 s	CH rock			
953 vs	—	trans CH wag			
859 s	—	CH_2 wag			
595 s	—	cis CH wag			

1648 cm^{-1} and an s-cis isomer C=C band at 1618 cm^{-1}. (See Section 6-3.2.) As in halogen substitution (other than fluorine), the attachment of a heavy atom such as S or Si next to the C=C moiety lowers the C=C stretching frequencies.

The $=CH_2$ and trans CH waggings are very characteristic in IR absorption for the vinyl compounds. The cis CH wag in vinyls, on the other hand, is weaker and is sensitive to mechanical interaction effects. The cis in-phase wag vibrational frequency varies from 711 cm^{-1} for vinyl fluoride ($H_2C{=}CHF$) to 515 cm^{-1} for trimethylvinylsilane ($H_2C{=}CH{-}Si(CH_3)_3$). Therefore, this particular vibrational mode is of little use in elucidating the presence of the C=C bond other than providing information on the substituent involved. The Raman intensity of three ethylenic CH waggings ranges from weak to undetectable; thus, their use as Raman group frequencies is marginal.

On further substitution, the C=C stretching frequency will vary from those of vinyl compounds. The direction of frequency shift depends on the nature of the substituents on the C=C bond. A considerable number of vibrational studies have been done on the subject of di-, tri-, and tetrasubstituted ethylenes. Some examples are vinylidene halides ($H_2C{=}CX_2$, X = F, Cl, Br) [75–78], *cis-* and *trans-*1,2,difluoroethylenes [79], *cis-* and *trans-*1-chloropropene [81], 1-chloro-2-methylpropene [80], and 1,1,2-trichloro-3,3-difluoro-1-propene [58].

6-11.

CONJUGATION WITH A C=C MOIETY

Conjugation of a C=C moiety with another C=C double bond splits the C=C stretching absorption into symmetric and antisymmetric C=C−C=C stretching bands. This is largely because of the in-phase and out-of-phase mechanical interaction between the stretching of two conjugated C=C groups. The symmetric mode is located at a higher frequency, while the antisymmetric mode is about 50 cm^{-1} lower, as shown in Table 6-6.

The relative IR and Raman intensity of the two C=C stretching bands is very dependent on the conformational structure, and hence the symmetry property, in the vicinity of the C=C−C=C skeleton. Thus, it is necessary to discuss briefly the rotational conformation of conjugated dienes in order to understand the IR and Raman group frequency of this type of compound. For a conjugated diene containing the $>C_1{=}C_2{-}C_3{=}C_4<$

Table 6-6.

The C=C Stretching Frequencies of 1,3-Butadiene and Its Derivatives

Compounds	Conformation	Symmetric Stretch		Antisymmetric Stretch		
		IR	R	IR	R	Ref.
$CH_2=CH-CH=CH_2$ (liq.)	s-trans	—	1638	1592	—	
$CH_2=CR-CH=CH_2$						
$R = -CH_3$ (liq.)	s-trans	1642 w	1645 s	1595 s	1595 vw	18
$R = -CH_2CH_3$ (liq.)	s-trans	1633 w	1645 s	1590 s	1590 vw	82
$R = -CH(CH_3)_2$ (liq.)	s-trans	1632 w	1640 s	1590 s	1580 vw	82
$R = -C(CH_3)_3$ (liq.)	s-trans	1642 w	1650 w	1596 w		82
$R = -C(CH_3)_3$ liq.	s-cis	1611 s	1610 s	1645 w		82
$R = -CH_2C(CH_3)_3$ (liq.)	s-trans	1642 w		1596 s		82
$CH_2=CCl-CH=CH_2$ (liq.)	s-trans	1632 m	1628 vs	1591 s	1581 w	83, 84
$CHR_1=CH-CR_2=CH-OC_2H_5$						
$R_1, R_2 = H$			1659 vs[b]		1588 w	88
			1644 vs[b]		1554 w	
$R_1 = H, R_2 = CH_3$			1654 vs		1553 w	88
$R_1 = CH_3, R_2 = H$			1657 vs		—	88
$R_1 = C_2H_5, R_2 = H$			1657 vs[b]		—	88
			1647 vs[b]		—	
$R_1, R_2 = CH_3$			1653 vs		—	88
R_1 = 1-*cyclo*-C_6H_9, $R_2 = H$			1649 vs		1590 w	88
R_1 = 1-*cyclo*-C_5H_7, $R_2 = H$			1651 vs		1545 w	88
$CH_2=C(CH_3)-C(CH_3)=CH_2$	s-trans	—	1625 vs	1603 vs	—	83
$CH_2=CCl-CCl=CH_2$ (liq.)	s-trans	1568 s				83
$HOOCCH=CH-CH=CHCOOH$	cis-cis[a]	1640 m		1595 vs		85
	trans-trans[a]	1640 s		1615 s		85
	trans-cis[a]	1640 s		1605 s		85
$(CH_3)_2C=CH-CH=C(CH_3)_2$		—	1668 vs	~1618 m	—	34c
$F_2C=CH-CH=CF_2$	s-trans	—	1765 vs	1717 s		86, 87
$Cl_2C=CH-CH=CCl_2$	s-trans	—	1609 vs	1550 s		86
$CCl_2=CCl-CCl=CCl_2$	gauche	1610 m	1608 vs	1560 m	1560 vs	83
$(CF_3)_2C=CX-C(CF_3)=CF_2$						
$X = -P(=O)(OEt)_2$		1740		1640		89
$X = -N(CH_3)_2$		1670		1650		89
$X = -NCH_2CH_2OCH_2CH_2$		1670		1635		89

[a] Conformation with respect to double bonds instead of single bond.
[b] Two bands observed for each stretching mode due to rotation of —OEt group.

skeleton, three rotational isomers can be found with different dihedral angles around the C_2-C_3 single bond. These isomers are described as s-trans (180°), s-cis (0°), and gauche (0°–180°) isomers. The designation "s" is referring to the C_2–C_3 single bond. In the s-trans configuration, the two conjugated double bonds are co-planar with the two C=C bonds, parallel and pointing away from each other. When one of the double bonds is rotated 180° around the C_2-C_3 axis, the s-cis form results. The gauche isomer occurs when the rotational angle is less than 180°, and the two double bonds are not co-planar in this configuration.

Chang *et al.* [90] suggested that the steric effect is a major factor in determining the stability of the three configurations of halogenated butadienes. In most cases, the s-trans isomer is the most favorable isomer. The s-cis configuration is favored only when there are simultaneous steric hindrances present on C_1 and C_3 or C_2 and C_4 atoms. Gauche isomers are found for perfluoro and perchloro butadienes, which have serious steric interaction between halogen atoms on C_1 and C_3 (or C_2 and C_4) and C_1 and C_4.

Take 1,3-butadiene as an example. Since there is no steric hindrance within the molecule, this compound exists exclusively as an s-trans isomer in liquid and solid states. As a consequence of its symmetry, C_{2h}, the C=C symmetric stretching mode of 1,3-butadiene is strongly Raman-active and polarized at 1638 cm^{-1} (liq.). This stretching mode is not observed in the infrared, however [18]. Conversely, the antisymmetric C=C—C=C stretching mode exhibits very strong IR absorption at 1592 cm^{-1} but is Raman-inactive [18].

Table 6-6 lists the C=C stretching frequency of selected substituted butadienes. With the exception of highly fluorinated compounds, the symmetric C=C stretchings on these compounds occur between 1668 and 1609 cm^{-1} (IR: w; R: S) and the antisymmetric vibration at 1610–1550 cm^{-1} (IR: s, R: w) for s-trans isomers generally. In some cases, the coexistence of s-trans and s-cis isomers at room temperature results in two bands for each stretching mode. The spectrum in these regions may be further complicated by the rotation of substituents on ethylenic carbons [16, 83]. For instance, the ethoxy butadienes listed in Table 6-6 present a total of four bands, instead of two, in the C=C stretching range.

6-12.

CONJUGATION WITH A C=O MOIETY (α, β-UNSATURATED CARBONYL MOIETIES)

As shown in Table 6-7, the C=C stretching frequency of acrylaldehyde

$$(CH_2{=}C{-}CH{=}O,\ \text{IR: } 1615\ cm^{-1})$$

and crotonaldehyde

$$(CH_3CH{=}CH{-}CH({=}O),\ \text{IR: } 1641\ cm^{-1})$$

are lowered by conjugation compared with those of monoalkylethylenes (1650–1638 cm^{-1}) and dialkylethylenes (1660–1640 cm^{-1}). Similar effects are observed for α,β-unsaturated ketones. Cottee *et al.* reported the

Table 6-7.

The C=O and C=C Stretching Frequencies of Some α,β-Unsaturated Carbonyl Compounds

Compound		C=O Stretch IR	C=C Stretch IR	R	Ref.
$H_2C{=}CH{-}C({=}O)H$ (CCl_4 soln.)	s-trans	1703 vs	1615 w		96
$CH_3CH{=}CH{-}C({=}O)H$ (CCl_4 soln.)	s-trans	1693 vs	1641 s		96
$H_2C{=}CH{-}C({=}O)CH_3$ (CCl_4 soln.)	s-trans	1687 vs	1648 w		
	s-cis	1707 vs	1618 m, sh		96
$CH_3CH{=}CH{-}C({=}O)CH_3$ (CCl_4 soln.)	s-trans	1674 vs	1645 m, sh		
	s-cis	1692 s	1632 s		96
$H_2C{=}CH{-}C({=}O)OH$ (liq.)		1705		1637	107
$H_2C{=}CH{-}C({=}O)OCH_3$ (CS_2 soln.)		1728 vs	1633 w	1633	69, 99
$H_2C{=}CH{-}C({=}O)OEt$ (CS_2 soln.)		1724 vs	1636 w		99
			1620 w		
$(CH_3)CH{=}CH{-}C({=}O)OCH_3$ (CS_2 soln.)		1727 vs	1663 m		99
		1680 vw	1655 vw, sh		
$(CH_3)CH{=}CH{-}C({=}O)OEt$ (CS_2 soln.)		1723 vs	1663 s		99
		1681 vw	1655 w, sh		
$H_2C{=}CH{-}C({=}O)Cl$ (liq.)		1761		1620	108

1647–1617 cm^{-1} range for the C=C stretching frequency of 18 α,β-unsaturated ketones [97], including open-chain and cyclic systems with one, two, or three substituents on the C=C double bond. Incidentally, C=C stretching frequencies of α,β-unsaturated esters fall in the same region as those of unconjugated ethylenes.

As with the conjugated dienes discussed in the previous section, the spectral features of α,β-unsaturated carbonyl compounds are closely related to the rotational configuration. For compounds containing a $C_1{=}C_2{-}C_3{=}O$ skeleton, two conformers (s-trans and s-cis) arisen from rotation about the $C_2{-}C_3$ single bond are possible. Vibrational studies carried out by Bowles *et al.* [98] suggested that acrylaldehyde ($CH_2{=}CH{-}CH{=}O$) and crotonaldehyde ($CH(CH_3){=}CH{-}CH{=}O$) exist as s-trans isomers at room temperature. Hence, only one C=C and one C=O stretching band are found. When the aldehydic hydrogen is replaced with an alkyl group, the steric hindrance between the alkyl and ethylenic groups tends to increase the potential energy of the s-trans configuration; therefore, some s-cis isomers may be observed at room temperature. For example, two C=C bands are reported in the infrared spectra of methyl vinyl ketone

$$(CH_2{=}CH{-}C({=}O)CH_3)$$

and ethylidenacetone

$$(CH(CH_3){=}CH{-}C({=}O)CH_3)$$

because of the coexistence of s-trans and s-cis conformers.

The differentiation between s-trans and s-cis isomers usually can be achieved by the spectral patterns of ν(C=O) and ν(C=C) bands [91–106]. Cottee *et al.* [97] found that while the C=C stretching frequency is of little diagnostic value in distinguishing s-trans and s-cis ketones, their IR intensities are quite characteristic for this purpose. The IR band of s-cis conformers in the C=C stretching region is usually strong, while those of a s-trans isomers appear to be weak. Furthermore, the frequency difference between the C=O and C=C stretchings is characteristic for the two conformers. When the >C=C—C=O moiety is not included in a five-membered ring, the difference between the C=O and C=C stretching frequencies is less than 60 cm^{-1} for the s-trans configuration and greater than 70 cm^{-1} for the s-cis isomer [97, 109]. In addition, Taylor [91] found that the C=O and C=C stretchings of s-trans and s-cis isomers respond to the solvent polarity in different ways. Collectively, the IR absorption intensity, IR C=O and C=C stretching frequency difference, and solvent polarity dependency of the C=O and C=C stretching frequencies provide valuable means for distinguishing the isomer involved.

6-13.

CONJUGATION WITH A PHENYL GROUP

Nyquist [30] recently reported the IR group frequencies of styrene

$$(C_6H_5{-}CH{=}CH_2),$$

48 ring-substituted styrenes (Ph—CH=CH_2; Ph: substituted phenyl group), 25 ring-substituted α-halostyrenes (Ph—CCl=CH_2), and α-alkystyrenes (Ph—CR=CH_2). For styrene derivatives, the phenyl and vinyl groups are co-planar only when there is no substituent at both positions 2 and 6 of the phenyl group. The C=C stretching frequency of these specific styrenes occurs in the region of 1637–1616 cm^{-1} in infrared spectra. The =CH_2 wagging mode absorbs infrared radiation in the range of 928–901 cm^{-1}. However, when there are substituents on both positions 2 and 6 of the phenyl group, the coplanar structure has too much steric hindrance. In the forced non-planar configuration, conjugation between the vinyl and phenyl groups is reduced. As a consequence, the C=C stretching band is found at 1644–1623 cm^{-1}, which is closer to the C=C stretching frequency of monalkylethylenes (1650–1638 cm^{-1}). The =CH_2 wagging frequency of these compounds is in the 947–918 cm^{-1} region.

Table 6-8.

The C=C Stretching Frequencies of Alkenes with the C=C Group Conjugated to a C≡N Moiety

Compound	IR	R
$CH_2=CH-CN$ (liq.)	1609 s	1607 s, p
cis-$CH_3CH=CH-CN$ (liq.)	1635 vs	1629 vs, p
trans-$CH_3CH=CH-CN$ (liq.)	1634 vs	1641 vs, p
cis-$C_2H_5CH=CH-CN$	1621 m	1628 vs
$(CH_3)_2C=CH-CN$	1635 s	1634 s, p

For α-halostyrenes and α-alkylstyrenes, the phenyl and isopropenyl groups are co-planar except for the 2-substituted compounds. The co-planar molecules exhibit IR bands in the 1634–1609 cm^{-1} region for the C=C stretching mode, and 899–877 cm^{-1} for the $=CH_2$ wagging mode. The frequencies of these two vibrational modes for the 2-substituted α-halostyrenes and α-alkylstyrenes are located in the 1645–1632 cm^{-1} and 899–911 cm^{-1} regions.

6-14.

CONJUGATION WITH A NITRILE (C≡N) GROUP

The vibrational analysis of unsaturated nitrile compounds, such as acrylonitrile ($H_2C=CH-CN$) [110], *cis*- and *trans*-crotononitriles ($CH_3-CH=CH-CN$) [111], *cis*-2-pentenenitrile (*cis*-Et$-CH=CH-CN$) [112], 2-methylpropenitrile ($H_2C=C(CH_3)-CN$) [113] and 3-methyl-2-butenitrile ($(CH_3)_2C=CH-CN$) [114] have been reported. The C=C stretching frequencies of these compounds are summarized in Table 6-8. In general, conjugation between the C=C and C≡N moieties results in a lowering of the C=C stretching frequency. Moreover, the presence of a CN group adjacent to the double bond enhances the IR intensity of the C=C stretching mode.

REFERENCES

1. R. Lespeiau and M. Bourguel, *Compt. Rend.* **190,** 1504, 1930.
2. M. Bourguel, *Compt. Rend.* **193,** 934, 1931.
3. M. Bourguel, *Compt. Rend.* **194,** 1736, 1932.
4. M. Bourguel, B. Gredy, and L. Piaux, *Compt. Rend.* **195,** 129, 1933.
5. M. Bourguel, *Compt. Rend.* **195,** 311, 1933.

6. M. Bourguel and L. Piaux, *Bull. Soc. Chim. Fr.* **5**, 2, 1958, 1935.
7. H. van Risseghem, B. Gredy, and L. Piaux, *Compt. Rend.* **196**, 938, 1933.
8. B. Gredy, *Bull. Soc. Chim. Fr.* **5**, 2, 1029, 1935.
9. B. Gredy, *Bull. Soc. Chim. Fr.*, **5**, 2, 1038, 1935.
10. B. Gredy, *Bull. Soc. Chim. Fr.* **5**, 2, 1951, 1935.
11. B. Gredy, *Bull. Soc. Chim. Fr.* **5**, 4, 415, 1937.
12. J. H. Hibben, "The Raman Effect and Its Chemical Application," Reinhold, New York, 1939, pp. 161–177.
13. N. Sheppard, *J. Chem. Phys.* **17**, 74, 1949.
14. H. W. Thompson and P. Torkington, *Trans. Faraday Soc.* **42**, 432, 1946.
15. N. Sheppard and D. M. Simpson, *Quart. Rev. London* **6**, 1, 1952.
16. L. J. Bellamy, "The Infrared Spectra of Complex Molecules," John Wiley & Sons, New York, 1975.
17. N. B. Colthup, L. H. Daly, and S. E. Wiberley, "Introduction to Infrared and Raman Spectroscopy," 3rd Ed., Academic Press, New York, 1990.
18. F. R. Dollish, W. G. Fateley, and F. F. Bentley, "Characteristic Frequencies of Organic Compounds," John Wiley & Sons, New York, 1974.
19. R. A. Nyquist, "The Interpretation of Vapor-Phase Infrared Spectra, Group Frequency Data," Sadtler Research Laboratories, 1984.
20. L. Piaux, *Ann. Chim.* **4**, 147, 1935.
21. K. W. F. Kohlrausch, "Raman Specktren," Akad. Berlag Becker, and Erler Kom.-Ges., Leipzig, pp. 302–319, 1943.
22. P. F. Gruzder, *Zh. Fiz. Khim.* **28**, 507, 1954.
23. L. M. Sverdlov, *Prov. Acad. Sci. USSR, Phys. Chem. Sect.* **112**, 109, 1957.
24. D. G. Rea, *Anal. Chem.* **32**, 1638, 1960.
25. B. L. Crawford, Jr., J. E. Lancaster, and R. G. Inskeep, *J. Chem. Phys.* **21**, 678, 1953.
26. D. C. McKean, M. W. Mackenzie, A. R. Morrison, J. C. Lavalley, A. Janin, V. Fawcett, and G. M. Edwards, *Spectrochim. Acta* **41A**, 435, 1985.
27. K. E. Sterin, V. T. Aleksanyan, and C. N. Zhizhin, "Raman Spectra of Hydrocarbon Compounds," 1976 (USSR), title only translated.
28. R. L. Arnett and B. L. Crawford, Jr., *J. Chem. Phys.* **18**, 118, 1950.
29. C. J. Pouchert, "The Aldrich Library of FT-IR Spectra," Aldrich Chemical Company, Inc., 1985.
30. R. A. Nyquist, *Appl. Spectrosc.* **40**, 196, 1986.
31. H. W. Thompson and P. Torkington, *Trans. Faraday Soc.* **41**, 246. 1945.
32. R. S. Rasmussen and R. R. Brattain, *J. Chem. Phys.* **15**, 120, 1947; **15**, 131, 1947; **15**, 135, 1947.
33. H. L. McMurray and V. Thornton, *Anal. Chem.* **24**, 318, 1952.
34. "The Sadtler Standard Raman Spectra," Sadtler Research Laboratories, 1976. (a) 3092R, (b) 3093R, (c) 1127R.
35. G. Varsanyi., "Assignments for Vibrational Spectra of Seven Hundred Benzene Derivatives," John Wiley & Sons, New York, 1975.
36. J. E. Kilpatrick and K. S. Pitzer, *J. Res. NBS* **38**, 191, 1947.
37. H. W. Schrotter and E. G. Hoffman, *Liebigs Ann. Chem.* **672**, 44, 1964.
38. N. Sheppard, *J. Chem. Phys.* **17**, 455, 1949.
39. C. M. Pathak and W. H. Fletcher, *J. Mol. Spectrosc.* **31**, 32, 1969.
40. W. C. Harris and I. W. Levine, *J. Mol. Spectrosc.* **39**, 441, 1971.
41. W. H. Snyder and H. S. Kimmel, *J. Mol. Struct.* **4**, 473, 1969.
42. R. D. Mclachlan and R. A. Nyquist, *Spectrochim. Acta* **24A**, 103, 1968.
43. G. H. Griffith, L. A. Harrah, J. W. Clark, and J. R. Durig, *J. Mol. Struct.* **4**, 255, 1969.

44. T. Torgrimsen, P. Klaeboe, and F. Nicolaisen, *J. Mol. Struct.* **20,** 213, 1974.
45. C. Sourisseau and B. Pasquier, *J. Mol. Struct.* **12,** 1, 1972.
46. A. L. Verma and P. Venkateswarlu, *J. Mol. Spectrosc.* **39,** 227, 1971.
47. B. Silvi and P. Perchard, *Spectrochim. Acta.* **32A,** 23, 1976.
48. J. R. Durig, J. F. Sullivan, and C. M. Whang, *Spectrochim. Acta* **41A,** 129, 1985.
49. G. A. Crowder, *J. Mol. Spectrosc.* **20,** 430, 1966.
50. T. Torgrimsen and P. Klaeboe, *J. Mol. Struct.* **20,** 229, 1974.
51. P. Klaeboe, T. Torgrimsen, and D. H. Christensen, *J. Mol. Struct.* **23,** 15, 1974.
52. S. Samdal, H. M. Seip, and T. Torgrimsen, *J. Mol. Struct.* **42,** 153, 1977.
53. Ø. Tongmo, Q. Shen, K. Hagen, and R. Seip, *J. Mol. Struct.* **71,** 185, 1981.
54. S. H. Schei and P. Klaeboe, *J. Mol. Struct.* **96,** 9, 1982.
55. S. H. Schei, *Spectrochim. Acta* **39A,** 1043, 1983.
56. G. A. Crowder, *J. Mol. Spectrosc.* **23,** 1, 1967.
57. G. A. Crowder, *J. Mol. Spectrosc.* **23,** 103, 1967.
58. P. Klaeboe, G. Neerland, and S. H. Schei, *Spectrochim. Acta* **38A,** 1025, 1982.
59. A. O. Diallo, *Spectrochim. Acta* **39A,** 613, 1983.
60. D. A. C. Compton, S. C. Hsi, and H. H. Mantsch, *J. Phys. Chem.* **85,** 3721, 1981.
61. D. A. C. Compton and W. F. Murphy, *Spectrochim. Acta* **41A,** 1141, 1985.
62. P. Torkington and H. W. Thompson, *Trans. Faraday Soc.* **41,** 236, 1945.
63. B. Bak and D. Christensen, *Spectrochim. Acta* **12,** 355, 1958.
64. M. de Hemptinne, *Trans. Faraday Soc.* **42,** 5, 1946.
65. J. Evans and H. Bernstein, *Can. J. Chem.* **33,** 1792, 1955.
66. O. W. Gullikson and J. R. Nielsen, *J. Mol. Spectrosc.* **1,** 158, 1957.
67. J. A. Rolfe and L. A. Woodward, *Trans. Faraday Soc.* **50,** 1030, 1954.
68. P. Torkington and H. W. Thompson, *J. Chem. Soc.* **303,** 1944.
69. W. R. Feairheller, Jr. and J. E. Katon, *J. Mol. Struct.* **1,** 239, 1967.
70. S. G. Frankiss, *Spectrochim. Acta* **22,** 295, 1966.
71. V. F. Kalasinsky, S. E. Rodgers, and J. A. S. Smith, *Spectrochim. Acta* **41A,** 155, 1985.
72. I. S. Ignatyev, A. N. Lazarev, M. B. Smirnov, M. L. Alpert, and B. A. Trofimov, *J. Mol. Struct.* **72,** 25, 1981.
73. J. R. Durig, W. J. Natter, and M. Johnson-Streusand, *Appl. Spectrosc.* **34,** 60, 1980.
74. J. Fabian, H. Kröber, and R. Mayer, *Spectrochim. Acta* **24A,** 727, 1968.
75. D. C. Smith, J. R. Nielsen, and H. H. Claassen, *J. Chem. Phys.* **18,** 326, 1950.
76. P. Joyner and G. Glockler, *J. Chem. Phys.* **20,** 302, 1952.
77. F. Winther and D. O. Hummel, *Spectrochim. Acta* **23A,** 1839, 1967.
78. J. R. Scherer and J. Overand, *J. Chem. Phys.* **32,** 1720, 1960.
79. N. C. Craig and J. Overend, *J. Chem. Phys.* **51,** 1127, 1969.
80. G. A. Guirgis, Y. S. Li, and J. R. Durig, *J. Chem. Phys.* **75,** 3274, 1984.
81. J. R. Durig and G. A. Guirgis, *J. Raman Spectrosc.* **13,** 160, 1982.
82. D. Craig, J. J. Shipman, and R. B. Fowler, *J. Am. Chem. Soc.* **83,** 2885, 1961.
83. G. J. Szaz and N. Sheppard, *Trans. Faraday Soc.* **49,** 358, 1953.
84. N. V. Tarasova and L. M. Sverdlov, *Russ, J. Phys. Chem.* **42,** 840, 1968.
85. P. Sohår and G. Varsânyi, *J. Mol. Struct.* **1,** 437, 1968.
86. F. A. Miller, W. F. Elbert, and W. Pingitore, *J. Mol. Struct.* **41,** 19, 1977.
87. R. M. Conrad and D. A. Dows, *Spectrochim. Acta* **21,** 1039, 1965.
88. S. M. Makin, B. K. Kruptsov, V. M. Medvedeva, and L. N. Smirnova, *J. Gen. Chem. USSR* **8,** 2492, 1962.
89. N. P. Gambaryan, D. P. Del'tsova, V. A. Livshits, and E. C. Ter-Gabriélyamn, *Bull. Acad. Sci., USSR, Div. Chem. Sci, Part II* **35,** 2147, 1987.
90. C. H. Chang, A. L. Andreasen, and S. H. Bauer, *J. Org. Chem.* **36,** 920, 1971.

91. P. J. Taylor, *Spectrochim. Acta* **32A,** 1471, 1976.
92. R. Mecke and K. Noack, *Spectrochim. Acta* **12,** 391, 1958.
93. R. Mecke, and K. Noack, *Chem. Ber.* **93,** 210, 1960.
94. R. L. Erskine and E. S. Waight, *J. Chem. Soc.* 3425, 1960.
95. K. Noack and R. N. Jones, *Can. J. Chem.* **39,** 2201, 1961.
96. A. J. Bowles, W. O. George, and D. B. Maddams, *J. Chem. Soc. B*, 810, 1969.
97. F. H. Cottee, B. P. Straughan, C. J. Timmons, W. F. Forbes, and R. Shilton, *J. Chem. Soc. B*, 1146, 1967.
98. A. J. Bowles, W. O. George, and D. B. Cunliffe-Jones, *J. Chem. Soc. B*, 1070, 1970.
99. W. O. George, D. V. Hassid, and W. F. Maddams, *J. Chem. Soc. Perkin II*, 400, 1972.
100. D. Smith and P. J. Taylor, *Spectrochim. Acta* **32A,** 1477, 1976.
101. D. Smith and P. J. Taylor, *Spectrochim. Acta* **32A,** 1503, 1976.
102. D. Smith and P. J. Taylor, *Spectrochim. Acta* **32A,** 1489, 1976.
103. J. E. Katon and P.-H. Chu, *J. Mol. Struct.* **78,** 141, 1982.
104. W. O. George and A. J. Porter, *J. Chem. Soc. Perkin II*, 954, 1973.
105. J. M. Landry and J. E. Katon, *Spectrochim. Acta* **40A,** 871, 1984.
106. C. Tétlez, R. Knudsen, and O. Sala, *J. Mol. Struct.* **67,** 189, 1980.
107. W. R. Feairheller, Jr. and J. E. Katon, *Spectrochim. Acta* **23A,** 2225, 1967.
108. J. E. Katon and W. R. Feairheller, Jr., *J. Chem. Phys.* **47,** 1248, 1967.
109. E. A. Braude and C. J. Timmons, *J. Chem. Soc.*, 3766, 1955.
110. F. Halverson, R. F. Stamm, and J. J. Whalen, *J. Chem. Phys.* **16,** 808, 1948.
111. J. R. Durig, C. K. Tong, C. W. Hawley, and J. Bragin, *J. Phys. Chem.* **75,** 44, 1971.
112. D. A. C. Compton and W. F. Murphy, *J. Phys. Chem.* **85,** 482, 1981.
113. J. Bragin, K. L. Kizer, and J. R. Durig, *J. Mol. Spectrosc.* **38,** 289, 1971.
114. D. A. C. Compton, W. F. Murphy, and H. H. Mantsch, *J. Raman Spectrosc.* **11,** 349, 1981.
115. S. K. Freeman, "Applications of Laser Raman Spectroscopy," John Wiley & Sons, New York, 1974, p. 112.
116. W. J. Potts and R. A. Nyquist, *Spectrochim. Acta* **15,** 679, 1959.
117. N. L. Owen and N. Sheppard, *Trans. Faraday Soc.* **60,** 634, 1964.
118. Y. M. Kawa, *Bull. Chem. Soc. Japan* **29,** 110, 1956.
119. N. B. Colthup, *Appl. Spectrosc.* **25,** 368, 1971.
120. J. K. Brown and N. Sheppard, *Trans. Faraday Soc.* **51,** 1611, 1955.
121. N. B. Colthup and M. K. Orloff, *Spectrochim. Acta* **27A,** 1299, 1971.
122. H. Yashida, Y. Furukawa, and M. Tasumi, *J. Mol. Struct.* **194,** 279, 1989.

CHAPTER 7

Acetylenes

7-1. INTRODUCTION

A large amount of literature appeared on the vibrational spectra of acetylenes during the 1960s and the early 1970s. The vibrational spectra of alkyl acetylenes were reviewed previously by Sheppard and Simpson [1]. More recently, a few acetylene vapor phase IR spectra were investigated by Nyquist [43]. More details on the characteristic IR and Raman frequencies were reported by Bellamy [2] and Dollish *et al.* [3], respectively. The characteristic frequencies of acetylenes arise from vibrations involving $\equiv$C—H stretching and bending, C$\equiv$C stretching, and $\equiv$C—C stretching and bending modes. The values of these vibrational frequencies are summarized in Table 7-1. More specific values for various compounds are listed in Tables 7-2 to 7-4.

The presence of monosubstituted acetylenes can be readily determined by the strong IR bands due to the $\equiv$C—H stretching and bending modes. Variable IR intensity is found for the C$\equiv$C stretching mode of monosubstituted alkynes. However, this C$\equiv$C stretching vibration becomes IR-forbidden when the C$\equiv$C moiety is symmetrically disubstituted. Fortunately, Raman spectrometry provides an easy answer to this problem, since the C$\equiv$C stretching is very strong and polarized in the Raman spectra. The spectral position and pattern of Raman bands usually provides enough information to identify the type of substitution.

Table 7-1.

Characteristic Frequencies of Acetylenes

Approximate Description of the Vibration	Frequency (cm^{-1})	IR	R
≡C—H str.	3340–3290 (soln.)	s	w, p
R—C≡C—H	3320		
GCH_2—C≡C—H[a]	3320–3310		
M—C≡C—H[b]	3303–3281		
≡C—H bend.	681–610 (soln.)	s	w
R—C≡C—H	640–625	s	m-w, dp
M—C≡C—H	707–675, 663–577		
C≡C str. monosubstituted	2130–2100	var	vs, p
R—C≡C—H	2130–2120	w	vs, p
M—C≡C—H[b]	2055–2019	s-m	vs, p
disubstituted	2250–2200	w	vs, p
	2325–2290 (Fermi resonance)	vw	var
R—C≡C—R′	2237–2230	o	vs, p
M—C≡C—CH_3[b]	2200–2170	s	vs, p
—C≡C—C≡C— asy. str.			
H—C≡C—C≡C—X[c]	2245–2175	s	
—C≡C—C≡C— sym. str.			
H—C≡C—C≡C—X[c]	2100–2000		
R—C≡C—C≡C—R′	2257–2251		vs, p
≡C—C str. monosubstituted	1000–940	w	m-w
disubstituted	1160–1005	w	m-w
C—C≡C—C antisym. str.	~1160		
sym. str.	842–758		
≡C—C bend			
CH_3—$(CH_2)_n$—C≡CH, $n = 0$–5	348–336	w	m-w
CH_2XC≡CH[c]	314–311, 186–311		
CHO—C≡C—H	261, 226		
COOH—C≡C—H	245, 218		

[a] G: H, R, SCH_2Ph, SPh, OPh, NR_2, F, Cl, Br, I. [b] M: P, As, Sb, Si, Ge, Sn.
[c] X: F, Cl, Br, I.

7-2.

THE C—H STRETCHING OF C≡C—H GROUPS

The acetylenic C—H stretching vibration is characterized by the sharp IR absorption band in the 3340–3290 cm^{-1} region (Table 7-2) when measured in CCl_4 solution. The IR intensity of this band is so strong that it is very distinct and observable even in the presence of the strong, broad O—H band in alcohols or acids (3500–2600 cm^{-1}). The weaker shoulder that often appears at the lower-frequency side of the C—H IR band can be explained by the Fermi resonance effect. This vibrational mode is Raman-active (weak, polarized) in some cases.

Table 7-2 provides a list of the C—H stretching frequencies for a variety of compounds in the solution state. Generally, the C—H stretching frequencies are sensitive to the substituent on the C≡C moiety and the

Table 7-2.

The ≡C—H Stretching Frequency (cm^{-1}) of Monosubstituted Alkynes in CCl_4 Solution, etc.

Compounds	Frequency (cm^{-1})	Reference
(a) H—C≡C—CH_3	3320	4
H—C≡C—CH_2R	3320	4
(b) H—C≡C—CH_2X, X = F	3322	4, 5
Cl, Br, I	3315	4, 5
OH, OCH_3, O—Ph	3320–3312	4
SCH_2—Ph, S—Ph	~3318	4
NR_2	3316–3313	4
H—C≡C—CF_3	3316	6
(c) H—C≡C—F	3355	7
H—C≡C—Cl	3340	7
H—C≡C—Br	3325	7
(d) H—C≡C—CH=CH_2	3305	1
H—C≡C—C(CH_3)=CH_2	3315	16
H—C≡C—CH=CH(CH_3)	3300	16
H—C≡C—COOH	3306	8
H—C≡C—C(=O)Cl	3301	9
H—C≡C—C(=O)H	3335 (vap.)	17
H—C≡C—Ph	3316	4
(e) H—C≡C—C≡C—H	3293	10
H—C≡C—C≡C—Cl	3295	11
H—C≡C—C≡C—Br	3328	11
H—C≡C—C≡C—I	3315	11
(f) SiH_3—C≡CH	3311 (vap.)	15
M(C≡C—H)$_3$ M = P, As, Sb	3303–3294	12
M(C≡C—H)$_4$ M = Si, Ge, Sn	3299–3293	13
CH_3M(C≡C—H)$_3$ M = Si, Ge	3283–3281 (liq.)	14

Table 7-3.

The C≡C Stretching Frequency (cm^{-1}) of Monosubstituted Alkynes

HC≡C−X	Frequency	Ref.	HC≡C−X	Frequency	Ref.
HC≡C−CH_3	2130	4	HC≡C−C(=O)H	2080 (liq.)	17
HC≡C−CH_2R	2121	4	HC≡C−C_6H_5	2113	4
HC≡C−CH_2OH	2120	4	HC≡C−CH=CH_2	2099 (liq.)	1
HC≡C−CH_2OCH_3	2119	4	HC≡C−C(CH_3)=CH_2	2099 (liq.)	16
HC≡C−CH_2O−Ph	2131-2121	4	HC≡C−CH=CH(CH_3)	2101 (liq.)	16
HC≡C−CH_2S−Ph	2122	4	SiH_3−C≡CH	2054 (vap.)	15
HC≡C−SCH_2−Ph	2120	4	P(C≡CH)$_3$	2052	12
HC≡C−CH_2NR_2	~2100	4	As(C≡CH)$_3$	2045	12
HC≡C−CH_2F	2148	4, 5	Sb(C≡CH)$_3$	2025	12
HC≡C−CH_2Cl	2131	4, 5	Si(C≡CH)$_4$	2055	13
HC≡C−CH_2Br	2126	4, 5	Ge(C≡CH)$_4$	2055	13
HC≡C−CH_2I	2128	4	Sn(C≡CH)$_4$	2037	13
HC≡C−CF_3	2255	6	CH_3Si(C≡CH)$_3$	2047	14
CH≡C−C(CH_3)$_3$	2105 (liq.)	19	CH_3Ge(C≡CH)$_3$	2046	14
HC≡C−CR_2OH	2115-2102	4	CH_3Sn(C≡CH)$_3$	2019	14
HC≡C−CR_2OR	2111-2103	4			
HC≡C−F	2255	7	HC≡C−C≡C−H	2175, 2005	10
HC≡C−Cl	2110	7	HC≡C−C≡C−Cl	2244, 2075	11
HC≡C−Br	2085	7	HC≡C−C≡C−Br	2238, 2100	11
HC≡C−COOH	2120	8	HC≡C−C≡C−I	2202, 2042	11
HC≡C−C(=O)Cl	2130	9			

physical state of the molecule. As observed for O−H vibrations, the ≡C−H stretching and bending modes are affected by weak hydrogen bond formation in the condensed phase. Consequently, it is not surprising that the gas-phase data (3370–3300 cm^{-1}) may shift 10–30 cm^{-1} upwards when compared to those frequencies in the liquid phase. Nyquist [46] reports 3340–3325 cm^{-1} for 1-alkynes in the vapor state. When a saturated carbon atom is adjacent to the HC≡C− moiety, the group frequency is in the 3320–3310 cm^{-1} region in solution.

Monoalkyl acetylenes absorb near 3320 cm^{-1}. This stretching frequency is reduced when a heteroatom is attached to the α-carbon atom (Table 7-2b). In addition, conjugation with C=O or C=C lowers the vibrational frequency to the 3316–3300 cm^{-1} region (Table 7-2d). The opposite shift is observed when a halogen atom is connected directly to the triple-bond carbon (Table 7-2c). An unusually large spectral shift to below 3300 cm^{-1} may result if the triple bond carbon is attached to an atom with vacant *d*-orbitals, such as P, As, Si, or Ge (Table 7-2f).

Table 7-4.

The C≡C Stretching Frequencies (cm^{-1}) of Disubstituted Acetylenes

X−C=C−Y	Frequency	Ref.	X−C=C−Y	Frequency	Ref.
$CH_3-C\equiv C-CH_3$	2233 (liq.)	19, 20	$Cl-C\equiv C-CH_2Cl$	2261	33
$CH_3-C\equiv C-R$,	2239–2233	1, 21	$Cl-C\equiv C-Cl$	2234	34
R: *n*-alkyl	(liq.)		$Cl-C\equiv C-Br$	2205	34
$CH_3-C\equiv C-CH_2Cl$	2242	22	$Cl-C\equiv C-I$	2177	34
$CH_3-C\equiv C-CH_2Br$	2242	22	$Cl-C\equiv C-CF_3$	2264	35
$CH_3-C\equiv C-CF_3$	2271	23	$Cl-C\equiv C-CN$	2195	36
$CH_3-C\equiv C-Cl$	2235	24, 25	$Br-C\equiv C-CH_2Br$	2226	33
$CH_3-C\equiv C-Br$	2239	26	$Br-C\equiv C-Br$	2185	34
$CH_3-C\equiv C-I$	2210	26	$Br-C\equiv C-I$	2154	37
$CH_3-C\equiv C-(C{=}O)Cl$	2235	9	$Br-C\equiv C-CF_3$	2235	35
$CH_3-C\equiv C-CHO$	2211	27	$Br-C\equiv C-CN$	2121	36
$R-C\equiv C-R'$	2235–2231	21, 28, 29	$I-C\equiv C-I$	2118	34
$R-C\equiv C-CH_2OH$	2228–2226	21	$I-C\equiv C-CF_4$	2208	35
$R-C\equiv C-CH(CH_3)OH$	2250	21	$I-C\equiv C-CN$	2128	36
$R-C\equiv C-C(CH_3)_2OH$	2239–2237	21	$CF_3-C\equiv C-CF_3$	2305	6
$R-C\equiv C-Br$	2219–2201	19, 29	$CN-C\equiv C-CN$	2119	38
$R-C\equiv C-I$	2212–2190	19, 29	$P(C\equiv C-CH_3)_3$	2199	39
$R-C\equiv C-CH{=}CH_2$	2236	16	$As(C\equiv C-CH_3)_3$	2189	39
$R-C\equiv C-C({=}O)CH_3$	2212–2205		$Sb(C\equiv C-CH_3)_3$	2168	39
$R-C\equiv C-C_6H_5$	2221–2202	21, 29,	$Si(C\equiv C-CH_3)_4$	2197	39
		31, 32	$Ge(C\equiv C-CH_3)_4$	2192	39
$Ph-C\equiv C-Ph$	2222	40	$Pb(C\equiv C-CH_3)_4$	2176	39
$CHO-C\equiv C-Cl$	2217	30	$CH_3-C\equiv C-C\equiv C-CH_3$	2264 (R)	42
$CHO-C\equiv C-Br$	2180	30	$R-C\equiv C-C\equiv C-R'$	2257–2251 (R)	42
$CHO-C\equiv C-I$	2159	30	$Ph-C\equiv C-C\equiv C-Ph$	2210–2205 (R)	41
				2125–2120 (IR)	

7-3.

THE C−H BENDINGS OF C≡C−H GROUPS

The characteristic IR bands in the 681–610 cm^{-1} region are attributed to the ≡C−H bending modes. For monoalkyl acetylenes, these vibrational modes occur in a narrower region of 640–628 cm^{-1}. The intensity is usually very strong in IR, while it is medium to weak in Raman. The ≡CH bending frequency tends to be increased or decreased when the substituent decreases or increases the total electron density on the ≡CH carbon [44]:

$$\mathrm{N{\equiv}C{-}C{\equiv}C{-}H \leftrightarrow \bar{N}{=}C{=}C{=}\overset{+}{C}{-}H} \qquad (663\ \mathrm{cm^{-1}}),$$

$$\mathrm{F{-}C{\equiv}C{-}H \leftrightarrow \overset{+}{F}{=}C{=}\bar{C}{-}H} \qquad (578\ \mathrm{cm^{-1}}).$$

An interesting phenomenon observed for the ≡C—H bending modes is that the frequencies are higher in the liquid state than in the gas phase. This is in contrast to the usual observation for stretching modes, but is typical of group bending modes involving hydrogen bonding. Nyquist [46] reports 639–629 cm^{-1} for 1-alkynes in the vapor state. A second feature of these vibrations is that their overtones give rise to a moderately intense IR band. However, these overtones fall in the very cluttered fingerprint region at 1300–1200 cm^{-1} for alkyl acetylenes (1257–1247 cm^{-1} in the vapor state) [46].

The out-of-plane and in-plane bending modes are degenerate for those molecules with axial symmetry about the C≡C—H group; therefore, only the spectral band will be observed. Examples of such compounds are CH_3—C≡CH (630 cm^{-1}, soln.) [4], F—C≡CH (578 cm^{-1}, vap.) [7], Cl—C≡CH (604 cm^{-1}, vap.) [7], and Br—C≡CH (618 cm^{-1}, vap.) [7].

A splitting of the ≡C—H bending frequencies will be observed when the compound has no axial symmetry about the C≡C moiety. Examples for the splitting of ≡C—H bending modes can be found in the spectra of propargyl derivatives (HC≡C—CH_2G, where G is any functional group other than the H atom) and in the spectra of acetylenes with planar substituents (e.g., C=C, C=O, or phenyl group) attached to the acetylenic carbon. The attachment of phenyl group or carbonyl group adjacent to the acetylenic carbon causes greater frequency splitting between the two bending vibrations. Some examples are the ≡C—H bending frequencies of HC≡C—C_6H_5 [4], HC≡C—C(=O)OH [8], and HC≡C—C(=O)H [17] at 648 and 611 cm^{-1} (soln.), 693 and 650 cm^{-1} (soln.), and 691 and 669 cm^{-1} (vap.), respectively. The splitting effect of propargyl derivatives has been correlated by Nyquist and Potts [4] to the electronegativity of the —G group. Generally, the splitting in the ≡C—H bending frequencies is proportional to the difference between the elctronegativities of the G group and the carbon atom. As a result of small differences in electronegativity, the separation is negligible or unresolvable in the cases of HC≡C—CH_2R, HC≡C—CH_2S—, and HC≡C—CH_2I (i.e., G = R, S—, and I, respectively). But the splittings increase to (a) 10–15 cm^{-1} for HC≡C—CH_2Br and HC≡C—CH_2Cl, (b) approximately 25 cm^{-1} for HC≡C—CH_2NR_2, and (c) 35–38 cm^{-1} for HC≡C—CH_2OR and HC≡C—CH_2F. The ≡C—H bendings occur in the 707–675 cm^{-1} and 663–577 cm^{-1} regions for acetylenes containing atoms such as Si, Ge, Sn, P, As, and Sb attached to the triple bond [12–14].

7-4.

C≡C STRETCHING

The polarized C≡C stretching band is very prominent in Raman spectra, while its IR intensity varies. This vibration is IR-active only in cases of monosubstituted or antisymmetric disubstituted acetylenes. Even so, the IR absorption of R—C≡CH type compounds is weak, and the intensity decreases progressively when the triple bond moves away from the terminal position. Thus, this band is barely observed in the IR spectrum of 2,10-dodecadiyne ($CH_3—C≡C—(CH_2)_6—C≡C—CH_3$). It appears to be virtually IR-forbidden when the C≡C group is located three or more positions from the end of the chain. Infrared intensity enhancement of the C≡C stretching band by an acetylinic halogen [7] or conjugation to C=C, C=O [17] and phenyl [18] groups has been reported.

The C≡C stretching frequency of acetylene (HC≡CH) occurs at 1961 cm^{-1} (liq.) [3]. This frequency is increased on substitution. For example, the vibrational frequency of $HC≡C—CH_3$ (2142 cm^{-1}, liq.) [4] is 181 cm^{-1} higher than that of acetylene (HC≡CH), and that of $CH_3—C≡C—CH_3$ (2313 cm^{-1}, liq.) [19] is 352 cm^{-1} higher than that observed for acetylene. In the IR (in the vapor state) 1-alkynes absorb weakly at 2162–2137 cm^{-1} [46]. Most organic monosubstituted acetylenes show C≡C Raman bands in the region of 2130–2100 cm^{-1}, and disubstituted acetylenes are indicated by a band in the 2250–2200 cm^{-1} region.

The substituent effect on C≡C stretching of disubstituted acetylenes is somewhat parallel to that of monosubstituted acetylenes. For example, the frequency of HC≡C—Br (2085 cm^{-1}) [7] is lower than that of HC≡C—Cl (2110 cm^{-1}) [7], and the same trend is observed in disubstituted compounds containing chlorine and bromine atoms next to the triple bond (Table 7-4). Monoalkyl (HC≡C—R) acetylenes show characteristic Raman bands in the regions of 2130–2120 cm^{-1} and dialkyl compounds (R—C≡C—R′) in those of 2325–2290 cm^{-1} and 2237–2229 cm^{-1}. The band in the 2325–2290 cm^{-1} region of dialkyl acetylenes arises from the Fermi resonance involving the interaction of the ≡C—C stretch overtone and the C≡C stretching. This band is also observed in many disubstituted alkynes. Such doubling of the Raman band in this region has been very useful in differentiating the type of substitution. The attachment of oxygen or sulfur atoms to the α-position has little effect on the frequency when compared with the values of alkyl acetylenes. However, α-halogenation raises the frequency to 2148–2126 cm^{-1}, and the connection of an NR_2 group to the α-carbon atom lowers the vibrational frequency to the 2100 cm^{-1} region. The effect of the halogen atom adjacent to the C≡C moiety on the spectral position

(Table 7-3) varies according to the C—X bond strength [2]. For example, a fluorine atom increases the frequency immensely to 2255 cm^{-1}, while a chlorine atom returns the frequency back to the normal region. A bromine atom, on the other hand, decreases the frequency to a value of 2085 cm^{-1}. Again, a large frequency lowering was reported when an atom with vacant *d*-orbitals (P, As, Sb, Si, Ge, Sn) is connected to the acetylenic group [12–14, 39]. Conjugation to a phenyl or C=C group causes frequency lowering, while conjugation to a C=O group shifts the spectral band in a much less predictable manner.

For compounds containing the —C≡C—C≡C— grouping, the vibrational motions are more suitably described as antisymmetric and symmetric —C≡C—C≡C— stretchings. The compound H—C≡C—C≡C—H has C≡C stretching bands at 2172 cm^{-1} (symmetric stretch) and 2005 cm^{-1} (antisymmetric stretch). The characteristic Raman band (symmetric stretch) of the R—C≡C—C≡C—R structure is at 2257-2251 cm^{-1} (strong, polarized) [42]. The Ph—C≡C—C≡C—Ph type compounds give rise to a characteristic, very strong Raman band at 2210–2205 cm^{-1} and a medium to weak IR band at 2125–2120 cm^{-1} (antisymmetric stretch) [41]. The H—C≡C—C≡C—X type compounds have strong IR bands at 2244-2202 cm^{-1} and weak IR bands at 2100–2037 cm^{-1}. On the basis of intensity, the higher frequency band was assigned to the antisymmetric C≡C stretch [11]. However, the reverse assignment is true for all other diacetylenes above. For these unsymmetrically substituted cases, the higher frequency band is probably mainly C—C≡C—X stretch, and the lower frequency band is probably mainly H—C≡C—X stretch, as in previous correlations.

7-5. THE C—C STRETCHING OF C≡C—C MOIETIES

The ≡C—C stretching bands (1161–758 cm^{-1}) are weak in IR and not simple characteristic frequencies in Raman. The identification of these vibrational bands is more difficult because other types of C—C stretching vibrations also occur in the same region and can interact. The spectral shift from 1000–940 cm^{-1} (monosubstituted acetylenes) to 1161–1005 cm^{-1} (disubstituted acetylenes) does provide a valuable characteristic doublet feature in the C≡C stretching region of dialkyl acetylenes, because of the Fermi resonance effect with the ≡C—C stretch overtone.

7-6. ≡C—C BENDINGS

Like the ≡C—H deformations, the two ≡C—C bending modes are degenerate when axial symmetry about the C≡C bond is present. In the case of no axial symmetry about the C≡C bond, the two ≡C—C bending modes are no longer degenerate, and the separation between two spectral frequencies is dependent upon the substitution involved. Only a single C—C deformation Raman band (348–336 cm^{-1}) is observed for straight-chain monoalkyl acetylenes ($CH_3—(CH_2)_n—C≡CH$, $n = 0–5$). In methyl alkyl acetylenes ($CH_3—C≡C—R$), a Raman band is seen at 380–366 cm^{-1}, along with a Raman band near 1380 cm^{-1} (CH_3 symmetric deformation) [45]. The α-halogenerated acetylenes ($CH_2X—C≡CH$; X = F, Cl, Br, I) [4] possess two deformation frequencies, at 314–311 cm^{-1} and 186–158 cm^{-1}, as a result of the loss of degeneracy and the high electronegativity of the halogen. The presence of a carbonyl group next to the C≡C bond lowers the bending frequencies, as exemplified by HC≡C—C(=O)H (261 and 226 cm^{-1}) [17] and HC≡C—C(=O)OH (245 and 218 cm^{-1}) [8].

REFERENCES

1. N. Sheppard and D. M. Simpson, *Quart. Rev.* **6,** 1, 1952.
2. L. J. Bellamy, "The Infrared Spectra of Complex Molecules," John Wiley and Sons, New York, 1975.
3. F. R. Dollish, W. G. Fateley, and F. F. Bentley, "Characteristic Raman Frequencies of Organic Compounds," John Wiley and Sons, New York, 1974.
4. R. A. Nyquist and W. J. Potts, *Spectrochim. Acta* **16,** 419, 1960.
5. J. C. Evans and R. A. Nyquist, *Spectrochim. Acta* **19,** 1153, 1963.
6. C. V. Berney, L. P. Cousins, and F. A. Miller, *Spectrochim. Acta* **19,** 2019, 1963.
7. G. R. Hunt and M. K. Wilson, *J. Chem. Phys.* **34,** 1301, 1961.
8. J. E. Katon and N. T. McDevitt, *Spectrochim. Acta* **21,** 1717, 1965.
9. E. Augdahl, E. Kloster-Jensen, and A. Rostad, *Spectrochim. Acta* **30A,** 399, 1974.
10. I. Freund and R. S. Halford, *J. Chem. Phys.* **42,** 4131, 1965.
11. P. Klaeboe, E. Kloster-Jensen, and S. J. Cyvin, *Spectrochim. Acta* **23A,** 2733, 1967.
12. F. A. Miller and D. H. Lemmon, *Spectrochim. Acta* **23A,** 1099, 1967.
13. R. E. Sacher and D. H. Lemmon, *Spectrochim. Acta* **23A,** 1169, 1967.
14. R. E. Sacher, W. Davison, and F. A. Miller, *Spectrochim. Acta* **26A,** 1011, 1970.
15. E. A. V. Ebsworth, S. G. Frankiss, and W. J. Jones, *J. Mol. Spectrosc,* **13,** 9, 1964.
16. T. G. V. Yakovleva and A. A. Petrov, *Opt. Spectrosc.* **11,** 320, 1961.
17. G. W. King and D. Moule, *Spectrochim. Acta* **17,** 286, 1961.
18. J. C. Evans and R. A. Nyquist, *Spectrochim. Acta* **16,** 918, 1960.
19. F. F. Cleveland and M. J. Murray, *J. Chem. Phys.* **11,** 450, 1943.
20. B. L. Crawford, Jr., *J. Chem. Phys.* **7,** 555, 1939.

21. B. Gredy, *Ann. Chim., XI Ser.* **4,** 5, 1935.
22. R. D. McLauchlan, *Spectrochim. Acta* **23A,** 1793, 1967.
23. E. C. Tuazon and W. G. Fateley, *J. Chem. Phys.* **53,** 3178, 1970.
24. A. G. Meister, *J. Chem. Phys.* **16,** 950, 1948.
25. D. W. Davison and H. J. Bernstein, *Can. J. Chem.* **33,** 1226, 1955.
26. R. A. Nyquist, *Spectrochim. Acta* **21,** 1245, 1965.
27. J. C. D. Brand and R. A. Powell, *J. Mol. Spectrosc.* **43,** 342, 1972.
28. M. Bourguel and P. Daure, *Bull. Soc. Chim., Fr.* **47,** 1349, 1930.
29. M. J. Murray and F. F. Cleveland, *J. Am. Chem. Soc.* **63,** 1718, 1941.
30. E. Lagset, P. Klaeboe, E. Kloster-Jensen, S. J. Cyvin, and F. M. Nicolaisen, *Spectrochim. Acta* **29A,** 17, 1973.
31. M. J. Murray and F. F. Cleveland, *J. Am. Chem. Soc.* **60,** 2664, 1938.
32. M. J. Murray and F. F. Cleveland, *J. Am. Chem. Soc.* **61,** 3546, 1939.
33. R. A. Nyquist, A. L. Johnson, and J.-S. Lo, *Spectrochim. Acta* **21,** 77, 1965.
34. P. Klaeboe, E. Kloster-Jensen, D. H. Christensen, and I. Johnsen, *Spectrochim. Acta* **26A,** 1567, 1970.
35. E. Augdahl, E. Kloster-Jensen, V. Devarajan, and S. J. Cyvin, *Spectrochim. Acta.* **29A,** 1329, 1973.
36. P. Klaeboe and E. Kloster-Jensen, *Spectrochim. Acta* **23A,** 1981, 1967.
37. D. H. Christensen, T. Stroyer-Hansen, P. Klaeboe, E. Kloster-Jensen, and E. E. Tucker, *Spectrochim. Acta* **28A,** 939, 1972.
38. F. A. Miller, R. B. Hannan, Jr., and L. R. Cousins, *J. Chem. Phys.* **21,** 110, 1953.
39. R. E. Sacher, B. C. Pant, F. A. Miller, and F. R. Brown, *Spectrochim. Acta* **28A,** 1972, 1361.
40. "The Sadtler Standard Raman Spectra," Sadtler Research Laboratories, 1976, 1526R.
41. "The Sadtler Standard Raman Spectra," Sadtler Research Laboratories, 1976, 2710R, 3741R, 3742R, 3743R.
42. A. G. Meister and F. F. Cleveland, *J. Chem. Phys.* **12,** 393, 1944.
43. R. A. Nyquist, "The Interpretation of Vapor-Phase Infrared Spectra, Group Frequency Data," Sadtler Research Laboratories, 1984.
44. N. B. Colthup, *Appl. Spectrosc.* **50,** 589, 1976.
45. K. W. F. Kohlrausch, "Ramanspektren," Akad. Verlagses, Leipzig, 1943, p. 321.
46. R. A. Nyquist, *Appl. Spectrosc.* **39,** 1088, 1985.

CHAPTER 8

The —C≡N and —N≡C Groups

8-1. INTRODUCTION

Nitriles (C—CN), cyanamides (N—CN), cyanates (—O—CN), thiocyanates (—S—CN), selenocyanates (—Se—CN), and isocyanides (C—NC) are the most commonly encountered compounds containing —C≡N (or —N≡C) triple bonds. These compounds have very valuable characteristic —C≡N (or —N≡C) stretching frequencies in a relatively unpopulated spectral region, which allows for easy identification. Group frequencies for —C≡N groups are listed in Table 8-1.

Table 8-1.

Characteristic Frequencies (cm^{-1}) of —C≡N and $-\overset{+}{N}\equiv\overset{-}{C}$ Groups

	cm^{-1}	IR	R	Reference
1. Nitriles (C—CN)				
C≡N str.				
R—CN	2250–2230	s	vs, p	13, 14, 43, 18–20, 45
$CH_3(CH_2)_n$—CN	2250–2245	s	vs, p	43
ring —CN	2242–2230	s	s, p	28, 29, 32, 43
XCH_2—CN (X = F, Cl, Br, I)	2260–2244	w	s, p	43, 45
—OCH_2—CN	2256–2245	w	s, p	43
XCH_2—CH_2—CN (X = O, Cl, Br)	2256–2250	s	s, p	43, 45

Table continued overleaf

Table 8-1—continued

		cm^{-1}	IR	R	Reference
RO—C(=O)—CH_2—CN		near 2260	w		43
>N—C(=O)—CH_2—CN		2270–2257	w		43
RR′C=CR″(CN)		2235–2215	s	vs	43
(R,R′,R″ = H, alkyl group)					
ArHC=CAr(CN)		2235–2215	w	w	45
R—C(=O)—CN		2225–2210	s	vs, p	43
EtO—C(=O)—CN		2246	m		43
benzonitriles		2240–2220	m-s	s, p	40, 41, 43, 45
polynuclear aromatic nitriles		2222–2210	m-s		43
C—C≡N bend.					
CH_3—CN		378	s	vs, dp	50
GCH_2—CN:	o.p.[a]	378–370		w	8, 10
	i.p.[a]	307–106		m-s, dp	16, 18
H_2C=CH(CN):	o.p.	362			22
	i.p.	242		s	
RR′C=CR″(CN):	o.p.	282–223	m	w	23–26
	i.p.	242–153		ms, dp	
2. Cyanamides (N—CN)					
C≡N str.		2225–2210	s	vs, p	43, 45, 53
3. Cyanate (—O—CN)					
C≡N str.		2256–2245	s	vs, p	54–56
4. Thiocyanates (—S—CN)					43, 57–60
C≡N str.					
R—S—SN		2157–2155	s	s, p	
Ar—S—CN		2174–2161	s	s, p	
S—C≡N bend.		513–453	w		
		416–405	m		
5. Selenocyanates (—Se—CN)					61, 62
C≡N str.		near 2152	s		
Se—C≡N bend.		420–400	vw		
		365–360	w		
6. Isocyanides (—C—N≡C)					
N≡C str.					
R—NC		2146–2134	s	s, p	43, 63–65
$C_6H_5CH_2$—NC		2152	s		43
RO—C(=O)—CH_2—NC		2166–2164	s		43
(R = Me, Et)					
Ar—NC		2125–2109	s	s, p	63

[a] o.p.: out-of-plane; i.p.: in-plane.

In general, the characteristic stretching frequency of the C≡N moiety usually occurs at above 2210 cm^{-1} when the group is attached to a light atom such as C, N, or O. This stretching frequency is lowered when a heavier atom (e.g., S, Se) is at the adjacent position. The presence of an isonitrile group (i.e., an isocyanide group) is characterized by a broad band below 2160 cm^{-1}. Although the rotational isomerism affects the X—C stretching frequencies of X—CN compounds (X = C, S, and Se) little rotational isomeric effect was observed on the relatively localized C≡N stretching mode and X—C≡N bending (X = C, S, Se).

8-2. NITRILE COMPOUNDS (C—CN)

Vibrational spectra of nitrile compounds are best characterized by the sharp C≡N stretching bands. Early vibrational studies of nitriles were reported by Reitz *et al.* (Raman) [1–3], Kitson and Griffith (IR) [4], and Hidalgo (IR) [5]. More recent vibrational analysis are available for the CN groups attached to saturated carbon atoms [6–21], C=C double bonds [22–25], rings [27–33, 42, 51], —C≡C—X (X = F, Cl, Br, I, CN) [34–37], or benzene rings [38, 39].

The sharp band in the region of 2270–2210 cm^{-1} is easily assigned to the characteristic C≡N stretching mode [40, 41]. The IR intensity of this vibrational mode varies considerably from very strong to very weak [43]. Such intensity variation is the direct result of the inductive effect arising from the neighboring environment or neighboring substituent [44]. In the solid-phase IR spectra of benzonitriles, the Christiansen effect is commonly observed [43, 45] if the sample is insufficiently ground.

The strong Raman scattering of the C≡N stretching mode is polarized in the majority of cases. The depolarization ratio ρ for alkyl substituted nitriles is about 0.05–0.10 [66]. We have examined the 84 Raman spectra reported in reference 45. Among these Raman spectra, only those compounds with structures of the types (Ar)HC=C(Ar)(CN) and $(C_6H_4X)CH_2$—CN (where X: strong electron withdrawing group) show weak Raman intensity.

8-2.1. Aliphatic Nitriles

The C≡N stretching mode gives rise to strong IR and Raman bands for alkyl nitriles in the 2250–2230 cm^{-1} region [13, 14, 18–20, 43, 45]. A narrower frequency range (2250–2245 cm^{-1}) is expected when the alkyl group is a straight chain ($CH_3(CH_2)_n-C\equiv N$) [43]. The CH_2 deformation in $R-CH_2-C\equiv N$ groups absorbs in the IR at 1435–1405 cm^{-1} [46].

The effect of branching can be summarized as follows: (a) The CN stretching frequency for secondary nitriles (i.e., $R_2CH-C\equiv N$) is located in the same range as a straight chain. The IR absorption of $(CH_3)_2CH-C\equiv N$ (2245 cm^{-1}) and especially of $(NC-CH_2CH_2)_2CH(CH_3)-CN$ (2249 cm^{-1}) are good examples. In the latter example, the primary and secondary $C\equiv N$ stretching bands are unresolvable, and only one IR band is observed [43]. (b) Further branching on the α-carbon atom appears to lower the CN vibrational frequency a little. For example, the IR spectrum of *t*-butylcyanitride ($(CH_3)_3C-C\equiv N$) is characterized by a strong $C\equiv N$ band at 2234 cm^{-1} [43]. Also, it should be noted that the $C\equiv N$ stretching band is not necessarily the strongest Raman band below 2600 cm^{-1} for nitriles containing a branched alkyl group. For example, vibrations involving quaternary and tertiary carbon atoms usually give rise to Raman intensities comparable with or even stronger than that of the $C\equiv N$ stretching mode [45].

8-2.2. Nitrile with Substituents on the α-Carbon

The presence of an electron-attracting atom such as Cl or O on the α-carbon next to the nitrile group can markedly reduce the IR intensity of the $C\equiv N$ band [46]. The band frequency is slightly higher than the frequency for alkyl nitriles. The Raman intensity is not markedly affected. The vibrational frequencies of $F-CH_2-CN$, $Cl-CH_2CN$, $Br-CH_2-CN$, and $I-CH_2-CN$ are located at approximately 2257 cm^{-1} (liq.; IR, vw), 2260 cm^{-1} (liq.; IR, w), 2254 cm^{-1} (liq.; IR, m), and 2244 cm^{-1} (liq.; IR, s), respectively [43]. Further halogenation on the α-carbon atom is best illustrated by the Raman bands of Cl_2CH-CN and Cl_3C-CN at 2253 cm^{-1} and 2250 cm^{-1}, respectively [40]. The IR intensities are weak. The attachment of an oxygen to the α position has an effect similar to that of the halogen. For example, the stretching frequency is increased to the 2255–2245 cm^{-1} region, and the IR intensity of this stretching frequency varies from medium-weak to very weak [43].

The attachment of a halogen or an oxygen atom to the β-carbon atom causes a small shifting of the $C\equiv N$ stretching frequency to the same characteristic frequency region as that of α-substituted nitriles. However, the $C\equiv N$ IR intensities of the β-halogenated or β-oxygenated nitriles are stronger than those of the corresponding α-substituted nitriles, and more like those of alkyl nitriles. Therefore, a medium $C\equiv N$ band is expected in the IR spectra [43, 45].

The presence of a carbonyl group next to the α-carbon atom produces a similar effect on the C≡N band as that observed in the IR spectra of the halogenated nitriles. For instance, the IR spectra of NC—CHRC(=O)NH_2 and NC—CH_2C(=O)NHR exhibit weak bands between 2270 and 2256 cm^{-1}, and those of the NC—CH_2C(=O)OR type of molecules generally display a very weak band near 2260 cm^{-1} [43].

The presence of two CN groups on the same carbon increases the CN stretching frequency a little from that of alkyl nitriles. Examples of such dicyano compounds are malononitrile ($CH_2(CN)_2$, 2273 cm^{-1}) [43], 1,1-dicyanocyclopropane (2263, 2257 cm^{-1}) [30], and 1,1,2,2-tetracyanoethane (2259 cm^{-1}) [43].

The stretching frequency of a C≡N group attached to a non-aromatic ring system is located in a lower region of 2242–2230 cm^{-1} [28, 29, 32, 43]. In this case, the frequency lowering was attributed to the partial conjugation of the C≡N group with the ring system [32].

8-2.3. Conjugated Nitriles

Conjugation to C=C and C=O groups lowers the C≡N frequency to 2235–2215 cm^{-1} and 2225–2210 cm^{-1}, respectively [43]. This CN frequency can be shifted to an even lower value if a strong electron-donating group such as an amino moiety (—NH_2 or —NR_2) is present on the β-carbon of the double bond. For instance, C≡N stretching modes of 3-dimethylaminoacrylonitrile, $(CH_3)_2$NCH=CH—C≡N, and 3-amino-crotononitrile, (CH_3)(NH_2)C=CH—C≡N, absorb IR radiation at the unusually low frequencies of 2195 cm^{-1} and 2180 cm^{-1}, respectively [43]. In addition, the C≡N IR absorption bands of these compounds are exceptionally broad compared to the sharp CN stretching bands normally observed.

The intensity of the C≡N mode is enhanced in IR and Raman spectra by the conjugated group in the majority of cases [40, 41, 43, 45]. However, exceptions exist for compounds with the structure of (Ar)HC=C(Ar)(CN) or EtO—C(=O)—CN. In the IR and Raman spectra of 21 molecules with the (Ar)HC=C(Ar)(CN) type structure [41], the characteristic C≡N stretching frequencies fall in a region similar to that of other conjugated C≡N groups (2235–2210 cm^{-1}). However, the IR and Raman intensities of these conjugated C≡N compounds are weak, especially when the aryl group contains a strong electron-withdrawing group such as —NO_2 or —Cl. In the case of EtO—C(=O)—CN, the medium IR band occurs at 2246 cm^{-1}, which is about 25 cm^{-1} higher than the absorption region of the R—C(=O)—CN type of molecules (2225–2210 cm^{-1}). This elevated C≡N stretching frequency of EtO—C(=O)—CN can be attributed to the presence of an oxygen atom on the α-carbon atom of the CN group.

8-2.4. Aromatic Nitriles

The characteristic frequency of C≡N stretching vibrations of benzonitriles falls in the range of 2240–2220 cm^{-1} [40, 41]. The CN stretching bands of polynuclear aromatic nitriles are observed in the region of 2222–2210 cm^{-1} [43].

Among the IR spectra of 80 benzonitriles published in *The Aldrich Library of FT-IR Spectra* [43], more than 65 spectra are characterized by a strong to medium band in the 2240–2220 cm^{-1} region. The vibrational frequency and IR intensity are correlated to the Hammett-type substitution parameters both experimentally and theoretically [47–49]. Electron-withdrawing groups, such as the nitro group, decrease the IR band intensity and increase the frequency, whereas electron-donating groups, such as the amino group, increase the IR intensity and decrease the frequency [46, 67].

The CN stretching band of benzonitrile (C_6H_5CN) occurs at 2230 cm^{-1}, and those of most monosubstituted benzonitriles are observed in the 2241–2225 cm^{-1} region [43]. Substitution of the aromatic hydrogen by an electron-withdrawing group, such as —F, —NO_2, —OH, or —CF_3, results in a C≡N stretching frequency shift to the higher limit of the characteristic spectral region. The presence of —Cl and —Br at meta and para positions also shifts the frequency to above 2230 cm^{-1}. Furthermore, successive substitution of electron-withdrawing or -donating groups can also shift the C≡N stretching frequency beyond the characteristic frequency region of 2240–2220 cm^{-1}. For example, 2-*N*-isopropylamino-4-methylbenzonitrile exhibits an IR band at 2208 cm^{-1}, and 2,3,5,6-tetrafluoro-4-cyanobenzonitrile demonstrates IR absorption at 2253 cm^{-1}. Other examples of benzonitriles exhibiting a C≡N band beyond the characteristic 2240–2220 cm^{-1} frequency region are *meta*-hydroxylbenzonitrile (2246 cm^{-1}), *ortho*- and *para*-aminobenzonitriles (2213 cm^{-1}), and *para-N,N*-dimethylaminobenzonitrile (2210 cm^{-1}) [43].

The IR intensity of benzonitriles varies from medium-weak to strong, with the electron-withdrawing or -donating properties of the substituents on the benzene ring. The Raman intensity of the C≡N band is enhanced by the conjugation of the aromatic ring. Nevertheless, the aromatic ring stretching and deformation modes often exhibit stronger Raman intensities than the CN stretching vibration [45].

8-2.5. Skeletal Vibrations for C—C≡N Compounds

The two C—C≡N bending modes occur in spectral regions slightly higher than the corresponding vibrations of the C—C≡C moiety. For acetonitrile

(CH_3—CN), the out-of-plane and in-plane bendings are degenerated. Therefore, only one band is observed at 378 cm^{-1} (IR: m, R: vs, dp) [50]. However, the degeneracy is removed when one or more hydrogen atoms on the methyl group are replaced by other functional groups. The removal of this degeneracy gives two vibrational bands that are recognized as the in-plane and out-of-plane bendings. For GCH_2CN (G = F, Cl, Br, CN, OCH_3) [8, 10, 16–18] type compounds, the out-of-plane bending frequency is observed in a very narrow region (378–370 cm^{-1}, R: s). This frequency region is very close to that found for acetonitrile. The in-plane bending frequencies, however, occur in a much wider spectral range (358–165 cm^{-1}) because of the coupling of this vibration with other in-plane vibrations. This vibrational mode gives rise to prominent IR and Raman intensities and is useful in confirming the type of substituent, G, since a heavier substituent usually shifts the in-plane bending frequency towards a lower value. For example, some of the mass-dependencies of the in-plane bendings are illustrated by the vibrational bands in the following: CH_3—CH_2CN (358 cm^{-1}, IR : w, R : s) [16], CH_2Cl—CH_2CN (307 cm^{-1}, IR : vs, R : s) [16], CH_2Br—CH_2CN (285 cm^{-1}, IR : s, R : vs) [16], $CH_2(CN)$—CH_2CN (224 cm^{-1}, IR : vs, R : vs) [17], F—CH_2CN (238 cm^{-1}, R : w) [10], Cl—CH_2CN (198 cm^{-1}, IR : s, R : m, dp) [10], Br—CH_2CN (172 cm^{-1}, IR : s, R : m, dp) [10], and I—CH_2CN (165 cm^{-1}, IR : s, R : w, p) [10].

The out-of-plane and in-plane C—C≡N bendings of H_2C=CH—C≡N occur at 362 cm^{-1} and 242 cm^{-1}, respectively [22], while the corresponding vibrational bands of its alkyl substituted derivatives fall in the far infrared regions of 282–223 cm^{-1} and 203–167 cm^{-1} [23–26]. In the case of conjugation to the C≡C group, the frequency of these bendings shifts to a higher value. For example, the strong IR and Raman bands in the 483–496 cm^{-1} region were assigned to the C—C≡N bending mode of X—C≡C—C≡N (X = Cl, Br, and I) [35].

The C—C—CN bending offers little practical application as a characteristic frequency, not only because of the wide spectral range (600–480 cm^{-1}), but also because of the poor IR and Raman intensities.

8-3.

CYANAMIDES ($>$N—C≡N)

There are a limited number of vibrational analyses reported for this class of compounds. The C≡N stretching band of H_2N—CN, the simplest cyanamide, is located at 2260 cm^{-1} [43, 52]. But alkylation on the N atom causes the C≡N frequency to be lowered for this vibrational mode. In dimethylcyanamide, one observes strong IR and Raman bands at 2220 cm^{-1} ($(CH_3)_2N-CN$) [43, 53]. IR spectra of seven dialkyl cyanamides, including cyclic and acyclic compounds, were reported in *The Aldrich Library of FT-IR Spectra*. The C≡N stretching absorption of these compounds is in the 2215–2200 cm^{-1} region [43]. Generally, these CN stretching bands are broader in IR spectra than those of the nitrile compounds. Hence, the CN absorptions of nitrile and cyanamide can be readily differentiated by their band shape, even though they are generally located in similar spectral regions.

The Raman spectra of dimethylcyanamide and diallylcyanamide, published in *The Sadtler Standard Raman Spectra* [45], reveal a sharp, polarized, and strong C≡N band that looks very similar to those of nitriles. Therefore, the Raman spectral band shape and intensity cannot be readily used as the basis to distinguish cyanamide and nitrile.

Skeletal vibrations of dimethylcyanamide are at 1058 cm^{-1} (antisymmetric C—N—C stretching; IR : s, R : not observed), 762 cm^{-1} (symmetric C—N—C stretching; IR : ms, R : vs, p), 620 cm^{-1} (N—C≡N in-plane bending, IR : m, R : w), and 530 cm^{-1} (N—C≡N out-of-plane bending, IR : mw, R : w) [53]. The symmetric C—N—C stretching band can be a good Raman group characteristic frequency, since a medium, polarized band near 810 cm^{-1} is also observed for diallylcyanamide [45].

8-4. CYANATES, THIOCYANATES, AND SELENOCYANATES

8-4.1. Cyanates (—O—C≡N)

The C≡N stretching frequencies of cyanate compounds are located in the region of 2256–2245 cm^{-1} [54–56]. The presence of a cyanate group is further characterized by the C—O stretchings at the 1123–1084 cm^{-1} (alkyl cyanates) and 1190–1100 cm^{-1} (aryl cyanate) regions [55, 56].

8-4.2. Thiocyanates (—S—C≡N)

The C≡N stretching of aliphatic thiocyanates (R—S—CN) exhibits strong IR and Raman bands at 2157–2155 cm^{-1} [43, 57–60], while C≡N stretching of aromatic thiocyanates (Ar—S—CN) displays strong IR and Raman bands at a higher value (2174–2161 cm^{-1}) [57]. Skeletal vibrations involving SCN groups are C—S(CN) and S—C(N) stretchings and two S—C≡N bending modes. Both kinds of S—C stretching are sensitive to rotational isomerism. Vibrational frequencies are within a broad region of 698–586 cm^{-1} [57]. Bendings occur at 513–453 cm^{-1} (IR : w) and 416–405 cm^{-1} (IR : m) [57].

8-4.3. Selenocyanates (—Se—C≡N)

The CN stretching band of alkyl selenocyanates is located near 2152 cm^{-1} [61, 62]. This class of compounds is also characterized by the Se—C(N) stretching at 522–520 cm^{-1} (IR : ms) and Se—C≡N bendings at 420–400 cm^{-1} (IR : vw) and 365–360 cm^{-1} (IR : W).

8-5. ISOCYANIDES (C—N≡C)

The IR spectra of 18 isocyanides were reported by Ugi and Meyr [63], and the vibrational analysis of methylisocyanide was carried out by Dadieu [64] and Duncan [65]. The —N≡C stretching frequency occurs below 2200 cm^{-1}, and the IR band width is broad. The characteristic spectral position for aliphatic isocyanides is in the 2146–2134 cm^{-1} region. This frequency is lowered to 2125–2109 cm^{-1} if the isocyanide is conjugated with an aryl group instead [63]. For the RO—C(=O)—CH_2—NC type of a structure, where R = Me or Et, the stretching band is located in the 2166–2164 cm^{-1} region [43]. It is worthy to note that the —NC absorption of benzyl isocyanide ($C_6H_5CH_2$—NC) is at 2152 cm^{-1} [43].

REFERENCES

1. A. W. Reitz and R. Scrabal, *Monatsh. Chem,* **70,** 398, 1937.
2. A. W. Reitz and R. Sabathy, *Monatsh. Chem.* **71,** 100, 1938.
3. A. W. Reitz and R. Sabathy, *Monatsh. Chem.* **71,** 131, 1938.
4. R. E. Kitson and N. E. Griffith, *Anal. Chem.* **24,** 334, 1952.
5. A. Hidalgo, *Compt. Rend.* **249,** 395, 1959.
6. N. E. Duncan and G. J. Janz, *J. Chem. Phys.* **23,** 434, 1955.
7. J. L. Duncan, *Spectrochim. Acta* **20,** 1197, 1964.
8. S. W. Charles, F. C. Cullen, G. I. L. Jones, and N. L. Owen, *J. Chem. Soc., Faraday Trans. II* **70,** 758, 1974.
9. R. G. Jones and W. J. Orville-Thomas, *J. Chem. Soc.,* 4632, 1965.
10. J. R. Durig and D. W. Wertz, *Spectrochim. Acta* **24A,** 21, 1968.

11. P. Klaeboe, *Spectrochim. Acta* **26A,** 87, 1970.
12. T. Fujiyama, *Bull. Chem. Soc. Jpn.* **44,** 1194, 1971.
13. C. J. Wurrey, W. E. Bucy, and J. R. Durig, *J. Phys. Chem.* **80,** 1129, 1976.
14. T. Fujiyama, K. Tokumaru, and T. Shimanouchi, *Spectrochim. Acta* **20,** 415, 1964.
15. J. R. Durig, C. M. Player, Jr., Y. S. Li, J. Bragin, and C. W. Hawley, *J. Chem. Phys.* **57,** 4544, 1972.
16. P. Klaeboe and J. Grundens, *Spectrochim. Acta* **24A,** 1905, 1968.
17. G. Schrumpf and S. Martin, *J. Mol. Struct.* **101,** 57, 1983.
18. S. W. Charles, F. C. Cullen, and N. L. Owen, *J. Mol. Struct.* **34,** 219, 1976.
19. T. Fujiyama, *Bull. Chem. Soc. Jpn.* **44,** 89, 1971.
20. C. J. Wurrey, Y. Y. Yeh, M. D. Weakley, and V. F. Kalasinsky, *J. Raman Spectrosc.* **15,** 179, 1984.
21. D. A. C. Compton, S. C. Hsi, and H. H. Mantsch, *J. Phys. Chem.* **85,** 3721, 1981.
22. F. Halverson, R. F. Stamm, and J. J. Whalen, *J. Chem. Phys.* **16,** 808, 1948.
23. J. R. Durig, C. K. Tong, C. W. Hawley, and J. Bragin, *J. Phys. Chem.* **75,** 44, 1971.
24. J. Bragin, K. L. Kizer, and J. R. Durig, *J. Mol. Spectrosc.* **38,** 289, 1971.
25. D. A. C. Compton and W. F. Murphy, *J. Phys. Chem.* **85,** 481, 1981.
26. D. A. C. Compton, W. F. Murphy, and H. H. Mantsch, *J. Raman Spectrosc.* **11,** 349, 1981.
27. J. Mallols, V. Tabacik, and S. Sportouch, *J. Raman Spectrosc.* **11,** 312, 1981.
28. G. Schrumpf, *Spectrochim. Acta* **39A,** 511, 1983.
29. C. J. Wurrey and A. B. Nease, in "Vibrational Spectra and Structure," Vol. 7 (J. R. Durig, Ed.), Elsevier, Amsterdam, 1979, Chapter 1.
30. R. W. Mitchell and J. Nakovich, Jr., *Spectrochim. Acta* **29A,** 1153, 1973.
31. R. J. Mureinik and C. W. F. T. Pistorius, *Spectrochim. Acta* **23A,** 1603, 1967.
32. D. A. Compton, J. L. Greene, M. A. S. Hazle, and J. G. Grasselli, *Spectrochim. Acta* **39A,** 541, 1983.
33. G. Davidovics, B. Gedoux, J. P. Aycard, and J. Pourcin, *J. Raman Spectrosc.* **11,** 262, 1981.
34. D. H. Christensen, I. Johnsen, P. Klaeboe, and E. Kloster-Jensen, *Spectrochim. Acta* **25A,** 1569, 1969.
35. P. Klaeboe and E. Kloster-Jensen, *Spectrochim. Acta* **23A,** 1981, 1967.
36. F. A. Miller and R. B. Hanna, *J. Chem. Phys.* **21,** 110, 1953.
37. F. A. Miller, D. H. Lemmon, and R. E. Witkowski, *Spectrochim. Acta* **21,** 1709, 1965.
38. P. Venkoji, *Spectrochim. Acta* **42A,** 1301, 1986.
39. S. Ram, B. K. Gupa, and J. S. Yadav, *Ind. J. Pure and Appl. Phys.* **19,** 1110, 1981.
40. F. R. Dollish, W. G. Fateley, and F. F. Bentley, "Characteristic Raman Frequencies of Organic Compounds," John Wiley & Sons, New York, 1974.
41. L. J. Bellamy, "The Infrared Spectra of Complex Molecules," John Wiley & Sons, New York, 1975.
42. G. Schrumpf and H. Dunker, *Spectrochim. Acta* **41A,** 841, 1985.
43. C. J. Pouchert, "The Aldrich Library of FT-IR Spectra," Aldrich Chemical Company, 1985.
44. S. Besnainou, B. Thomas, and S. Bratoz, *J. Mol. Spectrosc.* **21,** 113, 1966.
45. "The Sadtler Standard Raman Spectra," Sadtler Research Laboratories, 1976.
46. N. B. Colthup, L. H. Daly, and S. E. Wiberley, "Introduction to Infrared and Raman Spectroscopy," 3rd Ed., Academic Press, New York, 1990.
47. C. N. R. Rao and R. Venkataraghavan, *Can. J. Chem.* **39,** 1757, 1961.
48. L. Deady, A. R. Katritzky, R. A. Shanks, and R. A. Topsom, *Spectrochim. Acta* **29A,** 115, 1973.

49. T. Satto, M. Yamakawa, and M. Takasuka, *J. Mol. Spectrosc.* **90**, 359, 1981.
50. J. Zarembowitch and R. Maleki, *Spectrochim. Acta* **39A**, 43, 1983.
51. P. M. Green, C. J. Wurrey, R. Krishnamoorthi, and Y. Y. Yeh, *J. Raman Spectrosc.* **17**, 355, 1986.
52. W. H. Fletcher and F. B. Brown, *J. Chem. Phys.* **39**, 2478, 1963.
53. F. B. Brown and W. H. Fletcher, *Spectrochim. Acta* **19**, 915, 1963.
54. N. Groving and A. Holm, *Acta Chem. Scand.* **19**, 443, 1965.
55. H. Hoyer, *Chem. Ber.* **94**, 1072, 1961.
56. D. Martin, *Chem. Ber.* **97**, 2689, 1964.
57. G. L. Caldow and H. W. Thompson, *Spectrochim. Acta* **13**, 212, 1959.
58. E. Lieber, C. N. R. Rao, and J. Ramachandran, *Spectrochim. Acta* **13**, 296, 1959.
59. N. S. Ham and J. B. Willis, *Spectrochim. Acta* **16**, 279, 1960.
60. R. P. Hirschmann, R. N. Kniseley, and V. A. Fassel, *Spectrochim. Acta* **20**, 809, 1964.
61. E. E. Aynsley, N. N. Greenwood, and M. J. Sprague, *J. Chem. Soc.*, 2395, 1965.
62. W. J. Franklin, R. L. Werner, and R. A. Ashby, *Spectrochim. Acta* **30A**, 387, 1974.
63. I. Ugi and R. Meyr, *Chem. Ber.* **93**, 239, 1960.
64. A. Dadieu, *Monatsh. Chem.* **57**, 437, 1931.
65. J. L. Duncan, *Spectrochim. Actra* **20**, 1197, 1964.
66. S. K. Freeman, "Applications of Laser Raman Spectroscopy," John Wiley & Sons, New York, 1974, p. 75.
67. H. W. Thompson and G. Steele, *Trans. Faraday Soc.* **52**, 1451, 1956.

CHAPTER 9

Compounds Containing the Carbonyl Group

9-1. INTRODUCTION

Most carbonyl compounds have IR and Raman bands in the 1900–1550 cm^{-1} region [1–3]. The frequencies tend to be the same in the two techniques, but the intensities differ. In the infrared, C=O bands tend to be among the strongest in the spectra, whereas in the Raman, they are considerably weaker. When there are two C=O groups in a molecule with a center of symmetry (such as an *s*-trans oxalate or a carboxylic acid dimer), then the antisymmetric C=O stretch is only active in the IR and the symmetric stretch is only active in the Raman. Except for these cases, the C=O band can normally be observed in both techniques at about the same frequency, and factors that shift the C=O frequency can be observed in both IR and Raman.

The intensity of the C=O band in the IR makes the detection of this group easier in the infrared spectrum, particularly when it is present in small amounts. In addition, some side group bands show up better in the IR. These include the C—O band in esters (one or more bands in the 1300–1000 cm^{-1} region), the broad OH band near 3000 cm^{-1} in carboxylic acid dimers, or the CH_3(C=O) band near 1375 cm^{-1} in acetate esters and methyl ketones.

Raman intensities of C=O bands tend to be dependent mainly on resonance effects and are little affected by inductive effect [4]. Conjugation increases the C=O Raman band intensity in the order ester < ketone < aldehyde [4]. A unique characteristic of a Raman C=O band is the depolarization ratio ρ ($I_{\perp}/I_{\parallel}$), which has considerable interpretive value [4].

The C=O stretch band is not as sensitive to change in the group environment as the OH group, but it is more sensitive than many other groups. When spectra are compared of a C=O compound in different states [132], the highest frequencies are in the vapor state. Dilute solutions in nonpolar solvents such as CCl_4 or CS_2 have the highest liquid-state frequencies, but the frequencies, are still 5–20 cm^{-1} lower than those in the vapor state. The C=O frequency will be lower still for the condensed state or for solutions in more polar solvents such as $CHCl_3$. The C=O frequencies that change the most are those where the C=O bonds have the greatest polar interaction with the solvent [133]. If the C=O oxygen is intermolecularly hydrogen-bonded in the condensed state, the C=O frequency shift upon going to the vapor state can be considerably greater than that for non-hydrogen-bonded aldehydes, ketones, or esters, for example.

In the matrix isolation technique, the sample is trapped in a matrix of solid argon at a temperature of about 10 K and a high vacuum. Carbonyl frequencies in matrix isolation are intermediate between those in the vapor state and those in the solid or liquid state or solution [139].

General regions for carbonyl stretching frequencies in cm^{-1} are given in Table 9-1 where the data are taken from Colthup *et al.* [2]. General factors that shift C=O frequencies are discussed next, and then individual C=O groups are discussed.

9-2. FACTORS THAT SHIFT CARBONYL BANDS

9-2.1. Mass Effects

The C=O stretching frequency is relatively insensitive to a change in the mass of an atom directly attached to the C=O carbon, as long as that atom is carbon or a heavier atom [2, 5, 6]. If the attached atom is changed from a carbon to a hydrogen atom, the calculated mass effect would be to lower the C=O stretching frequency about 17 cm^{-1}, so that from the mass effect alone, an aldehyde should have a C=O frequency about 17 cm^{-1} lower than a ketone [5]. Since aldehydes have C=O frequencies about 15 cm^{-1} higher than ketones, this means that there is a change in the force constant

Table 9-1.

C=O Spectral Regions (in cm^{-1})

Dialkyl ketones	1725–1705	
Singly conjugated ketones	1700–1670	
Doubly conjugated ketones	1680–1640	
α-Chloro ketone (Cl near O) (cyclic equatorial)	ca. 1745	
α-Chloro ketone (Cl not near O) (cyclic axial)	ca. 1725	
Ketones in five-membered ring (unconjugated)	1750–1740	
o-Hydroxy aryl ketones	1670–1630	
1,3-Diketones, enol form	1640–1580	
Aliphatic aldehydes	1740–1720	
Aromatic aldehydes	1710–1685	
Formate esters	1725–1720	
Other saturated esters	1750–1735	
Conjugated esters (electron-attracting groups on the oxygen raise the C=O frequency)	1735–1715	
Lactones six-membered ring	1750–1715	
Lactones five-membered ring	1795–1740	
Carbonates noncyclic organic	1780–1740	
Carbonates in five-membered ring	1850–1790	
Carboxylic acid dimer	1720–1680	
Carboxylic acid monomer	1800–1740	
Carboxylic acid salt	1650–1540 1450–1360	
Amides and ureas (electron-attracting groups on the nitrogen raise the C=O frequency)	1695–1630	
Lactams five-membered ring	1750–1700	
Lactams four-membered ring	1780–1730	
Carbamates	1740–1683	
Anhydrides, noncyclic, unconjugated	1825–1815 1755–1745	higher band stronger
Anhydrides, noncyclic, conjugated	1780–1770 1725–1715	higher band stronger
Anhydrides, cyclic, unconjugated	1870–1845 1800–1775	lower band stronger
Anhydrides, cyclic, conjugated	1860–1850 1780–1760	lower band stronger
Acid chlorides, aliphatic	1810–1795	
Acid chlorides, aromatic	1785–1765 1750–1735	
Chloroformates, aliphatic	1785–1775	
Thiol esters, unconjugated	1710–1680	
Thiol esters, conjugated	1700–1640	

of the C=O bond. Also, since amides, esters, and acid chlorides have C=O frequencies different from those of ketones, this implies that the C=O force constants are not the same.

9-2.2. Bond Angle Effects

In a carbonyl compound ($X_2C{=}O$) where the attached atoms X are carbons or heavier atoms, the X atoms move only slightly during the C=O stretching vibration. When the C=O bond stretches, the carbon motion contracts the C—X bonds a little. The C—X restoring force on the carbon is in the C—X direction, but it has a component in the C=O direction that cooperates with the C=O restoring force in restoring equilibrium. This C—X restoring force component varies with the O=C—X angle, and therefore changes as the ring size changes in cyclic carbonyl compounds. It is this change in interaction between C—X and C=O bonds that causes most of the change in the C=O stretching frequency as the ring size is changed. There is little, if any, C=O force constant difference between cyclohexanone and cyclobutanone [5–7]. Cyclic carbonyl compounds have average C=O frequency shifts as follows, compared to the noncyclic analogues: seven-membered ring, $-8 \pm 3\ \text{cm}^{-1}$; six-membered ring, $+7 \pm 14\ \text{cm}^{-1}$; five-membered ring, $+37 \pm 11\ \text{cm}^{-1}$; and four-membered ring, $+76 \pm 7\ \text{cm}^{-1}$ [8]. For carbonyls in bicycle compounds where the carbonyl is part of two rings, the smallest ring containing the carbonyl is chosen.

9-2.3. Inductive Effects

The C=O group is polar because the oxygen tends to attract electrons:

$$X_2C{=}O \leftrightarrow X_2\overset{+}{C}{-}\overset{-}{O}.$$

As seen in the right-hand resonance form, the more the oxygen can attract electrons, the weaker the carbonyl bond becomes. If one attached X is changed from an alkane group to a chlorine atom, the increased electron-attracting ability of the chlorine reduces the polar form contribution and increases the double bond character of the carbonyl group, which raises the frequency. Acid chlorides (R—C(=O)—Cl, 1800 cm^{-1}) have higher C=O stretching frequencies than ketones ($R_2C{=}O$, 1715 cm^{-1}) mainly because of this effect.

In α-chlorocarbonyl compounds, Cl—CH_2—C(=O)—X, rotational isomers can exist because of rotation about the C—C(=O) bond. The

Cl—C—C=O dihedral angle can be about zero degrees (cis) or ±120° (gauche). In this cis isomer, the Cl is near to the C=O oxygen and tends to inhibit its tendency to attract electrons. This field effect raises the C=O frequency about 20 cm^{-1} relative to the gauche isomer, where the Cl is more distant from the oxygen [9, 10].

9-2.4. Mesomeric Effects

When a nitrogen, oxygen, or chlorine is attached to a carbonyl carbon, the nonbonding electrons on the attached atom X can rearrange and donate electrons to the carbonyl oxygen:

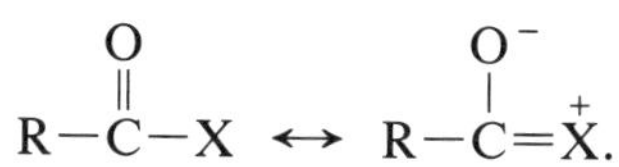

This weakens the C=O force constant depending on the electron donation tendency of X. This effect is slight in acid chlorides (X = Cl, 1800 cm^{-1}), but is more important in esters (X = OR, 1740 cm^{-1}). In amides (X = NH_2, 1665 cm^{-1}) it is a dominant effect, as the C=O frequency is lower than that of ketones (1715 cm^{-1}), even though the nitrogen is more electronegative than a carbon. The electron-donating tendency of X can be reduced if an electron-attracting group such as vinyl or phenyl is attached to it [11]:

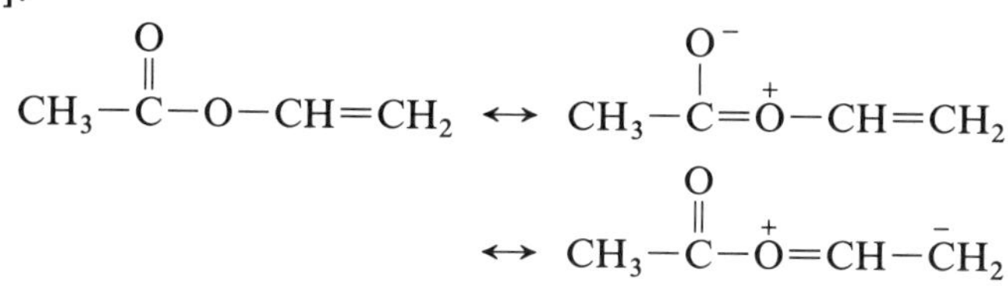

Vinyl acetate (1770 cm^{-1}) absorbs at a higher frequency than ethyl acetate (1740 cm^{-1}) because of this effect.

When the C=O bond is conjugated to a C=C bond or to a phenyl, the double-bond character of the carbonyl bond is reduced and the frequency is lowered 20–30 cm^{-1} [12]:

The lowering of the C=O frequency is even greater if there is an OH or NH on the β unsaturated carbon that can hydrogen-bond to the C=O:

This effect can be seen in *o*-hydroxyacetophenone (1640 cm^{-1}) and the enol form of acetylacetone types (ca. 1600 cm^{-1}) [2].

Conjugation effects are greatest when the unsaturated bonds (C=O and C=C) are coplanar, since this configuration allows the greatest overlap of the pi electrons. If steric effects prevent the coplanarity of the unsaturated bonds, the effects of conjugation are reduced. As an example, acetophenone (coplanar) has a C=O frequency of 1685 cm^{-1} [12], whereas 2,6-dimethylacetophenone (nonplanar) has a C=O frequency of 1705 cm^{-1} [4]. Ketones can be conjugated on both sides as in benzophenone (1653 cm^{-1}) or quinone (1650 cm^{-1}) [12].

9-3. ALDEHYDES (—CH(=O))

The characteristic frequencies of aldehydes are summarized in Tables 9-2 and 9-3. The presence of the aldehyde group is characterized by the aldehyde C—H stretching, the in-plane C—H bending (or rock), and the C=O stretching bands.

9-3.1. Aldehyde C—H Bands

The aldehyde C—H stretching mode of most aldehydes gives rise to a characteristic doublet in the region 2870–2695 cm^{-1} (Table 9-2). This doublet is due to a Fermi resonance effect between the aldehyde C—H stretching vibration and the first overtone of the in-plane C—H bending vibration (C—H rock). This is one of the best-known examples of Fermi resonance doubling in group frequency correlations [2]. When an overtone (or combination band) has a frequency close to that of a fundamental band of a different vibration, Fermi resonance may occur if they are of the proper types to interact. Fermi resonance results in a quantum-mechanical mixing of the overtone and fundamental. This causes the frequencies of the resulting two bands to diverge somewhat from the expected frequencies of the fundamental and the overtone before mixing. Most of the intensity of the band pair comes from the fundamental component. Therefore, if the two bands are of equal intensity, the fundamental and overtone are equally involved in each band. This occurs when the fundamental and overtone have close to the same frequency before mixing. If the intensities are unequal, the stronger band has more fundamental and less overtone

Table 9-2.

Characteristic Frequencies (cm^{-1}) of the Aldehyde C—H Group

Compound		(C—H str. + 2 × C—H rock)						C—H rock			
		cm^{-1}	IR	R	cm^{-1}	IR	R	cm^{-1}	IR	R	Ref.
CH_3—CHO	liq.	2844	m	—	2783	m	—	1400	m	—	12
RCH_2—CHO	liq.	2830–2815	m	—	2725–2713	m	mw, p	1400–1385	m	mw, p	3, 12–14
RR CH—CHO	liq.	2815–2805	m	—	2715–2700	m	mw, p	1400–1390	m	mw, p	3, 12–14
Me_3C—CHO	liq.	2808, 2709	m	—	2699	m		1400	m	—	12
ϕCH_2—CHO	liq.	2827	m	—	~2725	m		1388			12, 13
CF_3—CHO	vap.	2864	s	—				1384	s	—	18
CCl_3—CHO	liq.	2855	w	mw, p				1355	m	w, p	13–17
CBr_3—CHO	liq.	2844	m	mw	2675	vw	—	1354	m	w	12, 13, 15
CH_2=CH—CHO	liq.	2850, 2812[a]			2760, 2700			1423, 1363			12, 13, 19
CH_2=CR—CHO	liq.	2820–2810	mw	—	~2700	mw	—				12
Me—CH=CH—CHO	liq.	2816	mw		2733	mw	—	~1395	m	—	12, 19
RCH=CH—CHO	liq.	2820–2800	mw	—	2735–2720	mw	—	~1395			12
HC≡C—CHO	liq.	2869	—	w, p				1392	—	ms	20
Me—C≡C—CHO	liq.	2869	m	—				1395	m	w	21
Cl—C≡C—CHO	soln.	2860	m	w				1385	m	m	22
Br—C≡C—CHO	soln.	2860	m	m				1384	m	m	22
I—C≡C—CHO	soln.	2850	m	—				1385	m	—	22
ortho-Ar—CHO[b]	soln.	2900–2800	m	—	2765–2747	m	—	1410–1380	m	w, p	12–14, 23–25
Other Ar—CHO	soln.	2832–2812	m	—	2745–2720	m	—	1390–1380	m	w, p	12–14, 23–25
2-Furfurals	vap.	2812–2806			2750–2719			1400–1394			13
2-Thiophene	vap.	2810–2802			2750–2744			1392–1375			13

[a] C—H stretch band in Fermi resonance with both 1423 and 1363 cm^{-1} (four bands).

[b] Benzaldehydes substituted in the 2 position with halogen, nitro, methoxy, or hydroxy group.

Table 9-3.

Characteristic C=O Stretching Frequencies (cm^{-1}) of the —CHO Grouping

Compound	Vapor	IR	R	Solution	IR	R	Liquid (or Nujol)	IR	R	Ref.
R—CHO	1750–1740	vs		1740–1730	vs		1740–1725	vs	w	1, 12–14
CF_3-CHO	1788	s								18
CCl_3—CHO	1778	vs	w, p	1767	vs		1757	vs	ms, p	15, 16
CBr_3—CHO	1760			1748	vs		1745	vs	m, p	13, 15
H_2C=CH—CHO	1720						1696	vs		12
C=C—CHO				1705–1685	vs		1700–1680	vs		12
H—C≡C—CHO	1692	vs					1658			20
Me—C≡C—CHO	1699	vs					1666	vs	s	14, 21
Cl—C≡C—CHO	1694	vs		1670	vs	s				22
Br—C≡C—CHO	1692	vs		1670	vs	vs				22
I—C≡C—CHO				1676	s					22
Benzaldehydes	1728–1677	vs					1710–1630	vs	vs, p	12, 13
3- and 4-substituted and 3,4-disubstituted										
	1728–1715	s								
2-substituted and 2,3- and 2,4-disubstituted										
	1719–1700	s								
2-hydroxyl benzaldehydes										
	1682–1677						1680–1630	vs		

component, but the weaker band still has some fundamental component. If the weaker band is very weak, then the stronger band is nearly pure fundamental. Relative intensity changes in the two bands in a related series of spectra are caused by small shifts in the position of the overtone relative to the fundamental before mixing.

In the infrared spectra, the adehyde C—H stretch doublet bands are medium to weak in intensity. In the Raman spectra, these bands are medium to weak in alkyl aldehydes, but are even weaker in aromatic aldehydes. From the data available to us, it appears that the frequencies of the aldehyde C—H stretching bands are little affected by the sample states [3, 12–22], which implies the C—H does not hydrogen-bond.

In alkyl aldehydes, the C—H stretched doublet is at 2844–2805 and 2733–2699 cm^{-1} in the liquid state, as seen in Table 9-2. The lower frequency band is easier to detect because it is clear of other CH_2 bands in the 2850 cm^{-1} region. These aldehydic CH stretching frequencies of CH_3—CHO, RCH_2—CHO, R_2CH—CHO, and Me_3C—CHO in the liquid phase are located at 2733, 2725–2713, 2715–2700, and 2699 cm^{-1}, respectively. The aldehyde C—H rock bands are found in the region 1410–1380 cm^{-1}. In R—CH_2CHO compounds, this band is found near 1390 cm^{-1} between the CH_2(C=O) deformation band at 1415–1405 cm^{-1} and the C—CH_3 symmetric deformation band near 1378 cm^{-1} [12].

In Table 9-2, in conjugated aldehydes of the type RCH=CH—CHO (trans), the aldehyde CH stretching doublet and rock bands appear at 2820–2800, 2735–2720, and near 1395 cm^{-1} [12]. In acrolein, CH_2=CH—CHO, there is no aldehyde CH rock band in the usual 1410–1380 cm^{-1} region. Instead, there are two bands at 1423 and 1363 cm^{-1} [12, 13, 19]. This suggests that the aldehyde CH rocking vibration is interacting with the =CH_2 deformation (1403 cm^{-1} in CH_2=CH—C(=O)—CH_3 and 1395 cm^{-1} in CH_2=CH—C(=O)Cl) [12]. In the aldehyde CH stretch region of CH_2=CH—CHO, four bands are seen at 2850, 2812, 2760, and 2700 cm^{-1} (see Table 9-2), probably involving Fermi resonance with both 1423 and 1363 cm^{-1} bands. The compounds CH_2=CR—CHO (Table 9-2) are somewhat similar to acrolein. They show no CH rock band at 1410–1380 cm^{-1}, but show CH stretch bands near 2820 and 2700 cm^{-1} [12].

For benzaldehydes in CCl_4 solutions, the aldehyde CH stretch bands were found at 2832–2812 and 2745–2720 cm^{-1} if there were no substituents in the 2 position, ortho to the CHO group [24, 25]. For benzaldehydes substituted in the 2 position with such groups as halogen, nitro, methoxy, or hydroxy, the bands were somewhat higher, at 2900–2860 and 2765–2747 cm^{-1}, but some benzaldehydes with noninteracting ortho substituents had bands like those of the other isomers [24, 25]. This shift is in the wrong direction to be CH···X hydrogen bonding. The aldehyde CH rocking band in benzaldehydes is usually in the 1390–1380 cm^{-1} region [14, 24–26]. This band tends to be weak in the Raman spectrum, but can usually be seen in the IR as a medium-weak band.

9-3.2. Aldehyde C=O Bands

Aldehyde C=O bands are listed in Table 9-3. The aldehyde C=O band intensity is very strong in the IR. The frequencies for the vapor state are about 10 cm^{-1} higher than those for the liquid state [13, 140], except for cases involving intermolecular hydrogen bonding. For aliphatic aldehydes, the Raman intensity is weak and the depolarization ratio ρ is 0.24–0.28 [4]. In the liquid state, aliphatic aldehydes have C=O bands at 1740–1725 cm^{-1}. Unbranched aldehydes of the type RCH_2CHO have C=O bands in the narrow region 1729–1726 cm^{-1} [12]. Halogenation on the α-carbon increases the C=O frequency. Some C=O frequencies in CCl_4 are as follows: CH_3CHO, 1730 cm^{-1}; $ClCH_2CHO$, 1730 cm^{-1}; Cl_2CH—CHO, 1748 cm^{-1}; and Cl_3C—CHO, 1768 cm^{-1} [42]. Halogenated aldehyde C=O bands are somewhat intensified in the Raman spectra.

In α,β-unsaturated aldehydes C=C—CHO, both the C=O and the C=C stretching frequencies are lowered by conjugation. The C=O band is seen at 1700–1680 cm^{-1} in the condensed state, and the C=C band is seen about 40–45 cm^{-1} lower, near 1640 cm^{-1} [4, 12]. The IR intensity of the C=C band is somewhat enhanced. In general, conjugation increases the Raman intensity of the C=O band [4]. When the C=C—CHO group has only R or H on the C=C carbons, the Raman intensity of the C=O band (polarized) is stronger than that of the C=C band [14]. The Raman depolarization ratio ρ for this C=O band is 0.24–0.28 for most noncyclic cases [4]. When the C=C is part of a cyclohexene ring, the conjugated aldehyde C=O band has a ρ value of about 0.32 [4].

When the aldehyde group is on the end carbon of an already conjugated C=C group, a further change occurs. For aryl—C=C—CHO types, the C=O frequency is about 1680 cm^{-1} with ρ about 0.32–0.40, and for C=C—C=C—CHO types, the C=O frequency is about 1665 cm^{-1} with ρ about 0.40–0.45 [4, 14].

When the aldehyde group is conjugated to a C≡C group, the C=O band is lowered to about 1660 cm^{-1} in the condensed state, as seen in Table 9-2. Again, the C=O Raman band intensity is increased by conjugation [14, 20–22].

When the aldehyde group is conjugated in furfural and 2-thiophene carboxaldehyde, the C=O band is double because of two planar rotational isomers. The aldehyde C=O is *s*-cis or *s*-trans to the attached C=C in the five-membered ring. Furfural has the *s*-cis C=O band at 1692 cm^{-1}, $\rho = 0.30$, and the *s*-trans C=O band at 1672 cm^{-1}, $\rho = 0.19$ [4, 27]. The related 2-thiophene carboxaldehyde has the *s*-cis C=O band at 1675 cm^{-1}, $\rho = 0.28$, and the *s*-trans C=O band at 1652 cm^{-1}, $\rho = 0.20$ [4].

When the —CHO group is conjugated to an aromatic ring, the C=O frequency is lowered. For non-hydrogen-bonded benzaldehyde derivatives, the C=O frequencies for the condensed state are 1710–1660 cm^{-1}. In general, benzaldehydes with electron-withdrawing groups, such as $—CF_3$ or $—NO_2$, as substituents will give rise to a C=O band at the higher end of the spectral region, while those with electron-donating groups, such as more than one $—OCH_3$, as substituents will exhibit a C=O band near the lower end of the region. Hydroxybenzaldehydes have C=O bands in a lower region because of intra- and intermolecular hydrogen bonding. The majority of 2-hydroxybenzaldehydes have IR bands in the region 1675–1630 cm^{-1} (Nujol), while the 3- and 4-hydroxybenzaldehydes absorb in the region 1675–1645 cm^{-1} (Nujol). However, if these should also contain an $—NO_2$ substituent, the OH can bond to the NO_2 instead of the C=O and give rise to a C=O band at 1698–1671 cm^{-1} (Nujol). In the Raman spectra, aryl-CHO conjugation enhances the C=O stretching band, which in the majority of cases has an intensity similar to that of the ring stretching band near 1600 cm^{-1}. The depolarization ratio ρ is 0.21–0.29 [4].

9-4. KETONES

9-4.1. Open-Chain Alkanones

The methyl and ethyl alkanones, Me—C(=O)—R and Et—C(=O)—R, absorb IR radiation in the region 1737–1719 cm^{-1} in the vapor phase [13, 140]. Due to the polar nature of the carbonyl group, the C=O stretching frequency decreases as the sample phase changes from vapor to liquid phase. For the alkanones with no α-branching, the C=O stretching frequencies are located in the region of 1720–1712 cm^{-1} in liquid state [2]. See Table 9-4. A further frequency lowering is observed for solid samples, and the C=O stretching frequencies of 14-heptacosanone, $(n\text{-}C_{13}H_{27})_2C{=}O$, and 10-nonadecanone, $(n\text{-}C_9H_{19})_2C{=}O$, both appear at 1705 cm^{-1} as a doublet when measured in Nujol mull [12]. The C=O stretching frequencies of ketones in solution depend very much on the solvent. Bellamy has correlated a spectral region of 1725–1717 cm^{-1} for *n*-alkyl ketones in solution [1]. In the Raman spectra, the depolarization ratio ρ is about 0.21 [4].

Nyquist [13] observed a frequency lowering of the C=O stretching frequency of α-branched alkanones in vapor phase. Similar lowering is reported by Hirota *et al.* [28] for methyl-substituted acetones in CCl_4

Table 9-4.

The Characteristic C=O Stretching Frequencies (cm^{-1}) of Alkyl Ketones

Compound	Vapor	IR	Liquid	IR	R	Ref.
Alkyl ketones:						
Y—CH₂—CO—CH₂—Y′; Y, Y′ = H, R	1737–1730	vs	1720–1712	vs	ms	12–14
α-branched alkyl ketones	1730–1719	vs	1716–1708	vs	ms	12–14
α,β-Conjugated ketones:						
s-trans	1710–1691[a]	vs	1690–1675	vs	m	4, 19, 34, 37
s-cis			1700–1687	s	w	4, 19, 34, 37
Alkyl-aryl ketones			1705–1625	s	—	12, 47–49
Diaryl ketones			1670–1600	s	—	12
β-Diketones:						
keto form (doublet)			1740–1695			52, 56
enol form			1640–1570	vs	—	12, 52–56

[a] A correlation for 2-cyclohexene-1-ones.

solution. They reported that the "H-H" conformation, in which the C=O group is eclipsed to both α- and α'-hydrogen atoms, absorbs in the region of 1721–1716 cm^{-1}. The "H-Me" conformation, in which the carbonyl group is eclipsed to the α-hydrogen atom and the α'-methyl group, absorbs in the region 1713–1707 cm^{-1}. In the case of di-*t*-butylketone, only one rotational conformation, "Me-Me," is stable. This "Me-Me" conformation, in which the C=O group is eclipsed to both α and α'-methyl groups, exhibits a C=O IR band at a considerably lower value of 1686 cm^{-1} in CCl_4 solution. An additional effect besides α-branching is present here. If one methyl group in acetone is replaced with a *t*-butyl group, the C=O band is lowered by about 6 cm^{-1} [12], whereas if both methyl groups in acetone are replaced with *t*-butyl groups, the C=O band is lowered by 32 cm^{-1} in CCl_4 solution [6, 133]. For comparison, if both methyl groups in acetone are replaced with *i*-propyl groups, the C=O band is lowered by only 2 cm^{-1} [133]. In di-*t*-butyl ketone, steric effects increase the C—C(=O)—C bond angle, and this causes an additional lowering of the C=O wavenumber [6]. See Section 9-2-2.

Table 9-5.

The C=O Stretching Frequencies (cm^{-1}) of α-Halogenated, α-Methoxy, and α-Hydroxyl Ketones

Compound	cm^{-1}	Ref.
α-Halogenated ketones:		
$CH_3-CO-CH_2F$ (liq.)	1781	12
$CH_3-CO-CF_3$ (liq.)	1783	12, 30
$CF_3-CO-CF_3$ (liq.)	1806	31
$CH_3-CO-CH_2Cl$ (CS_2 soln.)	1750	29
$CH_3-CO-CHCl_2$ (liq.)	1742	12
$ClCH_2-CO-CH_2Cl$ (liq.)	1741	12
$Cl_2CH-CO-CHCl_2$ (liq.)	~1775	12
	1761	12
$Cl_2CH-CO-CCl_3$ (liq.)	1786	12
	~1740	12
$Cl_3C-CO-CCl_3$ (liq.)	1780	12
	~1750	12
$CH_3CHCl-CO-CH_3$ (liq.)	1725	12
α-Methoxy acetones:		
$CH_3-CO-CH_2OCH_3$ (liq.)	1731	12
$CH_3-CO-CH(OCH_3)_2$ (liq.)	1731	12
α-Hydroxyl ketones (liq.)	1725–1697	12

[a] The ν(C=O) of acetone, $CH_3-CO-CH_3$, is located at 1715 cm^{-1} in the liquid state.

9-4.2. α-Halogenated, α-Methoxyl, and α-Hydroxyl Ketones

The C=O stretching frequencies of select α-halogenated ketones are summarized in Table 9-5. In general, the effect of α-halogenation is to shift the C=O stretching frequencies of ketones toward a higher region. Frequently, more than one C=O band is observed for these compounds because of rotational isomerism [12, 13, 29]. From Table 9-5, it is apparent that the C=O stretching frequency increases progressively as the number of α-halogen atoms increases. According to this table, the effects of α-fluorination on the C=O stretching frequency appears to be more pronounced than those of α-chlorination and α-bromination.

The presence of a methoxy group on the α-carbon atom of methoxyacetones gives rise to a strong IR band near 1731 cm^{-1} in the liquid phase (see Table 9-5). This is about 15 cm^{-1} higher than the carbonyl stretching frequency of acetone (1715 cm^{-1}) measured as neat sample [12]. The characteristic C=O stretching frequencies of α-hydroxyl ketones are, however, located in a region of 1725–1697 cm^{-1}, which is similar to the region for alkyl ketones but lower than that for methoxyacetones.

9-4.3. Cycloalkanones

The carbonyl stretching frequencies of cycloalkanones and their derivatives are listed in Table 9-6. The change of sample state from vapor to liquid phase reduces the C=O stretching frequency by about 20 cm^{-1} because of molecular association. The data in Table 9-6 show that the C=O stretching frequency increases with decreasing ring size for cycloalkanones from C_9 to C_4. This frequency dependence upon the ring size was discussed in Section 9-2.2. For the small ring systems (C_4, C_5) the higher C=O stretching frequencies result from the decreased C—C(=O)—C bond angle at the carbonyl group, and not from a significant force-constant change (2, 5–7). The C—C(=O)—C bond angle for cyclohexanone is close to the normal sp^2, 120° bond angle, and therefore, its C=O stretching frequency is located at a value similar to those of open chain alkanones. As the ring size increases beyond C_6, the C—C(=O)—C bond angle increases a little from the normal value. Consequently, one observes a lowering in the C=O stretching frequency for C_7—C_{11} cycloalkanones. For cycloalkanones with ring size greater than C_{11}, the ring strain is released and the C=O stretching frequency returns to the normal value of open-chain cycloalkanones.

Table 9-6.

The C=O Stretching Frequencies (cm^{-1}) of Cycloalkanones and Their Derivatives [12, 13, 32]

Compound	Vapor	Liquid	Alkyl Substitution
Cyclobutanone (C_4)	1816 vs	1783 vs	
Cyclopentanone (C_5)	1765 vs	1746 vs	1751–1733[a]
Cyclohexanone (C_6)	1732 vs	1714 vs	1720–1705[a]
Cycloheptanone (C_7)	1721 vs	1702 vs	
Cyclooctanone (C_8)	1720 vs	1701 vs	
Cyclononanone (C_9)		1701 vs	
Cyclodecanone (C_{10})	1719 vs	1702 vs	
Cycloundecanone (C_{11})		1707 vs	
Cyclododecanone (C_{12})		1712 vs[b]	
Cyclotridecanone (C_{13})		1711 vs	
Cyclopentadecanone (C_{15})		1712 vs[b]	

[a] Including fused bicyclic systems containing five- or six-membered ring.
[b] Measured as Nujol mull.

In the Raman spectra, the C=O band of cyclobutanones have a ρ value of 0.4–0.5, whereas cyclopentanones, cyclohexanones, and cycloheptanones have ρ values in the 0.2–0.3 range [4].

Also listed in Table 9-6 are the characteristic C=O stretching frequencies of alkylated cyclohexanones and -pentanones. A deviation of the C=O stretching frequency from these ranges should be expected when α-halogenation, α-alkoxylation, or α,β-unsaturation occurs in the molecular structure. The influences of α-halogenation, α-alkoxylation, α,β-conjugation, etc., on the C=O stretching frequencies of cycloalkanones are parallel to those in the open-chain alkanones [1, 12, 13].

9-4.4. α, β-Unsaturated Ketones

In an olefin conjugated system, both the C=O and C=C stretching frequencies are lowered from those for the corresponding saturated moieties [1, 12, 13, 19, 32–46]. The characteristic C=O stretching frequencies of six 2-cyclohexene-1-ones are reported by Nyquist [13] to occur in the region of 1710–1691 cm^{-1} in the vapor phase. Bellamy [1] has correlated a region of 1690–1675 cm^{-1} for the C=O stretching frequencies of α,β-unsaturated ketones measured in solution. Our investigations on the IR spectra of six α,β-conjugated ketones reported by Pouchert [12] reveal a region of 1720–1655 cm^{-1} for the C=O stretching mode in liquid phase. The wide spectral range for the C=O stretching mode of α,β-conjugated ketones in the liquid state is likely due to the coexistence of different rotational isomers. For example, the IR spectra of liquid methyl and ethyl vinyl ketones (Me—CO—CH=CH_2 and Et—CO—CH=CH_2) [12] exhibit two C=O bands near 1700 cm^{-1} and 1685–1682 cm^{-1} due to the presence of *s*-cis and *s*-trans rotational isomers, respectively, where the C=O and C=C bonds rotate about the connecting single bond (*s*-) into a planar *s*-cis or *s*-trans configuration.

The characteristic C=O and C=C stretching frequencies and relative IR and Raman intensities can be used to differentiate the *s*-trans and *s*-cis rotational conformations. Typical C=O and C=C stretching frequencies for these can be found in Tables 9-7 and 6-5 [19, 34, 37]. The C=O stretching frequencies for the *s*-cis isomer are a little higher than those for the *s*-trans, isomer and the C=C stretching frequencies for the *s*-cis isomer are a little lower than those for the *s*-trans isomer. Typically, the C=O, C=C frequency difference is greater than 70 cm^{-1} for the *s*-cis isomer, but less than 60 cm^{-1} for the *s*-trans isomer [34]. As an example, CH_2=CH—CO—CH_3 in CCl_4 solution has C=O and C=C bands at 1707 cm^{-1} (IR : vs) and 1618 cm^{-1} (IR : m, sh) for the *s*-cis isomer, and bands at 1687 cm^{-1} (IR : vs) and 1648 cm^{-1} (IR : w) for the *s*-trans isomer [19]. The infrared intensities for the C=O band are somewhat stronger for the *s*-trans isomer, and the IR intensities for the C=C band are considerably stronger for the *s*-cis isomer [19, 34, 39]. In the Raman spectra, the C=O band is very weak for the *s*-cis isomer compared to that for the *s*-trans isomer. The Raman depolarization ratios ρ for *s*-trans isomers are like those for saturated acyclics, about 0.2. The ρ values for the *s*-cis acyclic isomers are about 0.4, and for fixed *s*-cis cyclic compounds, they are 0.42–0.52 [4].

These correlations were established using cyclic molecules where the *s*-cis or *s*-trans configuration is fixed. Then they were applied to acyclic

Table 9-7.

The Typical C=O and C=C Stretching Frequencies of Unconjugated and α,β-Unsaturated Ketones

	cm^{-1} (C=O)	A^a	cm^{-1} (C=C)	A^a
Unconjugated	1720–1712	1.00		
s-Trans conjugated	1690–1675	1.17	1644–1618	0.25
s-Cis conjugated	1700–1687	0.82	1624–1617	0.65

[a] Approximate relative IR intensity normalized against the C=O band intensity of saturated ketones.

compounds where both isomers are present. For example, in the Raman spectrum of 1-cyclohexenyl methyl ketone, the presence of a strong and polarized C=O band at 1666 cm^{-1} strongly suggests that the *s*-trans form is the predominant isomer for this compound. The compound 4-methyl-3-pentene-2-one ($Me_2C{=}CH{-}CO{-}Me$) exists predominantly as the *s*-cis isomer because of Me,Me steric hindrance in the *s*-trans isomer. It exhibits a weak Raman C=O band near 1688 cm^{-1} and a strong polarized C=C band [4].

There is likely to be some interaction between the C=O and C=C stretching vibrations so that during the C=O stretching mode, the C=C bond changes in length somewhat, and vice-versa. It has been suggested that the *s*-cis isomer interacts in this way a little more than the *s*-trans isomer, which causes the somewhat greater frequency difference in the *s*-cis isomer between the modes that are predominantly C=O and C=C stretch [44].

9-4.5. Aryl Ketones

Ketones conjugated to aromatic rings have the C=O frequencies lowered by conjugation. In the case of ketones, the C=O group can be conjugated on both sides, which has an additional lowering effect. The C=O frequencies for acetone, acetophenone, and benzophenone are 1715 cm^{-1}, 1685 cm^{-1}, and 1660 cm^{-1}, respectively [12]. Simple alkyl phenyl ketones have C=O bands in the condensed state about 1690–1680 cm^{-1} [12]. When the phenyl group has one or more substituents, the C=O frequency range is wider at 1705–1625 cm^{-1} [12, 47–49]. See Table 9-4. Electron-withdrawing substituents such as NO_2 raise the C=O frequency, and electron-donating groups such as NMe_2 lower the C=O frequency. Hydrogen-bonding OH groups, especially in the 2-position in aryl-alkyl ketones, lower the C=O frequencies [49]. The OH groups in the 2 position are intramolecularly hydrogen bonded to the C=O group and show a broad OH···O band in the IR around 3000 cm^{-1}. When the OH is in the 3 or 4 position, it is intermolecularly hydrogen bonded and shows an OH···O band in the 3400–3100 cm^{-1} region [12].

The C=O bands of aryl ketones and amides have roughly similar frequencies, but in the IR, the C=O band intensities for the aryl ketones are considerably lower than those for amides [50]. In the Raman, the C=O depolarization values are around 0.2 [4].

Substituted benzophenones are doubly conjugated ketones and have C=O bands in the region 1670–1600 cm^{-1} [12]. As in the aryl-alkyl ketones, the aromatic ring substituents shift the C=O band frequencies up or down depending on the electron-withdrawing or -donating characteristics or hydrogen bonding. Benzophenone with two *para*-dimethylamino groups on it has a C=O frequency shifted to 1600 cm^{-1}, which is superimposed on the ring band at 1600 cm^{-1} [12]. This is an extreme example of the electron-donating shift:

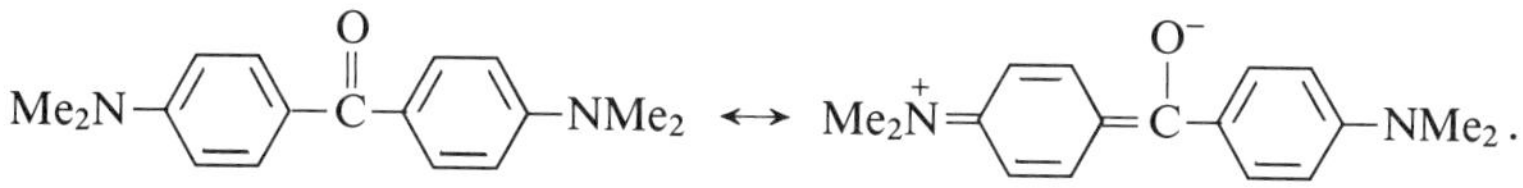

The quinones are also double conjugated. In the infrared the C=O groups have frequencies at 1650 cm^{-1} for quinone, 1661 cm^{-1} for naphthoquinone, and 1677 cm^{-1} for anthroquinone [12]. When these quinone types are substituted, the C=O bands in the IR are in roughly the same region as those for the benzophenones [2]. In the double ring compounds, 3,3′,5,5′-tetralkyl-4,4′-diphenoquinones, a strong IR band is seen involving out-of-phase C=O stretch mixed with ring vibrations in the region 1602–1586 cm^{-1} [134].

9-4.6. β-Diketones

When two ketone C=O groups are separated by one CH_2 group (*β*-diketones), the molecule can exist in the keto or enol form:

$$CH_3-\overset{\overset{\displaystyle O}{\|}}{C}-CH_2-\overset{\overset{\displaystyle O}{\|}}{C}-CH_3, \qquad CH_3-\overset{\overset{\displaystyle O\cdots H-O}{\|\qquad\quad |}}{C-CH=C}-CH_3.$$

In the keto form of acetyl acetone, the two C=O bonds give rise to C=O bands at 1727 cm^{-1} (symmetrical stretch) and 1707 cm^{-1} (antisymmetrical stretch) [51, 52]. These bands are usually weak because these molecules exist mainly in the enol form. See Table 9-4.

In the enol form, the intramolecularly hydrogen-bonded OH···O group absorbs in the region of 3000–2700 cm^{-1} and is very broad and rather weak in IR intensity. The double bonds in the OCCCO group are so delocalized that they are all best described as "bonds-and-a-half." The strongest IR band of most *β*-diketones is seen at 1640–1570 cm^{-1} (broad) [12, 52–56] and involves out-of-phase OCC stretching, in the same region as the out-of-phase $-CO_2Na$ stretch of carboxylic acid salts. The conjugated compound 1,3-diphenyl-1,3-propanedione exists entirely in the enol form and absorbs strongly at 1542 cm^{-1} in the IR [12].

9-5. ESTERS

Esters are characterized by a carbonyl band, usually in the 1750–1715 cm^{-1} region, which as usual for C=O bands is strong in the IR and weaker in the Raman spectra. In addition, there is a strong IR band (or band cluster) in the 1300–1150 cm^{-1} region involving (O=)C—O stretch. This band is the highest-frequency single bond stretch involving interacting C—C, C—O, and O—C bonds, but for simplicity, it is called the stretching of the C—O next to the C=O, because it tends to be characteristic for the type of ester derived from a particular acid [2]. A second IR band is seen usually near 1050 cm^{-1} for O—CH_3 or O—CH_2 groups and is called the O—C(H_2) stretch band because it tends to vary with the type of alcohol used to form the ester [2]. Again, however, this involves interaction with neighboring single bonds. Vibrational assignments have been given for methyl formate and methyl acetate [57–60, 141].

9-5.1. Ester Carbonyl Bands

Most simple esters to the type R—CO—OR have a C=O band in the 1750–1725 cm^{-1} region [3, 61]. Formate esters, HCO—OR, have a slightly lower C=O band at 1730–1715 cm^{-1} [3, 12, 61]. This is most likely caused by the mass effect of changing an atom next to the carbonyl from C to H without an appreciable C=O force constant change [2]. If a vinyl or a phenyl group is conjugated to the ester carbonyl, the C=O band appears at 1740–1715 cm^{-1} [3, 63–65]. The C=O force constant is decreased by the conjugation with the C=C or benzene ring. Strong electron-donating groups of hydrogen-bonding groups on the benzene ring can lower the C=O frequency further. *para*-Amino ethyl benzoate has a C=O band at 1685 cm^{-1} [12], and *ortho*-hydroxy or *ortho*-amino methyl benzoates (salicylates and anthranilates) absorb at 1679 and 1693 cm^{-1}, respectively [12]. See Table 9-8.

If a vinyl or phenyl group should be on the ester single-bonded oxygen, the C=O frequency is increased because the attached group competes with the carbonyl in withdrawing electrons from the single-bonded oxygen. Vinyl acetate (1762 cm^{-1}, liquid) can be compared with ethyl acetate (1742 cm^{-1}) [11, 12].

Saturated six-membered ring esters (δ-lactones) have a C=O frequency similar to those of the noncyclic analogues at 1750–1735 cm^{-1} [66]. Saturated five-membered ring esters (γ-lactones) have C=O frequencies somewhat higher at 1790–1760 cm^{-1} [67]. See Table 9-8.

Table 9-8.

C=O Stretching Frequencies (cm^{-1}) of Esters

Compound	νC=O (Vapor)	IR	νC=O (Liquid)	IR	R	References
Unconjugated esters:						
alkyl formates	1755–1750	s	1730–1715	s	m	3, 12, 61, 62
alkyl acetates	1770–1765	s	1750–1735	s	m	3, 12, 61, 62
propionates and up	1765–1755	s	1740–1725	s	m	3, 12, 61, 62
Conjugated esters:						
α,β unsaturated	1760–1740	s	1740–1715	s	m	12, 62, 65
aryl esters	1760–1740	s	1740–1715	s	m	12, 62, 65
salicylates, anthranilates	1720–1700	s	1695–1670	s	m	12, 62
Lactones:						
δ saturated	1780	s	1750–1735	s		12, 62, 66
γ saturated	1824–1816	s	1790–1760	s		12, 62, 67
>N−C(=O)−OR			1740–1683	s		69–71
RO−C(=O)−OR			1741–1739	s		72
RS−C(=O)−OR			1710–1702	s		72
Cl−C(=O)−OR			1780–1775	s		72
Cl−C(=O)−SR			1772–1766	s		72
RS−C(=O)−SR			1655–1640	s		72
>N−C(=O)−SR			1700–1650	s		72
R−C(=O)−SR			1700–1680	s		136
Ar−C(=O)−SR			1680–1640	s		136

When the atom attached to the C=O carbon is changed to an N, O, S, or Cl, further shifts occur because of force constant changes in the C=O bond. (See Table 9-8.) In carbamates (>N−CO−OR), the C=O band shifts with N substitution and hydrogen-bonding changes.

In the Raman spectra, the depolarization ratios ρ are about 0.09–0.12 for both saturated and α,β-unsaturated ester C=O bands, and somewhat higher at 0.15–0.21 for benzoates or cinnamates [4]. Saturated lactones in five- and six-membered rings have higher ρ values, at 0.15–0.17, than do noncyclic esters (0.09–0.12) [4], which mean that δ-lactones and noncyclic esters, which have similar C=O frequencies, can be distinguished by their ρ values [4].

9-5.2. Ester C−O Bands

Normal acetates $CH_3-CO-OR$ have a strong IR C−O stretching band at 1260–1230 cm^{-1} [61]. Acetates with $O-CH_2$ or $O-CH_3$ groups usually have a second IR O—C band at 1060–1035 cm^{-1}. Branching of this carbon (O—CHR_2, O—CR_3) raises this frequency somewhat [2].

Virtually all other normal saturated esters (formates, propionates, stearates, etc.) absorb strongly in the IR at 1210–1160 cm^{-1}, usually with multiple bands here [61].

Esters conjugated with a C=C group absorb strongly in the IR at 1300–1160 cm^{-1}, usually with multiple bands. Acrylates have bands near 1280 and 1200 cm^{-1}; methacrylates have bands near 1325, 1300, and 1180 cm^{-1}; maleates have bands near 1290, 1250, 1220 (strong), and 1162 cm^{-1}; and fumarates have bands near 1300, 1265, 1225 (weak), and 1175 cm^{-1} [63, 64].

Aryl conjugated esters such as benzoates or phthalates usually absorb strongly in the IR near 1280 and 1120 cm^{-1} [2].

9-5.3. Group on the Ester Oxygen

Methyl and ethyl esters can be distinguished in the IR [68]. The O—CH_3 group has a band near 2960 cm^{-1}, mw, (CH_3 antisymmetric stretch) and a band near 1440 cm^{-1}, m, (CH_3 symmetric deformation). The antisymmetric CH_3 deformation is near 1460 cm^{-1}, mw. The O—CH_2—CH_3 group has a band near 2980 cm^{-1}, m, (CH_3 antisymmetric stretch), 1475 cm^{-1}, w, (CH_2 deformation), 1455 cm^{-1}, w, (CH_3 antisymmetric deformation), 1400 cm^{-1}, w, (O—CH_2 wag) and 1375 cm^{-1}, m, (CH_3 symmetric deformation) [68]. In the O—CH_2—CH_3 group, the O—CH_2 wag and CH_3 symmetric deformation interact [2] to give the two bands mentioned, which look very much like the bands in this region from the *tert*-butyl group.

9-5.4. Other Ester Bands

The CH_3(C=O) group in acetate esters absorbs strongly in the IR at 1375–1365 cm^{-1} [2]. This is because of the CH_3 symmetrical deformation mode. In other saturated esters of the type R—CH_2—CO—OR, the CH_2(C=O) group absorbs in the IR near 1420 cm^{-1} because of CH_2 deformation [2].

Formate esters have Raman bands at 1383–1375 cm^{-1} (CH in-plane rock) and 775–750 cm^{-1} (O—C=O in-plane deformation) [3]. Acetate esters have Raman bands at 847–825 cm^{-1} (H_3C—C stretch) and 644–634 cm^{-1} (O—C=O in-plane deformation) [3]. In the IR, acetate esters show a weak doublet at 640–630 and 610–607 cm^{-1} (O—C=O deformation and C=O out-of-plane wag) [12, 61, 141].

Acrylates have a doublet in the C=C stretch region near 1640 and 1625 cm^{-1} with an intensity ratio in the IR of about 10 : 7, probably due to *s*-trans and *s*-cis rotational isomers. Other acrylate IR bands include 985 ± 5 cm^{-1} (=CH wag), 965 ± 5 cm^{-1} (=CH_2 wag), and 811 ± 3 cm^{-1} (=CH_2 twist). Methacrylate IR bands include 1640 cm^{-1} (C=C stretch), 939 ± 4 cm^{-1} (=CH_2 wag), and 815 ± 2 cm^{-1} (=CH_2 twist). Maleates have IR bands at 1645 cm^{-1} (C=C stretch) and 850 ± 50 cm^{-1} (*cis*-CH wag). Fumarates have an IR band at 979 ± 3 cm^{-1} (*trans*-CH wag) [63, 64].

In the aryl esters, the out-of-plane C=O wag mixes somewhat with the aryl CH wag vibrations to give the following strong IR bands: benzoates near 710 cm^{-1}, phthalates near 745 cm^{-1}, isophthalates and terephthalates near 730 cm^{-1} [2].

The Raman spectra of dialkyl phthalates have been studied [135], and the following Raman group frequencies in cm^{-1} were reported for the ester group: 1738–1728 vs, p, (C=O), 1294–1277 s, p, (C—C—O—C), and 1139–1118 m, —, (C—O—C—C). Raman bands in cm^{-1} chiefly involving the phthalate *ortho*-substituted ring were found at 3089–3079 m, p, 3045–3039 w, —, 1607–1603 s, dp, 1588–1583 m, p, 1497–1489 w, —, 1170–1163 m, p, 1047–1041 vs, p, 657–652 s, p, and 409–400 m, p. Aromatic rings are covered in Chapter 17.

9-6.

CARBOXYLIC ACIDS

Carboxylic acids —C(=O)—OH are characterized by the OH stretch bands, the C=O stretch bands, the C—O stretch bands, the OH in-plane bend bands, and the OH out-of-plane wag bands. These bands are all sensitive to the hydrogen-bonding state of the COOH group, which can exist as the unbonded monomer, the carboxylic acid dimer form, or sometimes in a hydrogen-bonded polymeric form. If there are other hydrogen-bond donor or acceptor groups, further variations in hydrogen bondings are possible.

The most common form for carboxylic acids is the hydrogen-bonded dimer:

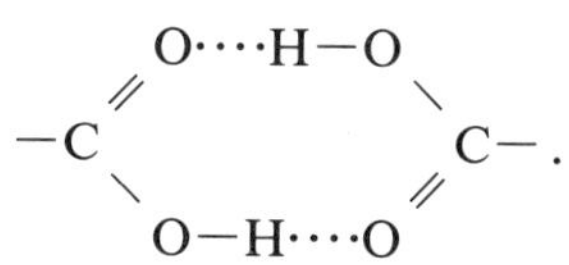

This form has a center of symmetry so that when the two monomer compounds vibrate in-phase, the vibration will be Raman-active only, and when they vibrate out-of-phase, the vibrations will be IR-active only. This is most notably seen in the dimer C=O stretching bands. In the Raman, the symmetrical C=O stretch is seen near 1650 cm^{-1}, and in the IR, the antisymmetrical C=O stretch is seen near 1700 cm^{-1}.

Vibrational assignments have been reported for acetic acid [73–75] and formic acid [142].

9-6.1. Carboxylic Acid OH Stretch

The carboxylic acid OH stretching bands are weak in the Raman, so IR data are generally used. In the hydrogen-bonded dimer form [76], the OH stretch band in the IR centers around 3000 cm^{-1} and is superimposed on the CH stretch bands, but is much broader than they are, having a half-intensity band width of roughly 600 cm^{-1}. On the low-frequency wing in the 2700–2500 cm^{-1} region, there are usually some weaker superimposed band shoulders assigned to overtones and combinations of bands near 1420 and 1300 cm^{-1} (due to OH bend and C—O stretch) enhanced by Fermi resonance with the broad OH stretch band [76].

In dilute CCl_4 solutions, some monomeric OH stretch absorption can be seen as a weak sharp band at 3550–3500 cm^{-1}, but the molecule still exists mainly as the hydrogen-bonded dimer [77].

In the vapor state, carboxylic acids can exist as dimers or monomers. At room temperature the dimer is predominant, but above about 150°C the monomer is predominant [76, 78]. Monomer bands are seen at 3585–3570 cm^{-1} [62, 72, 78], about 60 cm^{-1} higher than those from dilute solutions. See Table 9-9.

Table 9-9.

Vapor-Phase Monomeric Carboxylic Acid IR Bands (cm^{-1})

	Normal or α-Branched [62]	α,β-Unsaturated [62]	
OH stretch	3575–3570	3585–3580	m sharp
C=O stretch	1780–1770	1764–1760	s
COH in-plane bend	1385–1380	1365–1335	wm
C—O stretch	1180–1130	1150–1115	m
1100 cm^{-1} bands	1110–1080	1110–1070	m

9-6.2. Carboxylic Acid C=O Stretch

In the condensed state, most carboxylic acids exist as the dimer. In the IR, the dimer has a strong C=O band usually at 1740–1660 cm^{-1} [2]. See Table 9-10. In the Raman, the dimer has a weak C=O band usually at 1687–1625 cm^{-1} [3]. The Raman depolarization ratio is very low, about 0.05 [4]. In the vapor state, the dimer C=O band in the IR is usually near 1730 cm^{-1}, and the monomer C=O band can be seen usually near 1780–1760 cm^{-1} [62, 76] (see Table 9-9), about 20–25 cm^{-1} higher than those for dilute solutions.

As usual with C=O compounds, halogenation on the α-carbon tends to raise the C=O frequency. Trichloracetic acid in the solid state is in the dimer form and has an IR band at 1742 cm^{-1} [76] and a Raman band at 1687 cm^{-1} [79].

Carboxylic acid groups on aromatic rings in the condensed state absorb in the IR at 1710–1660 cm^{-1} [12, 80], and at 1686–1625 cm^{-1} in the Raman [81]. As usual, electron donor groups or hydrogen-bonding groups on the ring can lower the dimer C=O frequency. One must be on the lookout for alternate forms of hydrogen bonding other than the dimer form when hydrogen-bonding donor or acceptor groups are present. Monomers of aromatic acids in dilute solution absorb in the IR usually at 1760–1730 cm^{-1} [82].

When a hydroxyl or ether group can interact with the COOH group, alternate forms of hydrogen bondings are possible [2]. These groups may be in the same molecule with the COOH group, or the COOH molecule can be dissolved in water or alcohols or ethers. When the carboxyl OH group bonds to the ether oxygen, the carboxyl C=O is not hydrogen-bonded and appears in the ester region in *both* IR and Raman at 1760–1735 cm^{-1}. When the carboxyl C=O is hydrogen-bonded, but not dimerized as in alcohol-carbonyl bonding, a band is seen near 1730–1705 cm^{-1} in *both* IR and Raman. As can be seen, the carboxylic acid dimer is best characterized

Table 9-10.

Carboxylic Acid Dimer Bands (cm^{-1}) [76, 84, 85]

OH stretch	ca. 3000 very broad	IR : S
Overtones and combinations	2700–2500	IR : mw
C=O antisymmetric stretch	1740–1660	IR : vs
C=O symmetric stretch	1687–1625	R : w
C—OH in-plane bend	1440–1395	IR : m
C—O stretch	1315–1280	IR : s
OH···O out-of-plane wag	960–875	IR : m

by the Raman band at 1680–1640 cm^{-1} [85], where the IR has no band. Methoxyacetic acid has a very weak Raman dimer band at 1666 cm^{-1} and a relatively strong Raman at 1741 cm^{-1} [83] (like the IR band), indicating carboxyl-ether intramolecular hydrogen bonding [2]. Lactic acid is not dimerized, as it has no Raman band in the dimer region [83]. The C=O appears at 1744–1723 cm^{-1} in both the IR and Raman.

If a basic amino group is in the same molecule with a COOH group, the COOH group may bond with itself or may form an internal salt with the amine (zwitterion), as in amino acids. The IR spectra can distinguish these (see Section 9-6.5).

9-6.3. Carboxylic Acid C—O Stretch and OH Bend

In the infrared spectra, the best band for carboxylic acid dimers is the in-phase, out-of-plane OH···O wag at 960–875 cm^{-1} [84]. This IR band is only of medium intensity, but it is noticeably broader than other bands in the vicinity. If a non-dimeric form of hydrogen bonding is present for the COOH group, this band is at a lower frequency.

Two other IR band regions are found for COOH groups in the condensed state. A prominent band, or band cluster, appears in the IR at 1315–1200 cm^{-1} [2]. A narrower region, 1315–1280 cm^{-1}, in the IR has been assigned to dimers [84]. A somewhat weaker IR band is seen at 1440–1395 cm^{-1} [84]. These two bands near 1290 and 1410 cm^{-1} both involve interacting C—O stretch and C—O—H in-plane bending [84] (stretch-open and stretch-bend) [2]. However, because the 1315–1200 cm^{-1} band is in the same region where ester C—O stretch absorbs, we call this region mainly C—O stretch for simplicity and call the 1440–1395 cm^{-1} region mainly COH in-plane deformation. The carboxyl dimer bands are summarized in Table 9-10.

9-6.4. Long-Chain *n*-Aliphatic Acids

In the cystalline solid state, long straight-chain saturated carboxylic acids (such as stearic acid) and derivatives such as acid salts or amides have a regularly spaced progression of weak sharp bands in the IR between 1345 and 1180 cm^{-1} [86–88] that have been assigned to various interacting CH_2 wagging vibrations in the $(CH_2)_x$ part of the chain. These bands disappear in the liquid state or solution where the regular zig-zag configuration of the CH_2 chain in the crystal is lost.

A correlation has been made that relates the number of bands in the progression to the number of carbons in the acid [87]. If N is the number of bands in the 1345–1180 cm^{-1} region, then $2N$ is the number of carbons in acids with an even number of carbons, and $2N - 1$ is the number of carbons in acids with an odd number of carbons [87].

In acids, the acid C—O band is superimposed on the regularly spaced band progression, but in acid salts, the band progression region can be seen without interference.

Also, in the crystalline solid state the in-phase CH_2 rock IR band (near 723 cm^{-1} in liquids) is split and appears near 730 and 720 cm^{-1}.

9-6.5. Carboxylic Acid Salts

When a salt is made of a carboxylic acid, the C=O and C—O bonds of the acid are replaced by two equivalent (or nearly equivalent) C$\cdots$O "bond-and-a-half" bonds. These CO_2^- bonds interact out-of-phase and in-phase to give two bands. The antisymmetric CO_2 stretch band is usually seen at 1650–1540 cm^{-1}, strong in the IR and weak and depolarized in the Raman. The symmetric CO_2 stretch band is usually seen at 1450–1360 cm^{-1}, medium-weak in the IR and strong and polarized in the Raman [2, 3]. The antisymmetric X—CO_2 stretch is increased when the electron-withdrawing characteristic of X is increased [89]. Note that CH_3CO_2Na has a band at 1576 cm^{-1} and CCl_3CO_2Na has a band at 1675 cm^{-1} [12]. For a given acid salt, variations in the CO_2 stretch bands can occur when the positive counterion or the hydration state is varied.

Assignments for Raman and IR bands have been given for formate, acetate, and oxalate ions [90]. Also, some sodium salts of malonic, succinic, maleic, and fumaric acids in water solutions have been studied [91].

Acetate salts have IR bands at 1600–1550, s, 1450–1400, m, and weak bands near 1050, 1020, and 925 cm^{-1} [92]. Formate salts have IR bands near 2835, 1600, 1360, and 775 cm^{-1} [2]. Oxalate salts, $(O_2C{-}CO_2)^{2-}$, have two interacting CO_2 groups and have IR bands at 1620, 1320, 770, and 515 cm^{-1} [2]. Benzoate salts have bands at 1561–1552 (antisymmetric stretch) and 1427–1380 cm^{-1} (symmetric stretch) [93]. Amino acids in the zwitterion form ($H_3N{-}CH({-}X){-}CO_2$) have bands at 3100–2600 cm^{-1} (NH_3 stretch), 1665–1585 cm^{-1} (shoulder) (NH_3 antisymmetric deformation), 1605–1555 cm^{-1} (CO_2 antisymmetric stretch), 1530–1490 cm^{-1} (NH_3 symmetric deformation), 1425–1393 cm^{-1} (CO_2 symmetric stretch), and 1340–1315 cm^{-1} (CH deformation) [94, 95].

9-7. ANHYDRIDES

Anhydrides, —CO—O—CO—, of carboxylic acids have two carbonyl groups, which can vibrate in-phase or out-of-phase to give two bands. Noncyclic dialkyl-substituted anhydrides give rise to bands near 1820 cm^{-1} (symmetrical stretch) and 1750 cm^{-1} (antisymmetrical stretch). Conjugation lowers these bands to about 1775 and 1720 cm^{-1} [96, 97]. In these noncyclic anhydrides, the higher-wavenumber band is characteristically the more intense in the IR. See Table 9–11. As usual, halogenation of the α-carbon can raise the C=O frequency somewhat. An extreme example is CF_3—CO—O—CO—CF_3 (1879, 1811 cm^{-1}) [12].

Anhydrides that are part of a five-membered ring also give rise to two bands. Saturated five-membered ring anhydrides absorb near 1860 and 1780 cm^{-1}, and conjugated five-membered ring anhydrides absorb near 1850 and 1760 cm^{-1} [96, 97]. See Table 9-11. In these cyclic anhydrides, the lower-wavenumber band is characteristically the more intense in the IR. The ratio of the IR absorbances of the low (A_L) to those of the high (A_H) wavenumber C=O bands can be used to distinguish the types. A_L/A_H = 0.81–0.93 for unconjugated noncyclic anhydrides, and A_L/A_H = 6.3–11.1 for five-membered ring cyclic anhydrides. Six-membered ring anhydrides are intermediate, where A_L/A_H = 2.7 [96].

These IR intensity relationships are explained by the relative orientation of the two C=O bonds [96–99]. Since the dipole moment change is a vector, for vibrations involving two C=O bonds, the total dipole moment change is a vector sum, dependent on the angle. In five-membered ring anhydrides, the two C=O bonds are oriented about 150° from each other, so the in-phase stretch has a smaller dipole moment change than the out-of-phase stretch. Noncyclic anhydrides were originally thought to be planar and to have the C=O bonds nearly parallel. However, a vibrational analysis of acetic anhydride by Mirone *et al.* [98] showed that there were too many polarized Raman lines to be consistent with this form, and that carbonyl

Table 9-11.

Anhydride C=O IR Regions in cm^{-1} [96, 97]

Noncyclic:		
unconjugated	1825–1815 stronger	1755–1745 weaker
conjugated	1780–1770 stronger	1725–1715 weaker
Cyclic, five-membered ring:		
unconjugated	1870–1845 weaker	1800–1775 stronger
conjugated	1860–1850 weaker	1780–1760 stronger

oxygen mutual repulsion twists the two C—O bonds each about 30° so that the C=O bonds are oriented roughly 60° to each other in a nonplanar form. Now, the in-phase stretch has a larger dipole moment change than the out-of-phase stretch. Colthup [99] studied stearic anhydride and found a similar result for the liquid state, but for the crystalline solid state, the anhydride was planar and the C=O bonds were approximately parallel for this particular long-chain anhydride.

Multiple bands involving C—C—O and C—O—C stretch are seen in the IR usually at 1150–1000 cm^{-1} for noncyclic anhydrides and at 950–880 cm^{-1} for five-membered ring anhydrides [12].

The IR and Raman spectra of dimethyl dicarbonate have been studied [143]. Dimethyl dicarbonate (MeO—CO—O—CO—OMe) is an anhydride of methyl carbonic acid. In the liquid state, the in-phase C=O IR band is at 1830 cm^{-1} and the out-of-phase C=O IR band is at 1765 cm^{-1}. The relative C=O IR band intensity ratios suggest that, as with stearic anhydride [99], the C—O dihedral angle is about 30° for the liquid state and close to 0° for the solid state for this molecule [143].

9-8. AMIDES, LACTAMS, UREAS, AND IMIDES

Amides have a C=O band usually in the region 1695–1630 cm^{-1} that is strong in the IR and weak in Raman. Amides have NH_2 or NH groups that have various stretching and bending bands. These will be discussed in detail in Chapter 10, but are covered here as well. The C=O band in ureas is similar to that for amides, as is the C=O band for six-membered ring lactams. Imides, —CO—NH—CO— have C=O bands at somewhat higher frequencies, as do lactams in five- and four-membered rings.

9-8.1. Unsubstituted Amides

Unsubstituted amides, —CO—NH_2, in the hydrogen-bonded state have antisymmetric NH_2 stretch bands near 3350 cm^{-1} and symmetric NH_2 stretch bands near 3180 cm^{-1} [3, 12, 100, 101]. These bands shift to higher wavenumbers in solution, near 3520 and 3400 cm^{-1}, respectively [100, 101].

The C=O stretch band and the NH_2 scissors deformation band overlap in the hydrogen-bonded state. The C=O stretch is the stronger band in the IR and absorbs usually about 1680–1640 cm^{-1}, and the NH_2 deformation absorbs about 1640–1620 cm^{-1} [3, 12, 100, 101]. In solution, these two bands are usually seen at 1690–1670 cm^{-1} (C=O) and 1620–1614 cm^{-1} (NH_2 deformation) [3, 100, 101]. See Table 9-12.

Table 9-12.

Amide Bands in cm^{-1}

	Associated	Nonbonded	Reference
Unsubstituted amides:			
NH_2 antisymmetric stretch	3475–3350	ca. 3520	3, 100, 101
NH_2 symmetric stretch	3385–3160	ca. 3400	3, 100, 101
C=O stretch	1680–1640	1715–1675	3, 100, 101
NH_2 deformation	1640–1620	1620–1585	3, 100, 101
C—N stretch	1430–1390	1430–1390	3
NH_2 rock	1150–1100	1150–1100	3
NH_2 wag	750–600 broad		2
NCO in-plane def.	600–550	600–550	3
Monosubstituted amides:			
NH stretch, trans	ca. 3300	3490–3400	3, 100, 103–107
cis	ca. 3200		3, 103, 108
C=O stretch	1680–1630	1700–1650	3, 100, 103–107
C=O five-membered ring	1750–1700		2
C=O four-membered ring	1780–1730		2
CNH stretch-bend, trans	1570–1510	1550–1500	3, 100, 103–107
cis	1490–1400	ca. 1450	3, 108, 109
CNH stretch-open, trans	1310–1250	1250–1200	3, 100, 103–106
cis	1350–1310	ca. 1350	3, 108, 109
Disubstituted amides:			
C=O stretch		1680–1630	3, 100
NCO in-plane deformation			
H—CO—NR_2		ca. 650	3
R—CO—NR_2		620–590	3

Additional bands include C—N stretch usually at 1430–1390 cm^{-1}, NH_2 rock usually at 1150–1100 cm^{-1}, NH_2 wag at 750–600 cm^{-1} in the IR (very broad) [2], and in-plane NCO deformation at 600–550 cm^{-1} in the Raman [3].

9-8.2. Monosubstituted Amides

Noncyclic N-monosubstituted amides usually exist with the NH and C=O bonds in the trans configuration [100–107, 138, 144]. The hydrogen-bonded NH stretch is seen near 3300 cm^{-1} [3, 100], with a weaker band near 3100 cm^{-1} that is a Fermi resonance-enhanced overtone of the 1550 cm^{-1} band [102]. This band near 1550 cm^{-1} (strong in IR) involves both C—N stretch and C—N—H in-plane bend in the stretch-bend mode (sometimes called amide II). This band is very characteristic for noncyclic monosubstituted amides [103–106]. The stretch-open mode (sometimes called amide III) absorbs more weakly in the IR near 1310–1250 cm^{-1}. The C=O stretch band is very strong in the IR in the 1680–1630 cm^{-1} region. The out-of-plane NH wag absorbs broadly in the IR at 750–680 cm^{-1} [103–106].

In *N,N'*-dialkyl oxamides, the C=O bonds exist in the *s*-trans configuration, and a center of symmetry is present. The symmetrical C=O and CNH bands are seen in the Raman at 1695–1686 and 1567–1547 cm^{-1}, and the antisymmetrical C=O and CNH bands are seen in the IR at 1660–1643 and 1541–1508 cm^{-1} [137].

9-8.3. Lactams

Cyclic N-monosubstituted amides or lactams have the C=O and NH bonds forced into a cis configuration. In the hydrogen-bonded state, the NH stretch band appears near 3200 cm^{-1} with a weaker combination band near 3100 cm^{-1} [103, 108]. The C=O stretch band is strong in the IR and appears in six-membered ring lactams in the same region as in the noncyclic amides. Five-membered ring lactams absorb at 1750–1700 cm^{-1}, and four-membered ring lactams absorb at 1780–1730 cm^{-1} [2, 12]. There is no band in the 1570–1510 cm^{-1} region. The CNH stretch-bend mode for the cis or lactam configuration is lower, in the CH_2 deformation region, at 1490–1440 cm^{-1} [108, 109]. See Table 9-12.

9-8.4. Disubstituted Amides

Amides with no NH or NH_2 have a C=O band, strong in IR, usually at 1680–1630 cm^{-1} [100]. A band involving NCO in-plane bending is seen near 650 cm^{-1} for H—CO—NR_2 and at 620–590 cm^{-1} for R—CO—NR_2 [3].

9-8.5. Ureas

Ureas have C=O, NH_2, and NH bands fairly similar to those of amides. Approximate regions for substituted ureas are shown in Table 9-13. For the *N,N'*-dialkyl ureas, there are two CNH groups that interact to give two bands, near 1585 cm^{-1} and, more weakly in IR, near 1535 cm^{-1} [12].

9-8.6. Imides

Noncyclic imides (—CO—NH—CO—) have NH bands something like those in amides. In the solid state, the C—NH bands appear in the IR near 3250, 1507–1503, and 1236–1167 cm^{-1}. The C=O group absorbs strongly usually at 1740–1670 cm^{-1} [110–112].

Table 9-13.

Approximate IR Wavenumbers for Alkyl Ureas [12]

Group	$R_2N-CO-NR_2$	$RNH-CO-NHR$	$R_2N-CO-NH_2$	$RNH-CO-NH_2$
NH_2 antisymmetric stretch			3410	3430
NH stretch		3340		3350
NH_2 symmetric stretch			3200	3220
C=O stretch	1640	1625	1660	1650
NH_2 deformation			1610	1600
CNH stretch-bend		1585, 1535 w		1560

Cyclic imides in five-membered rings, such as phthalimides, maleimides, and succinimides, generally have two bands in the carbonyl region. The in-phase C=O band in the solid state appears at 1800–1735 cm^{-1} (IR : m, R : s), and the out-of-phase C=O band appears at 1750–1680 cm^{-1} (IR : vs, R : w) [111, 112, 131]. The hydrogen-bonded NH stretch band for non-N-substituted imides is seen at 3250–3150 cm^{-1} [131]. In the IR a band is seen at 1210–1140 cm^{-1} involving C—N—C stretch, which is strong in succinimides and weak to medium in phthalimides and maleimides [131]. As in the lactams, the cyclic imides do not have the 1505 cm^{-1} CNH band. In the Raman, phthalimides have an aromatic ring stretch band at 1630–1590 cm^{-1} R : m-s [131]. Another Raman band for N-substituted phthalimides is seen at 1030–990 cm^{-1} involving ring in-plane CH bending [131]

9-9.

ALIPHATIC ACID HALIDES

The characteristic vibrations of the —C(=O)—X group can be approximately described as the C=O and C—X stretching modes, the O=C—X in-plane bending, and the stretching mode of the C—C bond alpha to the —CO—X group. There are quite extensive literature references to the vibrational spectra of acid chlorides and acid bromides. Vibrational assignments of acetyl chloride [113–115] and bromide [114, 115], cyclopropane carbonyl chloride and bromide [116], and cyclobutane carbonyl chloride and bromide [117] have been reported. However, only a limited amount of literature is found for acid fluorides and iodides. The vibrational spectra of acetyl fluoride and iodide were examined by Ramsey and Ladd [115]. Berney has assigned the vibrational frequencies of trifluoroacetyl fluoride [118] and chloride [119] by IR and Raman techniques. See Table 9-14.

Table 9-14.

The Characteristic Frequencies (cm^{-1}) of Aliphatic Acid Halides

Approximate Description of Vibrational Modes		Frequency (cm^{-1})	IR	R
C=O stretching		1810–1775	vvs	m-w, p
C—X stretching:				
R—COX, X = F		832 (CH_3COF)		
Cl		731–565	s	m-w, p
Br		700–535	s	
I		542 (CH_3COI)		
RO—COCl	C—O—C antisymmetric stretch	1172–1134	vs	vw
	C—O—C symmetric stretch	850–770	s	s, p
	C—Cl stretch	690–680	s	o
OCX in-plane deformation:				
CF_3—COF		761	m	m
R—COCl		450–420	m	vs, p
RO—COCl		487–471	m	vs, p
—COCl		426	m	s
—COCl		437	m-s	vs, p
—COBr		350		s
—COBr		352	w	

9-9.1. C=O Stretching

As with the other C=O-containing compounds, the C=O group in acid halides has a weak Raman intensity. On the other hand, the highly polar nature of the C=O functional group gives rise to a very strong IR band in the 1810–1775 cm^{-1} region [12]. Acid bromides usually have higher C=O stretching frequencies than the corresponding acid chlorides. Conjugation to a carbon–carbon double bond lowers the C=O stretching frequency to 1780–1760 cm^{-1} [12] and often enhances its Raman intensity [14]. Frequently, a shoulder can be observed on the high-frequency side of the carbonyl band [12]. The origin of this shoulder is very likely the overtone of a strong band in the 1000–900 cm^{-1} range, which, in the case of acetyl chloride, is assigned to a vibrational mode involving the stretch of the C—C bond alpha to the CO—Cl group [113–115].

For α-branched or α-halogenated acyl halides, a doublet in the carbonyl stretch region is observed [12]. The origin of these spectral features is not easily understood and requires further study. One possible explanation involves the Fermi resonance of the C=O vibration with the overtone (or combination band) of the alpha C—C stretching vibration. Another possibility arises from the presence of the C=O absorption from different rotational isomers.

9-9.2. C—X Stretching

The C—X stretching mode of acid halides absorbs IR strongly in almost the same region as alkyl halides. But unlike those of alkyl halides, these polarized Raman bands of acid halides exhibit only medium to weak intensity [14]. The effect of rotational isomerism on the C—X stretching frequency was investigated by Katon and Feairheller [114]. There is only one rotational isomer for acetyl chloride (H trans to Cl); therefore, only one C—X stretching band is observed. In the case of propionyl halides (CH_3CH_2—CO—X), there would be two C—X stretching bands. For example, the trans form (C trans to Cl) of propionyl chloride absorbs at 693 cm^{-1}, and the gauche form at 569 cm^{-1} (C gauche to Cl, H trans to Cl).

For the rest of the saturated, straight-chain aliphatic acid halides, four rotational isomers are possible as a result of rotations about the α C—C bond and the β C—C bond. The α-trans form absorbs near 700 cm^{-1}, where the α-gauche form absorbs near 550 cm^{-1}. Further splitting of these two C—X stretching bands because of rotation about the β C—C bond is possible. As a consequence of rotations about both α and β C—C bonds, the spectra will present four C—Cl stretching bands. α-Branched or α-substituted compounds are expected to have complex spectral features in the C—X stretching region, since more rotational isomers are introduced.

The C—Br stretching frequencies of the —C(=O)—Br group fall in the 700–535 cm^{-1} region. In general, the C—Br stretching frequency is 15–40 cm^{-1} lower than the corresponding C—Cl frequency. The C—F stretching acetyl fluoride was reported to be at 832 cm^{-1}, and the C—I stretching of acetyl iodide at 542 cm^{-1} [115].

Trans and cis isomers are possible for α,β-unsaturated acyl halides [120]:

s-trans 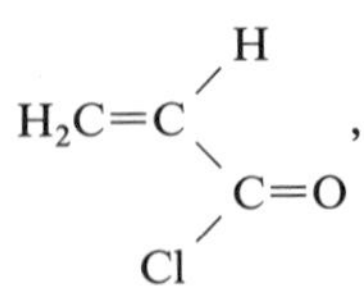 , *s*-cis 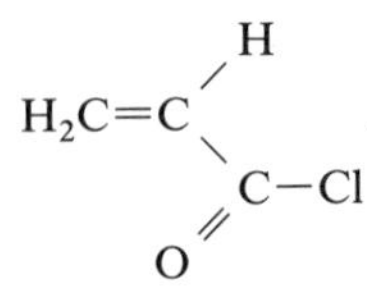 .

The separation between C—X absorption bands of cis and trans isomers can be as large as 100 cm^{-1}. For example, the *s*-trans isomer of acryloyl chloride (CH_2=CH—CO—Cl) has a C—Cl stretching band at 607 cm^{-1}, and that of the *s*-cis isomer is at 709 cm^{-1} [120]. The atoms trans to the Cl are H and C, respectively. Furthermore, frequency shifts of 10–20 cm^{-1} are observed for skeletal vibrations of different rotational isomers of acryloyl choride.

Compounds of RO—CO—Cl structures (chloroformates) show an intense IR band, which is totally unobservable in Raman, between 694 and 680 cm^{-1} because of the C—Cl stretching mode [12, 14, 121]. These chloroformates also have a strong IR band (weak in Raman) at 1172–1134 cm^{-1} involving C—O—C out-of-phase stretch [121] and another band, stronger in Raman [14] at 850–770 cm^{-1} (probably C—O—C in-phase stretch).

9-9.3. Deformation Vibrations

The most characteristic Raman band of acid chlorides arises from a vibration involving OCCl scissoring (in-plane bending) located in the 450–420 cm^{-1} region for R—CO—Cl [14]. This highly polarized band is extremely intense in the Raman spectra of many acid halides, and its peak intensity even exceeds the CH_2 stretching bands in the Raman spectra of $CH_3(CH_2)_{14}$—CO—Cl [14]. A frequency shift to 487–471 cm^{-1} for RO—CO—Cl [14, 121] and 511 cm^{-1} for CF_3—CO—Cl [119] has been observed; also, O=C—Cl scissoring frequencies of cyclobutanecarbonyl chloride [117] and cyclopropanecarbonyl chloride [116] fall in the region of 490–430 cm^{-1}. The OCF in-plane deformation frequency for CF_3—CO—F is at 761 cm^{-1} [118], while the OCBr scissoring frequency of cyclobutanecarbonyl bromide is at 350 cm^{-1} [117].

9-10. AROMATIC ACID HALIDES

The vibrational spectra of benzoyl halides (C_6H_5CO—X, where X = F, Cl, or Br) have been studied extensively [23, 122–130]. Characteristic frequencies of aromatic —C(=O)—X groups are listed in Table 9-15.

9-10.1. C=O and Ph—C Stretchings

The C=O stretching mode absorbs IR strongly in the 1800–1750 cm^{-1} region [12], and its Raman intensity is enhanced somewhat by the conjugated aromatic ring [14]. For benzoyl chlorides, an unequal intensity

Table 9-15.

The Characteristic Frequencies (cm^{-1}) of Aromatic Acid Halides

Approximate Description of the Vibrational Modes	Frequency (cm^{-1})	IR	R
C=O stretch	1800–1750	vs	var., p
Ph—C stretch	900–800	s	w
Overtones of Ph—C stretch	2 × (900–800)	s-m	var.
C—X stretch:			
benzoyl fluoride	1139		w
benzoyl chloride	673	s	s, p
3,5-dinitro-benzoyl chloride	604	w	
3,5-dibromo-4-methoxylbenzoyl chloride	669	s	
benzoyl bromide	657	s	s, p
4-methylbenzoyl bromide	604	s	
OCX deformation:			
benzoyl chloride	316	m	m, p
benzoyl bromide	263	m	s, p
4-methylbenzoyl bromide	235		m

doublet is usually seen in the C=O stretching region. This doublet is attributed to Fermi resonance between the C=O band and the overtone of the Ph—C stretching band in the 900–800 cm^{-1} region of the IR spectra [125, 128]. Our investigation aided by the Aldrich Library of FT-IR Spectra [12] indicates that the C=O and Ph—C stretching bands of *para*-substituted benzoyl chlorides are located at regions of 1780–1760 cm^{-1} and 891–872 cm^{-1}, respectively. Therefore, the Fermi resonance-enhanced overtone of Ph—C stretching bands often appears at the high-frequency side of the C=O stretching. The situation is reversed in the case of *ortho*-substituted benzoyl chlorides where the C=O stretching absorbs in the 1795–1765 cm^{-1} region and the Ph—C stretching occurs in the 885–860 cm^{-1} range. For *meta*-substituted benzoyl chlorides, the Ph—C stretching band occurs in a wider range (912–800 cm^{-1}). As a result, its overtone can appear as a shoulder on either side of the C=O stretching band (1767–1756 cm^{-1}).

9-10.2. C—X Stretching

There are controversial assignments on the C—Cl stretching mode of benzoyl chloride. Al-Jallo and Jalhoom [129] suggested that the strong IR band in the region of 900–800 cm^{-1} is due to the C—Cl stretching vibration.

Rao and Venkatoraghavan [125], on the other hand, designated this strong IR band to the Ph—C stretching mode. These two assignments were disputed by Condit *et al.* [130], who reassigned the 673 cm^{-1} vibrational band of benzoyl chloride to the C—Cl stretching vibration. Condit's assignment was supported by the Raman intensity, polarization, and the relative frequency shift (16 cm^{-1}) of benzoyl bromide. Varsanyi [23] further substantiated Condit's assignment by suggesting the C—Cl stretching frequencies of 3,5-dinitro-benzoyl chloride and 3,5-dibromo-4-methoxyl benzoyl chloride to be 683 cm^{-1} and 669 cm^{-1}, respectively.

9-10.3. C=O Deformation

The oxygen carbon out-of-plane deformation (γ OCX) is expected to be higher than the oxygen carbon in-plane deformation (δ OCX). For benzoyl chloride, the OCCl out-of-plane deformation occurs at 650 cm^{-1}, while the polarized Raman band at 318 cm^{-1} is attributed to the OCCl in-plane rock [130].

REFERENCES

1. L. J. Bellamy, "The Infrared Spectra of Complex Molecules," John Wiley & Sons, New York, 1975.
2. N. B. Colthup, L. H. Daly, and S. E. Wiberley, "Introduction to Infrared and Raman Spectroscopy," 3rd Ed., Academic Press, New York, 1990.
3. F. R. Dollish, W. G. Fateley, and F. F. Bentley, "Charateristic Raman Frequencies of Organic Compounds," John Wiley & Sons, New York, 1974.
4. S. K. Freeman, "Applications of Laser Raman Spectroscopy," John Wiley & Sons, New York, 1974.
5. S. Bratoz and S. Besnainou, *C.R. Acad. Sci.* **248,** 546, 1959.
6. J. O. Halford, *J. Chem. Phys.* **24,** 830, 1956.
7. N. B. Colthup and M. K. Orloff, *Chem. Phys. Lett.* **19,** 298, 1973.
8. H. K. Hall, Jr. and R. Zbiden, *J. Am. Chem. Soc.* **80,** 6428, 1958; **82,** 1215, 1960.
9. R. N. Jones, D. A. Ramsey, F. Herling, and K. Dobringer, *J. Am. Chem. Soc.* **74,** 2828, 1952.
10. L. J. Bellamy and R. L. Williams, *J. Chem. Soc. London*, 4294, 1957.
11. L. J. Bellamy, *Spectrochim. Acta* **13,** 60, 1958.
12. C. J. Pouchert, "The Aldrich Library of FT-IR Spectra," Aldrich Chemical Co., Vols. 1 and 2, 1985; Vol. 3, "Vapor Phase," 1989.
13. R. A. Nyquist, "The Interpretation of Vapor Phase Infrared Spectra, Group Frequencies," Sadtler Research Laboratories, 1985.
14. The Sadtler Standard Raman Spectra, Sadtler Research Laboratories, 1973–1976.
15. G. Lucazeau and A. Novak, *Spectrochim. Acta* **25A,** 1615, 1969.
16. J. R. Durig and W. J. Natter, *J. Raman Spectrosc.* **11,** 32, 1981.

17. G. Hagen, *Acta Chem. Scand.* **25,** 813, 1971.
18. C. V. Berney, *Spectrochim. Acta* **25A,** 793, 1969.
19. A. J. Bowles, W. O. George, and W. F. Maddams, *J. Chem. Soc.* (*B*), 810, 1969.
20. G. W. King and D. Moule, *Spectrochim. Acta* **17,** 286, 1961.
21. J. C. D. Brand and R. A. Powell, *J. Mol. Spectrosc.* **43,** 342, 1972.
22. E. Lagset, P. Klaboe, E. Kloster-Jensen, S. J. Cyvin, and F. M. Nicolaisen, *Spectrochim. Acta* **29A,** 17, 1973.
23. G. Varsanyi, "Assignments for Vibrational Spectra of Seven Hundred Benzene Derivatives," John Wiley & Sons, New York, 1974.
24. S. Pinchas, *Anal. Chem.* **29,** 334, 1957.
25. S. Pinchas, *Anal. Chem.* **27,** 2, 1955.
26. K. W. F. Kohlrausch and F. Koeppl, *Z. Physik, Chem.* **B24,** 370, 1934.
27. G. Allen and J. Bernstein, *Can. J. Chem.* **33,** 1055, 1955.
28. M Hirota, T. Hagiwara, and H. Satonaka, *Bull. Chem. Soc. Jpn.* **40,** 2439, 1967.
29. R. A. Nyquist, *Appl. Spectrosc.* **40,** 79, 1986.
30. J. R. Durig and J. S. Church, *Spectrochim. Acta* **36A,** 957, 1980.
31. C. V. Berney, *Spectrochim. Acta* **21,** 1809, 1965.
32. C. Laurence, M. Berthelot, and D. G. Morris, *Spectrochim. Acta* **39A,** 699, 1983.
33. V. Prelog, *J. Chem. Soc.*, 420, 1950.
34. F. H. Cottee, B. P. Straughan, C. J. Timmons, W. F. Forbes, and R. Shilton, *J. Chem. Soc.*, *B*, 1146, 1967.
35. R. Bartlet, M. Montange, and P. Arnaud, *Spectrochim. Acta* **25A,** 1081, 1969.
36. C. H. Chang, A. L. Andreasen, and S. H. Bauer, *J. Org. Chem.* **36,** 920, 1971.
37. R. Mecke and K. Noack, *Spectrochim. Acta* **12,** 391, 1958.
38. R. L. Erskine and E. S. Waight, *J. Chem. Soc.*, 3425, 1960.
39. K. Noack and R. N. Jones, *Can. J. Chem.* **39,** 2201, 1961.
40. K. Noack, *Spectrochim. Acta* **18,** 697, 1962.
41. K. Noack, *Spectrochim. Acta.* **18,** 1625, 1962.
42. W. P. Hayes and C. J. Timmons, *Spectrochim. Acta* **24A,** 323, 1968.
43. H. Junge, *Spectrochim. Acta* **24A,** 1965, 1968.
44. P. J. Taylor, *Spectrochim. Acta* **3A,** 1471, 1976.
45. D. Smith and P. J. Taylor, *Spectrochim. Acta* **32A,** 1477, 1976.
46. D. Smith and P. J. Taylor, *Spectrochim. Acta* **32A,** 1503, 1976.
47. R. N. Jones, W. F. Forbes, and W. A. Mueller, *Can. J. Chem.* **35,** 504, 1957.
48. R. N. Jones and E. Spinner, *Can. J. Chem.* **36,** 1020, 1958.
49. D. M. Hunsberger, *J. Am. Chem. Soc.* **72,** 5626, 1950.
50. H. W. Thompson and D. A. Jameson, *Spectrochim. Acta* **13,** 236, 1958.
51. N. B. Colthup and M. K. Orloff, *Spectrochim. Acta* **30A,** 425, 1974.
52. R. Mecke and E. Funk, *Z. Electrochim.* **60,** 1124, 1956.
53. R. S. Rasmussen, R. D. Tunnicliff, and R. R. Brattain, *J. Am. Chem. Soc.* **71,** 1068, 1949.
54. D. Hadzi and N. Sheppard, *Trans. Faraday Soc.* **50,** 911, 1954.
55. G. S. Hammond, W. G. Bordium, and G. A. Gutter, *J. Am. Chem. Soc.* **81,** 4682, 1959.
56. S. N. Ananchenko, I. V. Berezin, and I. V. Torgov, *Izv. Akad. Nauk, USSR* **9,** 1644, 1960.
57. J. K. Wilmhurst, *J. Mol. Spectrosc.* **1,** 201, 1957.
58. H. Susi and J. R. Scherer, *Spectrochim. Acta* **25A,** 1243, 1969.
59. E. M. Popov, G. A. Kogan, M. I. Struchkova, and V. N. Zheltova, *I. Struct. Chem., USSR* **12,** 61, 1971.
60. P. Matzke, O. Chacon, and C. Andrade, *J. Mol. Struct.* **9,** 255, 1971.

61. H. W. Thompson and P. Torkington, *J. Chem. Soc. London*, 640, 1945.
62. D. Welti, "Infrared Vapour Spectra," Heyden and Son, New York, 1970.
63. W. L. Walton and R. B. Hughes, *Anal. Chem.* **28,** 1388, 1956.
64. W. L. Walton and R. B. Hughes, *J. Am. Chem. Soc.* **79,** 3985, 1957.
65. J. L. Mateos, R. Cetina, E. Olivera, and S. Mesa, *J. Org. Chem.* **26,** 2494, 1961.
66. R. N. Jones and B. S. Gallagher, *J. Am. Chem. Soc.* **81,** 5242, 1959.
67. R. N. Jones, E. L. Angell, T. Ito, and R. J. D. Smith, *Can. J. Chem.* **37,** 2007, 1959.
68. A. R. Katritzky, J. M. Lagowski, and J. A. T. Beard, *Spectrochim. Acta* **16,** 954, 1960.
69. D. A. Barr and N. Hazeldine, *J. Chem. Soc. London*, 3428, 1956.
70. S. Pinchas and D. Ben Ishai, *J. Amer. Chem. Soc.* **79,** 4099, 1957.
71. A. R. Katritzky and R. A. Jones, *J. Chem. Soc. London*, 676, 1968.
72. R. A. Nyquist and W. J. Potts, *Spectrochim. Acta* **17,** 679, 1961.
73. W. Weltner, Jr., *J. Am. Chem. Soc.* **77,** 3941, 1955.
74. J. K. Wilmshurst, *J. Chem. Phys.* **25,** 1171, 1956.
75. M. Haurie and A. Novak, *J. Chim. Phys.* **62,** 137 and 146, 1965.
76. S. Bratoz, D. Hadzi, and N. Sheppard, *Spectrochim. Acta* **8,** 249, 1956.
77. J. D. S. Goulden, *Spectrochim Acta* **6,** 129, 1954.
78. D. Hadzi and M. Pintar, *Spectrochim. Acta* **12,** 162, 1958.
79. I. D. Poliakova and S. S. Raskin, *Opt. Spectrosc.* (*USSR*) **6,** 220, 1959.
80. M. St. C. Flett, *J. Chem. Soc.*, 962, 1951.
81. E. Hertz and H. Wittek, *Monatsch. Chem.* **74,** 271, 1943.
82. D. Peltier, A. Pichevin, P. Diazbo, and M. L. Josien, *C.R. Akaed. Sci.* **248,** 1148, 1959.
83. K. W. F. Kohlravsch, "Ramanspectren," Akad. Verlagsges. Leipzig, 1943.
84. D. Hadzi and N. Sheppard, *Proc. Roy. Soc.*, Ser A **216,** 247, 1953.
85. F. Gonzalez-Sanchez, *Spectrochim. Acta* **12,** 17, 1958.
86. R. N. Jones, A. F. McKay, and R. G. Sinclair, *J. Am. Chem. Soc.* **74,** 2575, 1952.
87. R. A. Meiklejohn, R. J. Meyer, S. M. Aronovic, H. A. Schuette, and V. W. Meloche, *Anal. Chem.* **29,** 329, 1957.
88. P. J. Corish and D. Chapman, *J. Chem. Soc. London*, 1746, 1957.
89. E. Spinner, *J. Chem. Soc.*, 4217, 1964.
90. K. Ito and H. J. Bernstein, *Can. J. Chem. 34,* 170, 1952.
91. C. B. Baddiel, C. D. Cavendish, and W. O. George, *J. Mol. Struct.* **5,** 263, 1970.
92. F. Vratny, C. N. R. Rao, and M. Dilling, *Anal. Chem.* **33,** 1455, 1961.
93. J. H. S. Green, W. Kynastron, and S. A. Lindsey, *Spectrochim. Acta* **17,** 486, 1961.
94. R. J. Koegel, J. P. Greenstein, M. Winitz, S. M. Bumbaum, and R. A. McCallum, *J. Am. Chem. Soc.* **77,** 5708, 1955.
95. K. Fukushima, T. Onishi, T. Shimanouchi, and S. Mizushima, *Spectrochim. Acta* **15,** 236, 1959.
96. W. G. Dauben and W. W. Epstein, *J. Org. Chem.* **24,** 1595, 1959.
97. L. J. Bellamy, B. R. Connelly, A. R. Phillpots, and R. L. Williams, *Z. Elektrochem.* **64,** 563, 1960.
98. P. Mirone, B. Fortunato, and P. Caniziani, *J. Mol. Struct.* **5,** 283, 1970.
99. N. B. Colthup, *Appl. Spectrosc.* **39,** 1030, 1985.
100. R. E. Richards and H. W. Thompson, *J. Chem. Soc. London*, 1248, 1947.
101. T. L. Brown, J. E. Regan, R. D. Schuetz, and J. Sternberg, *J. Phys. Chem.* **63,** 1324, 1959.
102. T. Mizazawa, *J. Mol. Spectrosc.* **4,** 163, 1960.
103. T. Mizazawa, T. Shimanouchi, and S. Mizushima, *J. Chem. Phys.* **24,** 408, 1956; **29,** 611, 1958.
104. R. D. S. Fraser and W. C. Price, *Nature* (*London*) **170,** 490, 1952.

105. C. G. Cannon, *Spectrochim. Acta.* **16,** 302, 1960.
106. M. Beer, H. B. Kessler, and G. B. B. M. Sutherland, *J. Chem. Phys.* **29,** 1097, 1958.
107. R. L. Jones, *Spectrochim. Acta* **23A,** 1745, 1967.
108. T. Mizazawa, *J. Mol. Spectrosc.* **4,** 155, 1960.
109. H. Brockman and H. Musso, *Chem. Ber.* **89,** 241, 1956.
110. J. Uno and K. Machida, *Bull. Chem. Soc. Jpn.* **34,** 545 and 551, 1961.
111. R. A. Abromovitch, *J. Chem. Soc. London*, 1413, 1951.
112. N. A. Borisevich and N. N. Khovratovitch, *Opt. Spectrosc.* (*USSR*) **10,** 309, 1961.
113. J. Overend, R. A. Nyquist, J. C. Evans, and W. J. Potts, *Spectrochim. Acta* **17,** 1205, 1961.
114. J. E. Katon and W. R. Feairheller, Jr., *J. Chem. Phys.* **44,** 144, 1966.
115. J. A. Ramsey and J. A. Ladd, *J. Chem. Soc., B*, 118, 1968.
116. J. E. Katon, W. R. Feairheller, Jr., and J. T. Miller, Jr., *J. Chem. Phys.* **49,** 823, 1968.
117. K. Kanai and J. E. Katon, *J. Mol. Struct.* **70,** 127, 1981.
118. C. V. Berney, *Spectrochim. Acta* **27A,** 663, 1971.
119. C. V. Berney, *Spectrochim. Acta* **20A,** 1437, 1964.
120. J. E. Katon and W. R. Feairheller, Jr., *J. Chem. Phys.* **47,** 1248, 1967.
121. H. A. Ory, *Spectrochim. Acta* **16,** 1488, 1960.
122. L. Kahovec and K. W. Kohlrausch, *Z. Phys. Chem.* **B38,** 119, 1937.
123. T. A. Hariharan, *Ind. J. Phys.* **26,** 114, 1952.
124. D. D. Thomson and J. F. Norris, *J. Am. Chem. Soc.* **58,** 1953, 1936.
125. C. N. R. Rao and R. Venkataraghavan, *Spectrochim. Acta* **18,** 273, 1961.
126. S. R. Singh, B. B. Lal, I. S. Singh, and M. P. Srivastava, *Indian J. Pure Appl. Phys.* **8,** 116, 1970.
127. N. S. Ham and H. B. Willis, *Spectrochim. Acta* **16,** 279, 1960.
128. E. Ortiz, J. F. Bertran, and L. Ballester, *Spectrochim. Acta* **27A,** 1713, 1971.
129. H. N. Al-Jallo and M. G. Jalhoom, *Spectrochim. Acta* **28A,** 1655, 1972.
130. D. A. Condit, S. M. Craven, and J. E. Katon, *Appl. Spectrosc.* **28,** 420, 1974.
131. P. T. McKittrick and J. A. Katon, *Appl. Spectrosc.* **44,** 812, 1990.
132. R. A. Nyquist, *Appl. Spectrosc.* **40,** 336, 1986.
133. R. A. Nyquist, *Appl. Spectrosc.* **44,** 433, 1990.
134. R. A. Nyquist, *Appl. Spectrosc.* **36,** 533, 1982.
135. R. A. Nyquist, *Appl. Spectrosc.* **26,** 81, 1972.
136. R. A. Nyquist and W. J. Potts,. *Spectrochim Acta* **15,** 514, 1959.
137. R. A. Nyquist, R. W. Chrisman, C. L. Putzig, R. W. Woodward, and B. R. Loy, *Spectrochim Acta* **35A,** 91, 1979.
138. R. A. Nyquist, *Spectrochim Acta* **19,** 509, 1963.
139. W. M. Coleman and B. M. Gordon, *Appl. Spectrosc.* **41,** 886 and 1159, 1987.
140. G. Song, X. Che, and D. Ding, *J. Mol. Struct.* **175,** 359, 1988.
141. J. Dybal and S. Krimm, *J. Mol. Struct.* **189,** 383, 1988.
142. Y. T. Chang, Y. Yamaguchi, W. H. Miller, and H. F. Schaefer, *J. Am. Chem. Soc.* **109,** 7245, 1987.
143. P. A. Lang and J. E. Katon, *J. Mol. Struct.* **172,** 113, 1988.
144. Y. V. Venkatachalapathi, D. F. Mierke, J. P. Taulane, and M. Goodman, *Biopolymers* **26,** 763, 1987.

CHAPTER 10

Compounds Containing —NH_2, —NHR, and —NR_2 Groups

10-1. INTRODUCTION

Compounds containing the —NH_2, —NHR, and —NR_2 moieties include aliphatic amines, aromatic amines, heterocyclic aromatic compounds, amides, thioamides, ammonium, salts, hydroxylamines, hydrazines, etc. For compounds containing NH_2, the characteristic group frequencies arise from antisymmetric NH_2 stretching, symmetric NH_2 stretching, NH_2 scissoring, NH_2 wagging, C—N stretching, and CCN bending modes. Likewise, the —NHR moiety is characterized by N—H stretching, N—H deformations, and C—N—C stretchings and bending vibrations. The —NR_2 functional group is characterized only by skeletal vibrations involving nitrogen atoms, and by modified CH_3—(N) and CH_2—(N) group vibrations.

The characteristic IR and Raman frequencies of —NH_2, —NHR, and NR_2 are summarized in Tables 10-1 to 10-3. The NH_2 and NH stretching bands appear in the same spectral region as OH stretching bands. Generally, hydrogen-bonded OH bands in the IR are much stronger than those of amino groups, but the reverse tends to be true in the Raman spectra, where OH bands are very weak.

Another similarity among the —OH, NH_2, and NH groups is their sensitivity to the presence of hydrogen bonding. The influence of hydrogen bonding is less pronounced on amino groups than on the OH group, however. Generally, H-bonding leads to a reduction in NH_2 (or N—H)

Table 10-1.
Characteristic Frequencies (cm^{-1}) of the $-NH_2$ Group

	cm^{-1}	IR	R	cm^{-1}
Antisymmetric NH_2 stretching:				
$R-NH_2$	3398–3381 (dil. soln.)			
PP′	3380–3361 (liq.)	mw	w	3422–3404 (vap.)
PS′	3370–3363 (liq.)	mw	w	3422–3400 (vap.)
PT′	near 3350 (liq.)	mw	w	near 3400 (vap.)
$Ar-NH_2$	3509–3460 (dil. soln.)			
	3500–3420 (liq.)	ms	vw	3548–3459 (vap.)
$C-C(=O)-NH_2$	near 3528 (dil. soln.)			
	3475–3350 (Nujol)			3565–3550 (vap.)
Symmetric NH_2 stretching:				
$R-NH_2$	3344–3324 (dil. soln)			
PP′	3310–3289 (liq.)	mw	mw, p	3361–3340 (vap.)
PS′	3285–3280 (liq.)	mw	mw, p	3340–3289 (vap.)
PT′	near 3280 (liq.)	mw	mw, p	near 3335 (vap.)
$Ar-NH_2$	3416–3382 (dil. soln.)			
	3420—3340 (liq.)	ms		3442–3360 (vap.)
$C-C(=O)-NH_2$	near 3413 (dil. soln.)			
	3385–3180 (Nujol)	s		3348–3430 (vap.)
NH_2 scissors:				
$R-NH_2$	1627–1590	m	vw	
$Ar-NH_2$	1638–1602	s	vw	
$C-C(=O)-NH_2$ (amide II band)				
	1620–1590 (soln.)	s	vw	
	1650–1620 (Nujol)	s	vw	1600–1576 (vap.)
NH_2 wag:				
$R-NH_2$				
PP′	854–778 (liq.)	s, br	vw	780–760 (vap.)
PS′	791–784 (liq.)	s, br	vw	799–779 (vap.)
PT′	897–848 (liq.)	s, br	vw	813–800 (vap.)
$Ar-NH_2$	700–600 (liq.)	m, br	vw	
$C-C(=O)-NH_2$	720–600 (Nujol)	m, br	vw	
C—N stretching and other characteristic skeletal stretching modes:				
$R-NH_2$				
PP′	1090–1040 (liq.)	s-m	s-m	
PS′	1040–1000 (liq.)	m-w	m, dp	
	1090–1080 (liq.)	m-w		
	1143–1130 (liq.)	s-m	m, dp	
	1163–1153 (liq.)	s-m	m, dp	
	1180–1170 (liq.)	s-m	m, dp	
PT′	1030–1000 (liq.)	m-w	w	
	1080–1060 (liq.)	m-w	w	
	1140–1110 (liq.)	m-w	w	
	1218–1195 (liq.)	ms	m-w	
	1245–1235 (liq.)	ms	m-w	
$Ar-NH_2$	1330–1260 (liq.)	s		1328–1228 (vap.) var.
$C-C(=O)-NH_2$ (amide III band)				
	1430–1390			
Skeletal deformation:				
$R-NH_2$ (CCN def.)	near 502			
	near 436			
$Ar-NH_2$ (CN bend.)	near 350 (i.p.)			
	near 200 (o.p.)			
$R-C(=O)NH_2$ (O=C—N def.)	600–500			

Table 10-2.

Characteristic Frequencies (cm^{-1}) of —NHR Groups

	cm^{-1}	IR	R	cm^{-1}	IR
N—H stretching:					
R—NHR	3320–3280 (liq.)	w	w, p	3360–3310 (dil. soln.)	w
Ar—NHR	near 3400 (liq.)	m	w, p	3458–3330 (vap.)	
				near 3450 (dil. soln.)	m
—C(=O)—NHR					
trans	3320–3270 (liq.)	m	w, p	3482–3460 (vap.)	
				3480–3440 (dil. soln.)	m
cis	near 3200 (liq.)	m		3180–3140 (dil. soln.)	m
overtone/or combination band				3100–3070	
C—N—H bend:					
Ar—NHR	c.a. 1510	ms	vw		
—C(=O)—NHR					
trans					
amide II	1570–1515 (solid)	s	w	1550–1489 (vap.)	
				1550–1510 (dil. soln.)	s
amide III	1300–1248	m	vs-s	1250–1230 (vap.)	
cis	1490–1440	m		1459–1450 (vap.)	
C—N stretching:					
R—NHR					
antisym. CNC str.:	1180–1130	ms	m		
sym. CNC str.:	900–850	var.	vs-s, p		
Ar—NHR	1350–1280	var.	var.		
—C(=O)—NHR					
trans	(see amide II and amide III bands under N—H bend)				
cis	1350–1310	ms			
C—N deformation:					
R—NHR	415–377				
$RCH_2C(=O)NHR$	near 630				

Table 10-3.

Characteristic Frequencies (cm^{-1}) of $—NR_2$ Groups

	cm^{-1}	IR	R
C—N stretching:			
$R—NR_2$	1250–1000		
	833–740		s, p
$HC(=O)NR_2$	870–820		s, p
$RC(=O)NR_2$	750–700		s, p
Skeletal deformation:			
$R—NR_2$	CNC def.	375–340	
$H—C(=O)NR_2$	O=C—N def.	near 650	
$R—C(=O)NR_2$	O=C—N def.	620–590	

stretching frequencies, an increase in NH_2 bending frequencies, a broadening of the spectral band, and an increase in IR intensity. The spectral frequency and pattern variations resulting from hydrogen bonding depend on the sample phase involved.

In general, NH_2 and N—H vibrational bands are more crucial in elucidating the type of NH_2 (or NHR) present, whereas the skeletal vibrational bands in Raman spectra are very useful for conformational studies, which are very important in determining the secondary structure of biological molecules.

The presence of rotational isomers due to rotation about the C—N and C—C bonds usually complicates the vibrational spectra, especially in the skeletal stretching and deformation regions. Recent vibrational studies on aliphatic amines have been carried out with emphasis on the conformational structures [1–15]. Infrared absorption bands due to NH_2 stretchings and scissoring are usually observed to be broad in condensed phase and, therefore, are not very useful in determining the rotational conformation. However, the N—H stretching frequencies of secondary arylamines [16] and secondary amides are correlated with the rotational structures [17, 18]. The NH frequencies of these compounds are discussed in Section 10-5.3. The skeletal stretchings and bending frequencies are more susceptible to the immediate environment. Therefore, a given compound with different rotational isomers usually has several skeletal vibrational frequencies.

It is necessary to define the classification of aliphatic amines before discussing the correlation of characteristic frequencies with the molecular structures. Unlike the previous discussions of alcohols and halides, where the terms "primary," "secondary," and "tertiary" are used to express the degree of branching on the α-carbon atom, these terms are now used to describe the number of hydrogen atoms attached to the nitrogen atom. To clarify this, let us describe a primary aliphatic amine ($R-NH_2$) as having three possible structures: RCH_2-NH_2, R_2CH-NH_2, or R_3C-NH_2. A secondary aliphatic amine can have the structures $RCH_2-NH-CH_2R$ or $R_2CH-NH-CH_2R$, etc. For simplicity, the letters P, S, and T are used to represent primary amine ($-NH_2$), secondary amine ($-NHR$), and tertiary amine (NR_2), respectively. Furthermore, the letters P′, S′, and T′ are used to describe the degree of branching on the α-carbon atom; the RCH_2-, R_2CH-, and R_3C- types of structure are represented by P′, S′, and T′, respectively. For example, the PP′ notation stands for the structure RCH_2-NH_2, while the PS′ symbol denotes the structure R_2CH-NH_2.

Vibrational analyses of open-chain amines [3–15, 30, 31], cyclic amines [11, 12, 32, 33], and allyl amine [1, 2, 14] have been reported. Raman and

IR spectra of aniline [34, 54] and its derivatives [35–53] have been studied extensively. Jakes and Krimm [92, 93] reviewed the extensive IR data on molecules with the amide group and reported the valence force field as well as normal coordinate analysis for the amide group. This study is very useful in understanding the form of the vibration. The vibrational spectra of thioamides have been studied [94—96, 112, 113].

Secondary amides such as the —C(=O)NH—R group exhibit nearly planar structure within the —C(=O)NH—C— moiety as a result of resonance stabilization [18]. The rotation about the C—N bond results in two isomers, namely those where the C=O and N—H bonds are cis and trans to each other [19]:

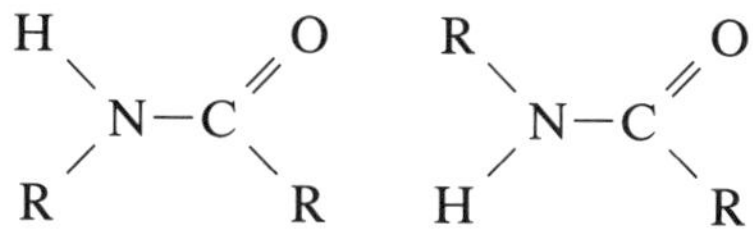

cis trans

It is possible to differentiate the two conformers by N—H stretching, N—H bending, and C—N stretching bands. Generally, the vibrational spectra of the cyclic amides with a ring size of more than eight atoms resemble those of the open-chain amides, while the spectral frequencies of cyclic amides with a ring size of eight atoms depend on the measuring condition [20b].

Since the cis and trans secondary amides exhibit different characteristic frequencies, it is necessary to discuss the situations in which these two isomers occur. For most open-chain secondary amides, the trans species are predominant [17, 19, 116]. For instance, liquid-phase *N*-methyl acetamide (CH_3—C(=O)NH—CH_3) contains less than 5% cis isomers at room temperature.

The relative abundance of the cis isomers increases to about 10% for *N*-methyl formamide (H—C(=O)NH—CH_3), in comparison to 0–5% for the previously mentioned *N*-methyl acetamide [20, 116]. This is because of the reduced steric hindrance when the CH_3— moiety on the carbonyl group of *N*-methyl acetamide is replaced by a less bulky hydrogen atom to form *N*-methyl formamide.

The cyclic amides with ring size less than eight atoms exist only as the cis isomer because of the ring structure. A few good examples of cyclic cis amides can be found in references 67, 97, and 103 through 107. As the ring size increases to more than eight atoms, the ring puckering motion allows the trans species to become the predominant form. The behavior of cyclic amides with ring size equal to eight is an intermediate situation.

10-2. NH_2 STRETCHING

The spectral positions of NH_2 antisymmetric and symmetric stretchings of aliphatic primary amines ($R-NH_2$), aryl primary amines ($Ar-NH_2$), and amides ($R-C(=O)-NH_2$) are given in Table 10-1. As a general trend, the vibrational bands are shifted by a H-bond interaction toward a lower frequency when the sample state changes from vapor phase, to dilute solution, to pure condensed phase. The attachment of $-NH_2$ to an alkyl (R−), aryl (Ar−), or carbonyl ($-C=O$) group can be differentiated by the spectral positions of NH_2 stretching modes. In the IR, the NH_2 and NH stretching bands increase in intensity as the base strengths decrease in the order amines, anilines, and amides.

10-2.1. Primary Aliphatic Amines ($R-NH_2$)

The NH_2 antisymmetric stretch mode occurs at 3422–3400 cm^{-1} [16a], 3398–3381 cm^{-1} [20a], and 3380–3350 cm^{-1} [21a] when measured by the IR method in vapor phase, dilute solution, and liquid phase, respectively, while symmetric bands in 3361–3289 cm^{-1} [16a], 3344–3324 cm^{-1} [20a], and 3310–3280 cm^{-1} [21a] are observed for vapor phase, dilute solution, and liquid phase samples. Branching on the α-carbon atom lowers the NH_2 stretching frequencies. The vibrational frequencies of antisymmetric and symmetric NH_2 stretching modes are further correlated with the types of amine (i.e., PP′, PS′, or PT′) (see Table 10-1), while the spectral pattern is relatively unperturbed [16a, 21a].

The characteristic IR pattern of liquid primary amines in this region consists of two medium bands of approximately equal intensity due to antisymmetric and symmetric NH_2 stretchings. It should be noted that the attachment of an −OR group to an α- or β-carbon atom will enhance the antisymmetric absorption [21e]. In addition, it is common to observe a shoulder on the low frequency side (3200–3160 cm^{-1}) of the symmetric NH_2 stretching band. This shoulder is attributed to the first overtone of NH_2 scissoring enhanced by Fermi resonance interaction with the NH_2 symmetric stretching [22]. It should be noted that the PS′ and PT′ types of amines may exhibit more H-bonding effect, for some of their NH_2 bands appear to be broader and stronger [21f]. The Raman band of the antisymmetric NH_2 stretching mode is depolarized and very weak, while the corresponding symmetric mode can be readily identified by a medium-weak polarized band in the expected spectral region [23, 24]. The overtone of NH_2 scissors shows very weak or no Raman scattering.

A useful empirical relationship (Eq. 10-1) between antisymmetric and symmetric NH_2 stretching wavenumbers, $\bar{\nu}$, of the primary amines was given by Bellamy and Williams [25] for primary amines containing two equivalent N—H bonds (i.e., the NH_2 group is free of H-bonding or the two N—H bonds take part in intermolecular hydrogen bonding in an identical fashion):

$$\bar{\nu}_{sym} = 345.53 + 0.876\,\bar{\nu}_{antisym}. \qquad (10\text{-}1)$$

Another empirical relationship (Eq. 2) was reported by Stewart [26]:

$$\bar{\nu}_{sym} = 0.98\,\bar{\nu}_{antisym}. \qquad (10\text{-}2)$$

If the relationships between the NH_2 stretch wavenumbers are not properly characterized by the above equations, the two NH bonds may not be equivalently hydrogen bonded.

10-2.2. Aryl Amines (Ar—NH_2)

The antisymmetric NH_2 stretching bands in different states are in the regions of 3548–3459 cm^{-1} (vap.) [16b], 3509–3460 cm^{-1} (dil. soln.) [20a], and 3500–3420 cm^{-1} (liq.) [21c, 70], while the corresponding symmetric vibrational band is situated at 3442–3360 cm^{-1} (vap.) [16b], 3416–3382 cm^{-1} (dil. soln.) [20a], and 3420–3340 cm^{-1} (liq.) [21c, 70]. Krueger [27] suggested an empirical equation,

$$\bar{\nu}_{sym} = 1023 + 0.682\,\bar{\nu}_{antisym}, \qquad (10\text{-}3)$$

to relate the antisymmetric and symmetric stretching wavenumbers for anilines. This equation is valid only for anilines containing equivalent N—H bonds.

In comparison with those of aliphatic amines, the NH_2 stretching frequencies of aromatic amines are increased. See Table 10-1. The IR spectra of liquid aromatic amines in this region have an NH_2 symmetric stretching band that is usually more intense than the antisymmetric stretch band. The overtone band near 3200 cm^{-1} in these cases appears to be better resolved in the IR spectra of aniline and its derivatives in the liquid state. A change in the spectral pattern is observed when the aromatic amines are measured in the solid state as mulls [21c]. The strong stretching bands, particularly the antisymmetric mode, are quite sharp and the relative intensities of the antisymmetric and symmetric modes are reversed as compared with the IR spectra measured in the liquid phase.

Data on naphthylamines are available where Ahuja and Singh [55] assigned the IR bands of 2-naphthylamine at 3390 cm^{-1} and 3310 cm^{-1} to antisymmetric and symmetric NH_2 stretching. Zabokrycka *et al.* [56] reported the NH_2 stretchings of solid 1-phenylazo-2-napthylamine to be located at 3434 cm^{-1} (s) and 3220 cm^{-1} (vw). The IR spectra (Nujol mull) of seven substituted naphthylamines reported in the *Aldrich Library of FT-IR Spectra* exhibits NH_2 stretching bands in the 3421–3370 cm^{-1} and 3324–3307 cm^{-1} regions [21g]. The spectral patterns of 1-amino-4-chloronaphthalene and 1-amino-4-bromonaphthalene are similar to those of arylamines measured as Nujol mull.

10-2.3. Primary Amides $R-(C=O)-NH_2$ and Thioamides $R-(C=S)-NH_2$

The free NH_2 antisymmetric and symmetric bands for amides in dilute solution were reported to be near 3528 cm^{-1} and 3413 cm^{-1}, respectively, in the infrared spectra [20b]. However, more recent studies by Nyquist suggested that the free NH_2 stretchings fall in the regions of 3565–3550 cm^{-1} (antisymmetric mode) and 3448–3430 cm^{-1} (symmetric mode) in the vapor phase [16d]. Wider spectral ranges are observed for solid-phase alkyl amides ($R-(C=O)NH_2$) and aryl amides ($Ar-(C=O)-NH_2$) measured as Nujol mull. The antisymmetric and symmetric stretchings of these two amides are observed in the 3475–3350 cm^{-1} and 3385–3180 cm^{-1} regions, respectively [21b, 21d].

The strong hydrogen-bond interaction ($NH\cdots O=C$) in the concentrated amide solutions and solid-phase amides, due to the presence of a carbonyl group, has a profound effect on the NH_2 vibrations. We found that the broad IR pattern within the 3500–3100 cm^{-1} region (see Table 10-1) is quite consistent for the 44 primary alkyl and aryl amides [21b, 21d]. In moderately concentrated solution, the coexistence of free NH_2 and various types of associated NH_2 species may give rise to more complicated absorption patterns.

In the case of thioamide, the NH_2 stretchings occur at slightly lower frequencies (3503 cm^{-1} and 3384 cm^{-1}) [28, 29]. The frequency dependence of thioamides on hydrogen bond formation is very similar to that of the amides [94].

10-3. NH_2 SCISSORS DEFORMATION

This vibrational mode appears in the 1638–1575 cm^{-1} region with strong to very strong IR intensity, but exhibits weak or no Raman scattering [21, 23]. In the case of aliphatic primary amines, this very strong IR band is normally found in the 1627–1590 cm^{-1} region with "band width at half maximum'' of approximately 60–100 cm^{-1} [21a]. Aryl amines show that NH_2 scissoring band in the region of 1638–1602 cm^{-1} [21c]. The possible interferences in this region are from the aromatic ring and C=C stretching modes.

As a consequence of hydrogen bonding, the scissoring frequency increases as a function of increasing concentrations. This is best illustrated by the NH_2 scissoring vibration of amides, which are also called amide II bands, in vapor phase (1600–1576 cm^{-1}), dilute solution (1620–1590 cm^{-1}), and solid state (1650–1620 cm^{-1}). The frequency shift of the NH_2 scissoring mode upon the change of sample phase (or concentration) has been applied to assign the C=O stretching and the NH_2 scissoring modes of amides. It has been found that the amide I band (mainly the C=O stretching) and the amide II band (mainly the NH_2 scissoring) of solid amides sometimes overlap, since the former (near 1650 cm^{-1}) is only slightly higher in frequency than the latter. However, if the sample is dissolved in nonpolar solvent for infrared study, the opposite effects of hydrogen bonding on the C=O stretchings and NH_2 scissors provide possibilities of resolving such overlapping. The weakening of hydrogen bonds upon dilution results in a higher C=O stretching frequency and a lower NH_2 scissoring frequency. The spectral shifts of the C=O stretching and NH_2 scissoring modes, upon dissolution, are often sufficient to allow resolution of the two otherwise nearly coincident bands.

10-4. NH_2 WAGGING MODE

With its strong and very broad features in the IR spectra, the NH_2 wagging mode in the 875–750 cm^{-1} region is very characteristic for liquid aliphatic primary amines. In this mode, the angle changes between the NH_2 plane and the N—C bond. The approximate "band width at half maximum intensity" is $\Delta\nu$ = 140–240 cm^{-1} for liquid samples [21a]. Primary arylamines and amides exhibit this IR band in a lower range (720–600 cm^{-1}) with medium intensity [21b, 21c, 21d]. The Raman intensity of this broad band is weak and usually overlapped with the CH_2 rocking bands, or skeletal vibrations. Nyquist [16a] reported the correlation between the vapor phase band position and the type of amine (i.e., PP′, PS′, or PT′) (see Table 10-1). Although our investigation of the liquid-phase IR spectra of amines suggests a trend similar to Nyquist's finding (see Table 10-1), one must be careful in exercising such correlations because multiple bands are quite common for this IR absorption [21].

10-5. N—H STRETCHING

Secondary amines and arylamines exhibit only a single characteristic IR and Raman band in the N—H stretching region (3500–3100 cm^{-1}). The secondary amides, however, present additional characteristic IR bands in this region, such as the 1550 cm^{-1} overtone band near 3100 cm^{-1}.

10-5.1. Dialkylamines (R—NHR) and Secondary Arylamines (Ar—NHR)

With the exception of piperidine and its derivatives (3361–3330 cm^{-1}, IR: weak), IR absorption from NH stretch in dialkylamines is too weak to be observed in the vapor-phase IR spectra [16c]. In dilute dialkylamine solutions, the weak NH stretch IR band is broadened a little and shifted to about 3320–3280 cm^{-1} by hydrogen bonding for liquid or solid samples [21a]. The polarized Raman band is weak for this vibrational mode [23]. Amines often contain a little water, which appears in the IR as a broad band superimposed on the sharper weak NH stretch band [21h]. In addition, weak combination bands may be confused with the weak NH stretch band. A very weak shoulder at 3200–3180 cm^{-1} appears in the condensed phase IR spectra of many secondary aliphatic amines. This is most likely a combination band. Its relative intensity seems to increase when the α- or β-carbon atom is branched. Two absorptions with approximately equal intensities in the regions 3320–3280 cm^{-1} (NH stretch) and 3200–3180 cm^{-1} (combination band) are observed for diisopropylamine and diisobutylamine [21i].

The N—H IR band is enhanced and shifted to a higher frequency for secondary arylamines (Ar—NH—R) [21c], while the polarized Raman band exhibits an intensity comparable to that of dialkylamines [23]. The spectral regions in vapor phase, dilute solution, and liquid phase are 3458–3338 cm^{-1} [16c], near 3450 cm^{-1} [20a], and near 3400 cm^{-1} [21c], respectively. Rotational isomers and intramolecular hydrogen bonding (if allowed by molecular structure) can yield more than one N—H band in the observed vapor phase or in the dilute solution. The substitution effect on the N—H frequency has been studied by Laurence and coworkers [110, 111].

10-5.2. Unsaturated Heterocyclic Secondary Amines

Pyrroles, indoles, and carbozoles exhibit N—H stretching frequencies with higher values (3530–3500 cm^{-1}) in the vapor state [16]. In Nujol mulls, these compounds absorb in the 3450–3380 cm^{-1} range. Further frequency lowering is expected upon substitution of (C=O)-containing groups on the pyrrole or indole ring. In addition to the usual NH···N type of hydrogen bonding, NH···π type hydrogen bonding was reported by Suwaiyan and Zwarich [59], Sanyal *et al.* [58], and Bernard-Houplain and Sandorfy [38] for indoles, and by Lautie and Novak [57] for crystalline pyrrole.

When measured as Nujol mull, the N—H stretching band of many secondary heteroaromatic amines, which contain more than one nitrogen atom in the aromatic ring, appears to be complicated in the region of 3200–2700 cm^{-1}. The compounds exhibiting this type of N—H···N= stretching band include pyrazoles, triazoles, imidazoles, benzimidazole, and benzotriazoles. Similar observations have been found in the IR spectra of indazole and its derivatives, for which unusually low N—H stretching frequencies (near 2820 cm^{-1}) are reported; these observations are attributed to a very strong hydrogen-bond interaction between the N—H group and the unsaturated N atom of the heterocyclic ring [63, 64]. The presence of the strong hydrogen bond between the N—H moiety and the unsaturated nitrogen atom is supported by x-ray diffraction for 5-amino-4H-imidazole derivatives [62]. Other than the strong hydrogen-bond interaction mentioned above, the IR absorption of these heterocyclic molecules may be further complicated by the presence of other types of hydrogen bonds, such as those between the N—H moieties and an aromatic ring or a saturated N atom. Furthermore, the amino-imino tautomerism may cause spectral complications in the N—H stretching region, as in the cases of 5-amino-4H-imidazole derivatives [60–62].

10-5.3. Secondary Amides and Thioamides

As discussed in the introduction section, most of the open-chain amides and cyclic amides with a ring size larger than eight atoms exist as trans isomers, while the cyclic amides with a ring size of less than eight atoms exist as cis isomers. In dilute solution, the NH stretch band of the trans species occurs at a higher frequency, at 3480–3440 cm^{-1} (free NH) [20, 116], than the corresponding cis isomer, at 3180–3140 cm^{-1} (bonded NH) [20]. The cis isomer forms a hydrogen-bonded dimer structure joined by two NH···O bonds. This cis form remains mostly H-bonded at a concentration of 0.003 mole/liter in CCl_4, where the trans form is mostly unassociated [114]. As with other NH vibrations, the band is broadened in the condensed phase, and the differentiation between the trans isomer near 3300 cm^{-1} and the cis isomer near 3200 cm^{-1} in this region of the vibrational spectra becomes somewhat more difficult. The Raman scattering of N—H stretching is weak and polarized.

A weak IR band in the region of 3100–3070 cm^{-1} for solid samples is characteristic for secondary amides with no heteroatomic substituent on either the N atom or the carbonyl group. This IR band vanishes upon dilution for the trans isomer and is attributed to the Fermi resonance-enhanced overtone of the 1550 cm^{-1} CNH (or amide II). In the case of the cis isomer, the IR band in solids persists on dilution. This 3100 cm^{-1} cis band is assigned to a combination [115] involving the amide C=O stretch near 1650 cm^{-1} and the cis CNH bend near 1450 cm^{-1} discussed in Section 10-6.2.

10-6. C—N—H BENDING

10-6.1. Amines

The C—N—H bending vibration in C_2NH groups changes the CNH angle. Secondary aromatic amines sometimes show an extra IR band about 1510 cm^{-1} near to the 1500 cm^{-1} aromatic ring band, involving CNH bending [122]. In secondary aliphatic amines, there are no definite bands for CNH bending above the 1470 cm^{-1} CH_2 and CH_3 bands [70].

10-6.2. Amides

For trans amides, there is extensive mixing between the C—N—H in-plane deformation and the C—N stretching modes of trans amides. Such vibrational coupling results in one band in the 1550–1510 cm^{-1} region (IR : s, R : w) [65, 66] (or amide II), where the C—N stretches while the CNH angle bends, and the other band in the 1300–1248 cm^{-1} region (IR : m, R : vs-s) (or amide III) where the C—N stretches while the CNH angle opens. Both bands are sensitive to deuteration on the amino nitrogen atoms and the presence of hydrogen bonding. The mixing of these two vibrational modes is essentially eliminated for the cis conformer [67]. It is thus adequate to describe these bands simply in the regions of 1490–1440 cm^{-1} and 1350–1310 cm^{-1} as mainly CNH in-plane bending and CN stretching, respectively [67]. The position of the 1350–1310 (C—N stretch) band is nearly unaffected when deuterated, and the frequencies are only slightly affected by hydrogen bonding. The different behaviors of these two bands between the trans and cis isomers and the N—H stretching frequency provide a good means of identifying the rotational configuration.

10-7. N—H WAGGING

Aliphatic secondary amines exhibit a strong broad IR band in the 785–720 cm^{-1} region that is attributed to the NH wagging vibration, where the NH bond bends relative to the CNC plane. The Raman intensity of this vibrational mode is very weak.

In the case of the secondary amides, this out-of-plane NH wagging band (also called the amide V band) is located in the 750–700 cm^{-1} region for associated open-chain amides [20, 21, 24], but shifts to a lower value when diluted by a solvent. For *N*-methylacetamide, the free N—H out-of-plane wag band was reported to be at 648 cm^{-1}. It was observed that the IR band eventually vanishes at very low concentrations of secondary amides or thioamides [20b]. The NH wag bands of cyclic cis amides were observed at a higher frequency, the 890–780 cm^{-1} region [21, 97]. Some empirical observations have shown that the position of the NH wag band is correlated with the secondary structure of polypeptides. The frequency of α-helical peptides is at about 618 cm^{-1}, while that of β-structural peptides is at about 710 cm^{-1} [98–102].

10-8. SKELETAL VIBRATIONS

10-8.1. C—N Stretching

For amines, the stretching modes involving the C—N skeleton usually give rise to one or more characteristic IR and Raman bands in the regions of 1250–1020 cm^{-1} and 900–720 cm^{-1}. These vibrational bands can be readily identified for small molecules when the C—O moiety does not spectrally interfere in these regions. In the case of the larger molecules, especially those containing branched alkyl groups, the assignment of the C—N stretching bands becomes difficult in the Raman spectra because of the presence of the stretching bands aroused from the carbon skeleton. The C—N vibrations are very sensitive to their environmental structures, and more than one band may result in the presence of other rotational isomers.

10-8.2. Amines

PRIMARY AMINES

The absorption frequency and the spectral patterns are correlated with the type of amine (primary, secondary, or tertiary) and the degree of branching on the alpha carbon atom. The C—N stretch of PP′ structure (CH_2-NH_2) occurs in the 1090–1040 cm^{-1} region with strong to medium IR and Raman intensities.

For the PS′ amines ($CH-NH_2$) in liquid phase, there is more than one characteristic C—N absorption [21]. These bands are located in the regions of 1180–1170 cm^{-1} (IR: s-m; R: m, dp), 1163–1153 cm^{-1} (IR: s-m; R: m, dp), 1143–1130 cm^{-1} (IR: s-m; R: m, dp), and 1040–1000 cm^{-1} (IR: m-w, R: m, dp; most near 1040 cm^{-1}) [21a, 23]. Another band in the 1090–1080 cm^{-1} region (IR: m-w) is also observed for some PS′ amines. It is quite likely that C—N stretching is involved in these vibrational bands to a different extent. The 1170 cm^{-1} band was assigned by Durig *et al.* [6] to a stretching mode involving the C—C bond for $(CH_3)_2CH-NH_2$. Since the IR spectra of tertiary carbon atoms exhibit characteristic absorptions in the regions of 1175–1165 cm^{-1} and 1170–1140 cm^{-1} due to the C_2CH-C stretchings [68], it seems appropriate to assign the bands in the 1180–1130 cm^{-1} region to a similar vibration of the C_2CH-N moiety where the IR intensity is enhanced by the polar NH_2 group. Bellamy assigned the band near 1040 cm^{-1} to the C—N stretching mode. This assignment has been confirmed by several investigators. A very characteristic polarized Raman band in 860–750 cm^{-1} (R: s, p) can be described as the symmetric C_2CH-N stretching mode.

The situation in PT′ type amines is similar to that in the PS′ amine structure. There appears a series of characteristic bands in the ranges of 1245–1235 cm^{-1} (IR: ms; R: m-w), 1218–1195 cm^{-1} (IR: ms; R: m-w), 1140–1110 cm^{-1} (IR: m-w, R: W), 1080–1060 cm^{-1} (IR: m-w; R: w), and 1030–1000 cm^{-1} (IR: m-w; R: w) [21a, 23]. Since compounds containing quaternary carbon atoms exhibit weak IR bands in the 1255–1245 cm^{-1} region and the 1250–1200 cm^{-1} region as a result of the CC_4 skeletal stretchings [68], the 1245–1235 cm^{-1} and 1218–1195 cm^{-1} bands of PT′ amines are most likely due to similar vibrations involving the C_3C-N moiety. The IR band near 1030 cm^{-1} was attributed to C—N stretching. The strongest Raman band observed in the spectrum of *t*-butylamine, $(CH_3)_3CNH_2$, is polarized and found at 752 cm^{-1} [23]. This band can be attributed to the symmetric C_3C-N stretching. Other PT′ amines exhibit this Raman band in between 760 and 720 cm^{-1} [23]. It is important to note that this is in the same spectral region as observed for the symmetric stretching modes of quaternary carbon atoms, CC_4.

SECONDARY AMINES

The antisymmetric stretch of the C—N—C moiety was correlated with a strong-to-medium IR band in the 1146–1132 cm^{-1} and 1191–1171 cm^{-1} regions for SP′P′ and SP′S′ amines, respectively [20a], while a similar Raman intensity in these regions was medium-to-weak. However, caution must be exercised in the above correlation because exceptions exist. For example, *N*-butyl-α-methylbenzylamine,

$$(C_6H_5)(CH_3)CH{-}NH{-}(CH_2)_3CH_3,$$

and *N*-isoproprópylbenzylamine,

$$(C_6H_5)CH_2{-}N{-}CH(CH_3)_2,$$

are both SP′S′ amines, but they exhibit IR bands near 1130 cm^{-1} and 1160 cm^{-1}, respectively [23a]. The IR bands of diisopropylamine

$$[(CH_3)_2CH{-}NH{-}CH(CH_3)_2]$$

and di-*sec*-butylamine

$$[(C_2H_5)(CH_3)CH{-}NH{-}CH(CH_3)(C_2H_5)],$$

at about 1180 cm^{-1} and 1165 cm^{-1}, are examples of the SS′S′ type of C—N absorption [21i]. The symmetric CNC stretching mode of secondary amines is characterized by a polarized Raman band in the 900–850 cm^{-1} region with medium intensity [24a].

TERTIARY AMINES

Although no characteristic skeletal vibration was reported, the IR spectra of tertiary amines with alkyl carbon number less than 5 exhibit rich skeletal information in the 1250–1000 cm^{-1} region due to the antisymmetric NC_3 stretchings [21a]. It is common to find more than one strong-to-medium IR band in this region. The characteristic Raman band falls in the 1070–1050 cm^{-1} region for antisymmetric NC_3 stretching and in the 833–740 cm^{-1} region for the symmetric stretching mode [24a].

10-8.3. Arylamines

The C—N stretch of arylamines couples strongly with the ring vibrations and occurs at a higher frequency than that of aliphatic amines. This vibration occurs in the region of 1330–1250 cm^{-1} for primary arylamines ($Ar{-}NH_2$), 1350–1250 cm^{-1} (usually two bands) for secondary arylamines (Ar—NH—R), and 1380–1265 cm^{-1} for tertiary arylamines ($Ar{-}NR_2$) [20, 70]. The Raman intensities vary. The IR intensities are usually medium-strong in aniline free bases, but are quite weak in aniline salts [70].

10-8.4. Amides

The C—N stretch (amide III band) for primary amides is located between 1430 and 1390 cm^{-1} (IR : s-m; R : m-w) [20b, 21b, 23, 24b].

The antisymmetric C—N stretching mode of secondary amides couples strongly with the N—H deformation and has been discussed in Section 6.2 of this chapter. In the Raman spectra of aliphatic secondary amides, the strong polarized band in the region of 900–800 cm^{-1} can be attributed to the symmetric CNC stretching mode [23b]. For example, the strongest Raman band of *N*-methyl acetamide is located at 884 cm^{-1} [23b].

Tertiary amides are best characterized by the symmetric CNC stretching band in Raman spectra. The vibration of $H-C(=O)-NR_2$ occurs at the 870–820 cm^{-1} region and is shifted to a lower value (750–700 cm^{-1}) when the carbonyl hydrogen is replaced by a heavier alkyl group (i.e., $R-C(=O)-NR_2$) [24b]. This ν_{sym}(CNC) Raman band is very pronounced for small tertiary amides; however, its existence is often obscured by the skeletal vibrations, especially for larger molecules. Unfortunately, the ν_{sym}(CNC) IR absorption is weak in the majority of cases.

10-9.

SKELETAL DEFORMATIONS

The skeletal deformations are sensitive to conformational configurations. Here Teixeira-Dias *et al.* [9] assigned the vibrational bands at 502 cm^{-1} and 436 cm^{-1} to CCN deformations for *n*-butylamine, while Durig *et al.* [6] assigned the bands in the region of 472–545 cm^{-1} and at 369 cm^{-1} to the same vibrations for $(CH_2)_2CHNH_2$. Vibrational studies on cyclopropylamine, cyclobutylamine, and cyclopentylamine suggested that the C—N in-plane bending mode is around 460–400 cm^{-1} and the out-of-plane bending mode is near 402–368 cm^{-1} [11–13]. The CNC bending of secondary and tertiary amines was correlated by Dollish *et al.* [24] with the Raman bands in the regions of 415–397 cm^{-1} and 375–340 cm^{-1}, respectively.

The bending modes of arylamines occur at lower frequencies. The frequencies at around 350 cm^{-1} and 200 cm^{-1} have been assigned to the in-plane and out-of-plane $C-NH_2$ deformations, respectively, for anilines [34, 36, 41, 45, 51–53].

In the case of amides, the characteristic bending is N—C=O in-plane deformation (amide IV band). The N—C=O bending mode is generally located at a higher frequency than the CCN deformation of amines, and a spectral range of 660–500 cm^{-1} was reported for primary amides. For secondary amides, the N—C=O bending frequencies are found to shift to near 630 cm^{-1} and near 780 cm^{-1}, respectively, for $R-CH_2-C(=O)NHR$ and HC(=O)NHR [24]. The tertiary amides containing the $H-C(=O)NR_2$ type of structure exhibit a N—C=O deformation band near 650 cm^{-1}, but this frequency is lowered to the 620–590 cm^{-1} region for tertiary amides with the $R-C(=O)NR_2$ type of structure [24].

10-10. THE $N-CH_2-$ AND $N-CH_3$ GROUPS

Similarly to the $-OCH_2-$ and $-OCH_3$ groupings previously discussed, the attachment of an amino nitrogen atom to the CH_2 or CH_3 group shifts the in-phase CH stretchings to lower frequencies near 2800 cm^{-1} in secondary and tertiary amines and anilines. This frequency displacement is helpful in detecting amino groups, especially for tertiary amines where the NH band is not present. See Table 10-4.

McKean *et al.* [117–120] have shown that the CH bonds in amino $N-CH_3$ and $N-CH_2$ do not have identical force constants. The CH bond that is trans to the nonbonding electron pair on the amino (or anilino) nitrogen is weaker than the rest. This causes the $(N)CH_3$ stretch bands to spread over a wider region than $(C)CH_3$ stretch bands. This effect disappears in amino salts or quaternaries and amides. Some characteristic CH_3 IR bands are as follows: $RNCH_3$, 2805–2780 cm^{-1}; $RN(CH_3)_2$, 2825–2810 and 2775–2765 cm^{-1}; $ArNCH_3$, 2820–2810 cm^{-1}; and $ArN(CH_3)_2$, 2810–2790 cm^{-1}. In secondary and tertiary amines, the NCH_2 group has IR bands at 2820–2760 cm^{-1} [123, 124]. These bands involve the in-phase CH_3 or CH_2 stretch, but with a larger amplitude for the weak CH bond, and complications from Fermi resonance with overtones of CH_3 and CH_2 bending overtones.

Table 10-4.

IR Bands (cm^{-1}) for $CH_3(N)$ and $CH_2(N)$ Groups in Amines and Anilines

Aliphatic $-N-CH_3$	2805–2780 cm^{-1}
Aromatic $-N-CH_3$	2820–2810
Aliphatic $-N-(CH_3)_2$	2825–2810, 2775–2765
Aromatic $-N-(CH_3)_2$	2810–2790
$N-CH_2$ in aliphatic secondary and tertiary amines	2820–2760

The (N)CH_3 symmetrical deformation band is seen near 1460–1400 cm^{-1} [20a, 121], which is higher than the region near 1380 cm^{-1} for $C-CH_3$ groups. The $N-CH_3$ antisymmetric deformation is seen near 1480–1440 cm^{-1} [121]. These bands can be obscured by $C-CH_2$ and $C-CH_3$ bands, but the absence of the 1380 cm^{-1} band in the IR spectrum indicates the absence of $C-CH_3$ groups.

10-11.

AMINE SALTS

10-11.1. Primary Amine Salts ($-\overset{+}{N}H_3$)

Primary amine salts ($R-\overset{+}{N}H_3$) absorb strongly between 3200 cm^{-1} and 2800 cm^{-1} because of antisymmetric and symmetric NH_3 stretchings [70]. Substitution of an aryl group for an alkyl group results in lower frequencies. For example, the hydrochloride salts of anilines exhibit NH_3 stretching frequencies that are lower than 3000 cm^{-1} [21]. Infrared spectra of primary amine salts are further characterized by strong-to-medium IR absorptions in the region near 2600 cm^{-1}. These absorptions are due to combination bands enhanced by Fermi resonance, involving the deformation vibrations. Furthermore, these amine salts exhibit a series of sharp IR bands in the 2800–2100 cm^{-1} region. The bands at the higher-frequency end (i.e., 2800 cm^{-1}) appear to have stronger IR intensity, while a gradual decrease in intensity is often observed for those IR bands toward the lower-frequency end (i.e., 2100 cm^{-1}). Another characteristic feature in the IR spectra of the hydrochloride salts of primary amines is the isolated band near 2000 cm^{-1} (medium) [21, 70]. This band has been assigned to a combination band of the NH_3^+ antisymmetric deformation band about 1580 cm^{-1} and the NH_3^+ torsional oscillation about 480 cm^{-1} [71]. Although it was suggested that the intensity of this band is related to the lowering of the C_{3v} symmetry in the ammonium ion ($R-\overset{+}{N}H_3$) [71], this relationship appears to be vague according to our examination of 40 primary amine salts reported in references 21a and 21c. Among the 22 aliphatic primary amine salts investigated, 18 compounds exhibited medium to weak IR bands in the region of 2070–2000 cm^{-1}. It appears to be less common in the IR spectra of primary arylamine salts, where only nine out of the 18 primary arylamine salts reported in reference 21c show this absorption. With the exception of aniline, the arylamine salts appear in a lower frequency region of 2000–1960 cm^{-1}. The infrared spectrum of aniline hydrochloride exhibits a medium band at 2014 cm^{-1}.

The primary amine salt spectral pattern can be readily differentiated from the C≡C or C≡N stretching on the basis of band shape. Generally, the stretching of the first appears to be broad and has multiple band structure, as described earlier in this section, while the stretching bands of the last two are sharp.

The antisymmetric and symmetric $\overset{+}{N}H_3$ deformations occur at the regions of 1625–1560 cm^{-1} and 1550–1505 cm^{-1}, respectively [69].

10-11.2. Secondary Amine Salts ($-\overset{+}{N}H_2-$)

Vibrations involving antisymmetric and symmetric $\overset{+}{N}H_2$ stretchings are characterized by strong multiple IR absorptions located in the 3000–2700 cm^{-1} region [21a, 21c, 70]. In addition, combination bands, enhanced by Fermi resonance, extend from 2700 cm^{-1} to 2300 cm^{-1}, with their intensities decreasing from high frequency to a lower value. The $\overset{+}{N}H_2$ deformation appears as a medium IR band in the region of 1620–1560 cm^{-1} [71, 21].

10-11.3. Tertiary Amine Salts ($>\overset{+}{N}H-$)

The strong multiband absorptions of tertiary amine hydrochloride salts in the 2700–2330 cm^{-1} region [70] are due to vibrations involving $\overset{+}{N}H$ stretching and overtones or combination bands in Fermi resonance [21a, 21c]. A narrower region was proposed for trimethylammonium halides [$(CH_3)_3\overset{+}{N}H\bar{X}$, X = Cl, Br, or I] [16, 72–74]. This spectral region of halide salts is unique because only a few compounds absorb IR strongly in this region.

The $\overset{+}{N}H$ stretching frequency depends on the counterion in the amine salt. For example, this frequency was found to increase progressively as the counterion changed from Cl^- to Br^- and to I^-. This NH^+ stretching frequency variation is attributed to the strength of hydrogen bonding due to $NH^+\cdots X$ interaction. For $(CH_3)_3\overset{+}{N}HSbCl_6^-$, the hydrogen bonding is less pronounced than in halide salts. The associated N—H moiety gives rise to a strong IR band at 3195 cm^{-1}, and the free N—H moiety absorbs near 3200 cm^{-1} [75].

10-11.4. Quaternary Amine Salts ($\overset{+}{N}R_4$)

Quaternary amine salts are simply characterized by the NC_4 skeletal vibrations, which are analogous to the skeletal stretchings of quaternary carbon atoms, CC_4. For $R-N(CH_3)_3$ types of amine salts, absorption bands were observed in the IR at 980–900 cm^{-1} [76], and CH_3 band are seen near 3020, 1485, and 1410 cm^{-1}.

The most extensively studied quaternary ammonium salts are tetramethylammonium salts. The antisymmetric and symmetric NC_4 stretchings occur at around 950 cm^{-1} and 750 cm^{-1}, respectively, for $(CH_3)_4N^+$ [77–85]. The two stretching modes of tetraethylammonium ion are located in the 1060–1030 cm^{-1} and 672–666 cm^{-1} regions [86, 87]. The spectral separation between these two stretching modes is larger than that for the tetramethylammonium ion.

10-12.

HYDROXYLAMINES ($-O-NH_2$) AND HYDRAZINES ($>N-N<$)

10-12.1. Hydroxylamines

Condensed-phase hydroxylamines exhibit NH_2 vibrations at slightly lower frequencies than do aliphatic primary amines. For example, the characteristic NH_2 frequencies of n-$C_nH_{2n+1}O-NH_2$ (n = 6, 8, 10) are in the regions of 3251–3237 cm^{-1} (NH_2 symmetric stretch) and 1595–1589 cm^{-1} (NH_2 scissoring). The $O-N$ stretching of hydroxylamine ($HO-NH_2$) occurs at 906 cm^{-1} [24], and the CON stretching of n-$RO-NH_2$ is in the 851–840 cm^{-1} region [88].

10-2.2. Hydrazine

For hydrazine (H_2NNH_2) in the liquid state, the vibrational frequencies of NH_2 stretchings (3336–3190 cm^{-1}), scissoring (1628 cm^{-1}), and wagging (882 cm^{-1}) fall in the same range as those for the aliphatic amines. The Raman spectrum of this compound exhibits very strong, polarized bands at 1111 cm^{-1} due to the $N-N$ stretching mode. The NH_2 absorptions of R_2N-NH_2 compounds (3371–3348 cm^{-1} and 3251–3139 cm^{-1}) are more separated ($\sim$180 cm^{-1}) than those of primary amines ($\sim$75 cm^{-1}) [89], and hence the bands are useful for structural diagnosis.

The vibrational analysis of methyl hydrazines has been studied [90, 91, 108, 109]. Ordinarily, the skeletal stretchings involving N atoms occur at about 1100–1000 cm^{-1} and near 800 cm^{-1}. On the formation of hydrazone ($\rangle C{=}N{-}NH{-}$) with benzaldehydes, the hydrazinic NH stretching takes place in the 3100–2500 cm^{-1} region in solid-state (KBr disc) IR spectra [114].

REFERENCES

1. J. R. Durig, J. F. Sullivan, and C. M. Whang, *Spectrochim. Acta* **41A**, 129, 1985.
2. B. Silvi and J. P. Perchard, *Spectrochim. Acta* **32A**, 23, 1976.
3. R. R. Durig and Y. S. Li, *J. Chem. Phys.* **63**, 4110, 1975.
4. A. S. Manocha, E. C. Tuazon, and W. G. Fateley, *J. Phys. Chem.* **78**, 803, 1974.
5. M. Tsuboi, K. Tamagake, A. Y. Hirakawa, J. Yamaguchi, H. Nakagawa, A. S. Manocha, E. C. Tuazon, and W. G. Fateley, *J. Chem. Phys.* **73**, 5177, 1975.
6. J. R. Durig, G. A. Guirgis, and D. A. C. Compton, *J. Phys. Chem.* **83**, 1313, 1979.
7. H. Wolff, D. Horn, and H. G. Rollar, *Spectrochim. Acta* **29A**, 1835, 1973.
8. M. G. Giorgini, M. R. Pelletti, G. Pliani, and R. S. Cataliotti, *J. Raman Spectrosc.* **14**, 16, 1983.
9. J. J. C. Teixeira-Dias, L. A. E. Batista de Carvalho, A. M. Amorin da Costa, I. M. S. Lampreia, and E. F. G. Barbosa, *Spectrochim. Acta* **42A**, 589, 1986.
10. A. M. Amorim da Costa, C. F. G. C. Geraldes, and J. J. C. Teixeira-Dias, *J. Raman Spectrosc.* **13**, 56, 1982.
11. V. F. Kalasinsky, G. A. Guirgis, and J. R. Durig, *J. Mol. Struct.* **39**, 51, 1977.
12. V. F. Kalasinsky and T. S. Little, *J. Raman Spectrosc.* **9**, 224, 1980.
13. V. F. Kalasinsky, D. E. Powers, and W. C. Harris, *J. Phys. Chem.* **83**, 506, 1979.
14. A. L. Verma and P. Venkateswarlu, J. Mol. Spectrosc. **39**, 227, 1971.
15. A. M. Amorim de Costa, L. A. E. Batista de Carvalho, J. J. C. Teixeira-Dias, E. F. G. Barbosa, and I. M. S. Lampreia, *Can. J. Chem.* **65**, 384, 1987.
16. R. A. Nyquist, "The Interpretation of Vapor-Phase Infrared Spectra, Group Frequency Data," Vol. I, Sadtler Research Laboratories, 1984. (a) Aliphatic primary amines: pp. 173–188. (b) Primary arylamines: pp. 189–210. (c) Secondary amines: pp. 211–224. (d) Amides: pp. 264–291.
17. H. E. Hallam and C. M. Jones, *J. Mol. Struct.* **5**, 1, 1970.
18. A. T. Tu, "Raman Spectroscopy in Biology, Principles and Applications," John Wiley & Sons , New York, 1982.
19. L. A. LaPlanche and M. T. Rogers, *J. Am. Chem. Soc.* **86**, 337, 1964.
20. L. J. Bellamy, "The Infrared Spectra of Complex Molecules," John Wiley & Sons, New York, 1975. (a) pp. 277–293. (b) pp. 231–262.
21. C. J. Pouchert, "The Aldrich Library of FT-IR Spectra," Aldrich Chemical Company, Inc., 1986. (a) Vol. I, pp. 279–396. (b) Vol. I, pp. 748–836. (c) Vol. I, pp. 1189–1324. (d) Vol. II, pp. 355–421. (e) Vol. I, p. 327, Spectra C and D and p. 330, spectra C and D. (f) Vol. I, p. 286, spectra A and B, and p. 288, spectra A through D. (g) Vol. I, pp. 1251–1254. (h) Vol. I, p. 316, spectrum C. (i) Vol. I, p. 296, spectrum C and p. 297, spectrum C.
22. C. Perchard and J. P. Perchard, *J. Raman Spectrosc.* **6**, 74, 1977.
23. The assignments are based on the related spectra reported in "The Sadtler Standard Raman Spectra," Sadtler Research Laboratories, 1976. (a) 3111R and 3627R. (b) 120R, 508R, and 1294R.

24. F. R. Dollish, W. G. Fateley, and F. F. Bentley, "Characteristic Raman Frequencies of Organic Compounds," John Wiley & Sons, New York, 1974.
25. L. J. Bellamy and R. L. Williams, *Spectrochim. Acta* **9**, 341, 1957.
26. J. E. Stewart, *J. Chem. Phys.* **30,** 1259, 1959.
27. P. J. Krueger, *Nature* **194,** 1077, 1962.
28. W. Walter and H. P. Kubersky, *J. Mol. Struct.* **11,** 207, 1972.
29. H. O. Desseyn, W. A. Jacob, and M. A. Herman, *Spectrochim. Acta* **28A,** 1329, 1972.
30. P. J. Krueger and J. Jan, *Can. J. Chem.* **48,** 3229, 1970.
31. N. O. Cherskaya, V. P. Gorelik, V. A. Shlyapochnikov, V. P. Ivshin, and T. N. Ivshina, *Bull. Acad. Sci., USSR, Div. Chem. Soc.* **35,** 1385, 1986.
32. R. A. Bell, B. E. Brown, M. Duarte, H. E. Howard-Lock, and C. J. L. Lock, *Can. J. Chem.* **65,** 261, 1987.
33. F. Christini, F. A. Devillanova, A. Diaz, F. Isaia, and G. Verani, *Spectrochim. Acta* **41A,** 487, 1985.
34. J. C. Evans, *Spectrochim. Acta* **16,** 428, 1960.
35. V. B. Singh, R. N. Singh, and I. S. Singh, *Spectrochim. Acta* **22,** 927, 1966.
36. J. H. S. Green, D. J. Harrison, and W. Kynaston, *Spectrochim. Acta* **27A,** 807, 1971.
37. M. Kasai, M. Hirota, Y. Hamada, and H. Matsuoka, *Tetrahedron* **29,** 267, 1973.
38. M. Bernard-Houplain and C. Sandorfy, *Can. J. Chem.* **51,** 1075, 1973.
39. L. Smetankine and J. Etchepare, *J. Mol. Struct.* **19,** 799, 1973.
40. J. A. Faniran and H. F. Shurvell, *Spectrochim. Acta* **31A,** 1127, 1975.
41. P. Venkatacharyulu, V. L. N. Prasad, Nallgonda, and D. Premaswarup, *Ind. J. Pure & Appl. Phys.* **19,** 1178, 1981.
42. J. A. Faniran, I. Iweibo, and R. A. Oderinde, *J. Raman Spectrosc.* **11,** 477, 1981.
43. T. G. Towns, *Spectrochim. Acta* **39A,** 801, 1983.
44. E. Steger, S. Garbe, and J. Klosowski, *J. Raman Spectrosc.* **14,** 194, 1983.
45. N. S. Sundar, *Can. J. Chem.* **62,** 2238, 1984.
46. A. R. Shukla, C. M. Pathak, N. G. Dongre, B. P. Asthana, and J. Shamir, *J. Raman Spectrosc.* **17,** 299, 1986.
47. R. D'Cunha, V. B. Kartha, and S. Gurnani, *Spectrochim. Acta* **39A,** 331, 1983.
48. J. A. Faniran and H. F. Shurvell, *Spectrochim. Acta* **38A,** 1155, 1982.
49. R. A. Kydd and S. Mah, *Spectrochim. Acta* **38A,** 1031, 1982.
50. R. B. Singh, N. P. Singh, and D. K. Rai, *Ind. J. Pure & Appl. Phys.* **19,** 740, 1981.
51. N. S. Sundar, *Spectrochim. Acta* **41A,** 905, 1985.
52. R. A. Yadav and I. S. Singh., *Spectrochim. Acta* **41A,** 191, 1985.
53. N. S. Sundar, *Spectrochim. Acta* **41A,** 1449, 1985.
54. A. T. Hutton and D. A. Thornton, *Spectrochim. Acta* **34A,** 645, 1978.
55. I. S. Ahuja and R. Singh, *J. Inorg. Nucl. Chem.* **35,** 302, 1973.
56. A. Zabokrycka, P. Klaeboe, B. N. Cyvin, S. J. Cyvin, and J. Brunvoll, *Spectrochim. Acta* **38A,** 357, 1982.
57. A. Lautie and A. Novak, *J. Chem. Phys.* **56,** 2479, 1972.
58. N. K. Sanyal, S. L. Srivastava, and S. R. Tripathi, *Spectrochim. Acta* **38A,** 933, 1982.
59. A. Suwaiyan and R. Zwarich, *Spectrochim. Acta* **42A,** 1017, 1986.
60. M. P. Sammes and A. R. Katritzky, *Adv. Heterocycl. Chem.* **35,** 409, 1983.
61. J. Bellanato, C. Avendano, M. T. Ramos, P. Smithverdier, F. Florencio, and S. Garcia-Blanco, *Spectrochim. Acta* **41A,** 99, 1985.
62. C. Avendano, M. T. Ramos, and J. Bellanato, *Spectrochim. Acta.* **41A,** 109, 1985.
63. E. Galvez, M. Martinez, G. G. Trigo, F. Florencio, J. Vilches, S. Garcia-Blanco, and J. Bellanato, *J. Mol. Struct.* **75,** 241, 1981.
64. S. Martinez-Carrera, *Acta Crystallogr.* **20,** 783, 1966.

65. I. Suzuki, *Bull. Chem. Soc. Jpn.* **35,** 540, 1962.
66. T. Miyazawa, T. Shimanouchi, and S. Mizushima, *J. Chem. Phys.* **29,** 611, 1958.
67. T. Miyazawa, *J. Mol. Spectrosc.* **4,** 155, 1960.
68. G. B. B. M. Sutherland and D. M. Simpson, *J. Chem. Phys.* **15,** 153, 1947.
69. B. Chenon and C. Sandorfy, *Can. J. Chem.* **36,** 1181, 1958.
70. N. B. Colthup, L. H. Daly, and S. E. Wiberley, "Introduction to Infrared and Raman Spectroscopy," 3rd Ed., Academic Press, New York, 1990.
71. E. Brissette and C. Sandorfy, *Can. J. Chem.* **38,** 34, 1960.
72. M. Meinander, G. Bergstrom, S. Forss, and F. Stenman, *J. Raman Spectrosc.* **8,** 265, 1979.
73. E. Moreno, M. Marraud, and J. Neel, *C. R. Acad. Sci. Paris C* **284,** 203, 1977.
74. M. Schlaak, M. Couzi, and P. V. Houng, *Ber. Biungsenges. Gesellsch.* **80,** 881, 1976.
75. R. F. Howe and M. J. Taylor, *Spectrochim. Acta* **43A,** 73, 1987.
76. A. S. Hume, W. C. Holland, and F. Fry, *Spectrochim. Acta* **24A,** 768, 1968.
77. A. M. Heyns and W. H. J. de Beer, *Spectrochim. Acta* **39A,** 601, 1983.
78. R. W. Berg, *Spectrochim. Acta* **34A,** 655, 1878.
79. R. W. Berg, *J. Chem. Phys.* **71,** 253, 1979.
80. R. W. Berg, *J. Chem. Phys.* **69,** 1325, 1978.
81. R. W. Berg, F. W. Paulsen, and N. J. Bjerrum, *J. Chem. Phys.* **67,** 1829, 1977.
82. K. M. Harmon, I. Gennick, and S. L. Madeira, *J. Phys. Chem., Ithaca* **78,** 2585, 1974.
83. G. Kabisch, *J. Raman Spectrosc.* **9,** 279, 1980.
84. G. Kabisch and M. Klose, *J. Raman Spectrosc.* **7,** 311, 1978.
85. W. van der Ohe, *J. Chem. Phys.* **62,** 3933, 1975.
86. W. H. J. de Beer and A. M. Heynes, *Spectrochim. Acta* **37A,** 1099, 1981.
87. M. A. Hooper and D. W. James, *Aust. J. Chem.* **24,** 1331, 1971.
88. W. O. George, J. H. S. Green, and M. J. Rix, *Spectrochim. Acta* **26A,** 2007, 1970.
89. D. Hadži, J. Jan, and A. Ocvirk, *Spectrochim. Acta* **25A,** 97, 1969.
90. D. W. E. Axford, G. J. J. Jane, and K. E. Russell, *J. Chem. Phys.* **19,** 704, 1951.
91. E. R. Shull, J. L. Wood, J. G. Aston, and P. H. Rank, *J. Chem. Phys.* **22,** 1191, 1954.
92. J. Jakes and S. Krimm, *Spectrochim. Acta.* **27A,** 19, 1971.
93. J. Jakes and S. Krimm, *Spectrochim. Acta* **27A,** 35, 1971.
94. W. Walter, T. Fleck, J. Voss, and M. Gerwin, *Justus Liebigs Ann. Chem.*, 275, 1975.
95. W. Walter and P. Staeglich, *Spectrochim. Acta* **30A,** 1739, 1974.
96. W. Walter and R. F. Becker, *Justus Liebigs Ann. Chem.*, 753, 187, 1971.
97. J. Bandekar and G. Zundel, *Spectrochim. Acta* **38A,** 815, 1982.
98. J. F. Rabolt, W. H. Moore, and S. Krimm, *Macromolecules* **10,** 1065, 1977.
99. W. H. Moore and S. Krimm, *Biopolymers* **15,** 2465, 1976.
100. T. Miyazawa, *J. Polymer Sci.* **362,** S62, 1962.
101. B. R. Malcolm, *Biopolymers* **16,** 2591, 1977.
102. J. Bandekar and S. Krimm, in "Peptides" (E. Gross and J. Meienhofter, eds.), Pierce Chemical Co., Rockford, Illinois, 1979, pp. 241–244.
103. H. Susi and J. S. Ard, *Spectrochim. Acta* **27A,** 1549, 1971.
104. C. L. Angell, *J. Chem. Soc.*, 504, 1961.
105. T. Shimanouchi, K. Kuratani, and S. Mizushima, *J. Chem. Phys.* **19,** 1479, 1951.
106. H. T. Miles, T. P. Lewis, E. D. Becker, and H. Frazier, *J. Biol. Chem.* **248,** 1115, 1973.
107. K. Fukushima, Y. Ideguchi, and T. Miyazawa, *Bull. Chem. Soc. Jpn.* **37,** 349, 1964.
108. J. R. Durig and W. C. Harris, *J. Chem. Phys.* **51,** 4457, 1969.
109. J. R. Durig, W. C. Harris, and D. W. Wertz, *J. Chem. Phys.* **50,** 1449, 1969.
110. C. Laurence and M. Berthelot, *Spectrochim. Acta* **34A,** 1127, 1978.
111. C. Laurence, M. Berthelot, and M. Lucon, *Spectrochim. Acta* **38A,** 791, 1981.

112. P. Geboes, H. Hofmans, H. O. Desseyn, R. Dommisse, A. T. H. Lenstra, S. B. Sanni, J. M. M. Smits, and P. T. Beurskens, *Spectrochim. Acta* **43A,** 35, 1987.
113. H. O. Dessyn, A. J. Aarts, and M. A. Herman, *Spectrochim. Acta* **36A,** 59, 1980.
114. I. Suzuki, M. Tsuboi, T. Shimanouchi, and S. Mizushima, *Spectrochim. Acta* **16,** 471, 1960.
115. T. Miyazawa, *J. Mol. Soectrosc.* **4,** 168, 1960.
116. R. A. Nyquist, *Spectrochim. Acta* **19,** 509, 1963.
117. D. C. McKean, J. L. Duncan, and L. Batt, *Spectrochim. Acta* **29A,** 1037, 1973.
118. D. C. McKean, *Chem. Soc. Rev.* **7,** 399, 1978.
119. D. C. McKean, J. E. Boggs, and L. Schäfer, *J. Mol. Struct.* **16,** 313, 1984.
120. D. C. McKean, *Croat. Chem. Acta* **61,** 447, 1988.
121. N. B. Colthup, *Appl. Spectrosc.* **34,** 1, 1980.
122. D. Hadzi and M. Skrbljak, *J. Chem. Soc. London*, 843, 1957.
123. R. D. Hill and G. D. Meakins, *J. Chem. Soc. London*, 760, 1958.
124. W. B. Wright, Jr., *J. Org. Chem.* **23,** 1362, 1959.

CHAPTER 11

The Nitro Group

11-1. NO_2 STRETCHINGS

Among the six vibrations involving the NO_2 group in nitrocompounds, the NO_2 stretchings are the two most useful group frequencies. The antisymmetric stretching mode is quite constant in spectral position, while the symmetric stretching mode is more prone to vibrational coupling [1]. Therefore, caution should be exercised in applying the NO_2 symmetric stretching mode as a characteristic vibration. Nevertheless, with the combination of their spectral positions and intensities, these two characteristic vibrations usually provide enough information for differentiating the unconjugated [2–6] and conjugated NO_2 groupings [3, 4, 7–9]. In addition, the atom (C, O, or N) adjacent to the NO_2 moiety can be determined by the vibrational spectra of the compounds [10–12]. The C—N stretching mode is used as a characteristic frequency of NO_2 groupings only for small nitroalkanes. However, the vibration is sensitive to the type of nitroalkane (i.e., primary, secondary, etc.) and to nitroalkanes' rotational conformations. Thus, these vibrations can be useful in conformational studies.

11-1.1. Nitroalkanes (R—NO_2)

The antisymmetric NO_2 stretching mode of nitroalkanes was found by Nyquist [3] to exhibit higher frequencies in the vapor phase (1621–1555 cm^{-1})

than in the liquid state (1601–1531 cm^{-1}). Usually, the $\nu_{as}(NO_2)$ frequency decreases by 18–32 cm^{-1} as the sample phase changes from vapor to liquid state. On the other hand, the frequency shift of the symmetric NO_2 stretching mode upon a change in the sample state is not as predictable as that for the corresponding antisymmetric mode. The $\nu_s(NO_2)$ frequencies of nitroalkanes can either decrease by as much as 10 cm^{-1}, or increase by as much as 22 cm^{-1}, when the measuring state changes from vapor (1397–1310 cm^{-1}) to liquid phase (1381–1310 cm^{-1}) [3]. In the IR spectra of unconjugated nitroalkanes, the intensity of the antisymmetric NO_2 stretching band is very strong, while the symmetric stretching band is somewhat weaker (strong-medium) [4, 13]. The peak intensity ratio of $\nu_{as}(NO_2)/\nu_s(NO_2)$ has been observed to vary from 5.9 to 1.8 in vapor-phase IR spectra [3]. Although there was no definite number reported for the intensity ratio in liquid-state IR spectra, the existence of an unconjugated NO_2 group can still be recognized by the spectral pattern of these two bands [4, 13]. The symmetric $\nu(NO_2)$ mode is predicted and found to be very strong and polarized in Raman spectra, while the antisymmetric stretching mode exhibits only medium to weak intensity [13]. Therefore, the relative intensities of the NO_2 antisymmetric and symmetric stretchings are reversed in Raman spectra.

Slovetskii *et al.* Have studied the correlation between the NO_2 stretching frequencies and the type of nitroalkanes (i.e., primary, secondary, or tertiary) by Raman spectroscopy [2]. Their results are summarized in Table 11-1 along with correlations from other references. It appears that the branching on the α-carbon atom lowers the vibrational frequencies in both antisymmetric and symmetric NO_2 stretching vibrations. Investigations on the available IR data suggest that these correlations are also appropriate in interpreting the IR absorption frequencies. The table shows the compounds used for this examination.

	$\nu_{as}(NO)_2$	$\nu_s(NO_2)$	Ref.
n-C_2H_5—NO_2 to n-C_6H_{13}—NO_2:	1562–1553 cm^{-1}	1387–1378 cm^{-1}	4, 14
Me_2CH—NO_2:	1552 cm^{-1}	~1360 cm^{-1}	4, 14, 15
Me_3C—NO_2:	~1540 cm^{-1}	1358 cm^{-1}	4

When a $-CH_3$ moiety appears on the α-carbon atom, the interaction of CH_3 symmetric deformation (near 1380 cm^{-1}) with symmetric NO_2 stretch will result in two strong IR bands near 1390 cm^{-1} and 1360 cm^{-1} [7].

Table 11-1.

The Group Frequencies (cm^{-1}) of $-NO_2$ Containing Compounds

Vibrational Mode	Compound	Vapor	IR	Liquid/Solid	IR	R
NO_2 antisymmetric stretch	$R-NO_2$	1621–1555	vs	1601–1531	vs	m-w
	$-CH_2-NO_2$			1560–1548	vs	m-w
	$-CHR-NO_2$			1553–1547	vs	m-w
	$-CR_2-NO_2$			1554–1534	vs	m-w
	$-C(NO_2)_2$			1587–1563	vs	m-w
	$-C(NO_2)_3$			1603–1597	vs	
	$Ar-NO_2$	1580–1530	vs	1555–1487	vs	
	$O-NO_2$			1640–1620	vs	
	$N-NO_2$			1630–1550	vs	
NO_2 symmetric stretch	$R-NO_2$	1397–1310	s	1310–1381	s	vs
	$-CH_2-NO_2$			1388–1376	s	vs
	$-CHR-NO_2$			1374–1366	s	vs
	$-CR_2-NO_2$			1354–1344	s	vs
	$-C(NO_2)_2$			1337–1327		
	$-C(NO_2)_3$			1307–1297		
	$Ar-NO_2$	1371–1325		1357–1318	vs	vs
	$O-NO_2$			1285–1270	vs	
	$N-NO_2$			1300–1250	vs	
C—N stretch	$R-NO_2$					
	$R-CH_2-NO_2$		gauche:	894–873	m-w	m-s
			trans:	914–898		m-s
	$R-CHR-NO_2$		gauche:	863–847	m-w	s
			trans:	908–868		m
NO_2 scissors	$R-NO_2$					
	$R-CH_2-NO_2$			618–609	w	m-w
	$R-CHR-NO_2$			627–619[a]	w	m
	$Ar-NO_2$			857–830	m	m-w
NO_2 wag	$R-NO_2$					
	$R-CH_2-NO_2$			643–614	w	w, sh
	$R-CHR-NO_2$			670–650	w	w-vw
NO_2 rock	$R-NO_2$					
	$R-CH_2-NO_2$			494–472		m-w
	$R-CHR-NO_2$			557–522		s

[a] 3-Nitropentane: 613 cm^{-1}.

11-1.2. Substituted Nitroalkanes

Table 11-2 illustrates the influences of substituents on NO_2 stretching frequencies. In this table, the NO_2 stretching frequencies of liquid nitromethane ($CH_3{-}NO_2$) are used as standards for comparison. The frequency shift from NO_2 stretchings of $CH_3{-}NO_2$ are summarized in the second and third columns for substituted nitroalkanes. The substitution of halogen atoms (i.e., F, Cl, Br), C=O, and NO_2 groups on the α-carbon atom results in a high frequency shift of the antisymmetric NO_2 stretching mode, while the symmetric NO_2 stretching frequencies are lowered by 8–29 cm^{-1}. Therefore, the NO_2 antisymmetric and symmetric stretchings bands are farther apart than those of unsubstituted nitroalkanes. On the other hand, the NO_2 stretching frequencies of most β-substituted nitroalkanes are only slightly affected. The exception is β-nitro substitution, which lowers the symmetric NO_2 stretching by 15 cm^{-1}. Nyquist pointed out that the frequency shift of substituted nitroalkanes is somewhat additive for antisymmetric and symmetric NO_2 stretchings in both vapor and liquid phases. As an example given in Table 11-2, the antisymmetric NO_2 stretchings frequency of $CH_3CCl_2{-}NO_2$ is estimated to be at 1587 cm^{-1}, which is only 6 cm^{-1} higher than the experimental value of 1581 cm^{-1}. Similarly, the symmetric NO_2 stretching frequency of this compound is calculated to be 1321 cm^{-1}, and the experimental value was found to be 1325 cm^{-1} (see Table 11-2).

11-1.3. Conjugation with a C=C Group

Conjugation with a C=C moiety lowers antisymmetric and symmetric NO_2 stretchings most probably through the resonance effect. For nitroethylene, the two vibrational modes were found at 1566 cm^{-1} and 1359 cm^{-1} by IR and Raman methods [16]. Brown [7] found that monoalkyl nitroethylenes absorb infrared radiation strongly in the regions from 1528 to 1520 cm^{-1} and from 1359 to 1347 cm^{-1} because of the antisymmetric and symmetric NO_2 stretching vibrations. He also reported that characteristic vibrational frequencies for di- and trialkyl nitroethylenes are in the regions of 1519–1511 cm^{-1} and 1355–1337 cm^{-1}, respectively.

The two strong IR bands at 1514 and 1344 cm^{-1} for β-nitrostyrene ($C_6H_5{-}CH{=}CHNO_2$) [4] can be attributed to the antisymmetric and symmetric NO_2 stretching modes. Our investigation [13a] on four ring-substituted β-nitrostyrenes indicates that the antisymmetric NO_2 stretchings of these compounds are located near 1530 cm^{-1}, and the symmetric NO_2 stretching falls in the region of 1352–1333 cm^{-1} [13a]. These vibrational

Table 11-2.

$\nu_{as}(NO_2)$ and $\nu_s(NO_2)$ Frequency Shifts of Substituted Nitrocompounds from Those of CH_3NO_2

CH_3NO_2 (liq., IR)	$\nu_{as}(NO_2)$: 1558 cm^{-1} $\nu_s(NO_2)$: 1375 cm^{-1}	
	$\Delta\nu_{as}(NO_2)$	$\Delta\nu_s(NO_2)$
α-substituent:		
$-H$	0	0
$-CH_3$	-5	-8
$-C_2H_5$	-5	-8
$-C_6H_5$		-8
$-C{=}O$	$+10$	-8
$-F$	$+17$	-23
$-Cl$	$+17$	-23
$-Br$	$+17$	-23
$-NO_2$	$+29$	-29
β-substituent:		
$-H$	0	0
$-OH$	-4	-6
$-Cl$	$+2$	-4
$-Br$	$+2$	-4
$-NO_2$	$+2$	-15

Example: $CH_3CCl_2-NO_2$

$\nu_{as}(NO_2)$	$\nu_s(NO_2)$	
1558	1375	
$+1\times(-5)$	$+1\times(-8)$	........1 (α-CH_3)
$+2\times(17)$	$+2\times(-23)$	........2 (α-Cl)
calc. 1587 cm^{-1}	calc. 1321 cm^{-1}	
obs. 1581 cm^{-1}	obs. 1325 cm^{-1}	

frquencies for β-nitrostyrenes are essentially in the same regions as those for monoalkyl nitroethylenes. The NO_2 stretchings of *m*-bis(2-nitropropenyl)-benzene,

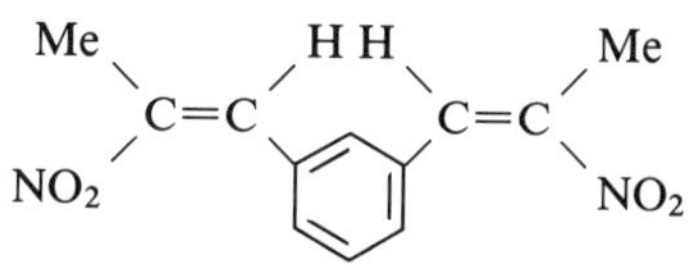

occur at approximately 1505 cm^{-1} and 1330 cm^{-1} [13b]. These frequencies are similar to those found for di- and trialkyl nitroethylenes. In the IR spectra of these ring-substituted β-styrenes, the symmetric NO_2 stretching band is observed to be as intense as the antisymmetric mode in the IR spectra. The relative intensities of the antisymmetric and symmetric NO_2 stretching bands in the Raman spectra, however, are similar to those observed for nitroalkanes [13a, 13b].

Nitroenamines, $(R_2N)CH{=}CH(NO_2)$, are found to be a special class of compounds as far as their vibrational spectra are concerned. Unlike the other nitrocompounds, the $\nu_{as}(NO_2)$ mode of nitroenamines exhibits a weak IR band in the region of 1530–1480 cm^{-1} and is easily confused with the aromatic ring or CH_2 deformations. Ostercamp and Taylor [17] have rationalized this anomalously weak $\nu_{as}(NO_2)$ band by the high degree of electronic polarization with the $>N{-}CH{=}CH{-}NO_2$ moiety due to the resonance effect. There are more irregularities about the vibrational spectra of nitroenamines that can be attributed to the same resonance effect. These include a weak C=C stretching band in the 1650–1600 cm^{-1} region of Raman spectra (IR : vs) and an unusually low NO_2 symmetric stretching frequency (1282–1229 cm^{-1}) with strong IR absorption but weak Raman intensity [17].

11-1.4. Aromatic Nitrocompounds

There exists a large collection of data on the vibrational spectra for aromatic nitrocompounds [9, 16, 18–28, 41, 42]. The NO_2 stretching frequencies of nitrobenzene occur at 1548 cm^{-1} and 1360 cm^{-1} for the antisymmetric and symmetric modes, respectively [16]. However, these frequencies in other substituted nitrobenzenes are affected by the substituents on the benzene ring through inductive effects, resonance effects, and hydrogen bond formation [12]. In general, the characteristic NO_2 stretching frequencies in the regions of 1587–1555 cm^{-1} and 1357–1318 cm^{-1} are assigned to the antisymmetric and symmetric NO_2 stretchings, respectively. Such frequency lowering, in both of these stretching modes, is attributed to the conjugation of NO_2 with the aromatic ring. As with the Raman spectra of nitroethylenes, the IR intensity of the symmetric NO_2 stretching is enhanced by conjugation, and the antisymmetric and symmetric stretching bands in the IR spectra are found to be very similar in peak intensity. The relative Raman intensities of nitrobenzenes, on the other hand, are very much similar to those found in the Raman spectra of nitroalkanes.

The normal coordinate analysis of some nitrobenzenes suggests that some vibrations of the benzene ring also participate in the so-called NO_2 symmetric stretching mode [20, 29]. Therefore, the NO_2 symmetric stretching mode of aromatic nitrocompounds is more complex than that of nitroalkanes. In the vibrational spectra of many *p*-nitroanlines or *p*-nitrophenols, two NO_2 symmetric stretching bands are observed [30–33]. It is found that intensities as well as spectral positions of this $\nu_s(NO_2)$ doublet are very dependent on the solvent polarity. Various theories, including self-association [32, 33] and Fermi resonance [31], have been proposed to explain the splitting of the $\nu_s(NO_2)$ band in these compounds.

The infrared $\nu(NO_2)$ frequencies of 33 nitropyridines were measured by Perjessy *et al.* [34]. They reported the NO_2 antisymmetric and symmetric stretching bands at 1543–1485 cm^{-1} and 1370–1336 cm^{-1}, respectively.

11-2.

C—N STRETCHING

11-2.1. Nitroalkanes

The CN stretching mode of small nitroalkanes is very strong and polarized in the liquid-state Raman spectra. This vibration occurs at 920 cm^{-1} for nitromethane and at 877 cm^{-1} for nitroethane [13c]. However, other primary nitroalkanes exist as mixtures of trans and gauche conformations due to the C—C rotation of the C—CNO_2 moiety; hence, two C—N stretching bands are observed in the region of 914–895 cm^{-1} (trans) and 894–868 cm^{-1} (gauche) [35]. In the case of small nitroalkanes, the C—N stretching band is intense and polarized in the Raman spectra, whereas the infrared absorption of this vibration is of medium to weak intensity. In the 1-nitropropane spectra the two polarized Raman bands at 895 cm^{-1} and 868 cm^{-1} can be assigned to the C—N stretchings of the trans and gauche isomers, respectively. The C—N stretching frequencies of secondary nitroalkanes appear in the lower regions of 908–868 cm^{-1} and 863–845 cm^{-1}. For 2-nitropropane, the C—N stretching of the trans isomer occurs near 900 cm^{-1} (IR : w, R : mw), and that of the gauche isomer at 850 cm^{-1} (IR : m, R : s). As the alkyl substituent on the NO_2 moiety becomes larger, the C—N stretching mode is often overwhelmed by other skeletal vibrations in the spectral region, and its assignment becomes uncertain.

11-2.2. Aromatic Nitrocompounds

The C—N stretching mode of aromatic nitrocompounds was originally correlated with a medium IR band in the region of 875–830 cm^{-1}. However, recent vibrational studies based on normal coordinate analysis revealed that

C—N stretching frequencies of aromatic nitrocompounds occur in the 1177–865 cm^{-1} region, because of the coupling between this vibrational mode and other ring vibrations. This is very similar to what happens with anilines. The 875–830 cm^{-1} band is reassigned to a vibration involving NO_2 scissoring, and this is discussed in the next section.

11-3. NO_2 DEFORMATIONS

The NO_2 deformation vibrations, which include NO_2 scissoring (in-plane), NO_2 wagging (out-of-plane), and NO_2 rocking (in-plane), usually occur at frequencies lower than 900 cm^{-1}. For the majority of compounds, the scissoring mode possesses the higher vibrational frequency of the three deformations, and it is by far the most characteristic deformation mode. The NO_2 rocking mode, on the other hand, has the lowest frequency for these three deformations.

According to the vibrational assignments of nitroalkanes [14, 6, 36], nitroethylene [16] and nitrobenzene [16, 21], the NO_2 deformation frequencies of unconjugated nitroalkanes are situated at lower frequencies than the corresponding vibrations of conjugated deformations. In the example of nitromethane, the NO_2 scissors, wagging, and rocking modes are observed at 656 cm^{-1} (liq.; IR : s; R : s, p), 608 cm^{-1} (liq; IR : m; R : w), and 482 cm^{-1} (liq.; IR : m; R : m), respectively [14]. Other *n*-nitroalkanes exhibit NO_2 scissoring bands in the region of 658–610 cm^{-1} [14]. The wagging mode occurs in a wider range of vibrational frequencies, 560–477 cm^{-1} [14]. Thus, these frequencies are less useful as characteristic frequencies. The NO_2 deformation frequencies of nitrocyclopropane are assigned by Holtzclaw *et al.* [36] as follows: scissoring (liq.; R : 771 cm^{-1}, w), wagging (liq.; R : 646 cm^{-1}, vw), and rocking (liq.; R : 491 cm^{-1}, m).

The NO_2 deformation occur at higher frequencies when conjugated to C=C or phenyl groups. For instance, the NO_2 scissoring of nitroethylene is located at 890 cm^{-1}, the NO_2 wagging at 714 cm^{-1}, and the rocking at 652 cm^{-1}. The NO_2 deformation frequencies of nitrobenzene are located in regions very close to those of nitroethylene, as its NO_2 scissoring is reported to be at 853 cm^{-1}, the wagging mode at 704 cm^{-1}, and the rocking vibration at 532 cm^{-1} [16]. The scissoring mode of other aromatic nitrocompounds is located in the region of 890–800 cm^{-1} (R : s-m; IR : s-w) [21–23]. In the case of 1,3,5-triamino-2,4,6-trinitrobenzene, the NO_2 scissoring frequency (IR : 782 cm^{-1}) is lower than the characteristic region [28].

11-4.

COVALENT NITRATES ($-O-NO_2$)

Only a limited amount of work has been reported on covalent nitrates. The antisymmetric and symmetric NO_2 stretching bands of covalent nitrates are located at 1640–1620 cm^{-1} and 1285–1270 cm^{-1}, respectively [11]. It should be noted that the antisymmetric NO_2 stretching may be confused with some substituted C=C moiety or aromatic ring stretchings. The symmetric NO_2 stretching band of primary nitrates usually appears as a single band, while the secondary nitrates, including monocyclic nitrates, exhibit a doublet structure in this spectral region. One exception for this correlation is secondary bicyclic nitrates, which display only a single ν_s band in their vibrational spectra. It is to be expected that this symmetric stretching mode is strong and polarized in the Raman spectra.

The O—N stretching frequency of covalent nitrates is observed at 870–855 cm^{-1} [11]. The NO_2 in-plane bending is found at 710–695 cm^{-1}, and the out-of-plane bending at 760–755 cm^{-1} [11].

11-5.

NITROAMINES ($>N-NO_2$)

The NO_2 symmetric stretching of the $>N-NO_2$ moiety occurs in a region very similar to that for the $-O-NO_2$ group. Infrared studies indicate that the characteristic $\nu_s(NO_2)$ of nitroamines is located at 1300–1250 cm^{-1} with very strong IR absorption [10, 37, 38]. The Raman intensity of this vibrational mode is expected to be strong and polarized. The antisymmetric NO_2 stretching is more sensitive to the environmental structure of the $N-NO_2$ moiety. For example, the alkyl-substituted nitroguanidines absorb infrared in the 1634–1605 cm^{-1} region due to $\nu_{as}(NO_2)$, while the arylguanidines absorb at lower frequencies (1587–1575 cm^{-1}) [37, 38]. Recently, Cherskaya *et al.* investigated the vibrational spectra of *N,N'*-dinitroalkylenediamines, $NO_2-NH-(CH_2)_n-NH-NO_2$ (n = 1-3) [39]. The $\nu_{as}(NO_2)$ vibration is assigned to bands between 1596 cm^{-1} and 1547 cm^{-1} (IR : vs, R : not observed), while the $\nu_s(NO_2)$ is found in the region of 1365–1290 cm^{-1} (IR : s-m, R : s-w). It is noteworthy that in-phase and out-of-phase $\nu_{as}(NO_2)$, as well as $\nu_s(NO_2)$, are possible for these compounds because of the presence of two nitro groups, but in this case, they are very weakly coupled and small frequency shifts are expected. Such mechanical interactions of the two NO_2 groups should be taken into consideration when examining the IR and Raman intensities of stretching modes and the polarization character of Raman bands. In these cases, the N—N stretching mode (1030–980 cm^{-1}) is found to be very strong and polarized in the Raman spectra, but weak in the IR.

These authors also reported bands for *N,N'*-dinitroalkenediamine salts, where the NO bonds have decreased double-bond character and the NN bond has increased double-bond character. The NO and NN bonds have rather similar force constants, so the $R{-}NNO_2^-$ ion somewhat resembles the NO_3^- ion. For $R{-}NNO_2$ salts, the band at 1310–1280 cm^{-1} (IR : s, R : w) was assigned to out-of-phase NO_2 stretch. A band at 1475–1430 cm^{-1} (IR : s, R : s, p) involves the in-phase stretching of the NO_2 bonds and NN contraction. Assignments were made on the basis of the IR and Raman intensities as well as the polarization character of the Raman bands [40].

REFERENCES

1. E. M. Popov and V. A. Shlyapochnikov, *Opt. & Spectrosc.* **15,** 174, 1963.
2. V. I. Slovetskii, V. A. Shylapochnikov, S. A. Shevelev, A. A. Fainzil'berg, and S. S. Novikov, *Izv. Acad. Nauk. USSR., Otd. Khim. Nauk* **330,** 1961.
3. R. A. Nyquist, "The Interpretation of Vapor-Phase Infrared Spectra, Group Frequency Data," Sadtler Research Laboratories, 1984, pp. 550–608.
4. C. J. Pouchert, "The Aldrich Library of FT-IR Spectra," Aldrich Chemical Company, Inc., 1985.
5. W. H. Lunn, *Spectrochim. Acta* **16,** 1088, 1960.
6. R. N. Hazeldine, *J. Chem. Soc.*, 2525, 1953.
7. J. F. Brown, *J. Am. Chem. Soc.* **77,** 6341, 1955.
8. R. R. Randle and D. H. Whiffen, *J. Chem. Soc.*, 4153, 1952.
9. M. St. C. Flett, *Spectrochim. Acta* **18,** 1537, 1962.
10. L. J. Bellamy, "The Infrared Spectra of Complex Molecules," John Wiley & Sons, New York, 1975, pp. 331–346.
11. R. A. G. Carington, *Spectrochim. Acta* **16,** 1279, 1960.
12. K. Nakanishi and P. H. Solomon, "Infrared Absorption Spectroscopy," Holden-Day, Inc., San Francisco, 1976, pp. 45–46.
13. "The Sadtler Standard Raman Spectra," Sadtler Research Laboratories, 1976. (a) 1528R, 1529R, 1537R, 1538R. (b) 1527R. (c) 296R, 297R.
14. D. C. Smith, C. Y. Pan, and J. R. Nielsen, *J. Chem. Phys.* **18,** 706, 1950.
15. J. R. Durig, J. A. S. Smith, Y. S. Li, and F. M. Wasacz, *J. Mol. Struct.* **99,** 45, 1983.
16. G. Varsanyi, S. Holly, and L. Imre, *Spectrochim. Acta* **23A,** 1205, 1967.
17. D. L. Ostercamp and P. J. Taylor, *J. Chem. Soc., Perkin Trans. II,* 1021, 1985.
18. B. J. Lindberg and B. Schroder, *Acta. Chem. Scand.* **24,** 3089, 1970.
19. O. Exner, S. Kovac, and E. Solcaniva, *Collect. Czech. Chem. Commun* **37,** 2156, 1972.
20. A. Kuwae and K. Machida, *Spectrochim. Acta* **35A,** 27, 1979.
21. G. Varsanyi, "Assignments for Vibrational Spectra of Seven Hundred Benzene Derivatives," John Wiley & Sons, New York, 1975.
22. S. Pinchas, D. Samuel, and B. L. Silver, *Spectrochim. Acta* **10,** 179, 1964.

23. N. S. Sundar, *Spectrochim. Acta* **41A,** 905, 1985.
24. J. H. S. Green, W. Kynaston, and A. S. Lindsey, *Spectrochim. Acta* **17A,** 486, 1961.
25. N. N. Sundar, *Spectrochim. Acta* **41A,** 1449, 1985.
26. M. Rangacharyulu and D. Premaswarup, *Ind. J. Pura & Appl. Phys.* **16,** 37, 1978.
27. A. M. C. Fernandez, A. M. T. Gonzalez, and A. G. Chozas, *J. Mol. Struct.* **143,** 557, 1986.
28. T. G. Towns, *Spectrochim. Acta* **39A,** 801, 1983.
29. N. Abasebogovic, L. Colombe, and P. Bleckmann, *J. Raman Spectrosc.* **6,** 92, 1977.
30. L. M. Epstein, E. S. Shubina, L. D. Ashkinadze, and L. A. Kazitsyna, *Spectrochim. Acta* **38A,** 317, 1982.
31. J. F. Bertran, M. Hernandez, and B. L. Serna, *Spectrochim. Acta* **38A,** 149, 1982.
32. M. Harrand, *J. Raman Spectrosc.* **4,** 531, 1975.
33. M. Harrand, *J. Raman Spectrosc.* **8,** 161, 1979.
34. A. Perjessy, D. Rasala, P. Tomasik, and R. Gawinecki, *Collect. Czech. Chem. Commun.* **50,** 2443, 1985.
35. G. Geiseler and H. Kesler, *Ber. Bunsenges. Phys. Chem.* **68,** 571, 1964.
36. J. R. Holtzclaw, W. C. Harris, and S. F. Bush, *J. Raman Spectrosc.* **9,** 257, 1980.
37. W. D. Kumler, *J. Am. Chem. Soc.* **76,** 814, 1954.
38. E. Lieber, D. R. Levering, and I. J. Patterson, *Anal. Chem.* **23,** 1594, 1951.
39. N. O. Cherskaya, V. P. Georelik, V. A. Shlyapochnikov, V. P. Ivshin, and T. N. Ivshina, *Bull. Acad. Sci. USSR, Div. Chem. Sci.* **35,** 1385, 1986.
40. N. O. Cherskaya, V. A. Shlyapochnikov, V. P. Ivshin, and T. N. Ivshina, *Bull. Acad. Sci. USSR, Div. Chem. Sci.* **35,** 1388, 1986.
41. J. H. S. Green and D. J. Harrison, *Spectrochim. Acta* **26A,** 1925, 1970.
42. J. H. S. Green and H. A. Lauwers, *Spectrochim. Acta* **27A,** 817, 1971.

CHAPTER 12

Double Bonds Containing Nitrogen Atoms

12-1. C-NITROSO COMPOUNDS

Most nitroso alkanes are unstable at room temperature. For example, primary and secondary nitroso alkanes easily undergo isomerization to form oximes ($\rangle C{=}N{-}OH$) [1]. Moreover, the Raman analysis of nitroso alkanes [65a] is hampered by photolysis in many cases. Consequently, the vibrational studies of these compounds are often carried out with the IR method either in the gas phase as transient species or by matrix isolation at very low temperatures [1, 3, 5–7]. Although tertiary C-nitroso compounds and nitroso benzenes are stable to isomerism, they dimerize in the solid state [1, 12]. Nevertheless, the vibrational spectra of monomeric tertiary nitroso alkanes and nitroso benzenes can be obtained by dissociating the dimer in dilute solution [1, 12].

12-1.1. The N=O Stretching of Monomeric C-Nitroso Compounds

Table 12-1 shows the N=O stretching frequency of some C-nitroso compounds. Generally, the N=O absorption of aliphatic nitroso compounds occurs in the 1621–1539 cm^{-1} region. Note that the spectral range (1621–1580 cm^{-1}) for α-halogenated C-nitroso compounds is located at the upper limit of the correlated range. A doublet arising from rotational

Table 12-1.

The N=O Stretching Frequency (IR) of C-Nitroso Compounds

Compound	Frequency (cm^{-1})	Ref.
Monomeric:		
Aliphatic nitroso compounds	1621–1539	1–9
$CH_3-N=O$	1564[a]	2–4
	1549[b]	
$Me_3C-N=O$	1574[a]	9
$R_2ClC-N=O$	1600–1580[a]	9
$CCl_3-N=O$	1621[a]	9
$CF_3-N=O$	1595[a]	5, 6
	1594[c]	
$CF_3CF_2CF_2-N=O$	1603[a]	6
$CH_2(OH)-N=O$	1559, 1555[b,e] IR : vs	7
cyclo-$C_6H_{11}-N=O$	1558[a]	8
1,4-dichloro-*cyclo*-$C_6H_9-N=O$	1570[a]	7
$N\equiv C-N=O$	1484[a]	13
Aromatic nitroso compound:	1523–1488	8–11
$Ph-N=O$	ca. 1523[a]	8, 10
p-$MeO-C_6H_4-N=O$	1508[d]	11

Dimeric C-Nitroso Compounds:

Aliphatic C-nitroso compounds:

trans: 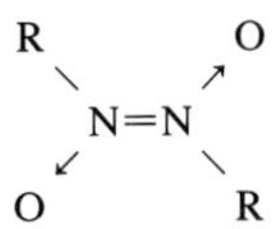

~1490 (R = *t*-Bu)
1290–1176

cis: 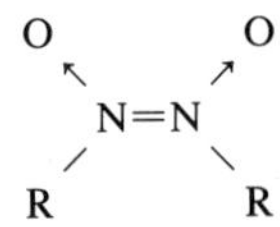

1420–1330
1344–1323

1, 12, 22

Aromatic C-nitroso compounds:

trans: 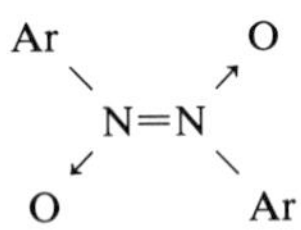

near 1440
1300–1253

cis: 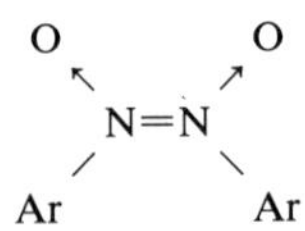

1407
1397–1389

1, 12, 22, 42, 43

[a] Gas phase.
[b] Matrix isolation.
[c] Condensed phase.
[d] Solution.
[e] Rotational isomer.

isomerism is observed in the N=O stretch region for some antisymmetrically substituted C-nitroso compounds. Taking $HOCH_2-N=O$ as an example, the IR bands at 1555 cm^{-1} and 1559 cm^{-1} are assigned by Müller *et al.* [7] to the N=O stretch mode of the trans and the cis conformer, respectively. Conjugation affects the N=O stretching mode by lowering the vibrational frequency. A good example is the N=O stretch absorption of N≡C−N=O, which occurs at 1484 cm^{-1} in the gas phase. In the case of nitrosobenzenes, the N=O stretching frequency is also lowered by conjugation to the region of 1513–1488 cm^{-1} [1, 8–11].

12-1.2. The N=O Stretching of Dimeric C-Nitroso Compounds (Azodioxy Compound)

As stated earlier, tertiary nitroso alkanes and nitroso benzenes are stabilized by dimerization in the solid phase. These dimers are the result when N atoms on each monomer interact to form a partial N=N double bond. Therefore, trans and cis geometrical isomers should be considered when dealing with this type of compound (see Table 12-1):

trans 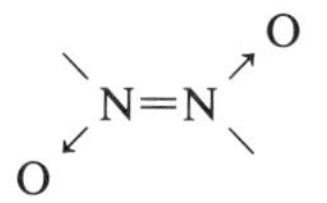 cis 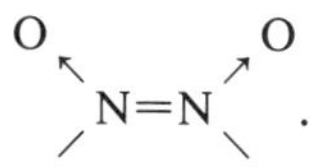 .

For dimeric nitrosobenzene, there is evidence that the π electrons are delocalized between the azodioxy bridge and the aromatic rings, even in the more sterically hindered cis configuration [42, 43]. As a consequence of the electron delocalization, the characteristic N→O frequency of dimeric nitroso compounds is lower than that of the corresponding monomers (Table 12-1). It should be noted that the N→O stretching frequencies of these dimers are in a spectral region similar to that of the azoxy compounds. This similarity becomes more obvious as one compares the structures of dimeric nitroso compounds, −N(O)=N(O)−, and azoxy compounds, −N=N(O)−. Therefore, it seems reasonable to predict that the N=N stretch frequencies of dimeric nitriso compounds are also located in a region near those of azoxy compounds (1480–1450 cm^{-1}). Indeed, the N=N stretching frequency of cis dimeric nitrosobenzene is observed at 1407 cm^{-1} (IR : vs, R : s) [43], and that of trans dimeric nitrosobenzene derivatives occurs near 1440 cm^{-1} (IR : —, R : vs) [42]. In the Raman spectrum for 2-methyl-2-nitrosopropane (nitroso-*tert*-butane), a medium-weak Raman band about 1490 cm^{-1} (which is not observed in the IR spectrum) may be attributed to the N=N stretching mode of the dimer. A strong Raman band near 1270 cm^{-1} involves the N→O in-phase stretch mode [65j].

12-1.3. The C—N Stretching

The C—N stretching mode of C-nitrosoalkanes is complicated by the coupling of other vibrations and appears as two IR bands near 1100 cm^{-1} and in the 860–750 cm^{-1} region [1, 4, 12]. A band involving C—N stretching in trifluoronitrosomethane (CF_3—N=O) is at a lower frequency (809 cm^{-1}, IR : vs, R : vs) than that of CH_3—N=O (870 cm^{-1}, IR : m) [5]. In the case of nitrosobenzenes, the interaction of the C—N stretching mode with ring vibrations increases the frequency to a higher value. Talberg [11] assigned the C—N stretching frequencies of nitrosobenzene and *p*-methoxynitrosobenzene by normal coordinate analysis to the absorptions at 1318 and 1342 cm^{-1}, respectively.

12-1.4. The C—N=O Bending

Another vibration involving the nitroso moiety is C—N=O bending. Barnes *et al.* [3] assigned the IR band of nitrosomethane at 574 cm^{-1} (m) to this vibrational mode.

12-2.

NITROSOAMINES (>N—N=O)

There is a marked drop in the N=O stretching frequency when the N=O group is attached to a nitrogen atom. For aliphatic nitrosoamines (R_2N—N=O), the N=O stretching frequency falls in the region of 1460–1425 cm^{-1} [14–16]. The N—N stretching frequency of aliphatic nitrosoamines is observed in the 1150–1030 cm^{-1} region [14–16], which is somewhat higher than that of nitroamines (R_2—N—NO_2) (1030–980 cm^{-1}). The Raman scattering arising from the symmetrical C_2N-skeletal vibrations gives rise to another prominent band. Taking dimethylnitrosoamine (Me_2N—NO) as an example, the strong Raman bands at 1442 cm^{-1} and 847 cm^{-1} are due to the N=O stretching mode and symmetrical C_2N— stretching mode, respectively [17].

In the case of aromatic nitrosoamines, Ar—N(R)—N=O, the N=O stretching frequency is raised to the region of 1500–1450 cm^{-1}, whereas the N—N stretching mode is lowered to the 1025–925 cm^{-1} region [16], which is very close to the N—N stretching frequency of aliphatic nitroamines (1030–980 cm^{-1}).

Table 12-2.

The Vibrational Frequencies (cm^{-1}) of Nitrosoamines

Compound	ν(N=O)	ν(N—N)	Other Vibrations	Ref.
$Me_2N—N=O$	1442		ν($—NC_2$): 847 (R : vs)	17
$R_2N—N=O$	1460–1425	1150–1030		16
Ar—(R)N—N=O	1500–1450	1025–925	ν(C—N): 1200–1160	16
$(NH_2)_2C=N—NO$	1504			68
HN=CX—NH—NH—N=O	1515			23
$(CF_3CH_2)_2N—NO$	1550			14
—C(=O)—HN—N=O	1527–1510			18

The N=O stretching mode of nitrosoguanidine was first assigned by Lieber *et al.* [23] in the IR to a band at 1653 cm^{-1}, which is of considerably higher value than those of other nitrosoamines (see Table 12-2). This assignment was later revised by Kumler [68], who attributed this IR band (1648 cm^{-1}) to the NH_2 bending vibration and reassigned the absorption at 1504 cm^{-1} to the N=O stretching mode. The N=O stretching frequency of *N*-nitroso-2,2,2-trifluoroethylamine, $(CF_3CH_2)_2N—NO$ (1550 cm^{-1}) [14], and *N*-nitrosoamides, —C(=O)NH—NO (1527–1510 cm^{-1}), is relatively higher than those of the aliphatic nitrosoamines [23, 18, 19]. Bellamy [19] rationalized that the frequency elevation of nitrosoguanidine, α-fluorinated nitrosoalkane, and *N*-nitroso amides is a factor of molecular geometry, which causes a certain spatial interaction between the N—N=O moiety and the substituents.

12-3. NITRITES (—O—N=O)

The presence of a neighboring oxygen atom increases the N=O stretch frequency of nitrites to 1680–1613 cm^{-1}, in comparison with the 1621–1488 cm^{-1} range of C-nitroso compounds. For most nitrites, the N=O stretch mode appears strong in both Raman and IR spectra [24a]. Furthermore, the rotation about the O—NO single bond introduces two strong N=O stretch bands in IR spectra due to C—O—N=O cis and trans isomers. In general, the N=O stretch modes of trans isomers (1680–1653 cm^{-1}) occur at higher frequencies than those of cis isomers (1625–1613 cm^{-1}) [20, 21].

Table 12-3.

The Characteristic Frequencies (cm^{-1}) of Nitrites (—O—N=O) [20]

	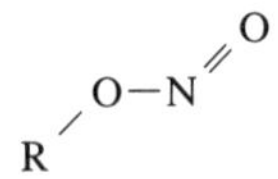 trans	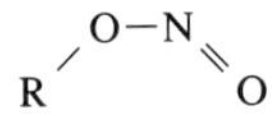 cis	$A_{(trans)}/A_{(cis)}$
N=O Stretch:			
Primary nitrites	1681–1669	1625–1613	0.95–3.5
Secondary nitrites	1667–1664	1618–1613	4.7–10
Tertiary nitrites	1655–1653	1613–1610	35–50
N—O Stretch:			
Primary nitrites	814–790		
Secondary nitrites	780–775		
Tertiary nitrites	764–751		
O—N=O Bend:			
Primary nitrites	625–565	691–617	
Secondary nitrites	605–594	688–678	

Branching on the carbon atom next to the —O—N=O moiety tends to reduce the N=O stretch frequency. Consequently, the N=O frequency decreases progressively as the adjacent alkyl group varies from primary through secondary to tertiary structures (see Table 12-3). Tarte [20] observed that the relative absorbance ratio, A_{trans}/A_{cis}, is very characteristic for each type of alkyl substituent. In the gas-phase IR spectra, the A_{trans}/A_{cis} ratio is 0.95–3.5 for primary alkylnitrites, 4.7–10 for secondary alkylnitrites, and 35–50 for tertiary alkylnitrite.

The presence of fluorine atoms in the vicinity of the —O—NO moiety appears to increase the N=O stretching frequency, and the strong IR bands of CF_3CH_2—O—NO at 1736 cm^{-1} and 1695 cm^{-1} were attributed to N=O stretchings of different conformations [19]. Since the fluorine atoms are quite remote from the —O—N=O moiety in this case, their influence on the N=O stretching frequency is more likely related to the spatial arrangement of the molecule than to the inductive effect [19].

12-4.

AZO COMPOUNDS (—N=N—)

Azo compounds are characterized by the N=N stretching mode in vibrational spectra. Most aliphatic azo compounds exist as the more stable trans form for C—N=N—C. However, the cis isomers are also observed in some cases, especially for aromatic azo compounds. Because of the symmetry property, the N=N stretch mode of symmetrically substituted trans isomer is usually IR-forbidden, but exhibits a very intense Raman band. The IR intensity of the N=N stretch band remains very weak even when the trans azo compound is antisymmetrically substituted. For cis azo compounds, however, the N=N band appears strong in the infrared spectrum.

12-4.1. The N=N Stretching Mode

Table 12-4 summarizes the N=N stretching frequency of selected azo compounds. The N=N stretch frequency of trans azo alkanes occurs in the 1576–1565 cm^{-1} region. Only a limited amount of studies have been done for cis azo alkanes. The N=N stretch frequency of this class of compounds is typified by the strong IR absorption of 1-pyrazoline derivatives in the 1550–1540 cm^{-1} region. See Table 12.4, (I).

The effect of conjugation results in a frequency lowering of the N=N mode. Therefore, compounds containing a carbonyl group next to the N=N moiety present a weak IR band near 1550 cm^{-1}. Upon conjugation with a phenyl group, the N=N stretch frequency is shifted to an even lower region. In Table 12-4, it is demonstrated that trans azobenzenes have an N=N stretch frequency at 1463–1380 cm^{-1} [22], which is about 150 cm^{-1} lower than that of trans azoalkanes. The N=N stretch frequency of cis azobenzenes, near 1511 cm^{-1}, however, is only about 50 cm^{-1} lower than that of 1-pyrozolenes (cis azoalkanes).

The difference between the frequency shifts of cis and trans azobenzenes from the corresponding azo alkanes can be explained by the steric hindrance effect. The steric interaction between the two phenyl groups in cis azobenzenes prevents coplanarity, which results in less conjugation and hence raises the N=N frequency to a region closer to that of the cis aliphatic azo compounds.

The assignment of the N=N stretch mode is more difficult for azo naphthols and azo naphthylamine. In the vibrational spectra of these substituted azo naphthalenes, the N=N band is overlaid by not only the aromatic C$\overset{..}{=}$C stretching bands but also the C=N stretching bands arising from the tautomerism effect [44, 45, 48]. A broad region of 1570–1360 cm^{-1} is expected for azo compounds containing naphthol and naphthylamine substituents next to the —N=N— moiety [22, 45–49].

Table 12-4.

The Characteristic Frequencies (cm^{-1}) of Azo, Azoxy, and Azothio Compounds

(I). Azo Compounds:

trans: 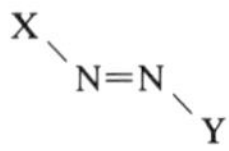

$-X^a$	$-Y$	N=N	IR	R	Ref.
−Me	−H	1550	vw	s	25, 26
−Me	−Me	1575	—	s	26, 27
$-CMe_2CN$	$-CMe_2CN$	1580			22
−R	−R	1576–1565	vw		35
−C(=O)Ph	−R	1550–1540	w		28
−C(=O)Me	−R	~1550	w		28
−COOR	−COOR	~1555			24
−NHPh	−Ph	1416			22
$-SCPh_3$	−Ph	1424		vs	29
$-SO_2Ph$	−Ar	1488–1423		s-vs	29
−Ph	−Ph	1442		vs	30–32
−Ar	−Ar	1463–1380			22

cis: 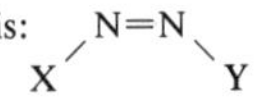

−X	−Y	N=N	IR	R	Ref.
−Ph	$-SCPh_3$	1492		m	29
−Ph	−Ph	1511	vs	w	31, 36
1-pyrazoline N=N		1548	vs		33
substituted 1-pyrazolines		1550–1540			34

(II). Azoxy Compounds:

trans 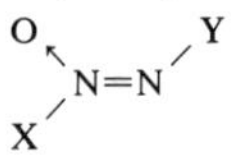

−X	−Y	N=N	IR	R	Ref.
−R	−R	1530–1495			38, 39
−Ph	−Ph	1413	vs	s	37

cis 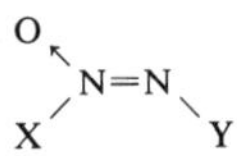

−X	−Y	N=N	IR	R	Ref.
−Ph	−Ph	1468	vs	w	36
Azoxybenzenes[b]		1490–1450			22, 40

(III). Azothiobenzenes:

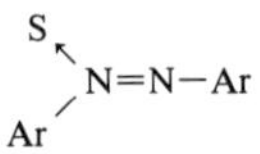

	N=N	Ref.
	1465–1445	41

[a] −Me: $-CH_3$, −Ph: $-C_6H_5$, −Ar: aromatic substituents.

[b] The geometry is not specified in the literature. However, these compounds are likely in the cis form, since the N=N frequencies are closer to that of *cis*-azoxybenzene than that of *trans*-azoxybenzene.

12-4.2. The Skeletal Vibrations

The vibrational analysis of azomethane reveals the symmetric C—N stretching frequency at 1186 cm^{-1} (IR : O, R : w) and the antisymmetric C—N stretching frequency at 1001 cm^{-1} (IR : s, R : —) [26, 27]. Furthermore, a very strong Raman band at 590 cm^{-1} (IR : —) is attributed to a mixing of symmetric C—N stretching and symmetric C—N=N bending modes. In the case of aromatic azo compounds, the vibration of the C—N group appears at around 1154 cm^{-1} [30].

12-5.

AZOXY COMPOUNDS (—N=N(→O)—)

The two characteristic absorptions of azoxy compounds are due to the antisymmetric N=N→O stretching (mainly N=N) and symmetric N=N→O stretching (mainly N→O). The N→O stretching frequency is correlated to an IR band in the region of 1335–1250 cm^{-1} [22, 50, 85, 86].

The presence of an oxygen atom on the N=N moiety largely enhances the IR intensity of the N=N stretch mode, especially in the case of trans isomers. Such enhancement is originated from the polar and symmetrical properties of the molecule in the vicinity of the N=N bond. The N=N stretching frequencies of aliphatic azoxy compounds are located in the 1530–1495 cm^{-1} region (Table 12-4, (II)). Like azo compounds, the cis azoxybenzene exhibits a higher N=N stretching frequency (1468 cm^{-1}) than does the corresponding trans isomer (1413 cm^{-1}), because of the reduced conjugation effect by steric hindrance in the cis configuration.

12-6.

AZOTHIO COMPOUNDS (—N=N(→S)—)

The N=N stretch band of azothiobenzenes is located in the spectral region of 1465–1445 cm^{-1}, which is essentially the same as that of azoxybenzenes (Table 12-4, (II), (III)). However, the N→S stretching frequency in the 1070–1058 cm^{-1} range [41] can be used to differentiate these two types of compounds, since the N→O stretching frequency of azoxybenzenes is located in a higher region (1335–1310 cm^{-1}).

12-7.

COMPOUNDS CONTAINING THE C=N GROUP

Compounds containing a C=N group are characterized by the C=N stretching band in IR and Raman spectra. The C=N stretching frequency, however, is expected to be phase-sensitive, since the C=N moiety is polar. In the presence of an —OH or a $—NH_2$ group, the frequency shift of the C=N stretch mode upon phase change is particularly pronounced because of hydrogen-bond interaction. In the case of the —CH=N— type of structure, the vibrational spectra are further characterized by the C—H stretching mode in a similar region to that of alkenes.

The IR intensity of the C=N stretching mode is often found between those of the C=O and C=C stretching modes. Although a symmetrically substituted C=C moiety can be easily distinguished from the C=N stretching by the IR intensity, the C=C stretch vibration remains a major interference source on assigning the C=N mode, especially in the case of unknown determination. Therefore, other vibrational bands should be consulted for structural elucidation. The assignment of the C=N stretch mode by IR spectroscopy is sometimes hindered when the C=N group is conjugated to an aromatic system. This difficulty arises from the fact that the C=N stretching frequency is lowered by the conjugation effect to a region where the spectral interference from aromatic ring stretchings (near 1600 cm^{-1}) is a consideration. Another spectral interference in IR spectra is due to the NH_2 scissoring mode in the 1640–1590 cm^{-1} region.

The C=N stretch band appears strongly in the Raman spectra. This Raman band is expected to be polarized when measured in solution or liquid phase [65c, 75]. Since the NH_2 scissoring mode is weak in Raman and the aromatic ring stretching Raman bands are depolarized, the assignment of the C=N stretch band with Raman spectroscopy is therefore more certain due to its strong and polarized nature.

12-7.1. Imines and Schiff Bases ($C_2C{=}N{-}$)

The C=N absorption of some selected imines and Schiff bases are listed in Table 12-5 [65b]. The IR spectra of methyleneimine ($H_2C{=}NH$) and ethylideneimine $CH_3CH{=}NH$ have been studied extensively [51–60]. In the case of methyleneimine, the C=N stretching mode occurs at 1638 cm^{-1} in the gas phase. When the molecule contains an antisymmetrically substituted C=N moiety, a splitting can be observed in the C=N stretching region due to cis and trans isomers. Hashiguchi *et al.* [58] reported a doublet at 1655 and 1651 cm^{-1} in the IR spectrum for the C=N stretch mode of ethylideneimine.

As demonstrated in Table 12-5, the C=N stretching frequency of the alkylated Schiff bases is located in the range of 1674–1649 cm^{-1}. A downward shift occurs when the C=N group is included as one of the ring bonds in a strained ring system. For pyrrolines, where the C=N moiety is included in a five-membered ring, the C=N stretching frequency is measured to be 1653–1613 cm^{-1} in CCl_4 solution [69]. An even lower frequency is observed as the ring strain increases. For example, the C=N stretch mode of 1-azetine (a four-membered ring) occurs at 1575 cm^{-1} (Table 12-5). Similar results are reported for chlorinated 1-azine derivatives, which have the C=N stretching band near 1580 cm^{-1} in the IR spectra of the neat sample [64].

Table 12-5.

The C=N Stretching Frequency (cm^{-1}) of Imines

Compounds	Frequency (cm^{-1})	Ref.	Note
(I). Nonconjugated Imines:			
$H_2C{=}N{-}H$	1638	51	gas
$(CH_3)HC{=}N{-}H$ trans:	1651	58	gas
cis:	1655	22	
$R_2C{=}N{-}H$	1646–1640		
$H_2C{=}N{-}CH_3$	1658	61	matrix isolation
$RHC{=}N{-}R$	1674–1665	80	
$R_2C{=}N{-}R$	1665–1645	22, 80	
$HC{=}N$ / $H_2C{-}CH_2$ (ring)	1575	67	
(II). Conjugation with C=C Bond and Phenyl Group:			
$(C_6H_5)RC{=}N{-}H$	1633–1620	22	
$(C_6H_5)_2C{=}N{-}H$	1600	66	
$H_2C{=}N{-}CH{=}CH_2$ [a]	1635 and 1618	62	gas
$(C_6H_5)HC{=}N{-}R$	1656–1629	22	
$(C_6H_5)HC{=}N{-}C_6H_5$	1637–1626	22	
$(C_6H_5)RC{=}N{-}R$	1650–1640	22	
$CH_3(C_6H_5)C{=}N{-}C_6H_5$	1640–1630	22	
$(C_6H_5)HC{=}N{-}CCl(C_6H_5)$	1605	64	neat

[a] See text. Most likely in-phase and out-of-phase interaction of C=N with C=C.

Conjugation of the C=N moiety with phenyl groups shifts the stretching frequency to the 1650–1600 cm^{-1}. There are disagreements on assigning the C=N stretch mode of 2-azabutadiene ($H_2C{=}N{-}CH{=}CH_2$). Amatatsu *et al.* [62] assigned the vapor-phase IR band at 1635 cm^{-1} to the C=N stretching mode, and the band at 1618 cm^{-1} to the C=C stretching mode for this particular compound. Guillemin *et al.* [63], however, attributed the higher-frequency band (1628 cm^{-1}) to the C=C stretch vibration and the lower-frequency band (1610 cm^{-1}) to the C=N stretch mode for the same molecule. Meyers [69] reported the IR spectra (CCl_4 solution) of some 1-pyrrolines, which consist of a cyclic C=N group conjugated by an exocyclic C=C bond. He assigned the IR band in the 1587–1560 cm^{-1} region to the C=N mode, and the band in the range 1664–1637 cm^{-1} to the C=C mode.

In the assignments of the authors just mentioned, the characteristic stretchings involving the C=C—N=C moiety are considered to be separately localized at the C=C and C=N bonds. However, it is very likely that C=C and C=N stretching modes will couple with each other, since these two vibrations exhibit similar stretching frequencies. By analogy with butadienes (C=C−C=C), glyoximes (HO−N=C−C=N−OH), and azines (−N=C−C=N−), it seems more reasonable to describe the characteristic stretchings of the C=N−C=C linkage as in-phase and out-of-phase C=C−N=C stretching modes. That, is, both C=C and C=N stretchings are involved in the IR bands of 2-azabutadiene at 1635 and 1618 cm^{-1}. Similarly, the IR bands in the 1587–1560 and 1664–1637 cm^{-1} regions are characteristic of C=C−N=C stretchings of the conjugated 1-pyrrolines.

Alkylaldehyde azines. $(R{-}CH{=}N{-})_2$ and $(R_2C{=}N{-})_2$, have anti-symmetric C=N stretch bands at 1663–1636 cm^{-1}. IR : m-s, and symmetrical C=N stretch bands at 1625–1608 cm^{-1}, R : s. Arylaldehyde azines, $(Ar{-}CH{=}N{-})_2$, have comparable bands at 1632–1606 cm^{-1} and 1563–1539 cm^{-1} in most cases [84].

12-7.2. Oximes (>C=N−OH)

Examples of the C=N stretching frequency of solid/liquid oximes are cited in Table 12-6. The IR absorption of the C=N mode is of medium-to-weak intensity, while the corresponding Raman band is strong and polarized [65c].

The C=N stretching frequency is expected to shift markedly upon the change of phase, since hydrogen bonding is significant for oximes. Bellamy [19] predicted a 20 cm^{-1} rise in the C=N stretching frequency for unassociated oximes as compared to the associated molecules. The formation of hydrogen bonds not only lowers the C=N stretch frequency, but also causes band splitting in the Raman spectra of some aromatic oximes [24b, 65h].

For aldoximes and ketoximes, the C=N stretching frequencies are generally located in similar regions. Geometrical isomerism appears to have little effect on the C=N frequency [19]. For aliphatic oximes, the C=N stretching appears in a region above 1650 cm^{-1}. The conjugation of the C=N group to a C=O group or an aromatic ring lowers the C=N stretching frequency to the 1650–1630 cm^{-1} region. For oximes containing a C=N bond exocyclic to a six- or five-membered ring, the C=N stretching frequencies are located in a higher region (see Table 12-6).

As with alcohols and phenols, the O−H stretching of oximes absorbs IR radiation strongly in the range of 3300–3100 cm^{-1} as a very broad band, but it contributes weak or no Raman scattering. The N−O stretching mode of oximes is correlated with one or more strong IR bands at 1030–870 cm^{-1} [22, 89].

Table 12-6.
The C=N Stretching Frequency (cm^{-1}) of Oximes

Compounds	C=N (cm^{-1})	Ref.
I. Aldoximes:		
(H)(R)C=N−OH	1670–1649	24a
(H)(CCl_3)C=N−OH	1619	73
(H)(Ph)C=N−OH[a]	1640–1630	65c
(H)(o-HO−C_6H_4)C=N−OH	1617	73
II. Ketoximes:		
$(CH_3)_2$NC=N−OH	1675	74
(R)(R′)C=N−OH	1666–1652	24b, 65c
(H_3C)($CH_3C(=O)$)C=N−OH	1635	65d
(R)(Ph)C=N−OH[a]	1640–1620	24a, 65c
$(C_6H_5)_2$C=N−OH	~1651	65e
Cyclohexanone oxime	1669	19, 65f
Cyclopentanone oxime	1685	19
d-Camphor oxime	~1685	65g

[a] Ph: phenyl or substituted phenyl group.

12-7.3. Glyoximes (HO−N=CX−CX=N−OH)

The coupling between the two C=N stretchings in glyoximes gives rise to symmetric and antisymmetric N=C−C=N stretchings. For glyoxime (HO−N=CH−CH=N−OH), Bardet and Alain [75] attributed the strong Raman band at 1636 cm^{-1} to the symmetric N=C−C=N stretching, and the weak IR band at 1610 cm^{-1} to the corresponding antisymmetric mode.

In the case of substituted glyoximes, Kahovec and Kohlrausch [73] correlated the strong Raman bands at 1630 and 1516 cm^{-1} to the C=N stretchings of methylglyoxime ($HO-N=CH-C(CH_3)=N-OH$), and those at 1645 and 1506 cm^{-1} to the same vibrations of dimethylglyoxime ($HO-N=C(CH_3)-C(CH_3)=N-OH$).

Interestingly, the relative intensities of antisymmetric and symmetric N=C—C=N stretching modes in IR and Raman spectra are very different from those observed for the C=C—C=C stretchings of conjugated butadienes. In the spectra of methylated glyoximes mentioned above, not only the symmetric N=C—C=N stretchings, but also the antisymmetric modes exhibit very weak IR intensity despite the fact that most glyoximes exist as the *s*-trans configuration in the vicinity of the N=C—C=N linkage [76]. The Raman intensity of the antisymmetric and symmetric N=C—C=N stretching modes, on the other hand, is prominent.

Based on the observation on these mentioned molecules, we propose (Table 12-7) the possible C=N stretchings for some selected glyoxime derivatives, for which IR and Raman spectra are available. As shown in Table 12-7, two Raman bands in the regions of 1650–1593 and 1610–1500 cm^{-1} are characteristic for substituted glyoximes.

12-7.4. Azines (>C=N—N=C<)

The vibrational analysis of solid formaldazine ($H_2C=N-N=CH_2$) is reported by Bondybey and Nibbler [70]. Similar to conjugated butadienes, this molecule exhibits three rotational conformations, namely, *s*-trans, *s*-cis,

Table 12-7.

The C=N Stretching Frequencies (cm^{-1}) of Glyoximes (HO—N=CX—CY=N—OH) in the Solid State

		N=C—C=N stretch						
—X	—Y		IR	R		IR	R	Ref.
—H	—H	1636 (sym. str.)	—	vs	1610 (antisym. str.)	vw	—	75, 76
—H	$-CH_3$	1630		s	1516		s	73
—H	$-NO_2$	1650	m	s	1608	w	s	76
$-CH_3$	$-CH_3$	1650	—	vs	~1512	—	m	65i, 73, 76
$-CH_3$	$-C_6H_5$	1641	—	s	~1500	vw	m	65i
$-C_6H_5$	$-C_6H_5$	1627	—	s	~1495?	m	m	65i, 76
1,2-cyclohexanedione, dioxime		1645	vw	w	1578	vw	s	65i
—Cl	—Cl	~1620	vw	—	1588	—	vs	65i, 76
—CN	—CN	1593	—	vs	1582	w	—	76

and gauche isomers. In the solid state, formaldazine exists exclusively as an *s*-trans isomer at subambient temperature. However, the *s*-trans and a small amount of the gauche conformers coexist at room temperature. For the *s*-trans formaldazine, a medium IR band at 1637 cm^{-1} is observed due to the antisymmetric (out-of-phase) C=N—N=C stretching vibration. The symmetric (in-phase) C=N—N=C stretching occurs strongly at 1612 cm^{-1} in the Raman spectrum [70].

The spectral correlation of the C=N stretching modes of azines has been discussed previously by several authors [19, 43a]. The antisymmetric C=N—N=C stretching modes of aldazines and ketazines are located in the region of 1663–1636 cm^{-1} (IR : s, R : w-O), and the corresponding symmetric C=N—N=C stretching vibration is situated in the 1625–1608 cm^{-1} range (IR : —, R : s). The effect of fluorination is illustrated by the C=N stretching frequencies of tetrafluoroformaldazine ($F_2C{=}N{-}N{=}CF_2$). For *s*-trans tetrafluoroformaldazine, King *et al.* [72] assigned the Raman band at 1758 cm^{-1} to the symmetric C=N—N=C stretching, and the strong IR band at 1747 cm^{-1} to the corresponding antisymmetric mode. Recently, Kolb *et al.* [71] reported two characteristic IR bands near 1635 and 1615 cm^{-1} for some ketazines with a structure such that the two N=C moieties are exocyclic to different five-membered rings.

In the Raman spectrum of solid formaldazine, the N—N stretching mode occurs at 1216 cm^{-1}(m) for the *s*-trans isomer. This vibration is not observed in IR due to its molecular symmetry. The vibrational spectra of formaldazine are further characterized by the CH_2 stretchings in the IR at 3056(m) and 2924(s) cm^{-1} and in the Raman at 3060(s) and 2930(vs) cm^{-1}.

12-7.5. Imidates and Imino Carbonates

The C=N stretching frequencies of imidates and imino carbonates are summarized in Table 12-8. According to the IR frequency measured by Mathis *et al.* [77], the C=N stretching frequency of imidates appears to be less affected by phase change, since the stretching frequencies measured in condensed phase and CCl_4 solutions are virtually the same.

Comparing the C=N absorption of benzimidic acid esters (Table 12-8, I(d)) and phenylacetimic acid esters (Table 12-8, I(c)), it is apparent that conjugation with a phenyl group lowers the C=N frequency by about 20 cm^{-1}. Pirkle and Sowin [79] have reported elevated C=N frequencies (1698–1675 cm^{-1}) for those imidates in which the C=N bonds are conjugated and exocyclic to a five-membered ring (see Table 12-8, I(e)).

For EtO—C(Ph)=N—NH—Ph types, Baccar *et al.* [78] assigned the IR band at 1595 cm^{-1} to the C=N stretching mode. When compared with

Table 12-8.

The C=N Stretching Frequency (cm^{-1}) of Imidates and Imino Carbonates

Compounds	C=N Stretch (cm^{-1})	Ref.
I. Imidates:		
(a) $RO(R')C{=}N{-}H$	1655–1652	24b
(b) $RO(Ar)C{=}N{-}H$	1645–1630	24b, 77
(c) $EtO(PhCH_2)C{=}N{-}Y^a$ Y = −Me, −Et, −*n*-Pr, −*n*-Bu, −*i*-Pr, −OEt	1676–1660	78
(d) $EtO(Ph)C{=}N{-}Y^a$ Y = −C(=O)Me, −COOMe, $-CH_2COOMe$, $-CH_2CN$, $-CH_2CH_2Br$	1665–1660	78
(e) Conj. imidates, exo C=N, five-membered ring	1698–1675	79
II. Imino Carbonates:		
$(EtO)_2C{=}NH$	1658	24b
III. Amidines:		
(a) $C_6H_5{-}NH(C_6H_5)C{=}NH$	~1620	65j
(b) $Me(Et{-}NH)C{=}N{-}Et$	1675 (monomer, dioxane soln.) 1592 (dimer, hexane soln.)	24b
(c) $R(C_6H_5{-}NH)C{=}N{-}C_6H_5$	1658–1632 (doublet) (chloroform soln.)	24b
(d) $Me(C_6H_5{-}NH)C{=}N{-}C_6H_5$	1640	65j
(e) $R(C_6H_5{-}NR')C{=}N{-}C_6H_5$	1633–1619 (single band) (chloroform soln.)	24b
IV. Guanidines:		
(a) $((CH_3)_2N)_2C{=}N{-}H$	1592	65k
(b) $(C_6H_5{-}NH)_2C{=}N{-}H$	~1635	65k
(c) $(C_6H_5{-}NH)_2C{=}N{-}C_6H_5$	~1630	65k

[a] Ph: substituted phenyl group.

the C=N stretching frequency of phenyl-substituted hydrazones or imidates, 1595 cm^{-1} is unusually low for the C=N stretch mode. It is well known that the aromatic ring stretchings of a monosubstituted phenyl group also exhibit medium IR bands near 1600 cm^{-1}.

12-7.6. Amidines (N—C=N)

Examples of the C=N stretching frequency of amidines are given in Table 12-8. For *N*-phenylbenzamidine (Table 12-8, III(a)) and *N,N'*-diphenylaceytamidine (Table 12-8, III(e)), the C=N stretch mode appears strongly in both IR and Raman spectra (65j). Some N-monosubstituted amidines (Table 12-8, III(c)) exhibit a doublet in the C=N region due to the presence of syn and anti rotational isomers. This doublet feature, however, is not observed for N,N-disubstituted amidines (Table 12-8, III(d)). The influence of hydrogen bonding on the C=N stretch frequency can be pronounced for amidines containing the —NH— group, such as the example given in Table 12-8, III(b). For this particular compound, the C=N frequency of the monomer occurs at 1675 cm^{-1} in dioxane solution. The formation of cyclic dimers in hexane solution, however, shifts the C=N frequency to a lower value of 1592 cm^{-1}.

12-7.7. Guanidines (N_2C=N—)

Guanidines usually absorb in the 1689–1500 cm^{-1} range because of the C=N stretching, NH deformation, and NH_2 deformation [22, 68]. Bellamy [19] quoted a higher region of 1718–1590 cm^{-1} for the C=N stretch absorption of guanidines. The interference of the NH as well as NH_2 deformations often causes an uncertainty in determining the C=N stretch mode in IR spectra. This difficulty is eliminated in Raman spectra, since the C=N stretching mode is expected to give rise to a prominent Raman band while the NH bendings are relatively weak. Examples of the C=N frequencies of guanidines in Raman spectra can be found in Table 12-8.

The C=N stretching band is markedly weakened in Raman spectra upon salt formation [65k]. This is due to the delocalization of the π electrons among the three CN bonds of the CN_3 moiety. Since the charge is distributed over three nitrogen atoms by delocalization, the NH_2 stretch frequencies of guanidines are less affected by salt formation than are the NH_2 stretch frequencies of amines. A strong Raman band is observed appearing consistently in the 1050–990 cm^{-1} region for the salts of guanidine $C(NH_2)_3^+$ [65k]. This band can be attributed to the symmetric CN_3 stretching mode of guanidine salts.

12-7.8. Hydrazone, Semicarbazone, Thiosemicarbazone, and Carbohydrazone

Hydrazones (>C=N−N<) with the >C=N−NH_2 structure absorb near 1650 cm^{-1}. Substitution on the terminal −NH_2 group, however, reduces the C=N stretching frequency to the 1644–1610 cm^{-1} region [19, 24b]. In the IR spectra of phenyl-substituted hydrazones, the assignments of the C=N stretching mode are interfered with by the IR bands due to aromatic ring stretchings [24b, 81]. The Raman spectroscopic method provides a more convenient way of characterizing the C=N linkage, since the C=N Raman bands are prominent in the majority of cases.

The C=N stretching modes of semicarbazones

(>C=N−NX−C(=O)−)

and thiosemicarbazones

(>C=N−NX−C(=S)−)

appear in a very similar spectral region. In general, the C=N stretch band is medium to weak in the IR spectra but strong in Raman [651]. For aliphatic semicarbazones and thiosemicarbazones, the C=N stretching frequency occurs in the region of 1655–1642 cm^{-1}. Conjugation of the C=N linkage in these compounds with a phenyl or carbonyl group lowers the C=N stretch frequency to the 1630–1610 cm^{-1} range [24b, 651].

The C=N stretching frequency of carbohydrazones

(>C=N−NX−C(=O)−NX−N=C<)

is expected to be found in a similar spectral region to that of the semicarbazones, since there is great similarity between the two types of molecules. Based on the Raman spectra available [651], a region of 1630–1610 cm^{-1} can be correlated to the C=N band of aryl carbohydrazones. Again, the Raman intensity of this vibrational mode is usually strong, while the IR absorption appears to be medium to weak in intensity.

12-7.9. Nitrones (>C=N(→O)−)

The nitrone group (C=N→O) has IR bands in the region 1620–1540 cm^{-1} involving the C=N stretch and in the region, 1280–1067 cm^{-1} involving the N→O stretch [86–88]. The N→O stretching frequencies of ketonitrones (C_2C=N(→O)−) are higher than those of aldonitrones (C−CH=N(→O)−), and in both of these, *N*-alkyl nitrones have higher N→O frequencies than do *N*-aryl nitrones [86].

12-8. COMPOUNDS CONTAINING THE N=S GROUP

The N=S stretching frequency of $C_6H_5{-}N{=}S(OMe)_2$ is at 1283 cm^{-1}, according to Dresdner and coworkers [82]. A recent study on the IR spectra of cyclic *N*-substituted imidosulfites [83] reveals that the N=S stretching vibration occurs in the range of 1333–1287 cm^{-1}. For cyclic imidosulfites, this N=S absorption often exhibits multiple-band structure due to rotational conformation and Fermi resonance.

REFERENCES

1. B. G. Gowenlock and W. Lüttke, *J. Chem. Soc., Quart. Rev. XII*, 321, 1958.
2. N. P. Ernsting and J. Pfab, *J. Chem. Soc., Faraday II* **74,** 2286, 1978.
3. A. J. Barnes, H. E. Hallam, S. Waring, and J. R. Armstrong, *Faraday Trans. II* **72,** 1, 1976.
4. J. P. Dognon, C. Pouchan, and A. Dargelos, *Chem. Phys. Lett.* **99,** 316, 1983.
5. H. F. Shurvell, S. C. Dass, and R. D. Gordon, *Can. J. Chem.* **52,** 3149, 1974.
6. J. Mason and J. Dunderdale, *J. Chem. Soc., London*, 754, 1956.
7. R. P. Müller, J. R. Huber, and H. Hollenstein, *J. Mol. Spectrosc.* **104,** 209, 1984.
8. W. Lüttke, *Z. Elektrochem.* **61,** 302, 1957.
9. P. Tarte, *Bull. Soc. Chim. Belg.* **63,** 525, 1954.
10. G. M. Bradley and H. L. Strauss., *J. Phys. Chem.* **18,** 1953, 1975.
11. H. J. Talbérg, *Acta Chem. Scand A* **33,** 289, 1979.
12. B. G. Gowenlock, H. Spedding, J. Trotman, and D. H. Whiffen, *J. Chem. Soc., London*, 3927, 1957.
13. E. A. Dorko and L. Buelow, *J. Chem. Phys.* **62,** 1869, 1975.
14. R. N. Hazeldine and B. J. H. Mattinson, *J. Chem. Soc., London*, 4172, 1955.
15. P. Tarte, *J. Chem. Phys.* **23,** 979, 1955.
16. R. L. Williams, R. J. Pace, and G. J. Jeacoke, *Spectrochim. Acta* **20,** 225, 1964.
17. V. J. M. G. Barredo and J. Goubeau, *Z. Org. Chem.* **251,** 2, 1943.
18. E. H. White, *J. Am. Chem. Soc.* **77,** 6008, 1955.
19. L. J. Bellamy, "The Infrared Spectra of Complex Molecules," 3rd. Ed., John Wiley & Sons, Inc., 1975, New York, pp. 331-346.
20. P. Tarte, *J. Chem. Phys.* **20,** 1570, 1952.
21. R. N. Haszeldine and J. Jander, *J. Chem. Soc.*, 691, 1954.
22. N. B. Colthup, L. H. Daly, and S. E. Wiberley, "Introduction to Infrared and Raman Spectroscopy," 3rd Ed., Academic Press, New York, 1990.
23. E. Lieber, D. R. Levering, and L. J. Patterson, *Anal. Chem.* **23,** 1594, 1951.
24. F. R. Dollish, W. G. Fateley, and F. F. Bentley, "Characteristic Raman Frequencies of Organic Compounds," John Wiley & Sons, New York, 1974. (a) p. 43. (b) pp. 134–142.
25. M. N. Ackerman, J. J. Burdge, and N. C. Craig, *J. Chem. Phys.* **58,** 203, 1973.
26. J. R. Durig, C. B. Pate, and W. C. Harris, *J. Chem. Phys.* **56,** 5652, 1972.
27. R. A. R. Pearce, I. W. Levin, and W. C. Harris, *J. Chem. Phys.* **59,** 1209, 1973.
28. T. R. Lynch, F. N. Maclachlan, and Y. Y. Siu, *Can. J. Chem.* **49,** 1598, 1971.

29. J. A. Carroll, D. R. Fisher, G. W. Canham, and D. Sutton, *Can. J. Chem.* **52,** 1914, 1974.
30. H. Hacker, *Spectrochim. Acta* **21,** 1989, 1965.
31. R. Kubler, W. Lüttke, and S. Weckherlin, *Z. Elektrochem.* **64,** 650, 1960.
32. R. G. Denning and J. Thatcher, *J. Am. Chem. Soc.* **90,** 5917, 1968.
33. J. R. Durig, J. M. Karriker, and W. C. Harris, *J. Chem. Phys.* **52,** 6092, 1970.
34. R. J. Crawford, A. Mishra, and R. J. Dummel, *J. Am. Chem. Soc.* **88,** 3959, 1966.
35. L. Spialter, D. H. O'Brien, G. L. Untereiner, and W. A. Rush, *J. Org. Chem.* **30,** 3278, 1965.
36. A. Gruger and N. Le Calvé, *J. Chim. Phys.* **69,** 743, 1972.
37. A. Gruger and N. Le Calvé, *Spectrochim. Acta* **28A,** 1253, 1972.
38. B. W. Langley, B. Lythgoe, and N. V. Riggs, *J. Chem. Soc., London*, 2309, 1951.
39. J. Jander and R. N. Hazeldine, *J. Chem. Soc.*, London, 919, 1954.
40. B. Witkop and H. M. Kissman, *J. Am. Chem. Soc.* **75,** 1975, 1953.
41. A. Foffani, G. Leandri, I. Zanon, and C. Carpanelli, *Tetrahedron Lett.*, No. 11, 21, 1959.
42. A. Gruger and N. Le Calvé, *Spectrochim. Acta* **31A,** 581, 1975.
43. A. Gruger, N. Le Calvé, and J. Fillaux, *Spectrochim. Acta* **31A,** 595, 1975.
44. D. Hadzi, *J. Chem. Soc.*, 2143, 1956.
45. P. J. Trotter, *Appl. Spectrosc.* **31,** 30, 1977.
46. I. M. Rao, D. Satyanarayana, and U. Agarwala, *Bull. Chem. Soc. Jpn.* **52,** 588, 1979.
47. D. Beteridge and D. John, *Analyst* **98,** 377, 1973.
48. P. M. Drozdzewski, *Spectrochim. Acta* **41A,** 1035, 1985.
49. A. Zabokrycka, P. Klaeboe, B. N. Cyvin, S. J. Cyvin, and J. Brunoll, *Spectrochim. Acta* **38A,** 357, 1982.
50. K. Nakanishi and P. H. Solomon, "Infrared Absorption Spectroscopy," 2nd Ed., Holden-Day, Inc., San Francisco, 1977.
51. Y. Hamada, K. Hashiguchi, M. Tsuboi, Y. Koga, and S. Kondo, *J. Mol. Spectrosc.* **105,** 70, 1984.
52. D. E. Milligan, *J. Chem. Phys.* **35,** 1491, 1961.
53. C. B. Moore, G. C. Pimentel, and T. D. Goldford, *J. Chem. Phys.* **43,** 63, 1965.
54. M. E. Jacox and D. E. Milligan, *J. Mol. Spectrosc.* **56,** 333, 1975.
55. M. Allegrini, J. W. C. Johns, and A. R. W. McKellar, *J. Chem. Phys.* **70,** 2829, 1979.
56. G. Duxbury, H. Kato, and M. L. Le Leere, *Faraday Dis.* **71,** 97, 1981.
57. G. Duxbury and M. L. Le Lerre, *J. Mol. Spectrosc.* **92,** 326, 1982.
58. K. Hashiguchi, Y. Hamada, M. Tsuboi, Y. Koga, and S. Kondo, *J. Mol. Spectrosc.* **105,** 81, 1984.
59. I. Stolkin, T.-K. Ha, and Hs. H. Gunthard, *Chem. Phys.* **21,**, 327, 1977.
60. F. J. Lovas, R. D. Suenram, D. R. Johnson, F. O. Clark, and E. Tiemann, *J. Chem. Phys.* **72,** 4964, 1980.
61. J. Hinze and R. F. Curl, *J. Am. Chem. Soc.* **86,** 5068, 1964.
62. Y. Amatatsu, Y. Hamada, and M. Tsuboi, *J. Mol. Spectrosc.* **123,** 276, 1987.
63. J. C. Guillemin, J. M. Denis, and A. Lablache-Combier, *J. Am. Chem. Soc.* **103,** 468, 1981.
64. A. Hassner, J. O. Currie, Jr., A. S. Steinfeld, and R. F. Atkinson., *J. Am. Chem. Soc.* **95,** 2982, 1973.
65. "The Sadtler Standard Raman Spectra," Sadtler Research Laboratories, Inc., 1976. (a) C-nitroso compounds: subject index 80 and spectra therein. (b) Imines and Schiff bases: subject index 71 and spectra therein. (c) See subject indexes 47 and 53, and spectra therein. (d) Spectrum no. 1863R. (e) Spectrum no. 2594R, (f) Spectrum no. 598R.

(g) Spectrum no. 2948R. (h) Spectrum no. 1809R. (i) Spectra no. 559R, 610R, 1426R, 2968R, 3415R. (j) Spectrum no. 3455R. (k) Spectra no. 4186R, 940R, 2660R. (l) Spectra no. 3155R, 3737R, 2816R.

66. A. Perrier-Datin and J.-M. Lebas, *Spectrochim. Acta* **25A,** 169, 1969.
67. Y. Amatatsu, Y. Hamada, and M. Tsuboi, *J. Mol. Spectrosc.* **123,** 267, 1987.
68. W. D. Kumler, *J. Am. Chem. Soc.* **76,** 814, 1954.
69. A. I. Meyers, *J. Org. Chem.* **24,** 1233, 959.
70. V. E. Bondybey and J. W. Nibbler, *Spectrochim. Acta* **29A,** 645, 1973.
71. V. M. Kolb, D. H. Hua, and W. L. Duax, *J. Org. Chem.* **52,** 3003, 1987.
72. S. T. King, J. Overend, R. A. Mitsch, and P. H. Ogden, *Spectrochim. Acta* **26A,** 2253, 1970.
73. L. Kahovec and K. W. F. Kohlrausch, *Monatsch. Chem.* **83,** 614, 1952.
74. W. C. Harris and S. F. Bush, *J. Chem. Phys.* **56,** 6147, 1972.
75. L. Bardet and M. Alain, *C. R. Acad. Sc., Paris* **217A,** 710, 1970.
76. N. O. Cherskaya, O. A. Rakitin, V. A. Shlyapochnikov, T. I. Godovikova, and L. I. Khmel'nitskii, *Bull. Acad. Sci., USSR, Div. Chem. Sci.* **53,** 2150, 1987.
77. R. Mathis, B. Maccar, J. Barrans, and F. Mathis, *J. Mol. Struct.* **7,** 355, 1971.
78. B. Baccar, R. Mathis, A. Secches, J. Barrans, and F. Mathis, *J. Mol. Struct.* **7,** 369, 1971.
79. W. H. Pirkle and T. J. Sowin, *J. Mol. Chem.* **52,** 3011, 1987.
80. J. Fabian, M. Legrand, and P. Poirier, *Bull. Soc. Fr.* **23,** 1499, 1956.
81. A. M. C. Fernandez, A. M. F. Gnzalez, and M. G. Chozas, *J. Mol. Struct.* **143,** 557, 1986.
82. R. D. Dresdner, J. S. Johar, J. Merritt, and C. S. Patterson, *Inorg. Chem.* **4,** 678, 1965.
83. P. Tisnes, C. Picard, J. D. Bastide, C. Zedde, L. Cazaux, P. Maroni, and G. Trinquier, *Spectrochim. Acta* **39A,** 943, 1983.
84. R. A. Nyquist, T. L. Peters, and P. B. Buddle, *Spectrochim. Acta* **34A,** 503, 1978.
85. J. Jander and R. N. Hazeldine, *J. Chem. Soc.*, 9A, 1954.
86. V. Baliah and V. Chandrasekharan Nair, *J. Indian Chem. Soc.* **66,** 42, 1988.
87. P. A. S. Smith and J. E. Robertson, *J. Am. Chem. Soc.* **84,** 1197, 1962.
88. J. Hamer and A. Macaluso, *Chem. Rev.* **64,** 474, 1964.
89. C. J. Pouchert, "The Aldrich Library of FT-IR Spectra," Aldrich Chemical Company, 1985.

CHAPTER 13

Cumulated Double Bonds

13-1. INTRODUCTION

Examples of cumulated double bonds are allenes (>C=C=C<), ketenes (>C=C=O), ketene imines (>C=C=N−), diazo compounds (>C=N=N), isocyanates (−N=C=O), isothiocyanates (−N=C=S), isoselenocyanates (−N=C=Se), azides (−N=N=N), and carbodiimides (−N=C=N−). In these X=Y=Z types of groups, the vibrations of the two double bonds are strongly coupled and result in out-of-phase and in-phase stretchings. Colthup *et al.* [1] noted that the out-of-phase stretch band of the group X=Y=Z is located in a region near that of the triple-bond stretching, and the in-phase stretching occurs in a region close to that of the single-bond stretchings. This statement is fully supported by the characteristic frequencies listed in Table 13-1 for various X=Y=Z types of compounds.

Since the in-phase mode exhibits stretching frequencies very similar to that of the single bond, coupling between the in-phase X=Y=X stretching and stretching vibrations of the neighboring single bonds is expected. As a result, the in-phase stretching is more sensitive to the substituents on the X=Y=Z group than the out-of-phase stretching mode. One may assume that the out-of-phase stretch of X=Y=Z should exhibit strong infrared absorption and weak Raman intensity, and vice versa. While this is largely true for most cumulated double bonds, many ketenes and diazo compounds

Table 13-1.

The Characteristic X=Y=Z Stretching Frequencies

Compounds	X=Y=Z Out-of-Phase Stretch	IR	R	X=Y=Z In-Phase Stretch	IR	R
>C=C=CH_2	1980–1901	vs	vw	1095–1076	vw	vs, p
>C=C=O	2197–2085	vs	var.	1420–1120	vs	s, p
>C=C=N–	2060–2000	s		~1233	s	s, p
>C=N=N	2113–2013	vs	m-mw, p	1389–1333	vs	vs-m, p
>N=C=O	2300–2250	vs, br	vw-O	1450–1400	w	s, p
>N=C=S	2200–2000	vs, br	m-mw, p	1090–925		s, p
–N=N=N	2169–2080	vs	ms-m, p	1343–1177	w	s, p
–N=C=N–	2152–2128	vs	w	~1460	w	s, p

appear to have strong IR and Raman bands in both regions. Splitting of the out-of-phase stretching band due to the rotational isomers [16, 17] and the Fermi resonance effect [39, 63, 65] were observed in some cases.

13-2. ALLENES (>C=C=C<)

The most characteristic frequencies of allenes arise from the out-of-phase and in-phase C=C=C stretchings. These vibrations are found in the regions of 2000–1900 cm^{-1} and 1130–1060 cm^{-1}, respectively [1].

Table 13-2 illustrates the stretching frequencies of selected allenes. A good review of the out-of-phase C=C=C stretching frequencies of 58 substituted allenes can be found in reference 4. Generally, both stretching frequencies are lowered by halogenation and conjugation (see Table 13-2). For $R_2P(O)$–HC=C=CH_2 and XC(=O)–HC=C=CH_2 types of allenes, an IR doublet was reported in the region for out-of-phase >C=C=CH_2 stretching [16, 17]. This phenomenon is attributed to the effect of rotational isomerism [17]. The symmetrical compound tetramethyl allene, $(CH_3)_2C=C=C(CH_3)_2$, has only weak IR absorption in the 2000–1900 cm^{-1} region.

The assignment of the in-phase C=C=C stretching mode is controversial. A previous study by Piaux and Gaudemar [11] designated the two strong-to-medium Raman bands at about 1130 cm^{-1} and 1100 cm^{-1} as the in-phase C=C=C stretching mode of mono *n*-alkyl allenes. More recently,

Table 13-2.
The $C{=}C{=}C$ Stretching Frequencies (cm^{-1}) of Allenes

Compounds	$C{=}C{=}C$ Out-of-Phase Stretch	IR	R	$C{=}C{=}C$ In-Phase Stretch	IR	R	Ref.
$H_2C{=}C{=}CH_2$	1957	s	—	1076	—	vs, p	2–5
$H_2C{=}C{=}HR$	1980–1945	ms		~1080	vw	s	4, 13, 14, 36
$RHC{=}C{=}CHR$	1955–1930	ms		~1080	vw	s	4, 36
$H_2C{=}C{=}CR_2$	1930–1915	ms		~1080	vw	s	4, 36
$H_2C{=}C{=}CHX$ (X = Cl, Br, I)	1963–1953	m	vs-w	1095–1076		vs, p	4, 6, 7
$H_2C{=}C{=}CHC_6H_5$	1925	s		1080, 1060			4, 10–12, 36
$(Me)(CR_2OH)C{=}C{=}CH{-}O(C{=}O)NR_2$	1980–1970 (IR)						8
$-CH{=}C{=}CH{-}(Me)C{=}CH-$	1927 (IR)						9
$CH{=}C(Me){-}CH{=}C{=}CH{-}(Me)C{=}CH-$	1901 (IR)						9
$H_2C{=}C{=}CR(C{=}O)R$	1945–1940 (IR)						15
$H_2C{=}C{=}CH(C{=}O)Y$ (Y: OH, NH_2, OR)	1930–1910 (IR)						16
$(Et)(Ph)C{=}C{=}CH(C{=}O)-$	~1930						15
$R_2P(O)HC{=}C{=}C(Me)R$	1960–1949 (IR)						17

investigation indicated that the $C{=}C{=}C$ in-phase stretching occurs at 1095–1060 cm^{-1} for most allenes (see Table 13-2). In the case of $CH_3CH{=}C{=}CH_2$, the prominent Raman bands at 1132 cm^{-1} (R : vs, p) and 1102 cm^{-1} (R : m, dp) are assigned to CH_3 and CH_2 wagging modes, respectively [14]. The in-phase $C{=}C{=}C$ stretchings of phenyl allenes are expected to occur at 1080 cm^{-1} and 1065 cm^{-1}.

The spectra of allenes are further characterized by the allenic C—H vibrations, which fall in the regions near those of ethylenic C—H vibrations. For allene ($H_2C{=}C{=}CH_2$) itself, the C—H stretchings, CH_2 scissoring, and $=CH_2$ wagging are at 3067 cm^{-1} and 2996 cm^{-1}, 1440 cm^{-1}, and 848 cm^{-1}, respectively [2–5]. The C—H vibrations of substituted allenes are situated at very similar regions as those of allene [3, 4, 6, 14]. The overtone of $=CH_2$ wagging near 1700 cm^{-1} is also observed in the infrared spectra.

The $C{=}C{=}C$ bending of allene occurs strongly at 356 cm^{-1} in the Raman spectrum [3]. However, much higher frequencies (625–592 cm^{-1} and 548–485 cm^{-1}) were reported for the haloallenes [6].

13-3.

KETENES ($>C{=}C{=}O$)

The two $C{=}C{=}O$ stretchings are located at a higher frequency region than those of allenes (see Table 13-1 and Table 13-3). Although the out-of-phase $C{=}C{=}O$ stretching occurs in the same regions as the $C{\equiv}C$ stretching of

Table 13-3.

The C=C=O Stretching Frequencies (cm^{-1}) of Ketenes

Compounds	C=C=O Out-of-Phase Stretch	IR	R	C=C=O In-Phase Stretch	IR	R	Ref.
$H_2C=C=O$	2153	vs	—[a]	1130	s	—	18, 19, 24, 25
$Me_2C=C=O$	2122	vs	—	1374	s	—	20
$(CF_3)_2C=C=O$	2197	vs	m, p	1420	s	s, p	21, 22
$(C_6H_5)_2C=C=O$	2105	vs	—	1120	w	—	22
$(Me_3Si)HC=C=O$	2112	vs	w	1270	vs	ms	23
$(Et_3Si)HC=C=O$	2115	vs	—	1275	s	—	23
$(Cl_3Si)HC=C=O$	2165	vs	—	1296	m	—	23
$(Me_3Si)_2C=C=O$	2085	vs	—	1295	m	—	23

[a] (—): Intensity not available.

monosubstituted acetylenes (2130–2100 cm^{-1}), these two absorptions can usually be differentiated by the infrared intensity. The C=C=O stretchings can be further differentiated from C≡C stretching by the presence of a medium-to-strong IR band originating from the ketene in-phase stretching at the region of 1420–1120 cm^{-1}.

Although the Raman intensities of these two C=C=O stretching modes are medium to strong in many cases, they are not necessarily the strongest bands in the Raman spectra. In the case of $(CF_3)_2C=C=O$, the strongest Raman band was found at 766 cm^{-1}, which is attributed to symmetric C_2C stretching of the $(CF_3)_2C=$ moiety.

Ketenes' stretching frequency and intensity are sensitive to substitution. For example, the characteristic C=C=O stretchings of $H_2C=C=O$ occur at 2153 cm^{-1} (IR : vs; R : w) and 1130 cm^{-1} (IR : s; R : vs) for out-of-phase and in-phase modes, respectively [18, 19, 24]. On methylation, the out-of-phase and in-phase C=C=O stretching bands of $(CH_3)_2$ C=C=O shift to 2134 cm^{-1} (IR : vs) and 1374 cm^{-1} (IR : vs), respectively [20]. Further fluorination on the methyl groups of the methylated ketene increases the out-of-phase and in-phase C=C=O stretchings to 2179 cm^{-1} (IR : vs; R : m, p) and 1429 cm^{-1} (IR : s, R : s, p), respectively [21, 22]. However, conjugation of a phenyl group to the C=C=O moiety lowers the stretching frequencies. $(C_6H_5)_2C=C=O$ is a good example, which exhibits a very strong IR band at 2105 cm^{-1} (out-of-phase) and a weak IR band at 1120 cm^{-1} (in-phase) [22]. In addition, the attachment of a R_3Si-group to the C=C=O moiety reduces the C=C=O stretching frequencies (see Table 13-3). However, the stretching frequencies of silylketones can be increased by chlorination. For example, the out-of-phase C=C=O stretching of $Cl_3Si-HC=C=O$ occurs at 2165 cm^{-1} (gas; IR : vs), and the in-phase mode at 1296 cm^{-1} (gas; IR : m) [23].

The C=C=O bending frequencies are expected to occur in the spectral regions close to those of C=C=C bendings. The C=C=O bendings of $(CH_3)_2C=C=O$ and $(CF_3)_2C=C=O$ are assigned to the IR bands in the regions of 676–668 cm^{-1} (IR : m-w) and 541–486 cm^{-1} (IR : ms).

13-4.

KETENE IMINES ($>C=C=N-$)

The C=C=N stretching frequencies of ketene imines are summarized in Table 13-4. Stevens and French [27] observed a characteristic IR band for $Y_2C=C=N-Y$ (where Y is alkyl or aryl) in the region of 2050–2000 cm^{-1}, which was assigned to the out-of-phase C=C=N stretching mode. For *N*-methyl ketene imine, out-of-phase stretching occurs at a somewhat higher value (2060 cm^{-1}), and in-phase C=C=N stretching exhibits a medium IR band at 1233 cm^{-1}.

Table 13-4.

Characteristic Stretching Frequencies (cm^{-1}) of Ketene Imines ($>C=C=N-$) and Diazo Compounds ($>C=N=N$)

Compounds	X=Y=Z Out-of-Phase Stretch	IR	R	X=Y=Z In-Phase Stretch	IR	R	Ref.
A. Ketene Imines							
$H_2C=C=N-CH_3$	2060	s	—[a]	1233	m	—	26
$Y_2C=C=N-Y$ (Y: −R or −Ar)	2050–2000 (IR)						27
$(Me-SO_2)_2C=C=N-Me$	2170 (IR)						1
B. Diazo Compounds							
$H_2C=N=N$	2102	vs	—	1170	s	—	28, 29
$RHC=N=N$	2049–2036	vs	—	1389–1333	s	—	30
$R_2C=N=N$	2032–2012	vs	—	1389–1333	vs	—	30
$R(C=O)-CH=N=N$	2100–2087	vs	(m, p)	1337[b]	vs	m, p	30, 33
$Ar(C=O)CH=N=N$	2133–2106	—	—	1353–1351	—	—	30, 31
$R(C=O)-CR=N=N$	2076–2057	vs	(mw, p)	1304[b]	s	vs, p	30, 32
$HC\equiv C-CH=N=N$	2060	vs	s	1168	vs		34
$(CN)_2C=N=N$	2147	vs	—	1246	s	—	35
$(CF_3)_2C=N=N$	2136	vs	w	1378	vs	w, p	21

[a] (—): Intensity not available.
[b] $R = CH_3$.

13-5.

DIAZO COMPOUNDS (>C=N=N)

The out-of-phase and in-phase C=N=N stretchings of diazo compounds fall in regions similar to those of C=C=O stretchings (see Table 13-1 and Table 13-4). For diazomethane ($H_2C{=}N{=}N$), the stretchings are at 2102 cm^{-1} and 1170 cm^{-1}. These stretching frequencies are susceptible to substitution. In the case of $(CF_3)_2C{=}N{=}N$, the presence of the fluorine atoms near the C=N=N moiety shifts the out-of-phase and in-phase C=N=N stretchings to the higher frequencies of 2136 and 1380 cm^{-1}, respectively [21]. Interestingly, the conjugation of the C=N=N grouping with other unsaturated moieties such as carbonyl and aromatic groups increases the out-of-phase C=N=N stretching frequency to a region higher than that of alkyl diazocompounds. For diazocarbonyl compounds, this unusual frequency enhancement is rationalized by the increasing triple-bond character due to the contribution of the resonance form (II), as in the equation

$$\underset{\text{(I)}}{O{=}C{-}CH{=}\overset{+}{N}{=}\overset{-}{N}} \longleftrightarrow \underset{\text{(II)}}{\overset{-}{O}{-}C{=}CH{-}\overset{+}{N}{\equiv}N.}$$

The significance of resonance form (II) is also demonstrated by the lowering of the C=O stretching frequency to 1647–1605 cm^{-1} for diazo ketones. A similar effect is observed for quinone diazines [73, 74].

13-6.

ISOCYANATES (—N=C=O)

There are only a limited number of studies on the vibrational analysis of this class of compounds. Table 13-1 compares the characteristic N=C=O stretching frequencies of isocyanates with other X=Y=Z moieties. Generally, the out-of-phase N=C=O stretching mode absorbs infrared strongly in the region of 2300–2250 cm^{-1}, while the Raman intensity of this vibrational mode is usually weak or undetectable [37–40, 47].

Stretching frequencies of N=C=O are generally not very sensitive to substitution. For example, conjugation with a phenyl group varies the out-of-phase stretching frequency by only a small amount. For aryl isocyanates, this frequency is in the region of 2282–2267 cm^{-1} (IR: strong and broad band) [41]. Attachment of the N=C=O moiety to a heavy atom (e.g., Si, Ge, or S) results in little variation of the N=C=O stretching frequency from those of alkyl isocyanates [38, 42–46].

The IR band of the N=C=O out-of-phase stretching mode can be differentiated from the C≡N stretching band (2270–2210 cm^{-1}) by its very intense and broad features. This absorption is often accompanied by shoulders on both sides of the IR band, and it is best illustrated by the IR spectra of aryl isocyanates [41, 47]. In the case of allyl isocyanate, such complex band structures were interpreted as various combination bands enhanced by Fermi resonance with the out-of-phase N=C=O stretch mode in the region near 2200 cm^{-1} [39].

The in-phase N=C=O stretching is assigned to a polarized strong Raman band in the 1450–1400 cm^{-1} region for methyl, ethyl, and isopropyl isocyanates [37–40]. Although the Raman spectrum of tertiary butyl isocyanate has been reported [42], the assignment of the in-phase N=C=O stretching mode is not certain. Nevertheless, a strong, polarized band observed at 1402 cm^{-1} is assigned to the in-phase N=C=O mode. The presence of a heavy atom adjacent to the N=C=O group caused a wider spectral region (1478–1357 cm^{-1}) for this vibration [38, 42–46]. The infrared intensity of the N=C=O in-phase stretch is weak in the majority of cases. However, the IR absorption in this region becomes strong to medium when two N=C=O groups attach to the same atom or adjacent atoms as a result of out-of-phase interaction between the two in-phase N=C=O stretchings [38, 43, 48].

The assignment of in-phase N=C=O stretching for phenyl isocyanate is controversial. Ham and Willis [50] attributed a strong IR band at 1108 cm^{-1} to this vibrational mode, while Stephenson *et al.* [49] assigned the polarized strong Raman band at 1443 cm^{-1} to the same motion. Furthermore, a polarized strong Raman band, which is not observed in the Raman spectra of other types of substituted benzene, is notable in the region of 1540–1500 cm^{-1} [47] for most ortho-substituted phenyl isocyanates. This polarized strong Raman band is likely to involve the in-phase N=C=O vibration.

The N=C=O bending modes exhibit a poorly defined band in the 650–600 cm^{-1} region for the liquid-state IR spectra of alkyl isocyanates, and a prominent band at 600–590 cm^{-1} for the solid-state IR spectra [37]. For the allyl isocyanate, the N=C=O bending is not observed in the IR spectra but shows up as a weak Raman scattering at 572 cm^{-1} [39].

13-7. ISOTHIOCYANATES (—N=C=S)

The out-of-phase N=C=S stretching of isothiocyanates gives rise to a very strong and broad IR absorption in the region of 2200–2000 cm^{-1}, which is approximately 100 cm^{-1} below the IR absorption of isocyanate groups. Similar to the IR spectra of organic isocyanates, the out-of-phase N=C=S stretching features a multiband structure. For alkyl isothiocyanates, a very strong band at 2150–2050 cm^{-1} with a medium shoulder at 2221–2150 cm^{-1} is the characteristic IR pattern [39, 50–54, 56, 57]. The multiple absorptions of aromatic isothiocyanates shift to the 2135–2015 cm^{-1} region [49, 50, 55]. The absorption complexity in this region was attributed to Fermi resonance of combination tones with the N=C=S out-of-phase stretching mode [39]. Interestingly, the Raman intensities of the N=C=S out-of-phase stretching modes are prominent and polarized in many cases [39, 49–51].

The out-of-phase and in-phase stretching frequencies are little affected by the attachment of a heavy atom such as Si or Ge. Several studies on silylisothiocyanates [38, 58, 59, 61] and germylisothiocyanate [60, 62] indicate that their characteristic N=C=S stretchings occur in regions near those of alkyl isothiocyanates. However, interaction between N=C=S groups should be considered when more than one NCS group is present on the same atom.

The in-phase N=C=S stretching mode of alkyl isothiocyanates was assigned to a polarized strong Raman band in the region of 690–650 cm^{-1} [50]. However, most authors preferred the polarized, strong Raman band at 1090–980 cm^{-1} for this vibrational mode [39, 51, 52, 56]. Recently, the normal coordinate analysis of methyl and ethyl isothiocyanates by Durig *et al.* [51, 52] revealed extensive mixing of N=C=S stretching and C—N stretching. Such vibrational coupling results in two strong polarized Raman bands in the 1090–1035 cm^{-1} and 690–650 cm^{-1} regions. For aromatic isothiocyanates, the in-phase stretching occurs in a lower frequency region, 936–925 cm^{-1}, with strong-to-medium IR absorption but weak Raman intensity [49, 50, 55].

13-8.

ISOSELENOCYANATES (—N=C=Se)

The infrared spectra of alkyl isoselenocyanates have been studied by Franklin and coworkers [53]. The out-of-phase and in-phase N=C=Se stretching modes of methyl isoselenocyanate (CH_3—CH=C=Se) occur at 2182 cm^{-1} and 983 cm^{-1}, respectively.

In the case of higher homologues, the out-of-phase stretching mode falls in the region of 2142–2100 cm^{-1}, while the in-phase stretching is displaced to a much lower region (559–504 cm^{-1}). In addition, rotational isomerism may produce a doublet for the N=C=Se stretching absorptions.

13-9.

AZIDES (—N=N=N)

Lieber *et al.* (63–65) studied the group frequencies of organic azides such as alkyl azides ($R-N_3$), aryl azides ($Ar-N_3$), acid azides ($-C(=O)-N_3$), carbamyl azides ($>NC(=O)-N_3$), vinyl azides ($H_2C=CRN_3$), and α-azido ethers ($R-X-CH_2-N_3$, X: O, S, or N) by infrared spectroscopy. Generally, the out-of-phase N=N=N stretchings fall in the region of 2169–2080 cm^{-1} (IR : vs), and the corresponding in-phase mode occurs in the region of 1343–1177 cm^{-1} (IR : w). Other bands with weaker IR intensities near 2400 and 2200 cm^{-1} are observed when an unsaturated moiety (e.g., Ar, C=C, or C=O) is adjacent to the N_3 group or when a heteroatom (O, S, or N) is attached to the carbon atom at the α position of the N_3 group. This splitting of the N=N=N out-of-phase bands is attributed to Fermi resonance of the N=N=N out-of-phase stretching with the combination tones of N=N=N in-phase and C—N stretchings, and other low-lying frequencies. Similar results were reported recently for 2-azido-1,3-butadiene [66, 68], 2,3-diazido-1,3-butadiene [67, 68], and 2,6-diphenyl azide [69].

The out-of-phase N=N=N stretching mode exhibits medium-to-strong Raman intensity, while the in-phase mode displays strong Raman intensity. Examples are methyl azides (R: 2104 cm^{-1}, ms, 1276 cm^{-1}, vs) [12] and 2-azido-1,3-butadiene (R: 2111 cm^{-1}, m, p, 1224 cm^{-1}, s, p) [66].

13-10.

CARBODIIMIDES (—N=C=N—)

The infrared spectra of R—N=C=N—R types of compounds has been explored by Khorana [70, 71] and by Meakins and Moss [72]. A review on the infrared and Raman spectra of carbodiimides was given by Mogul [75]. Generally, the N=C=N out-of-phase stretching of alkyl and aryl carbodiimides absorbs infrared strongly in the 2152–2128 cm^{-1} region. The IR intensity of this band is about 2.5 times as strong as the normal ketonic carbonyl absorption. The in-phase N=C=N stretching mode occurs near 1460 cm^{-1} with strong Raman scattering effect. Attachment of a Si atom to the N=C=N moiety shifts both frequencies toward the higher value. For silyl carbodiimides, the out-of-phase and in-phase stretchings are located at 2260–2180 cm^{-1} and near 1500 cm^{-1}, respectively [76, 77]. For germyl carbodiimide, NCN bands are at 2168–2140 cm^{-1} and 1419 cm^{-1} [78].

REFERENCES

1. N. B. Colthup, L. H. Daly, and S. E. Wiberley, "Introduction to Infrared and Raman Spectroscopy," 3rd Ed., Academic Press, New York, 1990.
2. R. C. Lord and J. Ocampo, *J. Chem. Phys.* **19,** 260, 1951.
3. R. C. Lord and P. Venkateswarlu, *J. Chem. Phys.* **20,** 1237, 1952.
4. J. H. Wotiz and D. E. Mancuso, *J. Org. Chem.* **22,** 207, 1957.
5. N. Sheppard and D. M. Simpson, *Quart. Rev.* **6,** 1, 1952.
6. R. A. Nyquist, Y. S. Lo, and J. C. Evans, *Spectrochim. Acta* **20,** 619, 1964.
7. R. A. Nyquist, T. L. Reder, F. F. Stec, and G. J. Kallos, *Spectrochim. Acta* **27A,** 897, 1971.
8. D. Hoppe, G. Gonschorrek, D. Schmidt, and E. Egert, *Tetrahedron* **43,** 2457, 1987.
9. W. Oroshnik, A. D. Mebane, and G. Karmas, *J. Am. Chem. Soc.* **75,** 1050, 1953.
10. L. Piaux and M. Gaudemar, *Compt. Rend.* **240,** 2328, 1955.
11. L. Piaux and M. Gaudemar, *Bull. Soc. Chim. Fr.* **23,** 794, 1956.
12. F. R. Dollish, W. G. Fateley, and F. F. Bentley, "Characteristic Raman Frequencies of Organic Compounds," John Wiley & Sons, Inc., New York, 1974, pp. 143–150.
13. T. L. Jacobs, R. Akuwie, and R. G. Cooper, *J. Am. Chem. Soc.* **73,** 1273, 1951.
14. G. L. Szasz, J. S. McCartney, and D. H. Rank, *J. Am. Chem. Soc.* **69,** 3150, 1947.
15. D. Bernard, A. Doutheau, and J. Gore, *Tetrahedron* **43,** 2721, 1987.
16. J. H. Wotiz and W. D. Celmer, *J. Am. Chem. Soc.* **74,** 1860, 1952.
17. R. Mathis, F. Mathis, N. Ayed, B. Baccar, and C. Charrier, *Spectrochim. Acta* **39A,** 233, 1983.
18. C. B. Moore and G. Pimentel, *J. Chem. Phys.* **38,** 2816, 1963.
19. W. H. Fletcher and W. T. Thompson, *J. Mol. Spectrosc.* **25,** 240, 1968.
20. W. H. Fletcher and W. B. Barish, *Spectrochim. Acta* **21,** 1647, 1965.
21. F. A. Miller and F. E. Kiviat, *Spectrochim. Acta* **25A,** 1577, 1969.
22. S. Nadzhimuldinov, N. A. Slovokhotova, and V. A. Kargin, *Russ. J. Phys. Chem.* **40,** 479, 1965.
23. A. N. Lazarev, T. F. Tenisheva, and L. L. Shchukovskaya, *Russ. J. Phys. Chem.* **43,** 949, 1969.
24. F. Halverson and V. Z. Williams, *J. Chem. Phys.* **15,** 552, 1947.
25. P. E. B. Buter, D. R. Eaton, and H. W. Thompson, *Spectrochim. Acta* **13,** 223, 1958.
26. Y. Amatatsu, Y. Hamad, and M. Tsuboi, *J. Mol. Spectrosc.* **123,** 476, 1987.
27. C. L. Stevens and J. C. French, *J. Am. Chem. Soc.* **76,** 4398, 1954.
28. C. B. Moore and G. C. Pimentel, *J. Chem. Phys.* **40,** 342, 1964.
29. C. B. Moore and G. C. Pimentel, *J. Chem. Phys.* **40,** 329, 1964.
30, P. Yates, B. L. Shapiro, N. Yoda, and J. Fugger, *J. Am. Chem. Soc.* **79,** 5756, 1957.
31. C. Laurence, M. Berthelot, L. L. Leveson, and C. W. Thomas, *Spectrochim. Acta* **38A,** 487, 1982.
32. C. Davidovic, F. Debu, C. Marfisi, M. Monnier, J. P. Aycard, J. Pourcin, and H. Bodot, *J. Mol. Struct.* **147,** 29, 1986.
33. A. Poetti, G. Paliani, M. G. Giorgini, and R. Cataliotti, *Spectrochim. Acta* **31A,** 1869, 1975.
34. F. K. Chi and G. E. Leroi, *Spectrochim. Acta* **41A,** 1759, 1075.
35. W. H. Smith and G. E. Leroi, *Spectrochim. Acta* **25A,** 1917, 1969.
36. L. J. Bellamy, "The Infrared Spectra of Complex Molecules," John Wiley & Sons, New York, 1975, pp. 69–71.
37. R. P. Hirschmann, R. N. Kniseley, and V. A. Fassel, *Spectrochim. Acta* **21,** 2125, 1965.

38. P. F. Koster, *Spectrochim. Acta* **24A,** 395, 1968.
39. T. Torgrimsen, P. Klaeboe, and F. Nicolaisen, *J. Mol. Struct.* **20,** 213, 1974.
40. J. F. Sullivan, D. T. Durig, J. R. Durig, and S. Cradock, *J. Phys. Chem.* **91,** 1770, 1987.
41. G. L. Caldow and H. W. Thompson, *Spectrochim. Acta* **13,** 212, 1958.
42. I. Kanesaka and K. Kawai, *Bull. Chem. Soc. Jpn.* **43,** 3298, 1970.
43. F. A. Miller and G. L. Carlson, *Spectrochim. Acta* **17,** 977, 1961.
44. J. R. Durig and J. F. Sullivan, *J. Mol. Struct.* **56,** 41, 1979.
45. J. Goubeau, E. Heabach, D. Paulin, and I. Widmaier, *Z. Anorg. Chem.* **300,** 194, 1959.
46. J. Prejer, *Rocz. Chem.* **41,** 647, 1947.
47. "The Sadtler Standard Raman Library," Sadtler Research Laboratories, 1976.
48. G. L. Carlson, *Spectrochim. Acta* **20,** 1781, 1964.
49. C. V. Stephenson, W. C. Coburn, and W. S. Wilcox, *Spectrochim. Acta* **17,** 933, 1961.
50. N. S. Ham and J. B. Willis, *Spectrochim. Acta* **16,** 279, 1960.
51. J. R. Durig, J. F. Sullivan, H. L. Heusel, and S. Cradock, *J. Mol. Struct.* **100,** 241, 1983.
52. J. R. Durig, H. L. Heausel, and J. F. Sullivan, *Spectrochim. Acta* **40A,** 739, 1984.
53. W. J. Franklin, R. L. Werner, and R. A. Ashby, *Spectrochim. Acta* **30A,** 1293, 1974.
54. R. N. Kniseley, R. P. Hirschmann, and V. A. Fassel, *Spectrochim. Acta* **23A,** 109, 1966.
55. P. G. Puranik and A. M. J. Rao, *Current. Sci.* **25,** 153, 1956.
56. E. Lieber, C. N. R. Rao, and J. Ramachandran, *Spectrochim. Acta* **13,** 296, 1959.
57. P. Kristian, O. Hritzova, and P. Talan, *Coll. Czech. Chem. Comm.* **42,** 671, 1977.
58. G. L. Carlson, *Spectrochim. Acta* **18,** 1529, 1962.
59. E. A. V. Ebsworth, R. Mould, R. Taylor, G. R. Wilkinson, and L. A. Woodward, *Trans. Faraday Soc.* **58,** 35, 1962.
60. G. Davidson, L. A. Woodward, K. M. Mackey, and P. Robinson, *Spectrochim. Acta* **23A,** 2383, 1967.
61. J. R. Durig, K. S. Kalasinsky, and V. F. Kalasinsky, *J. Mol. Struct.* **82,** 438, 1978.
62. J. R. Durig, Y. S. Li, and J. F. Sullivan, *J. Chem. Phys.* **71,** 1041, 1971.
63. E. Lieber, D. R. Levering, and L. J. Patterson, *Anal. Chem.* **23,** 1594, 1951.
64. E. Lieber, C. N. R. Rao, T. S. Chao, and C. W. W. Hoffman, *Anal. Chem.* **29,** 916, 1951.
65. E. Lieber, C. N. R. Rao, A. E. Thomas, E. Oftedahl, R. Minnis, and C. V. N. Nambury, *Spectrochim. Acta* **19,** 1135, 1963.
66. S. H. Schei, H. Riebe, C. J. Nielsen, and P. Klaeboe, *J. Mol. Struct.* **147,** 203, 1986.
67. C. J. Nielsen, P. Klaeboe, H. Priebe, and S. H. Schei, *J. Mol. Struct.* **147,** 217, 1986.
68. P. Klaeboe, C. J. Nielsen, H. Priebe, S. H. Schei, and C. E. Sjogren, *J. Mol. Struct.* **141,** 161, 1986.
69. A. El-Shahaway, *Spectrochim. Acta* **39A,** 115, 1983.
70. H. G. Khorana, *Can. J. Chem.* **32,** 261, 1954.
71. H. G. Khorana, *Chem. Rev.* **53,** 145, 1953.
72. G. D. Meakins and R. J. Moss, *J. Chem. Soc.*, 993, 1957.
73. R. J. W. Le Fevre, J. B. Sousa, and R. L. Werner, *J. Chem. Soc.*, 4686, 1954.
74. K. B. Whetsel, G. F. Hawkins, and F. E. Johnson, *J. Am. Chem. Soc.* **78,** 3360, 1956.
75. P. H. Mogul, *Nucl. Sci. Abstr.* **21,** 47014, 1967.
76. E. A. V. Ebsworth and M. J. Mays, *Spectrochim. Acta* **19,** 1127, 1963.
77. J. Pump, E. G. Rochow, and U. Wannagat, *Monatsch. Chem.* **94,** 588, 1963.
78. S. Cradock and E. A. Ebworth, *J. Chem. Soc.* (*A*), 1423, 1968.

CHAPTER 14

Organic Sulfur Compounds

14-1. INTRODUCTION

Depending on their molecular structure, organic sulfur compounds are characterized by vibrations of S—H, C—S, C=S, S=O, and $-SO_2-$ groups. The S=O and SO_2 stretch bands are strong in IR spectra, whereas SH and C—S stretch bands tend to be weak in the IR and are stronger in the Raman spectra. The C—S stretching frequency is quite consistent for many different types of compounds that contain the C—S moiety. Similarly to the alkyl halides discussed in Chapter 3, the conformations of alkyl sulfides can be specified as P_H, P_C, and S_{HH}, etc.; also, the correlation between the C—S stretching frequencies and the rotational isomers bears analogy with that for chloro compounds.

The C=S group has a lower stretching frequency than the C=O group and tends to interact with attached C—N, C—O, or C—C bonds, giving rise to more than one band involving C=S stretch, which complicates the interpretation.

14-2. THIOLS (R—SH AND Ar—SH)

Table 14-1 summarizes group frequencies of thiols. The presence of the —SH grouping is best characterized by the S—H and C—S stretchings

Table 14-1.

Characteristic S—H and C—S Stretch IR and Raman Frequencies (cm^{-1}) of Thiols

	cm^{-1}	IR	R	Ref.
S—H Stretch				
R—SH	2598–2584 (vap.)	vw		1
	2590–2580 (soln.)	w		1
	2590–2560 (liq.)	w	s, p	4a
Ar—SH	2608–2582 (vap.)	vw		1
	2590–2580 (soln.)	w		
	2580–2560 (liq.)	s-w	s, p	4a
C—S Stretch				
CH_3—SH	704	w	s, p	12
n-R—SH	660–650	w	s, p	4a, 14
Me_2CH—SH	623, 616			13
Me_3C—SH	587	—	vs, p	15, 16
$Cl(CH_2)_2$—SH	695 (T)	vs	s	17
	640 (G)		vs, p	
$Br(CH_2)_2$—SH	735 (T)	m	s	17
	662 (G)		m	
cyclo-C_6H_{11}—SH	734, 710[a] (equatorial)			18

[a] Doublet due to Fermi resonance.

in the Raman as polarized strong bands. The IR intensities of these two bands are weak in the majority of cases. However, the S—H stretch region (2590–2540 cm^{-1}) in the IR tends to be free of other bands. The CH_2—S group in thiols, sulfides, and disulfides has medium-intensity IR bands involving CH_2 wag at 1270–1220 cm^{-1} [123].

14-2.1. S—H Stretching

In the vapor phase, the S—H stretching were found in the regions of 2598–2584 cm^{-1} and 2608–2582 cm^{-1} for alkanethiols (R—SH) and benzenethiols (Ar—SH), respectively [1]. The S—H stretching frequency measured in dilute solution (2590–2580 cm^{-1}) is essentially located in the same range as that found for vapor-phase thiols [2]. In the liquid state, a lower spectral frequency (2590–2560 cm^{-1}) is observed as a result of self-association via hydrogen bonding [3, 4a]. This Raman band is very characteristic for the presence of the —SH moiety, since there is rarely any spectral interference in this region for most organic compounds. Unlike O—H stretching, the S—H stretching mode gives rise to a weak or sometimes barely detectable

IR band for aliphatic thiols (R—SH). Aryl thiols have S—H bands that are somewhat stronger in the IR. In the Raman spectra of both aliphatic and aromatic thiols, the S—H stretch band is strong [4a, 4d].

The cyclic —N=C(SH)— group usually exists in the keto form —NH—C(=S)—. This class of compounds usually shows very weak or undetectable S—H bands in the Raman and has definite N—H stretch bands in the IR.

Although the effects of the hydrogen bond interaction on the S—H vibrations are less significant than those found in alcohols and phenols, they should not be neglected. The S—H···S type of hydrogen bonding exists in dissolved and liquid thiols [5–10]. Such hydrogen bonding results in a lowering of the S—H stretching frequencies when the physical state of the thiols passes from vapor to liquid phase. Furthermore, the fundamental of the associated S—H species shows an intensity about 10–20 times greater than that of free S—H stretching in vapor phase [7, 8]. The enhancement of the S—H intensity in the IR by hydrogen bonding becomes very obvious when there is a second S—H group or an O—H moiety present in the same molecule, as in the case of alkanedithiols and hydroxylalkanethiols. The IR spectra of these compounds in liquid state display medium-to-strong S—H bands [11].

Intramolecular hydrogen bonding is also possible when allowed by molecular geometry. Good examples of such an interaction are 2-aminobenzenethiol and toluene-3,4-dithiol. The S—H stretchings of 2-aminobenzenethiol in different physical states were reported as follows: IR (vap.): 2560 cm^{-1} [1], Raman (liq.): near 2530 cm^{-1} [4b]. The S—H stretch mode of toluene-3,4-dithiol occurs at 2570 cm^{-1} in the vapor-phase IR spectrum [1]. The exceptionally low S—H frequencies, as compared to those of other benzenthiols (see Table 14-1), are attributed to the formation of the intramolecular hydrogen bond.

14-2.2. C—S Stretching

The C—S stretching mode usually appears as a polarized strong Raman band in the 735–590 cm^{-1} region [3], and the corresponding IR intensity is often weak [1, 4a, 11]. The correlation of the C—S stretching frequency to the rotational isomers is similar to that for the alkyl chlorides. For methanethiol, the three rotational isomers are equivalent, and its C—S stretching was assigned to a strong Raman band at 710 cm^{-1} by May and Pace [12]. Thiols with an alkyl group larger than CH_3 exhibit more than one conformer [13–19]. Hence, many thiols exhibit more than one C—S stretching band in this region (see table 14-1).

14-3. SULFIDES (Y—S—Y′; Y, Y′ = R AND/OR Ar)

The characterized C—S stretching frequencies of some sulfides are summarized in Table 14-2. Most dialkyl sulfides exhibit multiple bands in the C—S stretching region. Replacing an α-hydrogen atom or the alkyl group adjacent to the sulfur atom by other functional groups will cause some variation in the C—S stretching frequencies. Nevetheless, the C—S stretching frequencies remain in the characteristic region of 800–600 cm^{-1}, as illustrated in Table 14-2.

The antisymmetric and symmetric C—S—C stretchings of CH_3—S—CH_3 are located at 742 cm^{-1} (Raman : s) and 691 cm^{-1} (Raman : vs), respectively [20, 21]. For other disulfides containing the CH_3—S linkage, the CH_3—S

Table 14-2.

Characteristic C—S Stretching IR and Raman Frequencies (cm^{-1}) of Dialkylsulfides

Compounds	cm^{-1}	IR	R	Ref.
CH_3—S—CH_3	742	w	m, dp	20, 21
	691	m	s, p	
CH_3—S—	750–690	w	s, p	4e, 22
CH_3—S—R	730–720			4e, 23–27, 29, 35
—CH_2—S—	P_C: 750–710	w	s, p	4e, 22
	P_H: 700–640	w	s, p	
>CH—S—	S_{HH}: 660–630			22
Me_3C—S—	ca. 585	w	vs, p	4c, 29
(*cyclo*-C_6H_{11})$_2$—S	ca. 735, 696			53
CH_3—S—CH_2—CN	741	w		30
	728	w		
	715	sh		
	692	m		
$(PhCH_2)_2$—S	672	vs		31, 32
	650	ms		
CH_2=CH—S—CH_3	735	ms	s	33
	697	m	w	
	677	s	s, p	
$(CH_2{=}CH)_2$—S	730			34
	685			
Ph—S—Ph	701	sh	vs, p	35
CH_3—S—Ph	722	s	m	35, 54, 125
	689	s	s	

stretching mode is assigned to a band in the region of 750–690 cm^{-1}, since this Raman band is quite characteristic for these compounds [22]. An even narrower region of 730–720 cm^{-1} is expected for compounds having the $RCH_2{-}S{-}CH_3$ type of structure [23–27]. Empirically, the presence of the $(Me)_3C{-}S{-}$ linkage can be identified by a very strong Raman band near 585 cm^{-1} based on the fact that both $Me_3C{-}S{-}CMe_3$ and $Me_3C{-}S{-}CH_3$ exhibit a strong Raman band near this frequency [3, 4c, 111]. It is worth noting that these frequencies are very close to the C—S stretching frequency of $Me_3C{-}SH$ at 587 cm^{-1}.

For a $-C_\beta-C_\alpha-S-$ type of moiety, the conformational isomers arising from rotations about the $C_\beta-C_\alpha$ and $C_\alpha-S$ bonds give rise to different C—S stretching bands. Ohsaku [28] reported that the C—S stretching frequency is affected more by the conformation about the $C_\beta-C_\alpha$ bond than by the conformation about the $C_\alpha-S$ bond. As with the C—Cl stretching of alkylchlorides, the $C_\alpha-S$ stretching frequencies are correlated with the rotational isomers such as P_H, P_C, S_{HH}, etc. (see Table 14-2). It can be seen from Table 14-2 that the $C_\alpha-S$ stretching band occurs at a higher frequency for the P_C isomer than for the P_H form. The conformations about the $C_\alpha-S$ bond are usually specified as T and G for trans and gauche isomers, respectively. Although the $C_\alpha-S$ stretching frequency is less affected by the conformation about the $C_\alpha-S$ bond, the splitting of the $C_\alpha-S$ stretching band by this rotation is still observable in many cases. For example, the $CH_2{-}S$ stretching bands of $CH_3{-}CH_2{-}S{-}CH_3$ were observed at 678 cm^{-1} (Raman : m) and 653 cm^{-1} (Raman : vs) for T and G isomers, respectively [25].

It is noteworthy that, unlike the cyclic C—O—C or cyclic C—N—C moieties, C—S—C stretchings remain fairly localized even though included in a ring structure [40, 44]. As exemplified by Table 14-3, the cyclic C—S stretchings are expected in the same region as those of the open-chain C—S moiety.

Table 14-3.

The Characteristic C—S Stretching Frequencies (cm^{-1}) of Cyclic C—S Moieties

Compounds	Frequency (cm^{-1})	Ref.
Thiazolidines		36, 37
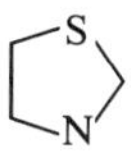	IR: 705 (m), 674 (s) R: 705 (vs), 674 (vs, p)	
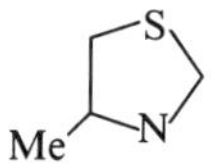	IR: 704 (w), 665 (m) R: 706 (vs), 664 (m)	

Table continues overleaf

Table 14-3.—continued

Compounds	Frequency (cm^{-1})	Ref.
Tetrahydrothiophene 	IR: 686 (m) R: 687 (vs, p)	38, 39
Tetrahydrothiopyran 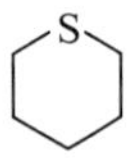	IR: 656 (m) R: 664 (vs, p)	4h, 11
1,4-Dithian 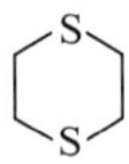	IR: 665 (s) R: 691 (vs), 665 (vs, p)	11, 40, 41
1,3-Dithiocyclopentane derivatives 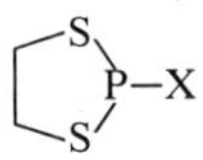	IR: 675–662 (m-vw), 652–640 (ms-w) R: 675–662 (vs-mw, dp), 652–640 (vs, p)	42
X = Cl, OMe, SMe, NMe_2, Ph		
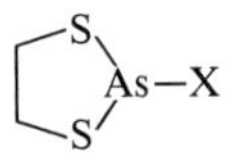	IR: 672–660 (ms-w), 656–648 (s-mw) R: 672–660 (m, dp), 656–648 (ms, p)	42, 43
X = Cl, I, NMe_2		
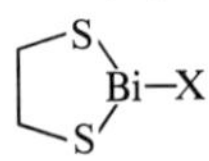	IR: 654 (w), 628 (w)	43
X = Cl, Br		
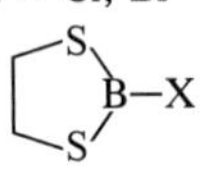	IR: 680–659 (ms-w) R: 680–659 (vs, p)	44
X = Cl, Br, Ph, NMe_2		
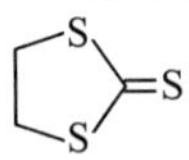	IR: 674 (m) R: 674 (vs, p)	45
Phenothiazen derivatives 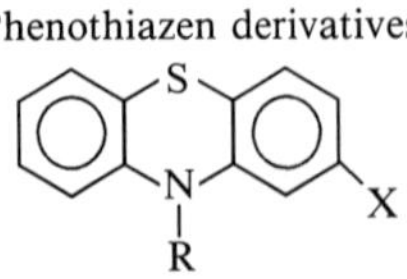	R: 685–675 (vs)	46
X = $-Cl$, $-CF_3$, R = $-CH_2CH_2CH_2N\langle$		

14-4.

DISULFIDES (Y—S—S—Y′; Y, Y′ = R AND/OR Ar) AND POLYSULFIDES

The vibrational spectra of dialkyl-, divinyl-, dibenzyl-, dicinnamyl-, and diaryldisulfides have been examined [35, 47–49]. Hayashi *et al.* [50, 51] reported the IR spectra of polymethylenedisulfide ($-(CH_2S_2)_n-$) and polyethylenedisulfide ($-(CH_2CH_2S_2)_n-$). Their results on the C—S and S—S frequencies are summarized in Table 14-4.

Variations in the S—S force constant and the S—S stretching frequency have been calculated as a function of the CSSC dihedral angle for CH_3—S—S—CH_3 using ab initio molecular orbital calculations [130].

Table 14-4.

The C—S and S—S Stretching Frequencies (cm^{-1}) of Disulfides

Compounds		cm^{-1}	IR	R	Ref.
C—S Stretch					
Me—S—S—Me		694	w	vs, p	4e, 52
Me—S—S—		700–680			4e, 52
$-CH_2$—S—S—	P_C:	750–700	w	s, p	4e, 52
	P_H:	670–630	w	s, p	
>CH—S—S—	S_{HH}:	640–590			52
Me_3C—S—S—CMe_3		566	vw	s, p	4e, 47
$PhCH_2$—S—S—CH_2Ph		662	vs	s	47
(*cyclo*-C_6H_{11}—S—$)_2$		ca. 735			53
		689			
(H_2C=CH—S—$)_2$		722			47
		602			
(PhCH=CH—S—$)_2$		662	m	s	47
(Ph—S—$)_2$		692	s	s	4e, 47
S—S Stretch					
(Me—S$)_2$		510	m	vs, p	4e, 52
(*n*-R—S$)_2$		527–522	w	m, p	47
		512–507	m-w	s, p	
(Me_3C—S$)_2$		544	sh	s, p	4e, 47
($PhCH_2$—S$)_2$		517		m	47
(Ar—S$)_2$		540–520		s-m, p	47
$(CH_2SS)_n$		488	w		50
		469	mw		
$(CH_2CH_2SS)_n$		506	w		51
		472	m		

14-4.1. C—S Stretching

There is a great similarity between the C—S stretchings of dialkyl sulfides and those of dialkyl disulfides. The effects of rotational isomerism on the C—S stretchings usually result in single or multiple C—S stretching bands in the 715–620 cm^{-1} region. The C—S stretching bands of dialkyl disulfides, like those of organic sulfides and thiols, are strong and polarized in the Raman spectra while the IR intensity is variable. Sugeta *et al.* [52] correlated the C_α—S stretching frequencies with rotational isomers about the C_β—C_α bond for dialkyl disulfides containing a —C_β—C_α—S—S— linkage. They found that the CH_3—S stretching frequency occurs in the 700–680 cm^{-1} region for compounds with a CH_3—S—S—R moiety. For the —CH_2—S—S linkage, the CH_2—S stretching mode of the P_H isomer is located in the 670–630 cm^{-1} region, while that of the P_C form is found at a higher region, 750–700 cm^{-1}. When the sulfur atom is attached to a secondary carbon atom, the C—S stretching frequency of the $>$CH—S moiety is situated in the 640–590 cm^{-1} region for the S_{HH} conformation. An example of disulfides with the sulfur atoms attaching to tertiary carbon atoms is Me_3C—S—S—CMe_3, which displays a C—S stretching band at 573 cm^{-1} (R : vs, p; IR: not observed) [4, 47]. One should note that this frequency is very close to that observed for Me_3C—SH at 587 cm^{-1} and those for Me_3C—S—CMe_3 and Me_3C—S—Me near 585 cm^{-1}.

14-4.2. S—S Stretching

The S—S stretching band is polarized and prominent in the Raman spectra, while the IR intensity is usually very weak because of its nonpolar nature. In the Raman spectra of di-*n*-alkyl disulfides, two S—S bands appearing in the regions of 527–522 and 512–507 cm^{-1} are attributed to different rotational isomers [3]. An increase in the S—S stretching frequency to 544 cm^{-1} is observed for $(Me)_3C$—S—S—$C(Me)_3$, which contains two tertiary alkyl groups next to the sulfur atoms. The S—S stretchings of diaryldisulfides were assigned to a single frequency in the region of 542–507 cm^{-1} [47]. A much lower frequency region was reported for polymethylenedisulfide and polyethylenedisulfide. Two weak IR bands in the 506–469 cm^{-1} region were assigned to the S—S stretching for these compounds [50, 51].

The S—S stretchings of polysulfides were found in a lower frequency region (510–450 cm^{-1}) compared with those of disulfides. The above correlation of the S—S stretching frequencies of polysulfides is obtained based on the vibrational studies on the molecules including R—S—S—S—R (R = CH_3 and C_2H_5) [66, 67] and sulfane derivatives (R—$(S)_n$—H and R—$(S)_n$—Cl; n = 2–4) [68].

14-5.

THIOCARBOXYLIC ACIDS AND THIOESTERS

14-5.1. S—H Stretching

Vibrational studies of thiocarboxylic acids (R—C(=O)—SH, R = H, Me, Et) and α-halogenated thiocarboxylic acids have been reported by Randhawa *et al.* [55, 56, 60, 61, 63] and Crowder *et al.* [57, 59, 62]. The S—H stretching mode of these compounds exhibits similar Raman characteristics (i.e., spectral range, intensity, and polarization) as those found for thiols (see Tables 14-1 and 14-5). The IR absorbance of thiocarboxylic acids, however, is several times stronger than that of thiols. Such enhancement in the IR absorption of the S—H stretching mode can be associated with hydrogen bond formation, since only weak absorbance is found for vapor-phase thiocarboxylic acids [1]. Like carboxylic acids, some

Table 14-5.

The Characteristic Frequencies (cm^{-1}) of Thiocarboxylic Acids and Thioesters

Compounds		cm^{-1}	IR	R	Ref.
R—C(=O)—SH					
S—H Stretch	vap.:	2600–2580	vw		55–63
	liq.:	2580–2540	vs-m	vs, p	55–63
C—S Stretch					
H—C(=O)—SH		~690			55
Me—C(=O)—SH		628		vs	56
Et—C(=O)—SH		712 (trans)			59
		598 (gauche)			
CH_2Cl—C(=O)—SH		710	vs	—	60
		590	—	s	
$CHCl_2$—C(=O)—SH		625	m	—	60
		565	—	m	
CH_2Br—C(=O)—SH		655		m	61
		565		m	
CF_3—C(=O)—SH		747	s	vs, p	62
CCl_3—C(=O)—SH		500	s	vs, p	63
R—C(=O)—S—R′					
C—S Stretch					
Me—C(=O)—S—Me	—S—CH_3:	726			64
	—C(=O)—S—:	628			
Me—C(=O)—S—Et	—S—CH_2:	680			64
	—C(=O)—S—:	626			

thiocarboxylic acids exist as cyclic dimers in vapor phase and dilute solution with an S—H···O type of hydrogen-bond interaction. Open-chain oligomers are formed in more concentrated solutions and in the liquid state [55, 57–59]. For thioacetic acid (Me—C(=O)—SH), it is possible to differentiate the type of hydrogen bonding by the S—H stretching frequency. For this particular compound, the free S—H stretching mode was observed to absorb infrared at 2588 cm^{-1}, while the cyclic dimer and open-chain oligomer absorb at 2530 cm^{-1} and 2570 cm^{-1}, respectively [58].

14-5.2. C=O Stretching

Nyquist and Potts [126] report that in solution, CH_3—C(=O)—SH has a C=O stretch band in the IR at 1712 cm^{-1}, and that aryl-C(=O)—SH compounds have IR bands at 1700–1690 cm^{-1} (strong). In the liquid state, CH_3—C(=O)—SH and thiobenzoic acid absorb strongly in the IR at 1695 and 1660 cm^{-1}, respectively [11].

14-5.3. C—S Stretching

The C—S stretching mode is generally observed as a strong Raman band in the region of 750–500 cm^{-1} for thioacids. In some cases, multiple C—S bands are found because of the rotational isomerism (see Table 14-5). The correlation between the C—S stretching frequencies and rotational isomers has been studied by several authors with an analogy to acid chlorides [60, 62].

14-5.4. C—S—H and O=C—S Deformations

The in-plane C—S—H bending is located in the region of 910–830 cm^{-1} (IR : s-m; R : m-w), while the corresponding out-of-plane bending falls in a lower region, 465–430 cm^{-1} (Raman : vs-m) [55–57, 63]. An exception to the above correlation is trichlorothioacetic acid. Randhawa *et al.* [63] have associated the out-of-plane C—S—H bending mode of trichlorothioacetic acid with a band at 616 cm^{-1} (IR : s; R : vw, dp).

Another characteristic skeletal deformation is due to O=C—S bending. The Raman intensity of this bending mode is quite prominent in the 460–430 cm^{-1} region [55–63]. Although the O=C—S bending and C—S—H out-of-plane bending modes occur in the same spectral region, the O=C—S bending vibration appears to exhibit a slightly lower frequency than the C—S—H out-of-plane bending in many cases.

14-5.5. C=S Stretching

The thiol-thione tautomerism has been proposed for thioacids:

$$\underset{\text{thiol}}{R{-}C({=}O){-}SH} \rightleftarrows \underset{\text{thione}}{R{-}C({-}OH){=}S}$$

While most thioacids are found existing exclusively in the thiol form, a small amount of thiones is detected for thioformic acid (H—C(=O)—SH) and trichlorothioacetic acid (CCl_3—C(=O)—SH) by the presence of a weak C=S stretching band near 1160 cm^{-1} in the IR and Raman spectra [55, 63, 65].

14-5.6. Thioesters (R—C(=O)—S—R′)

Two C—S stretching bands should be considered for thioesters (R—C(=O)—S—R′). For example, Crowder assigned the Raman bands of CH_3C(=O)—S—CH_3 at 628 cm^{-1} and 726 cm^{-1} to the C—S stretchings of C(=O)—S and S—CH_3 moieties, respectively [64]. In the case of CH_3—C(=O)—S—CH_2CH_3, the C—S stretchings of the —C(C=O)—S and S—CH_2— moieties were assigned to the Raman bands at 626 cm^{-1} and 680 cm^{-1}, respectively [64]. The C=O bands are strong in the IR at 1700–1680 cm^{-1} for R—C(=O)—S—R and at 1680–1640 cm^{-1} for aryl-C(=O)—S—R run in solution [126].

14-6. COMPOUNDS CONTAINING THE >C=S GROUP

14-6.1. Thioketones (Thiones) and Thiocarbonyl Halides (—C(=S)—X)

The vibrational analysis of tetramethylcyclobutane-1,3-dithione has been reported [69, 70]. The polarized strong Raman band of this compound at 1328 cm^{-1} (IR: not observed) was assigned to the symmetric C=S stretching mode, while the medium IR band at 1167 cm^{-1} (Raman: not observed) was attributed to the corresponding antisymmetric C=S stretching mode. The C=S in-plane bending mode appears at 390 cm^{-1} (IR: m, CCl_4 soln.) for this particular molecule in solution. Derivatives of thiobenzophenones have a medium-weak IR band at 1224–1207 cm^{-1} not seen in comparable benzophenones [127] that probably involves C=S stretching.

The C=S stretching mode of thiocarbonyl halides occurs in a region similar to that of thioketones. For thiophosgene ($Cl_2C{=}S$), the C=S stretching mode occurs strongly at 1120 cm^{-1} as a polarized Raman band [71, 72]. While bromination shows little effect on the C=S stretching frequency, as illustrated by BrClC=S [73], fluorination increases the C=S stretching frequency, and the Raman spectrum of FClC=S displays a strong band at 1242 cm^{-1} due to C=S stretching [74].

14-6.2. Dithioacids (R—C(=S)—SH) and Dithioesters (R—C(=S)—SR′)

The group frequencies of dithioacids and dithioesters may be illustated by dithioacetic acid and methyl dithioacetate. In the case of dithioacetic acid ($CH_3C({=}S){-}SH$), two S—H bands at 2568 and 2552 cm^{-1} were found in the IR spectrum when measured in CH_2Cl_2 solution. Only a single band at 2498 cm^{-1} was observed for the same vibrational mode in the neat liquid [75]. The variation in the spectral pattern of the S—H stretch mode upon the change of physical state can be explained by hydrogen bond formation. The C=S stretching mode is located at 1225 cm^{-1} (R : s), and the C—S stretching mode occurs at 585 cm^{-1} (R : vs). The C—S—H bending modes at 910 cm^{-1} (in-plane bend., R : s) and 450 cm^{-1} (out-of-plane bend., R : w) fall in regions similar to those reported for thiocarboxylic acids.

For methyl dithioacetate, the strong Raman band at 1203 cm^{-1} was attributed to the C=S stretching mode (IR: 1198 cm^{-1}, vs) [76]. The C—C—S antisymmetric and symmetric stretchings occur strongly at 857 cm^{-1} and 572 cm^{-1}, respectively, in Raman spectrum. Furthermore, a strong Raman band at 728 cm^{-1} was assigned to the C—S stretching mode of the $CH_3{-}S$ moiety.

14-6.3. Thioamides (—C(=S)—N〈)

In thioamides and thioureas, the NH and NH_2 bands are similar to those of amides and ureas. The NH_2—(C=S) group has bands near 3380, 3180, and 1630 cm^{-1}. The noncyclic —NH—(C=S) group has bands near 3170, 1530, and 1350 cm^{-1}. The cyclic —NH—(C=S) group also has a CNH band near 1550 cm^{-1}, whereas comparable cyclic lactams have no band in this region [123]. See Section 9-8.3.

The early studies on the C=S spectral correlation for thioamides encountered great difficulty due to vibrational mixings between the C=S stretching the other vibrational modes [3, 77, 78, 123]. However, based on the resonance structures of the thioamide functional group shown below, Desseyn *et al.* [79] threw some light on the correlation of spectral frequencies with certain vibrations of thioamides (see Table 14-6):

$$\rangle N{-}C{=}S \rightleftarrows \rangle \overset{+}{N}{=}C{-}\overset{-}{S}.$$

(I) (II)

The anomalously low C=S stretching frequency was attributed to the greater contribution of resonance form (II). For *N,N*-dimethylthioamide, such an assumption is supported by its greater dipole moment and the barrier to the internal rotation of the dimethylamino group, as compared with those of *N,N*-dimethyl amide, $Me_2N{-}C({=}O){-}$ [80]. The contribution of the more polar form (II) is also indicated by the unusually high C—N stretching frequencies (1550–1450 cm^{-1}). These assignments of the C=S stretching modes are supported by other authors [81–83]. Similar results were observed for $Y{-}C({=}S){-}NH(CH_3)$, where Y is $NH_2{-}C({=}O){-}$, $NH_2{-}C({=}S){-}$, $-CO_2K$, or $R{-}NH{-}C({=}S){-}$ [84–86]. In the IR and Raman spectra of the above compounds, the vibrations involving the C=S stretch mode are found to be prominent in the regions of 1092–1012 cm^{-1} and 891–779 cm^{-1}.

Table 14-6.

Some Characteristic Frequencies (cm^{-1}) of Thioamides [79]

	cm^{-1}	IR	R	
Primary thioamides:				
CS stretch	950–800	ms	ms	
NCS deformations	750–700	m	vs	
	500–400	m	vs	
Secondary thioamides:				
CS stretch	950–800	ms	ms	
NCS deformations	700–550	mw	vs	
	500–400	mw	s	
Tertiary thioamides:				
CN stretch	1563–1524	vs	m	
	1285–1210	s	w	
CC, CS, CNC stretch	1000–700	var.	var.	(multiple bands)
NCS deformation	626–500	m	vs	
	448–338	m-w	ms	

The mentioned structural resonance of thioamides is observed even when the —NH—C(=S)— moiety is included in a ring system, as in the case pyridine 2-thione [87, 89]:

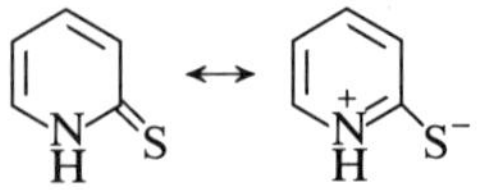

The vibrations involving C=S stretchings were found near 1140 cm^{-1} and 730 cm^{-1}. The polarized Raman band at 450 cm^{-1} (IR : s) can be assigned either to the C=S deformation [89] or to a mixing of ring deformation and C=S stretching [87].

Preti and Tosi [122] proposed a thione-thiol tautomerism to interpret the IR spectrum of benzoxazole-2-thione:

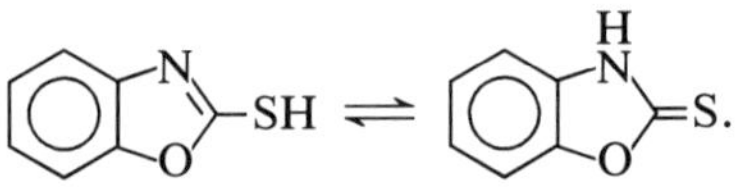

The thiol form on the left was a less preferable species, since the IR spectrum exhibits a strong N—H stretch band in the region above 3000 cm^{-1}. Furthermore, the S—H stretching band was not observed in the IR spectrum. We examined vibrational spectra of five compounds containing cyclic —N=C(SH)— groups (five-membered ring and six-membered ring). The result is similar to that reported by Preti and Tosi. In all cases, the NH stretching bands exhibit strong IR absorption and weak-to-medium Raman intensity, while the S—H stretching mode can be hardly observed. This finding suggests that the cyclic —N=C(SH)— moieties exist predominantly as the thione form of —NH—C(=S)—.

14-6.4. Thioureas (>N—C(=S)—N<)

As with thioamides, vibrational mixing and structural resonance are expected in the case of thioureas. Normal coordinate analysis was applied by Aitken *et al.* to study the vibrational spectra of thiourea. These authors found that the C=S stretching vibration of thiourea is involved in the three spectral bands at 1414 cm^{-1}, 733 cm^{-1}, and 487 cm^{-1} [90]. Among these bands, the 733 cm^{-1} absorption was found to involve mainly the C=S stretch mode. The unusually low C=S stretching frequency near 700 cm^{-1} is also seen in the spectra of *N,N'*-disubstituted thioureas [91]. The vibrational bands at 724 cm^{-1} and 688 cm^{-1} were reported for *N,N'*-dimethylthiourea and *N,N'*-diphenylthiourea, respectively, to contain more than 50% C=S stretching character. Interestingly, the vibrations involving the C=S stretch mode of a cyclic N—C(=S)—N moiety occur at frequencies similar to those of open-chain thiourea. For example, the vibrations involving the C=S stretching mode of pyrimidine-2-thione are located at 1210, 750, and 405 cm^{-1}:

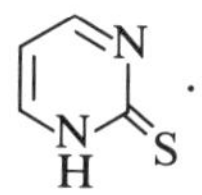

14-6.5. Trithiocarbonates (R—S—C(=S)—S—R)

The IR and Raman spectra of dimethyl trithiocarbonate

(Me—S—C(=S)—S—Me)

was reported by Herzög *et al.* [76]. The C—S stretch band of the CH_3—S moiety is at 722 cm^{-1} (IR : s, R : m). This frequency is in the same spectral region as those found for other CH_3—S containing compounds. Two very strong IR bands at 1078 cm^{-1} and 1033 cm^{-1} were attributed to the C=S stretchings of different rotational isomers. The strongest Raman band at 517 cm^{-1} (polarized) was assigned to the symmetric S—C—S stretching modes. Borch and coworkers [45] have studied the vibrations of ethylene trithiocarbonate,

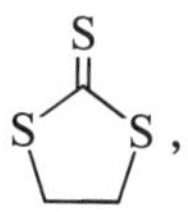

which is a five-membered ring trithiocarbonate, by IR (solid) and Raman (solution) techniques. The vibrational band at 674 cm^{-1} (IR : m, R : vs, p) was assigned to the CH_2—S stretching mode. Delocalization of the C=S double bond within the trithiocarbonate group, which in turn lowers the C=S stretching frequency as compared with those of thioketones, thioacids, and thioesters, was concluded for this compound. As in the case of thioamides, there are extensive mixings between the vibrations within the —S—C(=S)—S— moiety. For example, four bands at 1062 cm^{-1} (IR : s, R : vs, p), 882 cm^{-1} (IR : s, R : m, dp), 832 cm^{-1} (IR : s, R : w, dp), and 503 cm^{-1} (IR : m, R : vs, p) are associated with C=S stretching mode. Furthermore, the antisymmetric S—C—S stretching mode also contributes to the bands at 882 cm^{-1} and 832 cm^{-1}, while the symmetric S—C—S stretching contributes to the very strong Raman band at 503 cm^{-1} (polarized).

14-6.6. Thiazolidine-2-thiones (Compounds Containing Cyclic N—C(=S)—S Groups)

For thiazolidine-2-thione (following left),

vibrational bands involving C=S stretchings are found at 1085 cm^{-1} (IR : m) and 545 cm^{-1} (IR : m) [92, 93]. Similar frequencies were reported for its *N*-methyl analogue (above right), for which the strong IR bands at 1105 cm^{-1}, 981 cm^{-1}, and 595 cm^{-1} are associated with vibrations involving C=S stretching [94]. It can be seen that for these two thiazolidine-2-thiones, which contain cyclic N—C(=S)—S groups, the C=S stretching frequencies are similar to the C=S frequencies of compounds containing cyclic N—C(=S)—N and S—C(=S)—S groups. $CH_3NH—C(=S)—SCH_3$ has bands involving C=S stretch at 1164, 1034, 966, and 478 cm^{-1} [88].

14-6.7. Xanthates (R—O—C(=S)—S—X)

Sodium and potassium xanthate salts ($R—O—CS_2^-Na^+$) have strong IR bands at 1190–1175 cm^{-1}, involving COC out-of-phase stretch, and 1064–1021 cm^{-1}, involving CS_2 out-of-phase stretch [128, 129]. The in-phase CS_2 stretch is sensitive to rotational isomerization. A strong Raman band for in-phase CS_2 stretch is seen at 676–650 cm^{-1} for $(S_2)COCC$ trans, and at 626–620 cm^{-1} for $(S_2)COCH$ trans [129]. A medium-strong Raman band is seen at 477–446 cm^{-1} that involves COC deformation [129].

Other xanthate types including copper and zinc xanthates, *O,S*-dialkyl xanthate esters, and dixanthogens have IR bands involving the O—C=S group at 1250–1200 cm^{-1}, IR : s, 1140–1110 cm^{-1}, IR : m, and 1070–1020 cm^{-1}, IR : s [128]. Note that the IR band involving COC out-of-phase stretch is seen above 1200 cm^{-1} for copper and zinc complexes and below 1200 cm^{-1} for sodium and potassium salts.

14-7. COMPOUNDS CONTAINING S=O GROUPS

The characteristic S=O stretching frequencies of various S=O containing compounds are listed in Table 14-7. In general, the S=O stretching frequency is very characteristic for sulfoxides containing only hydrocarbon

Table 14-7.

The S=O Stretching Frequencies (cm^{-1}) of >S=O Groups

Compounds	cm^{-1}	IR	R	Ref.
R—SO—R	1070–1035	vs	m-w	3, 4f, 77
R—SO—Ar	1050–1040	vs	m-w	3, 4f
Ar—SO—Ar	1042–1035	vs	m-w	3, 4f
MeO—SO—OMe	1207	vs	vs, p	98
RO—SO—OR	1209–1198	s		3, 77
Me_2N—SO—NMe_2	1108	vs	vs	106
Me_2N—SO—Cl	1189	vs	m	106
RO—SO—Cl	1221–1214			3
F—SO—F	1308	vs	vs	107
Cl—SO—Cl	1233	vs		109
Br—SO—Br	1121		w	110

substituents; hence, a narrower region of 1070–1030 cm^{-1} should be expected. Conjugation does not affect the S=O stretching frequency, but the inductive properties of the substituents may lead to some frequency increase. This effect is fully demonstrated in Table 14-7. For example, the attachment of oxygen or halogen atoms to the S=O moiety results in a higher S=O stretching frequency, while hydrogen bond formation results in a low frequency shift of approximately 15–40 cm^{-1}. Splitting of the S=O stretching band due to isomerism is very common. In some cases, this frequency splitting allows the differentiation of different rotational isomers.

14-7.1. Sulfoxides, The C_2S=O Group

Vibrational analyses of dimethyl sulfoxide [95, 96] and diphenyl sulfoxide [35] have been reported. The splitting of the S=O band due to the rotational conformers about the C—S bonds has been studied by Oki *et al.* [97].

The S=O stretching mode absorbs IR strongly in the 1070–1030 cm^{-1} region for alkyl and aryl sulfoxides (i.e., YY′S=O; Y, Y′: R or Ar) and often appears as one of the strongest IR bands [3, 4f]. The S=O stretching mode exhibits poor Raman scattering efficiency, and it usually appears as a weaker band than the C—S stretching band in the Raman spectrum of a given organic sulfoxide. Taking CH_3—S(=O)—CH_3 as an example, the symmetric C—S—C stretching is 10 times stronger than the S=O stretch band (1042 cm^{-1}, liq.) in the Raman [4f, 95, 96]. Even the antisymmetric C—S—C stretch mode is stronger in the Raman than the S=O stretch mode [4g].

The C—S stretchings should be expected in a similar region as those found for dialkyl sulfides. For example, the antisymmetric and symmetric C—S—C stretchings of dimethyl sulfoxides are located at 689 cm^{-1} (IR : m; R : s, dp) and 672 cm^{-1} (IR : m, R : s, p), respectively [4g, 95, 96].

In the case of dimethyl sulfoxide, the vibrations involving C—S=O bending were assigned by Horrocks and Cotton to the bands at 382 cm^{-1} (IR : s; R : m, p) and 336 cm^{-1} (IR : m, R : s, dp) and the C—S—C bending mode of this molecule was assigned to the Raman band at 308 cm^{-1} (m, dp) [95].

14-7.2. Sulfites, The $O_2S{=}O$ Group

The attachment of two oxygen atoms to the S=O grouping results in an increase in the S=O stretching frequency. Vibrational spectra of dimethyl sulfite $(CH_3O)_2S{=}O$ were analyzed by Klaeboe [98]. In his work, the S=O stretching mode (IR: 1207 cm^{-1}, vs; R: 1207 cm^{-1}, vs, p), the antisymmetric O—S—O stretching mode (IR: 688 cm^{-1}, vs; R: 698 cm^{-1}, m, dp), and the symmetric O—S—O stretching vibration (IR: 733 cm^{-1}, vs, R: 732 cm^{-1}, s, p) in the liquid state were assigned. A more recent study by Huong and Raducanu using the IR technique suggested that the differentiation of rotational isomerism. about the two S—O single bonds is possible for dimethyl sulfite [99]. The S—O single bond rotational isomer with both CH_3 groups syn to the S=O bond absorbs IR at 1197 cm^{-1}, while the isomer with both CH_3 groupings anti or gauche to the S=O moiety absorbs at 1219 cm^{-1} in CCl_4 solution. Similar results were observed for $(CH_3O)ClS{=}O$, of which the IR band at 1198 cm^{-1} (CCl_4 soln.) and 1227 cm^{-1} (CCl_4 soln.) were attributed to the rotamers with the CH_3 grouping in the positions sym and anti to the S=O bond, respectively [99]. This type of band splitting due to conformational isomerism is also found in many five-membered, six-membered, and seven-membered cyclic sulfites [100–105]. Cazaux and coworkers [103] studied the S=O stretching band of 47 cyclic sulfites and suggested the following characteristic regions: chair with an equatorial S=O: 1257–1230 cm^{-1}; twisted boat: 1230–1209 cm^{-1}; chair with an axial S=O: 1212–1173 cm^{-1}.

14-7.3. The $N_2S{=}O$ Group

Paetzold and Rönsch [106] analyzed the IR and Raman spectra of $((CH_3)_2N)_2S{=}O$. The S=O stretching frequency of this compound is located at 1119 cm^{-1} (IR: vs; R: 1117 cm^{-1}, vs) in the liquid phase. The

Table 14-8.

The S—X Stretching Frequencies (cm^{-1}), X = N, O, F, Cl, Br

	Compounds	IR	R	Reference	Note
S—N Stretch	Me_2N—SO—NMe_2	678 vs	674 s	106	
		669 vs			
		654 vs	657 s		
		642 vs			
	Me_2N—SO—Cl	702 s	701 vs	106	
	Me—SO_2—NH_2	881 s		116	
	NH_2—SO_2—NH_2	931 s		117	$\nu_{antisym.}$ (N—S—N)
		904 s			$\nu_{sym.}$ (N—S—N)
	Me_2N—SO_2—NH_2	898 s	898 s	118	
		720 vs	714 vs		
	Me_2N—SO_2—Cl	716 s	718 vs, p	120	
	Cl—SO_2—N=C=O	741 vs	744 s	121	
S—O Stretch	MeO—SO—OMe	733 vs	732 s, p	98	$\nu_{sym.}$ (O—S—O)
		688 vs	698 m, dp		$\nu_{antisym.}$ (O—S—O)
S—F Stretch	F—SO—F	801 vs	801 m	107, 108	$\nu_{sym.}$ (F—S—F)
		721 vs	721 m		$\nu_{antisym.}$ (F—S—F)
S—CL Stretch	Cl—SO—Cl	492 vs		109	$\nu_{sym.}$ (Cl—S—Cl)
		455 vs			$\nu_{antisym.}$ (Cl—S—Cl)
	Me_2N—SO—Cl		394 vs	106	
	Me—SO_2—Cl	373 m	374 vs, p	119	
S—Br Stretch	Br—SO—Br		405 s, p	110	$\nu_{sym.}$ (Br—SO—Br)
			379 m, dp		$\nu_{antisym.}$ (Br—SO—Br)

existence of the rotational isomers gives rise to four very strong N—S—N stretching bands in the IR spectrum, at 678, 669, 654, and 642 cm^{-1}. However, only two bands, at 674 and 657 cm^{-1}, were found in the corresponding Raman spectrum (see Table 14-8).

14-7.4. Thionyl Halides, $X_2S=O$ (X = F, Cl, Br)

The effect of the halogen atoms on the S=O stretching frequency of these compounds is largely due to the inductive effect. For thionyl fluoride [107, 108], this vibrational mode occurs at 1308 cm^{-1} (IR : vs; R : vs), which is about 200 cm^{-1} higher than that of other $\rangle$S=O containing compounds. However, there is a progressive shift of the S=O stretching frequency towards a lower value as the halogen atom varies from fluorine, to chlorine, to bromine. Therefore, the S=O stretching band of thionyl chloride and thionyl bromide were found at 1251 cm^{-1} (IR : vs) [109] and 1121 cm^{-1} (R : w) [110], respectively.

The vibrational spectra of these compounds are further characterized by the antisymmetric and symmetric X—S—X stretching bands. These bands of thionyl halides are summarized in Table 14-8.

14-8. COMPOUNDS CONTAINING $-SO_2-$ GROUPS

The vibrational analysis of dimethyl sulfone (Me$-SO_2-$Me) [112], ethylmethylsulfone (Et$-SO_2-$Me) [113], phenylmethylsulfone (Me$-SO_2-$Ph) [114], diphenylsulfone (Ph$-SO_2-$Ph) [115], methanesulfonamide (Me$-SO_2-NH_2$) [116], sulfamide ($NH_2-SO_2-NH_2$) [117], *N,N*-dimethylsulfamide ($NH_2-SO_2-NMe_2$) [118], methanesulfonyl chloride (Me$-SO_2-$Cl) [119], dimethylaminosulfonylchloride ($Cl-SO_2-NMe_2$) [120], and $Cl-SO_2-$NCO [121] have been reported. These compounds are characterized by two distinct SO_2 stretching modes in their IR spectra. The Raman intensities of these two stretching modes are quite variable. Another vibrational band that is characteristic for the $-SO_2-$ grouping is due to the SO_2 scissoring mode.

14-8.1. SO_2 Stretchings

The characteristic $-SO_2-$ stretching frequencies of various types of $-SO_2-$ containing compounds are given in Table 14-9. As in the case of the S=O stretching vibration, the attachment of unsaturated functional groups to the $-SO_2-$ moiety has little effect on the SO_2 stretching frequencies [77]. Such stability of $-SO_2-$ stretching frequencies with respect to the neighboring unsaturated grouping is due to the fact that both S=O bonds are not co-planar with the adjacent unsaturated moiety. Hence, there is no conjugation effect (i.e., delocalization of π electrons) between the S=O bonds and the adjacent unsaturated grouping, and the $\nu(SO_2)$ frequencies remain in a constant region on stubsitution of unsaturated functional groups. Inductive effect, on the other hand, affects the two SO_2 stretching modes significantly [77]. In general, substituents with higher electronegativity shift SO_2 stretching frequencies toward higher values (see Table 14-9) [124]. The correlation between the $-SO_2-$ stretching frequencies and the inductive parameters, σ', of the substituents was discussed by

Table 14-9.

The Characteristic SO_2 Stretching Frequencies (cm^{-1})

Compounds	ν_{as} (SO_2)	IR	R	ν_s (SO_2)	IR	R	Reference
Sulfones:							
$R-SO_2-R$	1330–1295	vs	m-w	1152–1125	vs	vs, p	3, 21, 125, 112, 113
$Ar-SO_2-R$	1334–1325	s	w	1160–1150	s	w	77, 114
$Ph-SO_2-Ph$	1328	vs		1162	vs		115
Sulfonic Acids (Anhydrous):							
$R-SO_2-OH$	1352–1342	vs		1165–1150			125
Sulfonates:							
$R-SO_2-OR$	1358–1352		m	1172–1165		vs	3
$Ar-SO_2-OR$	1363–1338		w	1192–1185		vs	3
$Cl-SO_2-OR$	1406–1401		s-m	1191–1184		vs	3
Thiosulfonates:							
$R-SO_2-SR$	1334–1305		s-m	1128–1126		s	3
Sulfates:							
$RO-SO_2-OR$	1388–1372		s	1196–1188		vs	3
Sulfonamides:							
$Me-SO_2-NH_2$	1315	vs	m	1145	vs	vs	116
$(Me-SO_2-)_2NH$				1163		vs	3
$Ph-SO_2-NH_2$				1157		s	3
$H_2N-SO_2-NH_2$	1358	vs		1156	vs		117
$Me_2N-SO_2-NH_2$	1335	vs	m	1140	s	s	118
Me_2N-SO_2-Cl	1385	vs	m, dp	1177	vs	s, p	120
Organic Sulfonyl Halides:							
$Me-SO_2-F$	1401		m	1186		vs	3
$Ar-SO_2-F$	1412–1402		m-w	1197–1167		vs	3
$Me-SO_2-Cl$	1366	vs	w	1171	vs	s, p	119
$Ar-SO_2-Cl$	1384–1361		m-w	1184–1169		vs	3
$Cl-SO_2-NCO$	1412	vs	m	1182	vs	s	121
Sulfuryl Halides:							
$F-SO_2-F$	1497		w	1263		vs	3
$Cl-SO_2-Cl$	1414		s	1182		s	3

Nyquist [1]. Furthermore, the relationship between the $\nu_{as}(SO_2)$ and $\nu_s(SO_2)$ was found to be approximately linear [1].

The antisymmetric SO_2 stretching mode appears to be more sensitive to hydrogen bonding that the corresponding symmetric mode. When passing from gas phase to condensed phase, the $\nu_{as}(SO_2)$ frequency was found to decrease by 9–73 cm^{-1}, while only a 7–30 cm^{-1} decrease was observed for the $\nu_s(SO_2)$ mode. Similarly, the ν_{as} frequency measured in dilute solution is about 10–20 cm^{-1} higher than that obtained from the solid phase, while the $\nu_s(SO_2)$ frequency is fairly stable with respect to the change in physical states.

In anhydrous sulfonic acids, C—SO_2—OH, the SO_2 bands are like those in C—SO_2—OR [125]. In hydrated sulfonic acids (very common), the acid protonates H_2O to form the hydronium ion C—$SO_3^-H_3O^+$, and therefore somewhat resembles a sulfonic acid salt, C—$SO_3^-Na^+$ [125]. Sulfonic acid salts absorb in the IR at 1230–1120 cm^{-1} strong (antisym. str., two components) and 1080–1025 cm^{-1} medium (sym. str.) [123].

14-8.2. —SO_2— Bendings

The characteristic —SO_2— scissoring mode was previously correlated to a region of 610–545 cm^{-1} [77]. However, recent studies on the vibrational spectra of some —SO_2— containing compounds [112–121] suggested a lower frequency range of 586–504 cm^{-1} for this bending mode. The IR absorption of this vibration is strong in most cases, while the corresponding Raman intensity is variable. Another characteristic bending mode is due to the —SO_2— wagging in the 551–438 cm^{-1} range (IR : s, R : m-s). Both —SO_2— rocking and twisting modes are located in the region of 429–278 cm^{-1}.

14-9. S—X STRETCHINGS, X = O, N, F, Cl, Br

The S—X stretching frequencies of X—S(=O)—X′ and X—SO_2—X′ types of compounds are summarized in Table 14-8. In general, the S—X stretching frequency falls in the same region as the S—C stretching mode when the mass of X is comparable to that of the carbon atom, as in the case of S—N, S—O, and S—F stretchings. For S—Cl and S—Br stretchings, however, the S—X stretchings occur at a lower region of 492–379 cm^{-1} because the chlorine and bromine atoms are heavier. It is interesting to note that the symmetric X—S—X stretching modes of $(MeO)_2S{=}O$, $F_2S{=}O$, $Cl_2S{=}O$, and $Br_2S{=}O$ exhibit a higher vibrational frequency than the corresponding antisymmetric modes. This observation is the opposite of what was found for C—S—C stretching modes.

REFERENCES

1. R. A. Nyquist, "The Interpretation of Vapor Phase Infrared Spectra, Group Frequency Data," Vol. I, Sadtler Research Laboratories, 1984, pp. 465–470.
2. R. B. de Alencastro and C. Sandorfy, *Can. J. Chem.* **50,** 3594, 1972.

3. F. R. Dollish, W. G. Fateley, and F. F. Bentley, "Characteristic Raman Frequencies of Organic Compounds," John Wiley & Sons, New York, 1974, pp. 46–53.
4. "The Sadtler Standard Raman Spectra," Sadtler Research Laboratories, 1976. (a) Thiols: see subject index 61, 62. (b) *o*-Aminobenzenethiol: 229R. (c) Di-*t*-butylsulfide: 4182R. (d) 1046R, 1047R, 1051R, 1458R, 3219R. (e) Sulfides and disulfides: see subject index 75. (f) Sulfoxides and sulfones: see subject index 76. (g) 294R. (h) 1517R.
5. M.-L. Josien, P. Dizabo, and P. Saumagne, *Bull. Soc. Chim. Fr.*, 423, 1957.
6. M.-L. Josien, C. Castinel, and P. Sumagne, *Bull. Soc. Chim. Fr.*, 648, 1957.
7. R. A. Spurr and H. F. Byers, *J. Phys. Chem.* **62,** 425, 1958.
8. M. O. Bulanin, G. S. Denisov, and R. A. Pushkina, *Opt. Spectrosc.* **6,** 491, 1959.
9. J. G. David and H. E. Hallam, *Spectochim. Acta* **21,** 841, 1965.
10. S. H. Marcus and S. I. Miller, *J. Am. Chem. Soc.* **88,** 3719, 1966.
11. C. J. Pouchert, "The Aldrich Library of FT-IR Spectra," Aldrich Chemical Company, Inc., 1985, pp. 255–277, pp. 1173–1188.
12. I. W. May and E. L. Pace, *Spectrochim. Acta* **24A,** 1605, 1968.
13. D. Smith, J. P. Devlin, and D. W. Scott, *J. Mol. Spectrosc.* **25,** 174, 1968.
14. M. Hayashi, Y. Shiro, and H. Murata, *Bull. Chem. Soc. Jpn.* **39,** 112, 1966.
15. N. Sheppard, *Trans. Faraday Soc.* **46,** 527, 1950.
16. J. P. McCullough, D. W. Scott, H. L. Finke, W. N. Hubbard, M. E. Gross, C. Katz, R. E. Pennington, J. F. Messerly, and G. Waddington, *J. Am. Chem. Soc.* **75,** 1818, 1953.
17. M. Hayashi, Y. Shiro, M. Murakami, and H. Murata, *Bull. Chem. Soc. Jpn.* **38,** 1740, 1965.
18. D. W. Scott and G. A. Crowder, *J. Chem. Phys.* **46,** 1054, 1967.
19. J. P. McCullough, H. L. Finke, D. W. Scott, M. E. Gross, J. F. Messerly, R. E. Pennington, and G. Waddington, *J. Am. Chem. Soc.* **76,** 4796, 1954.
20. J. R. Allkins and P. J. Hendra, *Spectrochim. Acta* **22,** 2075, 1966.
21. G. Geiseler and G. Hanschmann, *J. Mol. Struct.* **11,** 283, 1972.
22. M. Ohsaku, *Bull. Chem. Soc. Jpn.* **48,** 707, 1975.
23. N. Nogami, H. Sugeta, and T. Miyazawa, *Bull. Chem. Soc. Jpn.* **48,** 2417, 1975.
24. N. Nogami, H. Sugeta, and T. Miyazawa, *Bull. Chem. Soc. Jpn.* **48,** 3573, 1975.
25. M. Sakakibara, H. Matsuura, I. Harada, and T. Shimanouchi, *Bull. Chem. Soc. Jpn.* **50,** 111, 1977.
26. M. Ohata, Y. Ogawa, H. Hatsuura, I. Harada, and T. Shimanouchi, *Bull. Chem. Soc. Jpn.* **50,** 380, 1977.
27. Y. Ogawa, M. Ohata, M. Sakakibara, H. Mutsuura, I. Harada, and T. Shimanouchi, *Bull. Chem. Soc. Jpn.* **50,** 650, 1977.
28. M. Ohsaku, *Bull. Chem. Soc. Jpn* **48,** 1037, 1975.
29. D. W. Scott, W. D. Good, S. S. Todd, J. F. Messerly, W. T. Berg, I. A. Hossenlopp, J. L. Lacina, A. Osborn, and J. P. McCullough, *J. Chem. Phys.* **36,** 406, 1962.
30. S. W. Charles, F. C. Cullen, and N. L. Owen, *J. Mol. Struct.* **34,** 219, 1976.
31. U. C. Joshi, M. Joshi, and R. N. Singh, *Ind. J. Pure & Appl. Phys.* **19,** 680, 1981.
32. R. J. H. Clark, C. D. Flint, and A. J. Hempleman, *Spectrochim. Acta* **43A,** 805, 1987.
33. J. Fabian, H. Kröber, and R. Mayer, *Spectrochim. Acta* **24A,** 727, 1968.
34. K. K. Georgieff and A. Dupré, *Can. J. Chem.* **37,** 1104, 1959.
35. J. H. S. Green, *Spectrochim. Acra* **24A,** 1627, 1968.
36. M. Guilliano, G. Mille, T. Avignon, and J. Chouteau, *J. Raman Spectrosc.* **7,** 214, 1978.
37. M. Guilliano, G. Davidovics, J. Chouteau, J. L. Larice, and J. P. Roggero, *J. Mol. Struct.* **25,** 329, 1975.

38. M. G. Giorgini, G. Paliani, and R. Cataliotti, *Spectrochim. Acta* **33A,** 1083, 1977.
39. J. R. Durig, J. F. Sullivan, T. S. Little, and D. T. Durig, *J. Mol. Struct.* **142,** 53, 1986.
40. O. H. Ellestad, P. Klaeboe, and G. Hagen, *Spectrochim. Acta* **28A,** 137, 1972.
41. P. Klaeboe, *Spectrochim. Acta* **25A,** 1437, 1969.
42. G. Davison and K. P. Ewer, *Spectrochim. Acta* **39A,** 419, 1983.
43. P. N. Gate, P. Powell, and D. Steele, *J. Mol. Struct.* **8,** 477, 1971.
44. G. Davison and K. E. Ewer, *J. Mol. Struct.* **74,** 181, 1981.
45. G. Borch, L. Henriksen, P. H. Nielsen, and P. Klaeboe, *Spectrochim. Acta* **29A,** 1109, 1973.
46. B. Kure and M. D. Morris, *Talanta* **23,** 398, 1976.
47. K. G. Allum, J. A. Creighton, J. H. S. Green, G. J. Minkoff, and J. L. S. Prince, *Spectrochim. Acta* **24A,** 927, 1968.
48. D. W. Scott, H. L. Finke, J. P. McCullough, M. E. Gross, R. E. Pennington, and G. Waddington, *J. Am. Chem. Soc.* **74,** 2478, 1952.
49. D. W. Scott, H. L. Finke, M. E. Gross, G. B. Guthrie, and H. M. Huffman, *J. Am. Chem. Soc.* **72,** 2424, 1978.
50. M. Hayashi, Y. Shiro, and H. Murata, *Bull. Chem. Soc. Jpn.* **39,** 1857, 1966.
51. M. Hayashi, Y. Shiro, and H. Murata, *Bull. Chem. Soc. Jpn.* **39,** 1861, 1966.
52. H. Sugeta, A. Go, and T. Miyazawa, *Chem. Lett.*, 83, 1972.
53. N. Sheppard, *Trans. Faraday Soc.* **46,** 429, 1950.
54. W. J. Balfour, K. S. Chandrasekhar, and S. P. Kyca, *Spectrochim. Acta* **42A,** 39, 1986.
55. H. S. Randhawa and C. N. R. Rao, *J. Mol. Struct.* **21,** 123, 1974.
56. H. S. Randhawa, W. Walter, and C. O. Meese, *J. Mol. Struct.* **37,** 187, 1977.
57. G. Crowder, E. Robertson, and K. Potter, *Can. J. Spectrosc.* **20,** 49, 1975.
58. R. Bicca de Alencastro and C. Sandorfy, *Can. J. Chem.* **51,** 1443, 1973.
59. G. Crowder, *J. Mol. Struct.* **32,** 207, 1976.
60. H. S. Randhawa and W. Walter, *J. Mol. Struct.* **38,** 89, 1977.
61. H. S. Randhawa and W. Walter, *Bull. Chem. Soc. Jpn* **51,** 1579, 1978.
62. G. A. Crowder, *Appl. Spectrosc.* **27,** 440, 1973.
63. H. S. Randhawa, C. O. Meese, and W. Walter, *J. Mol. Struct.* **36,** 25, 1977.
64. G. A. Crowder, *Appl. Spectrosc.* **26,** 486, 1972.
65. V. R. Engler and G. Gattow, *Z. Anorg. Allg. Chem.* **388,** 78, 1972.
66. R. Vogel-Högler, *Acta Phys. Austriaca* **1,** 311, 1948.
67. P. Donzelot and M. Chaix, *Compt. Rend.* **202,** 851, 1936.
68. F. Feher and W. Kruse, *Chem. Ber.* **91,** 2528, 1958.
69. H. F. Shurvell, R. D. Gordon, and R. B. Girling, *J. Raman Spectrosc.* **16,** 90, 1985.
70. P. K. Basu, U. C. Singh, K. N. Tantry, V. Ramamurthy, and C. N. R. Rao, *J. Mol. Struct.* **76,** 237, 1981.
71. H. W. Thompson, *J. Chem. Phys.* **6,** 748, 1938.
72. C. A. Frenzel, K. E. Blick, C. R. Bennett, and K. Niedenzu, *J. Chem. Phys.* **53,** 198, 1970.
73. J. L. Brema and D. C. Moule, *Spectrochim. Acta* **28A,** 809, 1972.
74. D. C. Moule and C. R. Subramaniam, *Can. J. Chem.* **47,** 1011, 1969.
75. H. S. Randhawa and W. Walter, *J. Raman Spectrosc.* **11,** 138, 1981.
76. K. Herzög, E. Steger, P. Rosmus, S. Scheithauer, and R. Mayer, *J. Mol. Struct.* **3,** 339, 1969.
77. L. J. Bellamy, "The Infrared Spectra of Complex Molecules," John Wiley & Sons, New York, 1975, pp. 394–410.
78. H. O. Desseyn and M. A. Herman, *Spectrochim. Acta* **23A,** 2457, 1967.
79. H. O. Desseyn, B. J. van der Veken, and M. A. Herman, *Appl. Spectrosc.* **32,** 101, 1978.

80. J. A. Le Poivre, H. O. Desseyn, and F. A. C. Alderweireldt, *Org. Magn. Reson.* **6,** 284, 1974.
81. P. Geboes, H. Hoffmans, H. O. Desseyn, R. Dommisse, A. T. H. Lenstra, S. Bamidele Sanni, J. M. M. Smith, and P. T. Beairskens, *Spectrochim. Acta* **43A,** 35, 1987.
82. W. Walter and P. Staglish, *Spectrochim. Acta* **30A,** 1739, 1974.
83. U. Anthoni and D. H. Nielsen, *J. Mol. Struct.* **116,** 175, 1984.
84. H. O. Desseyn, A. J. Aarts, and M. H. Herman, *Spectrochim. Acta* **36A,** 59, 1980.
85. H. O. Desseyn, A. Aarts, E. Esmans, and M. A. Herman, *Spectrochim. Acta* **35A,** 1203, 1979.
86. H. O. Desseyn, J. A. Le Doivre, and M. A. Herman, *Spectrochim. Acta* **30A,** 503, 1974.
87. D. N. Sathyanarayana and S. V. Kasmir Raja, *Spectrochim. Acta* **41A,** 809, 1985.
88. K. R. G. Devi, D. N. Sathyanarayana, and S. Manogaram, *Spectrochim. Acta* **37A,** 633, 1981.
89. R. Shunmugam and D. N. Sathyanarayana, *Spectrochim. Acta* **40A,** 757, 1984.
90. G. B. Aitken, J. L. Duncan, and G. P. McQuillan, *J. Chem. Soc. A*, 2695, 1971.
91. R. K. Ritchie, H. Spedding, and D. Steele, *Spectrochim. Acta* **27A,** 1597, 1971.
92. F. Cristiani, F. A. Devillanov, and G. Verani, *Spectrochim. Acta* **38A,** 1299, 1982.
93. F. A. Devillanova, K. R. Gayathry Devi, D. N. Sathyanarayana, and G. Verani, *Spectrochim. Acta* **35A,** 1083, 1979.
94. F. A. Devillanova, D. N. Sathyanarayana, and G. Verani, *J. Heterocyclic Chem.* **15,** 945, 1978.
95. W. D. Horrocks, Jr. and F. A. Cotton, *Spectrochim. Acta* **17,** 134, 1961.
96. M.-T. Forel and M. Tranquille, *Spectrochim. Acta* **26A,** 1023, 1970.
97. M. Oki, I. Oka, and K. Sakaguchi, *Bull. Chem. Soc. Jpn.* **42,** 2944, 1969.
98. P. Klaeboe, *Acta Chem. Scand.* **22,** 2817, 1968.
99. P. V. Huong and E. Raducanu, *J. Mol. Struct.* **23,** 81, 1974.
100. D. G. Hellier, J. G. Tillet, H. F. van Woerden, and R. F. M. White, *Chem. Ind. (London)*, 1956, 1963.
101. J. G. Pritchard and R. L. Vollmer, *J. Org. Chem.* **28,** 545, 1963.
102. P. C. Lauterbur, J. G. Pritchard, and R. L. Vollmer, *J. Chem. Soc.*, 5307, 1967.
103. L. Cazaux, J. D. Bastide, G. Chassaing, and P. Maroni, *Spectrochim. Acta* **35A,** 15, 1979.
104. K. L. Dorris, *Appl. Spectrosc.* **24,** 492, 1970.
105. H. Faucher, A. C. Guimaraes, and J. B. Robert, *Tetrahedron* **37,** 689, 1981.
106. R. Paetzold and E. Rönsch, *Spectrochim. Acta* **26A,** 569, 1970.
107. P. Bender and J. M. Wood, Jr., *J. Chem. Phys.* **22,** 1316, 1955.
108. J. K. O'Loane and M. K. Wilson, *J. Chem. Phys.* **23,** 1313, 1955.
109. D. E. Martz and R. T. Lagemann, *J. Chem. Phys.* **22,** 1193, 1954.
110. H. Stammreich, R. Forneris, and Y. Tavares, *J. Chem. Phys.* **25,** 1277, 1956.
111. D. W. Scott, W. D. Good, S. S. Todd, J. F. Messerly, W. T. Berg, I. A. Hossenlopp, J. L. Lacina, A. Osborn, and J. P. McCullough, *J. Chem. Phys.* **36,** 406, 1962.
112. T. Ubo, K. Machida, and K. Hanai, *Spectrochim. Acta* **27A,** 107, 1971.
113. A. H. Fawcett, S. Fee, M. Stuckey, and P. Walkden, *Spectrochim. Acta* **43A,** 797, 1987.
114. M. Bouquet, G. Chassaing, J. Corset, J. Favort, and J. Limouzi, *Spectrochim. Acta* **37A,** 727, 1981.
115. U. C. Joshi, M. Joshi, and R. N. Singh, *Ind. J. Pure & Appl. Phys.* **19,** 1226, 1981.
116. K. Hanai, T. Okuda, T. Uno, and K. Machida, *Spectrochim. Acta* **31A,** 1217, 1975.
117. T. Uno, K. Machida, and K. Hanai, *Spectrochim. Acta* **22,** 2065, 1966.

118. Y. Tanaka, Y. Tanaka, and Y. Saito, *Spectrochim Acta* **39A,** 159, 1983.
119. K. Hanai and T. Okuda, *Spectrochim. Acta* **31A,** 1227, 1975.
120. Y. Tanaka, Y. Tanaka, Y. Saito, and K. Machida, *Bull. Chem. Soc. Jpn.* **51,** 1324, 1978.
121. I. Kanesaka and K. Kawai, *Bull. Chem. Soc. Jpn.* **43,** 3298, 1970.
122. C. Preti and G. Tosi, *Spectrochim. Acta* **35A,** 577, 1979.
123. N. B. Colthup, L. H. Daily, and S. E. Wiberley, "Introduction to Infrared and Raman Spectroscopy," 3rd Ed., Academic Press, New York, 1990.
124. L. W. Daasch, *Spectrochim. Acta* **13,** 257, 1958.
125. S. Detoni and Hadzi, *Spectrochim. Acta* **11,** 601, 1957.
126. R. A. Nyquist and W. J. Potts, *Spectrochim. Acta* **15,** 514, 1959.
127. N. Lozac'h and G. Guilouzo, *Bull. Soc. Chim., Fr.*, 1221, 1957.
128. L. H. Little, G. W. Poling, and L. Leja, *Can. J. Chem.* **39,** 745, 1961.
129. N. B. Colthup and L. Porter Powell, *Spectrochim. Acta* **43A,** 317, 1987.
130. W. Zhao, J. Bandekar, and S. Krimm, *J. Am. Chem. Soc.* **110,** 6891, 1988.

CHAPTER 15

Organosilicon Compounds

15-1.

THE $-SiH_3$ GROUP

15-1.1. The SiH_3 Stretchings

Bellamy [1] has correlated the $-SiH_3$ stretching modes of monoalkylsilanes ($R-SiH_3$) to a strong IR band in the region of 2153–2142 cm^{-1}. See Table 15-1. The same vibrational modes for monoaromatic silanes were found in the region of 2157–2152 cm^{-1} [1]. In the majority of cases, the frequencies of the antisymmetric and symmetric $-SiH_3$ modes are so close as to be

Table 15-1.

The Group Frequencies of SiH_3 Moieties (cm^{-1})

		IR	Reference
The SiH_3 Stretchings:			
$R-SiH_3$	2153–2142	s	1, 22, 25
$Ar-SiH_3$	2157–2152	s	1, 25
$-C{\equiv}C-SiH_3$	2190–2170	s	1
The SiH_3 Deformations:			
	947–930		1, 2, 22, 25
	930–910		
The SiH_3 Rocking:			
	720–680		1, 2

indistinguishable. Although it was suggested that the presence of a conjugated triple bond with the $-SiH_3$ group can split the near-coincidence of the antisymmetric and the symmetric $-SiH_3$ stretchings, as in the case of 1,2-disilyl acetylene [1] (ν_{as}(SiH_3): 2190 cm^{-1}, ν_s(SiH_3: 2170 cm^{-1}), the IR study on silylacetylene ($H_3Si-C\equiv C-H$) by Ebsworth *et al.* [2] indicates that the antisymmetric and symmetric $-SiH_3$ modes are still superimposed for this silicon compound. The antisymmetric and symmetric SiH_3 stretching frequencies are expected to be considerably closer together than the comparable CH_3 frequencies because of the heavier mass of the silicon atom. The same is true for SiH_2 stretching frequencies [11].

15-1.2. The SiH_3 Deformations

By analogy with the $-CH_3$ deformations near 1460 cm^{-1} and 1370 cm^{-1}, the $-SiH_3$ deformations exhibit strong IR bands in the ranges of 947–930 cm^{-1} and 930–910 cm^{-1} that can be attributed to the $-SiH_3$ deformations [1, 25]. For example, the antisymmetric and symmetric $-SiH_3$ deformations of silylacetylene were observed at 946 and 935 cm^{-1}, respectively [2]. The $-SiH_3$ rocking mode has been correlated with an IR band in a region between 720 and 680 cm^{-1}. The SiH_3 rocking mode of silylacetylene was assigned to the strong IR band at 666 cm^{-1}.

15-2. THE $-SiH_2$ GROUP

Similar to those of the $-CH_2-$ moiety, the vibrations of a $-SiH_2-$ moiety are approximated as stretchings, scissoring, wagging, twisting, and rocking modes. The earlier data on the IR frequencies of $-SiH_2-$ moieties have been discussed by Bellamy [1]. More recently, the vibrational spectra of dimethylsilane [4], silacyclopentane [7], methyl methylsilyl sulfide [8], methyl methylsilyl ether [9], and ethyl methyl silane [10] have been reported. A summary of the group frequencies of the $-SiH_2-$ moiety is shown in Table 15-2.

15-2.1. $-SiH_2-$ Stretchings

Although the antisymmetric $-SiH_2-$ stretching mode of gas-phase silacyclopentane was observed by Durig and Wilis [7] using the IR technique to

Table 15-2.

The Group Frequencies of $-SiH_2-$ Moieties (cm^{-1})

SiH_2 Stretching	cm^{-1}	IR	R	Reference
$R-SiH_2-R$	2138–2117	s		1, 4, 5, 25
$SiH_2(CH_2)_4$ (ring)	2149	s	vs, p	7
$Ar-SiH_2-Ar$	2147–2130	s		25
$Ph-SiH_2-Ph$	2139	vs	m, p	3a, 18
$Me-SiH_2-SiH_2-Me$	2126	vs	vs, p	4
$Me-SiH_2-X$	2200–2140 (X = Me, F, Cl, Br, I)			6

SiH_2 Deformations	SiH_2 Scissors IR : s, R : w	SiH_2 Wag IR : s, R : w	SiH_2 Twist IR : s, R : w	SiH_2 Rock IR : s, R : w	Reference
$R-SiH_2-R$	950–930	900–840	690–560	540–480	1, 4, 5, 25
$Ar-SiH_2-Ar$	940–928	870–843			25
$Me-SiH_2-X$	980–940	955–875	740–630	520–460	6
$SiH_2(CH_2)_4$ (ring)	955	885	575	456	7

be at 2151.8 cm^{-1}, this IR band is not found in liquid-phase silacyclopentane. As a matter of fact, only one $-SiH_2-$ band is reported in many instances for liquid samples, and only the symmetric $-SiH_2-$ stretching mode was found. While the IR intensity of the SiH_2- band appears to be quite strong, the polarized Raman band assignable to the SiH_2- stretching mode exhibits intensity varying from medium to very strong.

The earlier studies indicate that the $-SiH_2-$ stretching modes of dialkylsilanes ($R-SiH_2-R$) are located in the region of 2138–2117 cm^{-1} [1, 4, 25]. The presence of a second Si atom on the $-SiH_2-$ group does not cause a significant shift in the $-SiH_2-$ stretching frequencies, as illustrated by the vibrational spectra of dimethyldisilane ($CH_3SiH_2SiH_2CH_3$), which exhibit a distinct vibrational band at 2126 cm^{-1} (IR : vs, R : vs, p) in the liquid phase [4]. The substitution of the alkyl groups by two aryl groups appears to increase the $-SiH_2-$ stretching frequencies slightly, and the $-SiH_2-$ stretching frequencies of diarylsilanes ($Ar-SiH_2-Ar$) are found in a region of 2147–2130 cm^{-1} [25]. For example, the IR spectrum of diphenylsilane (3a) exhibits a strong band at 2139 cm^{-1} that can be attributed to the $-SiH_2-$ stretching mode. The relative Raman intensity of the $-SiH_2-$ stretching appears to be polarized and of medium intensity. Halogenation on the Si atom induces a significant shift of the $-SiH_2-$ stretching frequencies toward higher values, as was found for difluorosilane and monochlorosilane.

15-2.2. The $-SiH_2-$ Deformations

For compounds with the $R-SiH_2-R$ type of linkage, the vibrational frequencies of the $-SiH_2-$ scissoring mode, the $-SiH_2-$ wagging mode, the $-SiH_2-$ twisting mode, and the $-SiH_2-$ rocking mode were reported to be located in the regions of 950–930 cm^{-1}, 900–840 cm^{-1}, 690–560 cm^{-1}, and 540–480 cm^{-1}, respectively (4). Halogenation on the Si atom appears to shift the above correlated frequency regions upwards somewhat, as demonstrated in Table 15-2.

15-3.

THE Si—H GROUP

15-3.1. The Si—H Stretching Mode

The Si—H stretching frequencies are sensitive to the inductive effect of the substituents. The frequency increases as the sum of the electronegativities of the substituents increases [13]. The characteristic frequencies of the Si—H compounds are listed in Table 15-3. In IR spectra, the Si—H stretching

Table 15-3.

The Group Frequencies of SiH Vibrations (cm^{-1})

Compounds	Si—H Stretch	IR	R	Si—H Bending	IR	R	Reference
$H-SiR_3$	2110–2094			842–800	s		1, 3b, 11
$H-SiR_2Ar$	2115–2103			842–800	s		1, 3d, 11
$H-SiRAr_2$	2125–2115			842–800	s		1, 3d, 11
$H-SiAr_3$	2132–2112	vs	m[a]	842–800	s		1, 3c, 11
$H-SiMe_2Cl$	2171	s		842[b]	s		3e
$H-SiMeCl_2$	2215	s		837[b]	s		3e
$H-SiF_3$	2282						1
$H-SiBr_3$	2236						1
$H-Si(OMe)_3$	2203						1
$H-Si(OEt)_3$	2194	s		880/860[b]	s		3f
$H-SiMe_2OSiMe_2$	2129	s		834[b]	m		3g
$-(SiHMe-O)_n-$	2166	s		838[b]	s		3h
$-(SiHPh-O)_n-$	2180	s					19
1-Hydrosilatrane ($H-Si(OCH_2CH_2)_3N$)	2088	s	vs	935	vs	sh	12, 16, 17

[a] The relative Raman intensity is slightly lower than that of the 1590 cm^{-1} band of the aromatic stretching mode.
[b] Tentative assignment.

mode appears in the region of 2110–2089 cm^{-1} for trialkylsilanes [1, 3b, 11, 25]. The substitution of the alkyl groups on the Si atom by aryl groups tends to increase the Si—H stretching frequencies, and a region of 2132–2112 cm^{-1} [1, 25] was reported for triarylsilanes. The Si—H stretching frequencies of some monoaryl- and diarylsilanes are intermediate [25].

Lee Smith [13, 15] and Thompson [14] both have reported formulas that allow the prediction of the Si—H frequencies from the parameters that are characteristic of the substituents on the Si atom. The (XYZ)SiH frequency in dilute CCl_4 solution can be calculated by a summation of wavenumber components, one for each of the three substituents. A selection of these components is given as follows in cm^{-1} [13], if there are no Si—O—Si groups: F 760.8, Cl 752.8, Br 745.3, OMe 734.4, OEt 732.0, H 724.8, Ph 708.7, Me 705.9, Et 699.1. For example, $(F_3)SiH$ stretch is calculated as $3 \times 760.8 = 2282.4$ cm^{-1} (observed 2282 cm^{-1}). The SiH frequencies in siloxanes are calculated by a slightly more complex function [13].

As illustrated in Table 15-3, halogenation and the presence of an alkoxy group will also, although to a lesser extent, increase the Si—H frequency. In some cases, the electronic structure of the Si atom can be probed by the Si—H frequency [16]. The exceptionally low Si—H stretching frequency of 1-hydrosilatrane in comparison with other alkoxy-substituted Si—H compounds has been attributed to the formation of an intramolecular bond between the N atom and the Si atom in the cyclic structure of the molecule [12, 16, 17].

15-3.2. The Si—H Deformation

The Si—H bending occurs in a lower region, 842–800 cm^{-1}, for alkyl and aryl silanes [11] with strong IR intensity [3b–3d]. A strong IR band is found in the spectra of chlorinated silanes [3e], 1,1,3,3,-tetramethyldisiloxane [3g], and polymethylhydrosiloxane [3h] near the higher-frequency end of the above region (see Table 15-3). In Table 15-3, note the extraordinarily high value of the Si—H bending frequency of 1-hydosilatrane at 935 cm^{-1}, which is about 100 cm^{-1} higher than the suggested region of 842–800 cm^{-1}, due to the unusual N→Si bond of the molecule [12, 16, 17].

15-4. THE Si—C STRETCHING

The Si—C stretching frequencies are reported in the range of 900–500 cm^{-1} [1]. Based on the available data, it appears that the Raman intensity of the

Table 15-4.

The Si—C Stretching Frequencies (cm^{-1})

Compounds		Frequency (cm)	IR	R	Reference
$Si(CH_3)_4$		860			1
$-Si(CH_3)_3$	sym. str.	715–680			1
	antisy. str.	660–485			1
Me—Si—Me		730–650			1
—O—SiMe (end group)		850–840			1
(side chain)		800–770			1
$R-SiH_3$	P_C	750–740			20
	P_H	690–685			20
R—Si—R′ (R,R′ = Me, Et; Et, Et; or Me, Pr):					
$Si(H_2)-CH_3$	str.	730–710	m-s	s-vs	10
$Si(H_2)-C(H_2)$	str.	760–610			10
	P_C	above 700	m-s	m-vs	
	P_H	below 700	mw-w	vvs	
$CH_2=CH-SiH_3$		624	m	m	21
$H_3C-Si-Si-CH_3$					4
$-(Si-O)_n-$, siloxanes		near 800[a]			19

[a]See text for further discussion.

Si—C stretching mode is very distinct, while the IR intensity can vary from weak to strong.

The Si—C stretchings are found to have considerable vibrational coupling with other vibrations of the neighboring moieties [10], and the frequencies are very dependent on the nature of the substituents [1]. The typical Si—C frequencies are summarized in Table 15-4. The coupling of the Si—C stretching vibration with other neighboring vibrations can be illustrated by the 800 cm^{-1} IR band of the open-chain dimethyl siloxanes. This IR band of dimethyl siloxane was assigned by Hummel [19] to the C—Si—C stretching mode, but Smith [15] assigned it to the methyl rocking mode. Colthup *et al.* [11] pointed out that there is probably vibrational coupling between the Si—C stretching mode and the CH_3 rocking mode for the Si—CH_3 group.

Like those of dialkylsulfide, the rotational conformations of dialkylsilanes may be classified as P_C and P_H conformations. More than one band in the vibrational spectra of dialkylsilanes can be attributed to the Si—C stretching modes due to the rotational isomers. The C—Si stretching frequencies of dialkylsilanes are found to be correlated with the rotational conformations. Matsuura and coworkers [10] have correlated the Si—C stretching frequencies of dialkylsilanes, based on a normal coordinate analysis of three dialkylsilanes, to be above 700 cm^{-1} for the P_C isomers, and below 700 cm^{-1} for the P_H conformers. The IR intensities appear to be medium-to-weak, and the Raman intensities are from medium-strong to strong.

For monoalkylsilanes, the Si—C stretching frequencies of the P_C conformation are reported to be in the region of 750–740 cm^{-1}, while those of the P_H conformation are found in the region of 690–685 cm^{-1} [20].

The SiCC bending of vinylsilane is located at 284 cm^{-1} with weak Raman intensity [21].

15-5. Si—C GROUPS

15-5.1. Si—R

The CH_3 in the Si—CH_3 group has a very strong, sharp IR band at 1280–1255 cm^{-1} (symmetric deformation) and a weak band near 1410 cm^{-1} (antisymmetric deformation). The $SiCH_3$ rock (plus probably some Si—C stretch interaction) has IR bands at 870–750 cm^{-1} [24]. See Table 15-5.

The Si—C_2H_5 group has IR bands at 1250–1220, 1020–1000, and 970–945 cm^{-1}. The group Si—CH_2—R has a weak band near 1410 cm^{-1} (CH_2 deformation), and a medium-intensity IR band at 1250–1200 cm^{-1} (CH_2 wag) [24].

Table 15-5.

Si—C Groups [24]

	cm^{-1}	Assignment
$SiCH_3$	ca. 1410 IR : w	CH_3 antisym. str.
	1280–1255 IR : s	CH_3 sym. str.
	870–750 IR : s	involves CH_3 rock
Si—CH_2—R	ca. 1410 IR : w	CH_2 def.
	1250–1200 IR : m	CH_2 wag
Si—CH=CH_2	1615–1590 IR : m	C=C str.
	1410–1390 IR : m	=CH_2 def.
	1020–1000 IR : m	=CH wag
	980–950 IR : m	=CH_2 wag
Si-phenyl	ca. 1430 IR : m	ring str.
	1125–1100 IR : s	ring + Si—C str.
	ca. 730 IR : m	ring CH wag
	ca. 700 IR : m	ring bend

15-2.2. Si—CH=CH$_2$

The Si—CH=CH$_2$ silicon-vinyl group has IR bands at 1615–1590 cm^{-1} (C=C stretch), 1410–1390 cm^{-1} (CH$_2$ deformation), 1020–1000 (=CH wag), and 980–950 cm^{-1} (=CH$_2$ wag) [24].

15-5.3. Si-Phenyl

When a silicon atom is directly attached to a phenyl group, a substituent-senstitive ring mode, which involves some Si—C stretch, is seen in the IR at 1125–1100 cm^{-1}. Other phenyl IR bands are seen near 1430, 730, and 700 cm^{-1} [24].

15-6.

THE Si—O STRETCHING

The IR freqencies of some organosilicon compounds containing the Si—O linkages are listed in Table 15-6.

Smith [15] has studied the correlation between the Si—O vibrational frequencies and the inductive effects of the substituents on the silicon atom. In general, a higher Si—O stretching frequency is observed for compounds with substituents of higher electronegativity on the silicon atom [15, 16].

The Si—O—C stretching frequencies of alkoxysilanes (Si—OR) are located in the region of 1100–1000 cm^{-1}.

The Si—O—Si stretching frequencies of siloxanes are characterized by the very strong IR complex in the region of 1100–1000 cm^{-1} [1, 19]. Long-chain siloxanes have two broad IR bands near 1090 and 1020 cm^{-1}. The possible spectral interference for the assignment of these vibrational modes is the Si—O—S stretching mode of alkoxysilanes.

The Si—O—Si stretching frequencies of cyclic siloxanes are correlated with the ring size [15, 23]. The Si—O—Si stretching frequencies of the (Si—O)$_3$ cyclotrisiloxanes are located in the region of 1020–1010 cm^{-1}, while the cyclotetrasiloxanes have bands near 1080 cm^{-1}. The Si—O—Si stretching frequency of a cyclopentasiloxane, which is located near 1100 cm^{-1}, is the highest frequency found for cyclosiloxanes. A gradual decrease of the Si—O—Si stretching mode from 1100 cm^{-1} was reported for cyclosiloxanes larger than pentamers. For cyclodecasiloxanes, the Si—O—Si stretching frequency is near 1050 cm^{-1}.

The Si—O—H group has a strong IR band involving Si—O stretch at 920–830 cm^{-1} [24]. The OH stretch band is like that in alcohols, 3700–3200 cm^{-1}, for free or hydrogen-bonded OH [24].

Table 15-6.
The Si—O Stretching Frequencies (cm^{-1})

	cm^{-1}	Reference
Alkoxysilanes:		
$Si-OCH_3$	1110–1107	1
	1100–947	
Si—OEt	1100–1075	1, 15
	1060–1040	
	970–940	
Si—O(*i*-Pr)	1060–1025	15
Si—OH	920–830	15
Si—OAc	955–935	15
Si—OPh	970–920	15
Siloxanes:		
Si—O—Si	1100–1000	1, 19
Open-chain siloxanes	1093–1076	1
	1055–1024	
Cyclotrisiloxanes	1020–1010	1, 15
Other cyclosiloxanes	1100–1050	1, 15
Cyclic aromatic siloxanes:		
$(SiPh_2O)_3$	1015	
$(SiPh_2O)_4$	1100	

15-7.

THE Si—Si STRETCHING

Durig and Church [5] have reported that the Si—Si stretching mode of disilane ($H_3Si-SiH_3$) gives rise to a strong Raman band at 423 cm^{-1}. Similar results were reported by Ohno *et al.* [4] for dimethyldisilane ($CH_3SiH_2-SiH_2-CH_3$). These authors found that dimethyldisilanes give rise to two strong and polarized Raman bands at 424 and 413 cm^{-1} that can be attributed to the Si—Si stretching mode.

15-8.

Si-HALOGEN

The Si—Cl group absorbs in the IR at 625–420 cm^{-1}, with subregions for $SiCl_3$ at 620–570 cm^{-1} IR : s and 535–450 cm^{-1} IR : m, for $SiCl_2$ at 600–535 cm^{-1} IR : s and 540–460 cm^{-1} IR : m, and for SiCl at 550–470 cm^{-1} [24]. See Table 15-7.

Table 15-7.
Si-Halogen [24]

	cm^{-1}
SiF_3	980–945 IR : s 910–860 IR : m
SiF_2	945–915 IR : s 910–870 IR : m
SiF	920–820 IR : m
$SiCl_3$	620–570 IR : s 535–450 IR : m
$SiCl_2$	600–535 IR : s 540–460 IR : m
SiCl	550–470 IR : m

The SiF group absorbs in the IR at 1000–800 cm^{-1}, with subregions for SiF_3 at 980–945 cm^{-1} IR : s and 910–860 cm^{-1} IR : m, for SiF_2 at 945–915 cm^{-1} IR : s and 910–870 cm^{-1} IR : m, and for SiF at 920–820 cm^{-1} [24].

Halosilanes react readily with moisture to form siloxanes [24].

REFERENCES

1. L. J. Bellamy, "The Infrared Spectra of Complex Molecules, Vol. I, 3rd Ed., John Wiley & Sons, New York, 1975, p. 374.
2. E. A. V. Ebsworth, S. G. Frankiss, and W. J. Jones, *J. Mol. Spectrosc.* **13,** 9, 1964.
3. C. J. Pouchert, "The Aldrich Library of FT-IR Spectra," Aldrich Chemical Company, Vol. 2, 1985. (a) p. 1120. (b) pp. 1098 and 1103. (c) p. 1120. (d) p. 1122. (e) p. 1101. (f) p. 1092. (g) p. 1091.
4. K. Ohno, M. Hayashi, and H. Murata, *Spectrochim. Acta* **39A,** 373, 1983.
5. J. R. Durig and J. S. Church, *J. Chem. Phys.* **73,** 4784, 1980.
6. E. A. V. Ebsworth, M. Onysyschuk, and N. Sheppard, *J. Chem. Soc.*, 1543, 1958.
7. J. R. Durig and J. N. Wilis, Jr., *J. Mol. Spectrosc.* **32,** 320, 1969.
8. K. Taga, K. Ohno, and H. Murata, *J. Mol. Struct.* **67,** 199, 1980.
9. K. Ohno, K. Taga, and H. Murata, *J. Mol. Struct.* **55,** 7, 1979.
10. H. Matsurra, K. Ohno, T. Sato, and H. Murata, *J. Mol. Struct.* **52,** 13, 1979.
11. N. B. Colthup, L. H. Daly, and S. E. Wiberly, "Introduction to Infrared and Raman Spectroscopy," 3rd Ed., Academic Press, New York, 1990.
12. I. S. Ignat'ev, A. H. Lazarev, S. G. Shevchenko, and V. P. Baryshok, *Bull. Acad. Sci. USSR, Div. Chem. Sci.* **35,** 1375, 1986.
13. A. L. Smith and N. C. Angelotti, *Spectrochim. Acta* **15,** 412, 1959.
14. H. W. Thompson, *Spectrochim. Acta* **16,** 238, 1960.
15. A. L. Smith, "Infrared Group Frequencies for Structure Determination in Organosilicon Compounds," in "Chemical, Biochemical, and Industrial Applications of Infrared Spectroscopy," John Wiley & Sons, New York, 1985.

16. A. L. Smith, *Appl. Spectrosc.* **41,** 1101, 1987.
17. M. Imbenotte, G. Palavit, and P. Legrand, *J. Raman Spectrosc.* **14,** 135, 1983, p. 73.
18. The Sadtler Standard Raman Spectra, Sadtler Research Laboratories, 1197R, 1600R and 1197R, 1976.
19. D. O. Hummel, "Applied Infrared Spectroscopy," Section 2.2 in "Polymer Spectroscopy" (D. O. Hummel, ed.), Verlag Chemie, Weinheim/Bergstr., 1974, p. 147.
20. H. Murata, H. Matsuura, K. Ohno, and T. Sato, *J. Mol. Struct.* **52,** 1, 1979.
21. V. F. Kalasinsky, S. E. Rodgers, and J. A. S. Smith, *Spectrochim. Acta* **41A,** 155, 1985.
22. D. F. Ball, T. Carter, D. C. McKean, and L. A. Woodward, *Spectrochim. Acta* **20,** 1721, 1964.
23. C. W. Young, P. C. Servais, C. C. Currie, and M. J. Hunter, *J. Am. Chem. Soc.* **70,** 3758, 1948.
24. A. L. Smith, *Spectrochim. Acta* **16,** 87, 1960.
25. R. N. Kniseley, V. A. Fassel, and E. E. Conrad, *Spectrochim. Acta* **15,** 651, 1959.

CHAPTER 16

Organophosphorus Compounds

16-1.

PH VIBRATIONS

16-1.1. PH and PH_2 Stretching Vibrations

The PH and PH_2 stretching vibrations give rise to strong, sharp IR and Raman bands in the region 2505–2222 cm^{-1} for all types [1–8]. In the PH_2 group, the antisymmetrical and symmetrical stretching bands are usually not resolved. In $C_2H_5PH_2$, the two bands are seen at 2304 and 2298 cm^{-1} in the IR gas-phase spectra, and at 2295 and 2288 cm^{-1} in the IR solid-state spectra [7]. The small wavenumber separation for the PH_2 stretch bands (compared with CH_2) results from a smaller HPH bond angle (93.5°) and a heavier central atom (P). In isopropyl phosphine, the PH_2 stretching bands have a spacing that is even smaller than in ethyl phosphine and is complicated by trans and gauche rotational isomerization [8].

There is a rough tendency for the PH stretching frequency (in Table 16-1) to increase when there is an increase in the sum of the electronegativities of the substituents on the phosphorus. Thomas [1] gives the following cm^{-1} ranges for various types of PH groups: phosphonic derivatives including O—PH(=O)—O, at 2457–2331 cm^{-1}; phosphinic derivatives including O—PH(=O)—C, at 2500–2310 cm^{-1}; phosphine oxides, C—PH(=O)—C, at 2505–2278 cm^{-1}; phosphines, C—PH—C, at 2335–2222 cm^{-1}; and sodium salts, at 2410–2285 cm^{-1}. Hydrobromide salts are hydrogen-bonded. The compound $(C_6H_5)_3PH^+Br^-$ absorbs weakly and broadly near 2170 cm^{-1} [9].

Table 16-1.

P—H Stretch Frequencies (cm^{-1}) [1]

	PH Stretch (cm^{-1})
$(RO)_2P(=O)H$	2450–2380
$(RO)_2P(=S)H$	2406–2392
RO, R > P(=O)H	2355–2310
HO, Ar > P(=O)H	2440–2381
$R_2P(=O)H$	2380–2278
$R_2P(=S)H$	2336–2288
RPH_2	2320–2275
R_2PH	2300–2270
Ph_2PH	2280

16-1.2. PH and PH_2 Bending Vibrations

The PH_2 has an HPH scissors bending or deformation vibration that gives rise to a band at 1090–1075 cm^{-1} in aliphatic and aromatic phosphines [7, 8, 10]. This band is medium-strong in the IR and medium-weak in the Raman. The PH_2 wag vibration gives rise to a band at 840–810 cm^{-1} [7, 8, 10]. This band is medium-strong in the IR and medium-weak in the Raman.

The PH wag gives rise to a band at 990–885 cm^{-1} [10]. This band is weak in the IR of dialkyl phosphines, but is medium-strong in the IR of diphenyl phosphine at 885 cm^{-1} [10]. In phosphine oxides, $C_2P(=O)H$, where C is part of an alkyl or phenyl group, a medium-intensity PH wag IR band is seen at 990–965 cm^{-1}, which shifts to 950–910 cm^{-1} in the comparable phosphine sulfides, $C_2P(=S)H$ [10]. Phosphorus acid esters of the type $(RO)_2P(=O)H$ have a very strong IR band at 980–960 cm^{-1} [10, 11]. This band is close to the nearby very strong C—O—P bands (particularly in P—O—C_2H_5), but has been shown to involve PH wag [11], which probably interacts with the C—O—P stretching vibration somewhat [10].

16-2.

THE P=O GROUP

The P=O stretching vibration [1–3, 9, 10, 12, 13] gives rise to a strong band in the IR and a medium band in the Raman. Most P=O compounds have a band in the region 1320–1140 cm^{-1}, but P=O compounds with fluorine or OH substituents extend the range to 1415–1085 cm^{-1} [1]. In the gas phase, $F_3P{=}O$ absorbs at 1415 cm^{-1}, and hydrogen-bonded types such as (HO)P=O absorb broadly as low as 1085 cm^{-1}. The P=O stretching frequency in $X_3P{=}O$ compounds varies chiefly with the sum of the inductive effects of the three substituents X on the P=O group [1]. Unlike the C=O group, it is little affected by mesomeric effects such as conjugation, or by being part of a strained ring. Bell *et al.* [12] established a straight-line plot of P=O stretching frequencies versus the sum of the Pauling electronegativities of the substituents on the P=O. Thomas and Chittenden [13] reworked this relationship and included data from over 900 compounds [1] to derive a relationship for the P=O stretch wavenumber $\bar{\nu}(P{=}O)$:

$$\bar{\nu}(P{=}O) = 930 + 40 \sum \Pi. \tag{1}$$

The group electronegativities (or more realistically, shift constants) were renamed the phosphorus inductive constants, or PI constants, symbolized by Π, the summation of which ($\sum \Pi$) appears in Eq. (1) [1, 13]. Thomas [1] has listed a large number of PI constants for various groups attached to the P=O, a selection of which appears in Table 16-2. Compounds containing the P—OH group have not been included for use in Eq. (1).

In compounds containing the P—OH group, the P=O band is broadened and lowered in frequency by hydrogen bonding and is often complex in

Table 16-2.

Phosphorus Inductive (PI) Constants for Substituent Groups [1]

CH_3	2.1	OC(=O)—	3.4	NHCHO	2.7
CH_2	2.0	$O{-}P^{V}$	3.3	NHP^{V}	2.1
CH	1.8	$O{-}P^{III}$	2.9	NHCOOR	2.4
C=C	2.4	NH_2	1.85	NCl_2	3.3
C=O	2.5	NHR	2.0	N=C=S	3.3
CR_2Cl	2.7	NR_2	2.4	N=C=O	3.6
$CRCl_2$	2.9	C≡CR	2.7	Br_3	8.2
CCl_3	3.0	C≡N	3.5	Br_2	5.5
CF_3	3.6	N=CHOR	2.3	Br	3.1
Aromatic	2.4	N=CRNR	2.7	Cl_3	9.0
OCH_3	2.9	$N{=}P^{V}$	2.6	Cl_2	6.3
OCH_2	2.85	$N\Big\langle\begin{matrix} CH_2 \\ \vert \\ CH_2 \end{matrix}$	2.2	Cl	3.4
OCH	2.75	N_3	3.0	F_3	12.0
OCH_2CH_2Cl	3.0	S—alkyl	2.4	F_2	8.0
OC=C	3.1	S—aryl	2.5	F	3.9
OC_6H_5	3.0	$S{-}P^{V}$	2.9	H	2.5

structure. Thomas [1] suggests a range of 1335–1087 cm^{-1} for most of these compounds. The limits are uncertain because of the band complexity.

In the IR spectra of P=O compounds containing the P—O—P group, such as phosphinic anhydrides (O=P—O—P=O) or related compounds, no P=O frequency doubling is noted when the anhydrides are symmetrical [1]. (Carboxylic anhydrides show a C=O doublet.) The P=O frequency (or frequencies in unsymmetrical cases) can be calculated by Eq. (1) if no P—OH is present [1].

A few examples of liquid-phase Raman P=O wavenumbers for vibrationally analyzed molecules include: $P(=O)F_3$, 1395 cm^{-1} [14]; $P(=O)Cl_3$, 1290 cm^{-1} [14]; $P(=O)Br_3$, 1261 cm^{-1} [14]; $CH_3IP(=O)F_2$, 1346 cm^{-1} [15]; and $(CH_3)_2P(=O)OCH_3$, 1210 cm^{-1} [16]. The IR gas-phase wavenumbers are 16–50 cm^{-1} higher [14–16], at 1418, 1326, 1277, 1366, and 1260 cm^{-1}, respectively.

16-3.

THE P=S GROUP

A vibration involving P=S stretching is found in the region 750–535 cm^{-1} [9, 10, 17–19], but unlike the P=O band, it is not always easy to identify. The P=S band varies in intensity, and the wavenumber shifts do not correlate with the substituent electronegativity. The complication involved is that the basic P=S stretching vibration is roughly estimated to have a frequency of about 675 cm^{-1} [19], and this may be near the symmetrical PX_3 stretching frequency for various types of X group substituents. Therefore, the P=S stretching vibration is expected to interact mechanically (to varying degrees) with the symmetrical PX_3 stretching vibration, giving rise to two bands. Each involves (to some extent) both the P=S stretch and the symmetrical PX_3 stretch, in-phase and out-of-phase. Some examples are listed in Table 16-3, where all the substituents are alike in a given molecule. All bands are polarized in the Raman spectra, and the lower frequency band in $X_3P=S$ is intense. The $X_3P=S$ bands that involve more symmetrical PX_3 stretch (third cm^{-1} column) have been assigned by comparison with the symmetrical PX_3 stretch in $X_3P=O$ and X_3P molecules [19]. The other $X_3P=S$ band, which is distinctly different from those in the $X_3P=O$ and X_3P spectra, is assigned to the vibration involving more P=S [19] and is shown in the last column of Table 16.3.

The Raman spectrum is very useful in the identification of the P=S band [18]. Of the two bands for $X_3P=S$ in Table 16-3, the lower-frequency band in each case is the in-phase $X_3P=S$ stretch. It is usually the stronger Raman

Table 16-3.

P=S Stretching Frequencies and Symmetrical PX_3 Stretching Frequencies (cm^{-1})

Substituent X	X_3P PX_3 Symmetric Stretch	$X_3P{=}O$ PX_3 Symmetric Stretch	$X_3P{=}S$ PX_3 Symmetric Stretch	$X_3P{=}S$ P=S Stretch	Reference
C_2H_5	699	692	686	535	19
OC_2H_5	731	734	785	610	19
SC_2H_5	—	558	526	685	19
F	890	875	894	695	14
Cl	507	486	435	753	14
Br	308	340	300	730	14

band, but is usually the weaker IR band. For example, in the IR spectra of $R_3P{=}S$ or $(RO)_3P{=}S$ compounds, the "P=S" band may be weak because the R—P=S or RO—P=S bonds are vibrating in-phase. However, if there is at least one Cl or S substituent, the "P=S" band is usually medium-to-strong in the IR because the Cl—P=S or S—P=S bonds are vibrating out-of-phase [10].

Examples of liquid-phase Raman P=S wavenumbers for vibrationally analyzed molecules include $(CH_3)_2P({=}S)C_2H_5$, 562 cm^{-1} [20], and $CH_3OP({=}S)Cl_2$, 719 and 702 cm^{-1} (rotational isomers) [21].

In Table 16-3, note that the "P=S" stretching frequency may be lower or higher than the "PX_3" symmetrical stretching frequency. When the "P=S" band has the lower frequency, the P=S group is vibrating in-phase with the PX_3 group to some extent, which lowers the "P=S" frequency. When the "P=S" band has the higher frequency, the P=S group is vibrating out-of-phase with the PX_3 group to some extent, which raises the "P=S" frequency. Therefore (compared to the P=S group), relatively high-frequency PX substituents such as P—R or P—OR lower the "P=S" frequency, and relatively low-frequency PX substituents such as P—Cl or P—SR raise the "P=S" frequency [10, 19]. The closer the frequencies of the PX_3 and P=S vibrations are before interaction, the larger will be the magnitude of the shift [10].

Substituent group shift constants, Δ, can be used to approximate some "P=S" frequency shifts [10]. These Δ shift constants vary with both substituent electronegativity and the effects of mechanical interaction. Surprisingly, these can be used even for mixed substituents in (XYZ)P=S compounds [10]. The P=S stretching frequency $\bar{\nu}(P{=}S)$ can be roughly approximated by the following equation [10], with empirical Δ values for each substituent:

$$\bar{\nu}(P{=}S) = 675 + \sum \Delta. \tag{2}$$

Table 16-4.

Frequencies (cm^{-1}) Involving P=S Stretch [1–3, 9, 10, 17–19]

Substituents on P=S			cm^{-1}	Substituents on P=S			cm^{-1}
OR	OR	Cl	672–648	R	R	R	599–535
OR	OR	SH	671–649	R	R	Ar	585–578
OR	OR	SR	663–645	R	Ar	Ar	638–599
OR	OR	H	658–636	Ar	Ar	Ar	663–627
OR	OR	F	650–630	F	F	SR	737–699
OR	OR	OR	670–600	F	F	H	ca. 705
OR	OR	R	650–589	F	F	F	695
OR	Cl	F	ca. 685	F	F	OR	677–604
OR	Cl	R	641–619	SR	SR	SR	700–685
OR	F	R	629–616	SR	R	OR	644–621
Cl	Cl	SR	ca. 728	Cl	Cl	Cl	753
Cl	Cl	OR	739–677	Cl	Cl	NR_2	675–625
C	Cl	R	685–650	Cl	NR_2	NR_2	620–600
R	R	Cl	653–593	NR_2	NR_2	NR_2	570–550
R	R	H	ca. 587	NHR	OR	Cl	675–625
R	R	OR	608–564	NHR	OR	OR	ca. 640

Δ values in cm^{-1}; Cl +23, S +10, H 0, OR −23, CH_2 −50, phenyl −13, F +7, NR_2 −40, NHR 0.

Some examples of bands that involve P=S are shown in Table 16-4. Thomas [1] often gives two regions that may involve P=S. Using the approach of Hooge and Christen [19] and Popov *et al.* with the Raman [18] and the above relationship, one of these regions is selected as the "P=S" frequency as discussed in this section and is listed in Table 16-4.

16-4.

THE P—OH GROUP

Compounds containing the P(=O)—OH group have a strong IR band at 1040–909 cm^{-1} [1, 22, 23] involving P—O stretch. This band has been found to be sensitive to the summation of the inductive effects of the other groups on the phosphorus besides the OH group [1, 22]:

$$\bar{\nu}[\text{P—O(H)}] = 650 + 40 \sum \Pi, \qquad \Pi(\text{P=O}) = 3.5 \quad \text{and} \quad \Pi(\text{P=S}) = 2.5. \tag{3}$$

In this equation, $\sum \Pi$ stands for the summation of the phosphorus inductive constants in Table 16-2, with the addition of the PI constants for P=O and P=S given above.

Compounds with two OH groups on the phosphorus, such as $RP(=O)(OH)_2$, are not included in the above equation. These acids have prominent IR bands at 1030–972 and 950–917 cm^{-1} [1, 22].

Care should be taken not to confuse these with other phosphorus groups that have bands in this region, including P—O—P, P—O—C_2H_5, and $O=PH(OR)_2$ types.

The O=P—OH and $O=P(OH)_2$ acids have very broad IR bands involving OH stretch extending over the whole region of 2725–1600 cm^{-1}, with maxima in the regions of 2725–2525, 2350–2080, and 1740–1600 cm^{-1} [1–3, 24, 25]. In $O=P(OH)_2$ acids with two OH groups per phosphorus, the 1700 cm^{-1} band, when present, is weaker than the 2600 and 2200 cm^{-1} bands, whereas in phosphinic acids, $R_2P(=O)OH$ (or other acids with one OH per phosphorus), the 1700 cm^{-1} band is the strongest of the three [1, 3, 24, 25]. This band pattern involves the overtone of the P—OH in-plane bending near 1230 cm^{-1} (a weak band usually obscured by P=O bands) and the overtone of the P—OH out-of-plane bending near 900 cm^{-1} (a weak band usually obscured by P—O bands) in Fermi resonance with the OH stretch [24, 25]. Related OH bands have been found in selenic and sulfinic acids [24, 25].

Acids of the type S=P—OH have diffuse bands at 3100–3000 and 2360–2200 cm^{-1} (OH stretch + OH deformation overtone, in Fermi resonance) and 932–909 cm^{-1} (P—O stretch) [1].

16-5.

SALTS OF ACIDS CONTAINING PHOSPHORUS

Salts of acids containing the PO_2^- group have two strong IR bands in the region 1323–1092 cm^{-1} (out-of-phase stretch) and 1164–995 cm^{-1} (in-phase stretch) [1, 10]. The two wavenumbers increase as $\sum \Pi$ increases (Table 16-5) if the cation is kept constant [1]. Acid salts of the type ($R_2PO_2^-$ or $R(H)PO_2^-$ absorb in the lower part of the two regions, whereas salts of the type $(RO)_2PO_2^-$ absorb in the upper parts of the regions [10]. See Table 16-5.

Salts of acids containing the PO_3^{2-} group absorb strongly in the IR at 1242–970 cm^{-1} out-of-phase stretch (two components), and medium-strong in the IR at 1021–893 cm^{-1} in-phase stretch [1, 10].

Salts of acids with the PS_2^- group have IR bands at 742–556 cm^{-1} (out-of-phase stretch) and 575–476 cm^{-1} (in-phase stretch) [1]. The two wavenumbers increase as $\sum \Pi$ increases, if the cation is kept constant [1]. Salts with the POS^- group have bands at 1240–1050 cm^{-1} (PO) and 658–545 cm^{-1} (PS) [1, 10]. See Table 16-5.

Table 16-5.

Stretching Frequencies (cm^{-1}) for Ions Containing Phosphorus [1][a]

$>PO_2^-$	1323–1093	1164–995
$-PO_3^{2-}$	1242–970	1021–893
PO_4^{3-}	1100–1000	
$>PS_2^-$	742–556	575–476
$>P\langle^{O-}_{S-}$	1240–1050	658–545

[a] For PO_2^- and PS_2^- ions. $\bar{\nu}$ increases as $\sum\Pi$ increases (for sodium salts).

16-6.

THE P—O—P GROUP

The P—O—P group has an out-of phase stretch vibration that gives rise to a strong IR band in the region 1025–900 cm^{-1} [1, 2, 22, 26]. Thomas and Chittenden [1, 22] report that P—O—P anhydrides with attached P—O or P—C bonds absorb at 988–908 cm^{-1} whereas if P-halogens are attached, the band appears at 1025–926 cm^{-1}. Also, if the attached P=O is replaced by a P=S or P=Se, the P—O—P frequency is lowered [1, 22]. A trivalent anhydride P—O—P frequency is somewhat lower than that of the corresponding pentavalent frequency [1, 22].

16-7.

THE P—O—C GROUP

Compounds containing a P—O—R group give rise to a band usually within the region 1088–920 cm^{-1} [1–3, 10, 17, 23]. These bands are usually very strong in the IR and medium-weak in the Raman. They involve out-of-phase P—O—C stretch, but with more O—C stretch than P—O stretch. As seen in Table 16-6, P—O—CH_3 groups have a band in the 1088–1015 cm^{-1} region [1], whereas P—O—CH_2—CH_3 groups have two prominent bands in the regions 1042–1008 and 982–920 cm^{-1} (O—C and C—C stretch) [1]. The P—O—CH_3 or P—O—C_2H_5 groups have another IR band, usually somewhat weaker, at 845–725 cm^{-1} thought to involve in-phase P—O—C stretch, or simply mainly P—O stretch [1, 10]. The P—O—R groups with

Table 16-6.

P—O—C, P—O—P, and P—O—H IR Bands (cm^{-1}) [1]

P^{V}—O—CH_3	1088–1018	O—C str.
P^{III}—O—CH_3	1034–1015	O—C
P—O—C_2H_5	1042–1008, 982–920	O—C—C
P—O—CH_2R	1042–987	O—C
P—O—CHR_2	1018–950	O—C
P—O—(CH_3 or C_2H_5)	845–725	P—O
P—O—Ar	1242–1110	O—Ar
P^{V}—O—Ar	996–905	P—O
P^{III}—O—Ar	875–830	P—O
P—O—P	1025–900	P—O—P antisym. str.
P—OH	1040–909	P—O
$P(OH)_2$	1030–972, 950–917	P—O
OH in P—OH	2725–2525, 2350–2050, 1740–1600	

R groups larger than C_2H_5 usually do not show much IR absorption in the 800 cm^{-1} P—O region, but they have strong C—O absorption in the 1042–950 cm^{-1} region, as seen in Table 16-6.

In the P—O—CH_3 and P—O—C_2H_5 groups, the methyl and ethyl groups have internal vibrations listed in Table 16-7 [11, 16, 21]. In particular, the P—O—CH_3 band at 1190–1170 cm^{-1} and the P—O—C_2H_5 band at 1167–1155 cm^{-1} are among the most distinctive. Both involve methyl rock.

Table 16-7.

CH_3 and C_2H_5 Bands (cm^{-1}) in $POCH_3$ and POC_2H_5 Groups [11, 16, 21]

P—O—CH_3	3026–3015	IR : w, R : w	CH_3 stretch
	3002–2993	IR : w, R : w	CH_3 stretch
	2956–2948	IR : m, R : m	CH_3 stretch
	2852–2844	IR : mw, R : m	CH_3 stretch
	1462–1454	IR : m, R : m	CH_3 antisym. def.
	1450–1439	IR : m, R : w	CH_3 sym. def.
	1190–1170	IR : ms, R : w	CH_3 rock
P—O—C_2H_5	ca. 2990	IR : m	C_2H_5 stretch
	2945	IR : w	C_2H_5 stretch
	2910	IR : w	C_2H_5 stretch
	1481–1470	IR : mw	OCH_2 def.
	1447–1445	IR : mw	CH_3 antisym. def.
	1396–1394	IR : m	OCH_2 wag
	1371–1370	IR : w	CH_3 sym. def.
	1167–1155	IR : m	CH_3 rock
	1105–1095	IR : mw	CH_3 rock

The P—O—aromatic group has strong IR bands in two main regions [1, 2, 10, 22, 34, 35]. A band in the 1242–1110 cm^{-1} involves O—Ar stretch as in aromatic ethers. A second band involving mainly P—O stretch appears at 996–905 cm^{-1} in pentavalent phosphorus compounds and at 875–830 cm^{-1} in trivalent phosphorus compounds. See Table 16-6.

16-8.

P—CH_3, P—CH_2, and P—PHENYL GROUPS

The P—CH_3 group has CH_3 antisymmetrical deformation bands at 1450–1395 cm^{-1}, IR : m, R : w, and a CH_3 symmetrical deformation band at 1346–1255 cm^{-1}, IR : m, R : w [1–3, 31]. Also, the P—CH_3 rock has a band at 977–842 cm^{-1} IR : s, R : w [1, 10, 31]. Within the regions given, pentavalent P types tend to have higher frequencies than trivalent P types [1, 31]. Examples are given in Table 16-8.

The P—CH_2—CH_3 or P—CH_2—R groups have a characteristic (P)CH_2 deformation band at 1440–1400 cm^{-1} IR : m, R : m, which can usually be seen when other alkane groups are present. Thomas [1] reports that the

Table 16-8.

P—C Groups

	cm^{-1}		Assignment	Reference
P—CH_3	1450–1395	IR : w, R : w	CH_3 antisym. def.	1–3
	1346–1255	IR : m, R : w	CH_3 sym. def.	1–3, 31
	977–842	IR : ms, R : w	CH_3 rock	1–3, 31
P—CH_3 Examples:				
CH_3PF_2	1417–1405	1292	889–859	27
$(CH_3)_2P(=O)OCH_3$	1420	1310	875	28
$(CH_3)_2P(=S)C_2H_5$	1420	1299	923	29
P—CH_2	1440–1405	IR : m, R : mw	CH_2 def.	10
P—CH_2 Examples:				
$C_2H_5PH_2$	1420			7
$C_2H_5P(=O)Cl_2$	1402			30
$C_2H_5P(=S)(CH_3)_2$	1402			29
P—phenyl	1130–1090	IR : m, R : w	ring mode + P—C	10
Examples:				
$P(Ph)_3$	1092			33
$As(Ph)_3$	1075			33
$Sb(Ph)_3$	1063			33

group $P-C_2H_5$ has two weak IR bands at 1282–1227 cm^{-1} (CH_2 wag and twist) when no P=O band is near this region. When a P=O falls in this region, P=O splitting may occur because of CH_2 interaction [30] and may indicate the presence of $O=P-C_2H_5$ [1, 11].

The P—C stretch of the above examples is found at 754–634 cm^{-1} [1, 2, 31, 32]. An example is $P(CH_3)_3$, with P—C stretch bands at 717 and 652 cm^{-1} [32]. However, P—C is not very characteristic, as it interacts with other bonds on the phosphorus. It is usually medium-weak in the IR and is usually strong in the Raman.

When a phenyl group is directly attached to a phosphorus, one of the substituent-sensitive phenyl vibrations involves some P—C stretch and is seen at 1130–1090 cm^{-1} IR : m, R : w [10]. This band in the IR is medium-weak in phosphines but is medium-strong in phosphine oxides, sulfides, or quaternaries [10]. Clark *et al.* [33] report that $P(Ph)_3$, $As(Ph)_3$, and $Sb(Ph)_3$ have phenyl substituent-sensitive bands at 1092, 1075, and 1063 cm^{-1}, respectively, and a second phenyl substituent-sensitive set at 683, 667, and 653 cm^{-1}, respectively. This second substituent P—phenyl band also involves some P—C stretch and is a medium-weak IR band in the region 730–680 cm^{-1} (see Chapter 17), near the usual 750 and 690 cm^{-1} IR phenyl bands [10].

16-9. PN GROUPS

Compounds containing $P-NH_2$, P—NH—R, and P—NH—Ar have bands involving some P—N stretch bands that are seen in the general region 1102–789 cm^{-1}, which is medium-strong in the IR [1, 10]. These P—N groups show some similarity to the P—O groups, but are somewhat weaker in the IR [10], and assignments are more difficult. These bands are listed in Table 16-9, along with $(P)-NH_2$ bands at 3425–3012 cm^{-1} and 1575–1538 cm^{-1} (stretch and deformation, respectively), a (P)—NH—R band at 3400–3125 cm^{-1} (NH stretch), (P)—NH—Ar bands at 3226–3115 cm^{-1} (NH stretch), and 1316–1274 cm^{-1} (N—Ar stretch), and various internal bands for $PN(CH_3)_2$, $PN(C_2H_5)_2$, and $PN(i\text{-}C_3H_7)_2$ [1, 10].

Cyclic phosphonitrilic compounds, $(X_2P=N)_n$, have strong IR bands (Table 16-9) involving PN stretch that are seen at 1297–1157 cm^{-1} for the cyclic trimer, and 1419–1180 cm^{-1} for the cyclic tetramer [1, 36–38]. Mixed FBr and FCl trimers had very strong IR bands at 1271–1223 cm^{-1} [36]. The cyclic trimer $(Cl_2P=N)_3$ [37], the cyclic tetramer $(Cl_2P=N)_4$ [38], and the

Table 16-9.

PN Frequencies (cm^{-1})

	cm^{-1}	Assignment
P—NH_2	3425–3012	NH_2 stretch
	1575–1538	NH_2 bend
	976–922	P—N stretch
P—NH—R	3400–3125	NH stretch
	1144–1076	C—N stretch
	1102–789	P—N stretch
P—NR_2	1053–885	
P—$N(CH_3)_2$	1316–1270, ca. 1190, and 1064	
P—$N(C_2H_5)_2$	ca. 1210 and 1175	
P—$N(i\text{-}C_3H_7)_2$	ca. 1200, 1183, 1160, and 1129	
(O=)P—NH—Aryl	3226–3115	NH stretch
	1316–1274	N—Ar stretch
	971–922	P—N stretch
P=N		
phosphonitrilic trimer	1297–1157	
phosphonitrilic tetramer	1419–1180	
noncyclic	1500–1141	

noncyclic polymer $(Cl_2P{=}N)_n$ [39] all have bands involving PN stretch at 1221 for the trimer, 1297 and 1275 for the tetramer, and 1281 and 1228 cm^{-1} for the polymer. These bands are strong in the IR and weak in the Raman.

Noncyclic P=N compounds of the types $(RO)_2P{=}N{-}C_6H_5$ and $(RO)(R)P{=}N{-}C_6H_5$ have a strong IR band at 1385–1325 cm^{-1} [40]. Thomas [1] lists a region of 1500–1141 cm^{-1} for all noncyclic P=N compounds.

16-10.

P—F, P—Cl AND P—S GROUPS

The P—F stretching vibration gives rise to strong IR bands in the region 905–760 cm^{-1} [1, 2, 9]. The P—F stretch can interact with P—C, P—O, or P—N stretch because of the similarities of the frequencies, which can complicate the assignment.

The P—Cl stretching vibration gives rise to strong IR bands in the region 607–400 cm^{-1} [1, 2, 9]. Within these P—X regions, the P—X stretch is

Table 16-10.

P—F, P—Cl, and P—S Frequencies (cm^{-1}) [1, 2, 9]

PF	905–760
PF_2	966–730
PF_3 (with =O or =S)	982–840, 894–875
PCl	565–544
PCl_2	607–488, 567–400
PCl_3 (with =O or =S)	578–494, 507–435
P—S—P	542–447
(O=)P—S—C	613–517, 582–510
(S=)P—S—C	557–497, 544–472
P—SH	548–522, 526–490
(S=)P—SH	2480–2300 cm^{-1}, H-bonded (SH stretch)
	2590–2550 cm^{-1} free SH, dilute solution

lowest for trivalent phosphorus, somewhat higher for P=S compounds, and higher still for P=O compounds [9]. Also, multiple halogens on the phosphorus have antisymmetrical and symmetrical PX_2 stretching bands. The compound $C_2H_5P(=O)Cl_2$ has PCl_2 antisymmetric stretch at 546 cm^{-1} IR : s, R : m, and PCl_2 symmetric stretch at 493 cm^{-1} IR : s, R : vs [30]. The P—X stretch bands are listed in Table 16-10.

The P—S stretching bands are found in the general region 613–440 cm^{-1} [1], which is in the same region as the P—Cl stretching bands. In acids of the type $(RO)_2P(=S)SH$ or $R_2P(=S)SH$, the hydrogen-bonded SH stretching vibration gives rise to a broad band at 2480–2300 cm^{-1} [10]. In dilute solution, the free SH stretching band absorbs sharply at 2590–2550 cm^{-1} [10].

Note that P-halogen compounds hydrolyze easily to make P—OH groups, which are sometimes seen as impurities in P-halogen compounds.

REFERENCES

1. L. C. Thomas, "Interpretation of the Infrared Spectra of Organophosphorus Compounds," Heyden, New York, 1974.
2. D. E. C. Corbridge, "The Infrared Spectra of Organophosphorus Compounds," in "Topics in Phosphorus Chemistry," Vol. 6 (M. Grayson and M. Griffiths, Eds.), John Wiley, London, 1969.
3. L. W. Daasch and D. C. Smith, *Anal. Chem.* **23**, 853, 1951.
4. L. J. Bellamy and L. Beecher, *J. Chem. Soc.*, pp. 475 and 1701, 1952.
5. L. J. Bellamy and L. Beecher, *J. Chem. Soc. London*, 728, 1953.
6. D. E. C. Corbridge, *J. Appl. Chem.* **6**, 456, 1956.

7. J. R. Durig and A. W. Cox, Jr., *J. Chem. Phys.* **63,** 2303, 1975.
8. J. R. Durig and A. W. Cox, Jr., *J. Phys. Chem.* **80,** 2493, 1976.
9. R. A. Nyquist and C. D. Craver, in "The Coblentz Society Desk Book of Infrared Spectra" (C. D. Craver, Ed.), The Coblentz Society, Kirkwood, Missouri, 1977.
10. N. B. Colthup, L. H. Daly, and S. E. Wiberley, "Introduction to Infrared and Raman Spectroscopy," 3rd Ed., Academic Press, New York, 1990.
11. R. A. Nyquist, *Spectrochim. Acta* **25A,** 47, 1969.
12. J. V. Bell, J. Heisler, H. Tannenbaum, and J. Goldman, *J. Am. Chem. Soc.* **76,** 5185, 1954.
13. L. C. Thomas and R. A. Chittenden, *Spectrochim. Acta* **20,** 467, 1964.
14. R. A. Nyquist, *Appl. Spectrosc.* **41,** 272, 1987.
15. G. H. Pieters, B. J. van der Veken, A. J. Barnes, T. S. Little, W. Y. Zhoa, and J. R. Durig, *Spectrochim. Acta* **43A,** 657, 1987.
16. B. J. van der Veken, R. L. Odeurs, M. A. Herman, and J. R. Durig, *Spectrochim. Acta* **40A,** 563, 1984.
17. R. C. Gore, *Discuss. Faraday Soc.* **9,** 138, 1950.
18. E. M. Popov, T. A. Mastryukova, N. P. Rodinova, and M. I. Kabachnik, *Zh. Obshch. Khim* **29,** 1998, 1959.
19. F. N. Hooge and P. J. Christen, *Rec. Trav. Chim. Pays-Bas* **77,** 911, 1958.
20. J. R. Durig and T. J. Hizer, *J. Mol. Struct.* **145,** 15, 1986.
21. R. A. Nyquist, *Spectrochim. Acta* **23A,** 1499, 1967.
22. L. C. Thomas and R. A. Chittenden, *Spectrochim. Acta* **20,** 489, 1964.
23. R. A. McIvor and C. E. Hubley, *Can. J. Chem.* **37,** 869, 1959.
24. S. Detoni and D. Hadzi, *Spectrochim. Acta* **20,** 949, 1964.
25. D. Hadzi, *Pure Appl. Chem.* **11,** 435, 1965.
26. R. A. McIvor, G. A. Grant, and C. E. Hubley, *Can. J. Chem.* **34,** 1611, 1956.
27. J. R. Durig, A. E. Stanley, and M. R. Jalilian, *J. Raman Spectrosc.* **10,** 44, 1981.
28. B. J. van der Veken, R. L. Odeurs, M. A. Herman, and J. R. Durig, *Spectrochim. Acta* **40A,** 563, 1984.
29. J. R. Durig and T. J. Hizer, *J. Mol. Struct.* **145,** 15, 1986.
30. J. R. Durig, C. G. James, and T. J. Hizer, *J. Raman Spectrosc.* **21,** 155, 1990.
31. L. C. Thomas and R. A. Chittenden, *Spectrochim. Acta* **21,** 1905, 1965.
32. M. Halmann, *Spectrochim. Acta* **16,** 407, 1960.
33. R. H. Clark, C. D. Flint, and A. J. Hempleman, *Spectrochim. Acta* **43A,** 805, 1987.
34. R. A. Nyquist, *Appl. Spectrosc.* **11,** 161, 1957.
35. A. C. Chapman and R. Harper, *Chem. Ind. (London)*, 985, 1962.
36. P. Clare and D. B. Sowerby, *Spectrochim. Acta* **37A,** 883, 1981.
37. P. C. Painter, J. Zarian, and M. M. Coleman, *Appl. Spectrosc.* **36,** 265, 1982.
38. J. Zarian, P. C. Painter, and M. M. Coleman, *Appl. Spectrosc.* **36,** 272, 1982.
39. M. M. Coleman, J. Zarian, and P. C. Painter, *Appl. Spectrosc.* **36,** 277, 1982.
40. M. I. Kabachnik, V. A. Gilyarov, and E. N. Tsvetkov, *Izv. Akad. Nauk, SSSR*, 2135, 1959.

CHAPTER 17

Aromatic and Heteroaromatic Rings

17-1.

BENZENE RINGS

The infrared and Raman spectra of benzene derivatives have been extensively studied in the literature [1–19]. A benzene ring, monosubstituted with a halogen, has a total of 30 modes of vibration, which are illustrated in Fig. 17-1 [1–3]. Six of these are substituent-sensitive in that the substituent atom moves appreciably or the substituent bond is significantly stretched or bent. All the rest have wavenumbers within about 30 cm^{-1} or so of a given average wavenumber and can be considered to be group frequencies. However, not all are useful, because many may be weak or absent in the IR or Raman spectra. Each mode has its own rules about IR or Raman intensity, or wavenumber shift with mass, or inductive or mesomeric effects of the substituents. This makes the substituted benzenes some of the most complex groups encountered.

The C—H stretching vibrations of substituted benzenes usually give rise to IR and Raman bands in the 3100–3000 cm^{-1} region [5, 20–22]. The phenyl group, with five aryl C—H bonds, often has a triplet of IR bands in this region [20], whereas the Raman usually has only one strong band at 3070–3030 cm^{-1} [19]. During the C—H stretching vibration, the C and H atoms move radially in opposite directions. In all other vibrations below 3000 cm^{-1}, the C—H bond hardly changes in length, which means that any carbon and its hydrogen that move radially, move in the same direction.

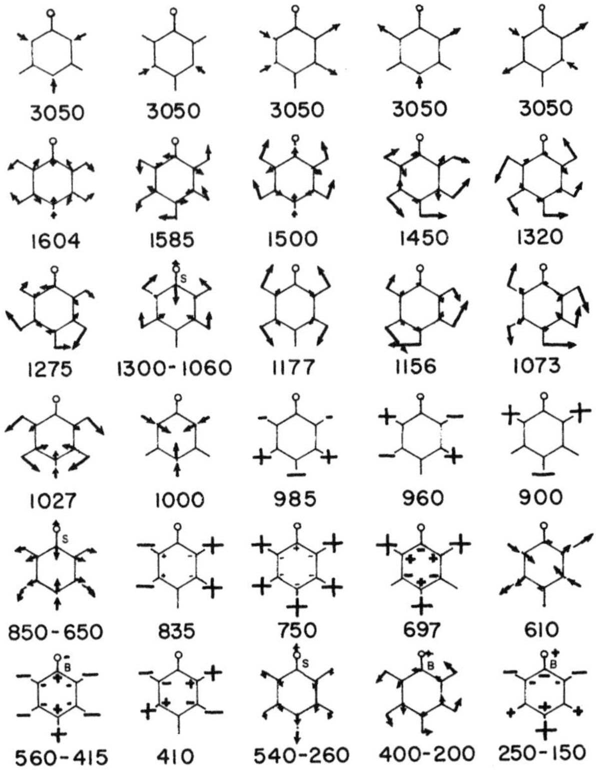

Fig. 17-1. Approximate vibrational modes of monosubstituted benzenes [1–3]. Substituent-insensitive modes have wavenumber regions usually within $\pm 30\,\mathrm{cm}^{-1}$ of the wavenumbers (cm^{-1}) given. Substituent-sensitive modes have wider regions as indicated.

Many bands in the 1600–1000 cm^{-1} region involve in-plane C—H bending vibrations that interact (sometimes strongly) with various ring CC vibrations. Above about 1200 cm^{-1}, a tangentially moving carbon and its hydrogen usually move in opposite directions, while below about 1200 cm^{-1} they usually move in the same direction [1, 23].

In benzene derivatives, out-of-plane C—H bending (or wag) vibrations usually give rise to bands at 1000–700 cm^{-1}. Above about 700 cm^{-1}, a carbon and its hydrogen usually move out-of-plane in opposite directions, whereas below about 700 cm^{-1}, they usually move in the same direction [3, 23].

Table 17-1 lists the major substituent-insensitive mono-, di-, and tri-substituted benzene IR and Raman bands. These will be discussed in detail in later sections.

Table 17-1.

Major Substituent-Insensitive Bands for Mono-, Di-, and Some Trisubstituted Benzenes

Substitution	cm^{-1}	IR[a]	R[a]
Mono, di, and tri	3100–3000	mw	s
Mono, di, and tri	1620–1585	v	m
Mono, di, and tri	1590–1565	v	m
Mono, ortho, and meta	1510–1470	v	vw
Para	1525–1480	v	vw
Mono, ortho, and meta	1465–1430	m	vw
Para	1420–1400	m	vw
Mono	1082–1065	m	vw
Ortho	1055–1020	m	s
Mono	1040–1016	m	m
Para	1023–1003	m	—
Mono, meta, and 1,3,5	1010–990	vw	vs
Meta, 1,2,4, and 1,3,5 (lone H)	935–810	m	—
Para and 1,2,4 (two adjacent H)	880–795	s	—
Meta and 1,2,3 (three adjacent H)	825–750	s	—
Ortho and mono (four and five adjacent H)	805–728	s	—
Mono, meta, and 1,3,5	710–665	s	—
Para	650–630	—	m
Mono	630–605	w	m

[a] Note: v is variable intensity; vw or vs is very weak or very strong intensity; s, m, and w are strong, medium, and weak intensity.

17-2. THE 1620–1400 cm⁻¹ RING BANDS

The CC bonds in benzene are all equivalent and are roughly a "bond-and-a-half." Because of this equivalence, the coupled CC bonds all interact, and the ring is best considered as a whole. Colthup *et al.* [23] used a mechanical standing-wave type description for ring vibrations where nodal lines separate the parts of the ring that vibrate out-of-phase with each other. As seen in Fig. 17-2a, there are two vibrations that involve "quadrant" stretch. Two nodal lines separate the ring into quadrants, and the ring CC bonds that are predominantly in one quadrant stretch, while those in the neighboring quadrants contract. As seen in Fig. 17-2b there are two vibrations that involve "semicircle" stretch. One nodal line separates the ring into semicircles, and ring CC bonds that are predominantly in one semicircle stretch, while those in the other contract. A ring bond that is bisected by a nodal line does not change in length.

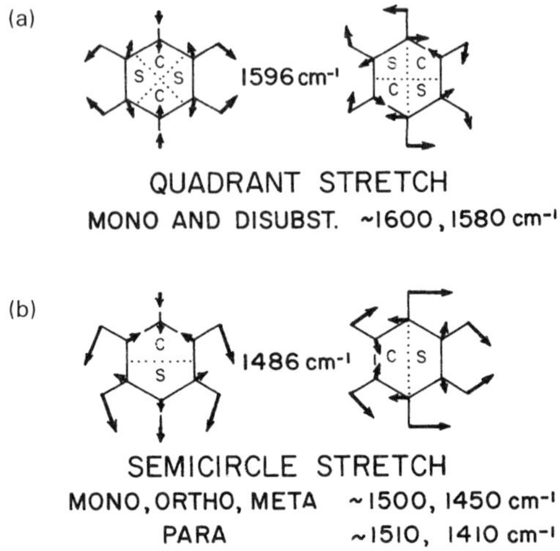

Fig. 17-2. (a) Quadrant and (b) semicircle stretch modes of benzene, each doubly degenerate (S is stretch, C is contract). For approximate modes for substituted benzenes, replace moving hydrogens with nearly stationary substituents at position 1 (top) for mono, 2,3 for ortho, 2,6 for meta, and 1,4 for para [1, 2].

As seen in Fig. 17-2, the quadrant stretch modes interact somewhat with in-plane CH bend, and this interaction is strong in the two semicircle stretch modes. Tangentially moving carbons and their hydrogens move in roughly opposite directions. Radially moving carbons and their hydrogens move in nearly the same direction. Substituents are nearly motionless.

Unsubstituted benzene at equilibrium has a center of symmetry. As seen in Fig. 17-2, one can start at any atom and go in a straight line through the center and an equal distance beyond, where one will find a symmetrically equivalent atom. In the vibrationally distorted molecule, the atoms move to the tips of the displacement vectors. As seen in Fig. 17-2a, the vibrationally distorted benzene performing either quadrant stretch vibration retains the center of symmetry. (Vector tip pairs are seen on the opposite sides of the center.) This means these vibrations in benzene, or in centrosymmetrically substituted benzenes, are IR-inactive. Therefore, symmetrically substituted para- (1, 4), tetra- (1, 2, 4, 5), or hexasubstituted benzenes show no quadrant-stretch IR absorption bands near 1600 and 1580 cm^{-1}. These are all Raman-active, however [5, 19].

The semicircle stretch vibrations Fig. 17-2b are antisymmetric with respect to the center of symmetry. In substituted benzenes, these are IR-allowed but are Raman-inactive in symmetrical para-, tetra-, or hexasubstitution and show only weakly in non-centrosymmetrically substituted benzenes [5 19].

In rings with a three-fold axis of symmetry, such as equivalently substituted 1,3,5-trisubstituted benzene, hexasubstituted benzene, or benzene itself, the two quadrant stretch modes are doubly degenerate and have the same frequency. The same is true for the two semicircle stretch modes. In mono- or disubstituted benzenes, there is no three-fold axis of symmetry, so the degeneracy is broken and the components come at separate frequencies.

17-2.1. The 1620–1565 cm^{-1} Bands

In mono- and disubstituted benzenes, the two quadrant stretch components usually give rise to bands at 1620–1585 and 1590–1565 cm^{-1}. The second band is often weaker in the IR than the first, but is sometimes enhanced by ring conjugation or halogen substitution. In the Raman spectra, the 1620–1565 cm^{-1} bands are medium in intensity [5, 19]. In the IR, the intensities are variable [4]. When the ring is para-substituted with identical substituents, a center of symmetry is present that is retained during the quadrant stretch vibrations, making these vibrations IR-inactive. However, if the substituents para to each other are different, these modes are IR-allowed. The greater the difference between the electron-donating or -withdrawing character of the para substituents, the stronger the IR bands will be [4]. Meta-substituted rings have IR band intensities in this region that vary with the sum (rather than the difference) of the electronic effects of the substituents, and ortho-substituted aromatics are intermediate [4]. Monosubstituted benzenes have weak IR intensities for weakly interacting substituents such as CH_2Cl, but stronger IR intensities for electron donors or withdrawers [4]. The wavenumbers for substituents such as Cl or S are at the lower end of the given regions, not because of the heavier mass (these atoms are nearly stationary) but because the C—Cl or C—S stretching and bending force constants (involved to some extent) are somewhat lower.

17-2.2. The 1525–1400 cm^{-1} Bands

The two semicircle stretch components in Fig. 17-2 are strongly mixed with in-plane C—H bend [1, 2]. In monosubstituted or ortho and meta disubstituted benzenes, these have IR bands at 1510–1470 and 1465–1430 cm^{-1}. For para-substituted benzenes, the IR bands are seen at 1525–1480 and 1420–1400 cm^{-1} [4]. For a para isomer, the band near 1500 cm^{-1} is usually 10–20 cm^{-1} higher than those for the other isomers. The IR intensity of the 1500 cm^{-1} band is strong for electron donors and weak when these are not present. For C=O substituents, the 1500 cm^{-1} band can be very weak [4]. The Raman intensity is very weak [5, 19]. The second band near 1450 cm^{-1} (or lower for para) is often obscured by CH_3 and CH_2 bands. Its IR intensity is more or less independent of the nature of the substituent [4]. Its Raman intensity is very weak [5, 19].

17-3.

THE 1300–1000 cm^{-1} REGION

17-3.1. The 1000 cm^{-1} Band

Figure 17-3 shows approximate vibrational modes for mono- and disubstituted benzenes in the 1300–1000 cm^{-1} region [1, 2]. In Fig. 17-3, first row, the band at 1010–990 cm^{-1} is usually the strongest band in the Raman

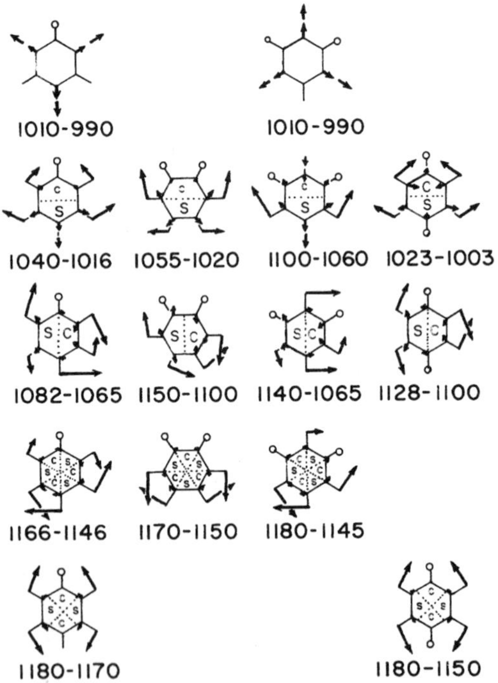

Fig. 17-3. Approximate vibrational modes for mono- and disubstituted benzenes in the 1200–1000 cm^{-1} region [1, 2].

Table 17-2.

Aromatic Bands in the 1200–1000 cm^{-1} Region

Mono	IR R	Ortho	IR, R	Meta	IR, R	Para	IR, R
1010–990	vw, vs			1010–990	vw, vs		
1040–1016	m, m	1055–1020	m, s	1100–1060	w, w	1023–1003	m, —
1082–1065	m, vw	1150–1100	m, vw	1140–1065	mw, w	1128–1100	w, —
1166–1146	w, w	1170–1150	w, w	1180–1145	w, w		
1180–1170	w, w					1180–1150	—, m

spectra of mono-, meta-, and also 1,3,5-trisubstituted benzenes [1, 2, 4–10]. In the IR spectra of mono- and meta-substituted benzenes, this shows as a weak, very sharp band that can often be picked out because of the well-defined narrow region 1000 ± 10 cm^{-1}. See Table 17-2, top row. In this mode, the 2, 4, and 6 carbons, with their hydrogens, move radially in-phase, and the 1, 3, and 5 carbons (with one or more substituents) are virtually stationary [1, 2], which makes this mode virtually substituent-independent for mono-, meta-, and 1,3,5-substituted benzenes.

This mode is derived from mixing the 992 cm^{-1} and 1011 cm^{-1} modes in unsubstituted benzene. In the 992 cm^{-1} mode, all the carbons move in-phase radially, which is the in-phase ring stretching (or "breathing") mode. In the 1011 cm^{-1} mode, all the carbons move radially also, but alternate carbons move outwards and inwards, which is the in-plane ring bending by sextants. For mono-, meta-, and 1,3,5-trisubstitution, these modes belong to the same symmetry species and can mix to give the mode described above. For ortho and para isomers, these modes belong to different symmetry species and cannot mix.

17-3.2. The 1150–1003 cm^{-1} Bands

In this region, mono- and disubstituted benzenes have two bands that, like the 1500 cm^{-1} bands (Fig. 17-2), involve semicircle stretch, strongly mixed with in-plane CH bending. In this lower region, tangentially moving carbons and their hydrogens move more or less in the same direction. See Fig. 17-3, rows 2 and 3, and Table 17-2, rows 2 and 3 [1, 2, 4–10, 19].

The Raman spectra of mono and ortho have a medium-strong band near 1027 and 1035 cm^{-1}, respectively. In the IR, a phenyl group has a characteristic triplet, two medium bands near 1073 and 1027 cm^{-1} and a very weak band at 1000 ± 10 cm^{-1}. In the IR, this region should be used with discretion, since many non-aromatic bands appear in this region [23].

17-3.3. The 1200–1145 cm^{-1} Bands

Bands are found in this region that involve mainly in-plane CH bend. These are illustrated in Fig. 17-3, rows 4 and 5, and Table 17-2, rows 4 and 5 [1, 2, 4–10, 19]. The tangentially moving hydrogens and their carbons move in roughly the same direction [1, 2]. The CH in-plane bending modes in Fig. 17-3, row 4, have a small amount of CC sextant stretch interaction, and those in row 5 have a small amount of quadrant stretch interaction. Like the aromatic bands at 1150–1003 cm^{-1}, the 1200–1145 cm^{-1} bands in the infrared spectra are often obscured by bands from the substituents.

17-4.

THE ARYL C—H WAG VIBRATIONS

The out-of-plane C—H wag vibrations give rise to the most useful IR bands, used in distinguishing different types of aromatic ring substitution. These are usually too weak in the Raman to be useful. The hydrogens on the substituted benzenes are separated into categories of adjacent hydrogens, since wagging adjacent CH bonds are coupled to each other, but are much less strongly coupled to other wagging C—H bonds that are separated by non-wagging substituents [23].

A monosubstituted benzene has five adjacent hydrogens, ortho has four adjacent hydrogens, meta has three adjacent hydrogens and one lone hydrogen, and para has two pairs of two adjacent hydrogens. The adjacent hydrogen group has as many wagging vibrational phases as there are C—H bonds. The mode that gives rise to the most intense IR band is the in-phase C—H wagging vibrations, found in the mean regions of 750 cm^{-1} for five and four adjacent hydrogens, 782 cm^{-1} for three adjacent hydrogens, 817 cm^{-1} for two adjacent hydrogens, and about 860 cm^{-1} for a lone hydrogen wag [4]. The overall regions are summarized in Table 17-3 [23, 24].

Table 17-3.

Adjacent Hydrogen Wag IR regions [23, 24]

Adjacent Hydrogens	Substitution Types	cm^{-1}
Five or four adjacent H	(mono or ortho)	805–728
Three adjacent H	(meta or 1,2,3)	825–750
Two adjacent H	(para, 1,2,4, or 1,2,3,4)	880–795
One lone H	(meta, or 1,2,4, or 1,3,5, or 1,2,3,5, or 1,2,4,5, or penta)	935–810

The aryl adjacent hydrogen correlations can be extended to include naphthalenes and pyridines, to be discussed later [23]. In pyridines, the ring nitrogen has no hydrogen on it, and it is counted as a substituted carbon in the correlation.

As with any correlation, the adjacent hydrogen wag correlations can be complicated by other substituent vibrations that have bands in the same region. The out-of-plane vibrations of aryl—NO_2 groups, aryl—CO_2X type groups (as in benzoic acids or esters), and aryl—CO—N types are particularly troublesome. The out-of-plane NO_2 or CO_2 vibrations, and various aryl—CH wag vibrations, interact to give strong IR bands at 750–700 cm^{-1} and weaker IR bands at 920–780 cm^{-1}. The wavenumber within the regions depends on the substitution [23]. In addition, nitroaromatics usually have an in-plane NO_2-ring vibration somewhere near 840 cm^{-1} that confuses the interpretation. These groups should be treated with caution when using the adjacent hydrogen correlations, and spectra of closely related compounds should be used for comparison. The aryl—CHO and the aryl—C(=O)R groups do not cause interaction effects. The C=O wag in ketones is lower than the aryl—CH wag region, and the out-of-plane CHO vibration (aldehyde CH wag) is higher [23].

The aryl C—H out-of-plane wag correlations in Table 17-3 have rather wide ranges, but for specific substituents, much closer predictions can be made [24]. It turns out that an individual aryl—CH wagging force constant decreases as the electron density on the CH carbon increases, as indicated by a molecular orbital calculation [24]. A substituent that changes the electron density on an ortho carbon changes the para carbon about half as much, and a meta carbon about one-third as much, but with a reversed sign [24]. The adjacent CH in-phase wag frequencies can be predicted by an appropriate summation of the effects of the substituents, to be discussed next.

17-4.1. The 900 cm^{-1} Phenyl Band

The five-adjacent C—H in-phase wag frequency near 750 cm^{-1} of a phenyl group is somewhat sensitive to substituent mechanical interaction effects and correlates rather poorly with the electron-donating or -withdrawing characteristics of the substituent [23]. There is a much weaker IR band near 900 cm^{-1} for a phenyl group that also involves aryl—CH wag where the two ortho hydrogens wag in one direction, the para hydrogen wags in the opposite direction, and the meta hydrogens move very little [3]. See Fig. 17-1.

The frequency of this $\bar{\nu}$(900 cm^{-1}) band varies over the region 934–864 cm^{-1}, and it correlates very well with the electron-donating or -withdrawing

Table 17-4.

Wavenumbers of the Monosubstituted Benzene Band near 900 cm^{-1} [24]

Substituent	cm^{-1}	Substituent	cm^{-1}
(CO)Cl	934	NCO	905
NO_2	934	i-C_3H_7	905
(CO)OR	933	NH_3Cl	904
SO_2	930	C_2H_5	904
CF_3	926	Br	903
$(CO)CH_3$	926	Cl	902
C≡N	925	SH	896
(CO)H	923	CH_3	896
$(CO)NH_2$	922	F	896
CO_2Na	919	SCH_3	893
C≡CH	916	$O(CO)CH_3$	890
CH_2Br	916	OH	887
C=C	909	OCH_3	883
CH_2OH	909	NH_2	880
$NH(CO)CH_3$	906	$NHCH_3$	869
t-C_4H_9	906	$N(CH_3)_2$	864

characteristics of the substituents [24]. A listing of some examples [24] is found in Table 17-4, where it can be seen that electron-withdrawing groups, such as NO_2, raise the $\bar{\nu}(900\ cm^{-1})$ frequency, but electron-donating groups, such as $N(CH_3)_2$, lower the frequency.

17-4.2. Adjacent Aryl CH Wag Frequency Shifts

Most of the aryl CH in-phase wag frequencies can be calculated as a linear combination of appropriate $\bar{\nu}(900\ cm^{-1})$ phenyl frequencies from Table 17-4 [24]. The calculation procedures are given in Table 17-5 [24]. In most cases, the substituents are equally weighted, but in three cases, namely 1,2,4-, 1,2,3-, and 1,2,3,5-substituted benzenes, unequal weighting is necessary from theory [24], as given in Table 17-5. For aryl—NO_2 or aryl—CO_2X groups, the calculations give the higher-frequency, weaker IR band, resulting from the group interaction described earlier.

The only out-of-phase wag given in Table 17-5 is the ortho 940 cm^{-1} band where the two aryl hydrogens next to the substituents wag in one direction, and the two others wag in the opposite direction. This is a much weaker band that the ortho four-adjacent CH in-phase wag band near 750 cm^{-1}, but it correlates much better with the electronic effects of the substituents. Note the standard deviations given for these two in Table 17-5.

Table 17-5.

Wavenumber Calculation for Aryl CH Wags [24]

CH Wag Type	Sum[a] $\bar{\nu}$(900)	Multiplied by	Plus or Minus	Observed Range, cm^{-1}	Std. Dev. cm^{-1}
meta lone H	$\bar{\nu}1 + \bar{\nu}3$	0.9	−757	935–810	8
meta 3 adj. H	$\bar{\nu}1 + \bar{\nu}3$	0.5	−120	825–740	6
ortho 4 adj. H	$\bar{\nu}1 + \bar{\nu}2$	0.5	−149	805–735	9
ortho 940 cm^{-1}	$\bar{\nu}1 + \bar{\nu}2$	0.5	+36	975–905	5
1,3,5 tri	$\bar{\nu}1 + \bar{\nu}3 + \bar{\nu}5$	0.7	−1040	930–810	6
1,2,4 tri, lone H	$\bar{\nu}2 + \bar{\nu}4$	0.7	−393	930–820	9
1,2,3 tri	$\bar{\nu}1 + \bar{\nu}3$	0.6	−303	825–750	9
1,2,3,5 tetra	$\bar{\nu}1 + \bar{\nu}3 + \bar{\nu}5 + \bar{\nu}5$	0.6	−1309	925–805	9
1,2,4,5 tetra	$\bar{\nu}1 + \bar{\nu}2 + \bar{\nu}4 + \bar{\nu}5$	0.6	−1280	930–840	9
1-naphthalenes					
3 adj. H	$\bar{\nu}1$	0.4	+432	805–775	2
2-naphthalenes					
lone H	$\bar{\nu}2$	0.6	+309	875–823	3
2 adj. H	$\bar{\nu}2$	0.3	+544	825–800	4

[a] In the first row, for example, $\bar{\nu}1 + \bar{\nu}3$ stands for the sum of the $\bar{\nu}$(900) wavenumbers from Table 17-4 for phenyls whose substituents are like those on positions 1 and 3 of the aromatics considered. This sum is to be multiplied by 0.9, and then subtract 757 to calculate the meta lone H wag wavenumber.

17-4.3. The Five-Adjacent and Two-Adjacent CH Wag Bands

The five-adjacent CH wag of phenyl and the two-adjacent CH wag of para are not included in Table 17-5. The five adjacent CH wag in mono-substitution seems to be more sensitive to mechanical interaction effects of the substituents, and these are superimposed on the electronic effects. The two-adjacent CH wag in para-, 1,2,4 tri-, and 1,2,3,4-tetrasubstitution are also affected in the same way [23]. Because of this similarity, the two-adjacent CH in-phase wag wavenumber can be calculated approximately as a linear combination of two five-adjacent CH wag phenyl wavenumbers, where the phenyls have substituents like those in the 1 and 4 positions of para-disubstitution, 1,2,4-trisubstitution, or 1,2,3,4-tetrasubstitution [23]:

$$\bar{\nu}(2\text{ adj H}) \cong 0.6[\bar{\nu}1(5\text{ adj H}) + \bar{\nu}4(5\text{ adj H})] - 75.$$

A few examples of $\bar{\nu}$(5 adj H), which are wavenumbers for five-adjacent in-phase CH wag for various phenyl substituents, are as follows: CH_3 728, CH_2R 743, CHR_2 759, CMe_3 763, Cl 741, O(H or C) 752, N(H or C) 752, $CH{=}CH_2$ 775, CO—R 761, CO(O or N) 805 (weak), NO_2 793 [23].

17-4.4. The 2000–1650 cm^{-1} Summation Bands

For the most part, the aryl—CH wag discussion has focused on the in-phase wag of adjacent hydrogens, since the in-phase wag gives rise to the most intense IR band. Two out-of-phase wags were discussed. In the $\bar{\nu}(900)$ vibration for monosubstituted benzenes, the two ortho hydrogens move one way, and the para hydrogen moves the other way. In the $\bar{\nu}(940)$ vibration for ortho, the two hydrogens next to the substituents move one way, and the other two move the other way. There are as many different CH wag vibrations as there are hydrogens for a given substitution. All of these are shown in Fig. 17-4 [3], where a standing-wave description is used [23]. In

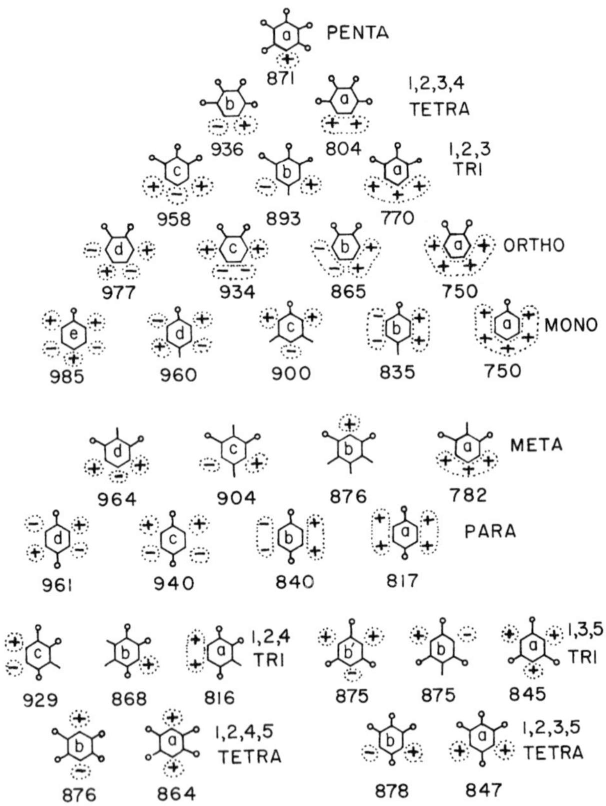

Fig. 17-4. The out-of-plane CH wag modes of substituted benzenes. Plus and minus indicate motion out of the plane of the page. Wavenumbers (cm^{-1}) are usually within ±30 cm^{-1} of the values given [3, 26].

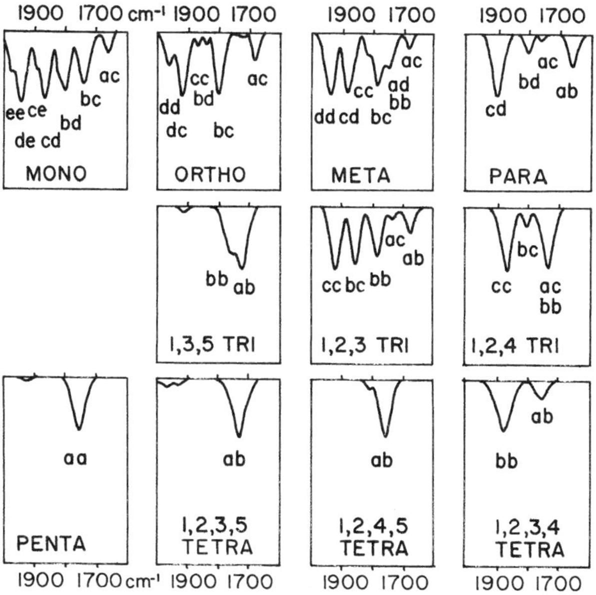

Fig. 17-5. Summation band patterns for substituted benzenes. The double letters (ab) indicate binary combinations of fundamentals given in Fig. 17-4 [25, 26].

the "a" mode, all the hydrogens wag in-phase. In the "b" mode, half the hydrogens move one way, and the other half move the other way, etc. In Fig. 17-4, anti-nodal regions enclose hydrogens that wag in-phase. Only the "a" modes are strong in IR, and all the rest involve out-of-phase wag and are weak or sometimes inactive in the IR.

In the 2000–1650 cm^{-1} region of the IR spectra of aromatic compounds, there is a pattern of weak bands that tend to be characteristic for a particular substitution [25]. For these bands to be seen clearly, the sample thickness should be somewhat greater than usual. These are illustrated in Fig. 17-5. If the electron-donating or -withdrawing properties of the substituents change, the pattern tends to shift as a whole, but it usually tends to remain recognizable [25] (but not always in extreme cases).

These bands have been assigned [26] to binary combinations and overtones of aryl CH wag modes illustrated in Fig. 17-4. In Fig. 17-5, the band assignments are indicated by two letters (such as bc) that refer to the two appropriate fundamentals (in Fig. 17-4) involved in the combination. These combination bands and overtones are jointly referred to as summation bands [26]. While these bands are weak as a whole, they are relatively strong in IR intensity for summation-type bands.

17-5. RING DEFORMATION MODES

17-5.1. The 700 cm^{-1} Band

In the IR spectra of mono-, meta-, and 1,3,5-substituted benzenes, there is a strong band in the region 710–665 cm^{-1}. Most of the monosubstituted benzenes absorb at 697 ± 11 cm^{-1}, and most of the meta-substituted benzenes absorb at 690 ± 15 cm^{-1} [4]. This band is weak or absent in the Raman spectra. This mode involves out-of-plane ring bending by sextants, as seen in Fig. 17-6, third row. The 2,4,6 carbons and hydrogens move in-phase, out-of-plane in one direction, and the 1,3,5 carbons move the other way, while the substituents and hydrogens on the 1,3,5 carbons are nearly motionless [3]. This band is strong in the IR for mono-, meta-, and 1,3,5-substituted benzenes, but is IR-inactive for ortho or para isomers with identical substituents. In 1,2,3- and 1,2,4-trisubstituted benzenes, this mode absorbs weakly in the 730–685 cm^{-1} region. Ortho and para isomers may absorb here weakly if the two substituents differ sufficiently [23].

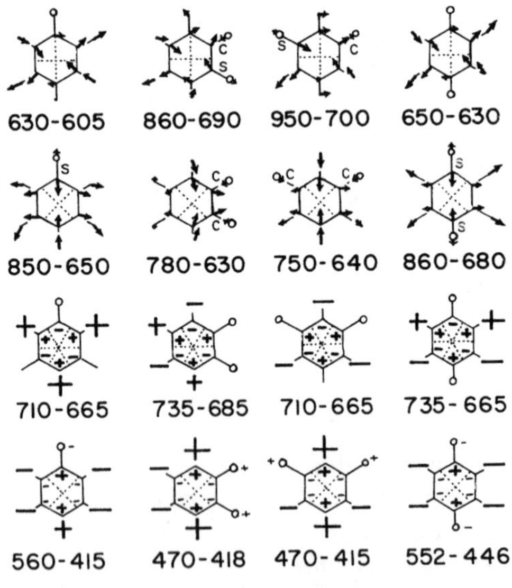

Fig. 17-6. Approximate quadrant in-plane bending modes, sextant out-of-plane bending modes, and IR-active quadrant out-of-plane bending modes [1–3]. Regions given in cm^{-1}.

17-5.2. The 650–605 cm^{-1} Bands

In the Raman spectra, there is a characteristic medium-intensity depolarized band at 630–605 cm^{-1} for monosubstituted benzenes and 650–630 cm^{-1} for para-disubstituted benzenes [5, 19, 26]. In the IR, these bands are weak for mono, and absent for para [27]. These are illustrated in Fig. 17-6, top row, for mono and para. These are one component of the in-plane ring bend by quadrants, where the 2,5 carbons move outward and the 3,6 carbons move inward. The 1,4 carbons move tangentially and do not stretch the C—X bonds. For ortho and meta isomers this mode becomes substituent-sensitive.

17-5.3. The 400 cm^{-1} Band

One component of the out-of-plane ring bend by quadrants has a characteristic frequency near 400 cm^{-1} for mono- and para-substituted benzenes. This is shown in Fig. 17-1. In this vibration, a two-fold axis of symmetry is retained that goes through the non-moving 1,4 carbons and substituents, making the vibration substituent-insensitive. However, the IR bands are usually forbidden or very weak, and the Raman intensities are very weak. A very weak Raman band is seen for monosubstituted benzenes at 420–390 cm^{-1} [5, 19].

17-6. SUBSTITUENT-SENSITIVE BANDS

17-6.1. The 1300–1060 cm^{-1} Aryl Bands

For monosubstituted benzenes, there is a substituent-sensitive mode where the 1,3,5 carbons move radially and the 2,4,6 carbons move only a little [1, 2, 4–10]. See Fig. 17-1. Because the substituted carbon number 1 is moving radially, it changes the length of the C—X substituent bond, and this C—X interaction makes this phenyl mode very substituent-sensitive. It is this mode that gives rise to phenyl bands that are called aryl—N in aniline, 1277 cm^{-1}, aryl—O in phenol, 1230 cm^{-1}, or aryl—Cl in chlorobenzene, 1083 cm^{-1}. For phenyl groups on Si, P, or S atoms, this band is seen in the 1130–1090 cm^{-1} region. Some examples are listed in the first column of Table 17-6, which lists Raman data from Kohlrausch [5]. These are all medium-strong in the Raman, and all except the 1208 cm^{-1} band in toluene are medium-strong in the IR.

Table 17-6.

Substituent-Sensitive Modes[a] in Phenyl Groups [5]

Substituent	1300–1060 cm^{-1}	850–650 cm^{-1}	560–415 cm^{-1}	540–260 cm^{-1}	400–200 cm^{-1}	250–150 cm^{-1}
NH_2	1277	818	504	531	385	233
OH	1230	812	507	532	—	240
F	1218	807	496	519	368	241
CH_3	1208	786	464	521	344	216
Cl	1083	702	467	418	297	196
SH	1097	698	461	413	280	188
Br	1071	673	460	315	254	181
I	1060	654	450	266	220	166

[a] See Fig. 17-1.

In this same region, disubstituted benzenes also have substituent-sensitive bands. *meta*-Dimethyl benzene (*m*-xylene) has a vibrational mode similar to that for the phenyl band above, involving in-phase aryl—C_2 stretch, medium-strong in the Raman at 1250 cm^{-1}. This shifts to 1123 cm^{-1} in the Raman for *meta*-dichloro benzene [1, 2, 5, 19].

Ortho and para dialkyl benzenes have strong Raman bands at 1230–1215 and 1230–1200 cm^{-1}, respectively, that involve in-phase aryl—C_2 stretch, somewhat mixed with the in-phase ring contract [19]. In ortho and para dichloro benzenes, these shift to 1129 and 1108 cm^{-1}, respectively [1, 2, 5, 19].

In the IR, substituent-sensitive aryl bands are found at 1270–1100 cm^{-1} involving aryl—F stretching (strong). Para Cl and Br benzenes absorb at 1096–1089 and 1073–1068 cm^{-1}, respectively. Meta Cl and Br benzenes absorb at 1078–1074 and 1073–1065 cm^{-1}, respectively. Ortho Cl and Br benzenes absorb at 1057–1034 and 1042–1028 cm^{-1}, respectively, in the IR [28, 29].

17-6.2. Quadrant In-Plane Bending Modes

Figure 17-6 (top two rows) shows some in-plane quadrant bending modes [1–3], all of which are substituent-sensitive except the 650–605 cm^{-1} modes for mono and para in the top row (previously discussed in Section 17-5.2). The other in-plane quadrant bend modes change the length of the aryl—X bonds, out-of-phase or in-phase for disubstitution, as shown in Fig. 17-6. (S is stretch and C is contract.) In the IR, these have variable intensities and can be confused with the adjacent hydrogen wag modes discussed previously, which is one reason why wag frequency shift calculations were given in Table 17-5.

In the monosubstituted benzenes, one quadrant in-plane bending mode interacts with the aryl—X stretch. When the C—X bond length increases and the distance between ring carbons 1 and 4 decreases, then a band appears at 850–650 cm^{-1}. See Fig. 17-6 (row 2), and Table 17-6 (column 2). If the substituent is an *n*-alkyl group, this band is seen at 780–730 cm^{-1} [19]. When the C—X and the carbon 1,4 spacing increase together, in-phase, a Raman band appears in the region 540–260 cm^{-1} as listed in Table 17-6 (column 4). See Fig. 17-1. For this band, Dollish *et al.* [19] list 521 cm^{-1} for toluene, 500–490 cm^{-1} for *n*-alkyl groups, and 462–458 cm^{-1} for CHR_2 substituents.

In ortho and meta isomers, the modes illustrated in Fig. 17-6, top two rows, give rise to some IR bands that are found in the same region as the aryl CH wag bands. Some examples are listed in Table 17-7 [30, 31]. For this halogen series, the ortho IR intensities are both medium and the meta IR intensities are strong and weak for the higher and lower frequency bands. For other substituents, the IR intensities are variable.

In the Raman, the out-of-phase modes for ortho and meta in Fig. 17-6 (top row) have usually very weak intensities for identical substituents. The quadrant in-plane bend modes, mixed with in-phase aryl—X_2 stretch in Fig. 17-6 (row 2), give rise to a strong Raman band at 780–630 cm^{-1} for ortho and meta isomers and 860–680 cm^{-1} for para [5, 19]. Ortho-, meta-, and para-dialkyl benzenes have strong Raman bands at 740–715, 740–700, and 830–780 cm^{-1}, respectively [19]. Some examples for para-substituted benzenes are *para*-xylene, 803 cm^{-1}; *para*-chlorotoluene 795 cm^{-1} and *para*-dichlorobenzene 748 cm^{-1}, all very strong in the Raman [19].

Table 17-7.

Quadrant In-Plane Bending IR Bands [1, 2, 30, 31]

	Ortho		Meta	
Substituents	Out-of-Phase[a] Medium	In-Phase Stretch Medium	Out-of-Phase Strong	In-Phase Stretch Weak
FF	838	760	950	733
FCl	825	679	880	681
ClCl	738	660	782	665
ClBr	720	647	757	653
BrBr	700	640	720	650
BrF	821	655	858	663

[a] Out-of-phase stretch refers to the fact that the two CX bonds stretch in-phase or out-of-phase during these aromatic ring vibrations.

In the Raman spectra, 1,2,3-trisubstituted benzenes have a strong band at 670–500 cm^{-1}, and 1,2,4-trisubstituted benzenes have a strong band at 750–650 cm^{-1} [19]. The 1,3,5-trisubstituted benzenes have the very strong Raman band near 1000 cm^{-1} discussed earlier. For 1,3,5-trialkyl benzenes, prominent Raman bands are seen at 580–510 and 520–480 cm^{-1} [19].

17-6.3. Quadrant Out-of-Plane Bending IR-Active Mode

The IR-active quadrant out-of-plane bending modes of mono- and di-substituted benzenes are illustrated in Fig. 17-6, bottom row. These are somewhat substituent-sensitive and can interact with out-of-plane bending modes of the substituents. These give rise to medium-intensity IR bands at 560–418 cm^{-1} for mono (Table 17-6, column 3), 470–418 cm^{-1} for ortho, 490–415 cm^{-1} for meta, 552–446 cm^{-1} for para, and 476–428 cm^{-1} for 1,2,4-trisubstituted benzenes [27]. These bands are very weak in the Raman [19].

17-7. FUSED-RING AROMATICS

Fused-ring aromatics such as naphthalenes, anthracenes, phenanthrenes, and even benzene rings fused to five- or six-membered-ring heterocyclics [30] have aryl—CH wag vibrations that give rise to strong IR bands that tend to follow the adjacent hydrogen correlations given for benzene derivatives [23]. Many of these also have ring stretch bands in the 1600–1500 cm^{-1} region that are often similar to those for benzene derivatives [23].

17-7.1. Naphthalenes

In the IR, alkyl-substituted naphthalenes have IR bands near 1600 cm^{-1} (a doublet), 1520–1505 cm^{-1}, and 1400–1390 cm^{-1} [32, 33]. The adjacent hydrogen wag IR correlations are as follows: 1-naphthalenes, 805–775 cm^{-1} (three adjacent H) and 780–760 cm^{-1} (four adjacent H); 2-naphthalenes, 875–823 cm^{-1} (lone H), 825–800 cm^{-1} (two adjacent H), and 760–735 cm^{-1} (four adjacent H) [23, 24, 34, 35]. When more highly substituted naphthalenes are included, the regions are 905–835 cm^{-1} (lone H), 847–799 cm^{-1} (two adjacent H), 820–730 cm^{-1} (three adjacent H), and 800–726 cm^{-1}

(four adjacent H) [32]. In the three- and four-adjacent hydrogen wag region, two bands are usually seen in the IR, one of which is probably a ring bending mode [32].

In the Raman spectra, a characteristic strong band is seen at 1390–1370 cm^{-1} for mono- and most disubstituted naphthalenes [36–38]. Raman bands for 1-naphthalenes are seen at 1580–1560 m, 1385–1370 vs, 1080–1020 w, 880–810 w-m, 720–650 m-s, and 535–512 cm^{-1} w [34, 36, 37, 39, 40]. Raman bands for 2-naphthalenes are seen at 1585–1570 m, 1390–1380 vs, 775–765 s, and 525–515 cm^{-1} ms [36–40]. The Raman spectra of 10 dimethyl naphthalenes have been reported [19, 41]. In the region within 1640–1570 cm^{-1}, most of these have a doublet.

17-7.2. Anthracenes and Phenanthrenes

Alkyl-substituted anthracenes have IR bands at 1640–1620 cm^{-1} and usually a band near 1550 cm^{-1}. There is no band near 1500 cm^{-1} [23, 33]. When one or both 9,10 ring positions have a lone hydrogen present, a relatively strong lone CH wag band is seen at 900–875 cm^{-1} in the IR [23, 33]. When at least one four-adjacent hydrogen group is present, a strong IR band is seen usually at 750–730 cm^{-1}.

Alkyl-substituted phenanthrenes have IR bands near 1600 cm^{-1} (often double), and a 1500 cm^{-1} band that distinguishes them from anthracenes [23, 33]. When the two-adjacent hydrogens on the middle ring are present, an IR band is seen near 820 cm^{-1}, and when at least one four-adjacent hydrogen group is present, a strong IR band is seen usually at 750–730 cm^{-1} [23, 33].

The Raman spectra of a few substituted anthracenes have been reported and have intense Raman bands at 1414–1384 and 426–390 cm^{-1} [19, 42, 43].

17-8. AROMATIC HETEROCYCLES, SIX-MEMBERED RINGS

The six-membered ring aromatic heterocycles are benzene-type rings where one or more aryl CH units are replaced with nitrogens. In benzene rings, of course, there are six "bond-and-a-half" CC bonds resulting from the delocalization of the three conjugated C=C bonds. When the six membered aromatic heterocycles have all double bonds in the ring like the benzenes, then the spectra have many benzene-like features, which will be discussed.

A major complication can occur, however, when there are OH or SH groups ortho or para to a ring nitrogen. In such cases, tautomerization can occur (and usually does), forming a C=O or C=S bond [44, 45]. An example is cyanuric acid:

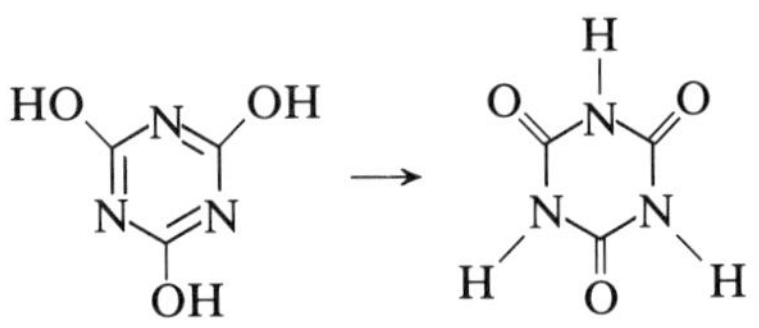

In the keto form, there are fewer than three double bonds in the ring and one or more double bonds (C=O or C=S) external to the ring. Heterocycles in the keto form have major differences in their spectra and should be considered separately from heterocycles in the normal benzene-like form. In the keto form for hydroxy substitution, a carbonyl band can be seen in the IR, usually in the 1720–1640 cm^{-1} region, and the NH···O can be seen absorbing broadly in the IR at 3200–2600 cm^{-1}. The tautomeric form (keto or enol) of OH- or SH-substituted aromatic heterocycles is best determined by comparison with the spectrum of a similar molecule where the hydrogen is replaced with alkyl group that fixes the structure:

−N=C(OR)− −NR−C(=O)−
−N=C(SR)− −NR−C(=S)−.

When an OH group is on a pyridine in the 3 position, tautomerization cannot occur. The strongly hydrogen-bonded OH···N is seen in the IR at 2800–2300 cm^{-1}, with a second weak broad component near 1800–1700 cm^{-1} [30], which is a Fermi resonance-enhanced overtone of the weak out-of-plane OH···N wag band near 900–850 cm^{-1}.

Sodium salts of OH substituents (ONa) and NH_2 substituents are usually in the enol or normal ring form [23].

17-8.1. Pyridines

Substituted pyridines have aryl-type CH stretch bands in the 3100–3000 cm^{-1} region. Monosubstituted pyridines have correlatable IR and Raman bands listed in Table 17-8 [19, 33, 36, 46–49]. The vibrational modes have many similarities to those of the benzenes. In the IR, the adjacent hydrogen correlation can be used, counting the nitrogen (which has no hydrogen) as a substituted carbon. There is one intensity variation. In 3-substituted pyridines, there is a lone CH between the nitrogen and the substituted

Table 17-8.

Bands for Monosubstituted Pyridines

	2-Substituted			3-Substituted			4-Substituted		
Assignment	cm^{-1}	IR	R	cm^{-1}	IR	R	cm^{-1}	IR	R
Quadrant stretch	1620–1570	s	ms	1595–1570	m	ms	1605–1565	s	ms
Quadrant stretch	1580–1560	s	m	1585–1560	s	m	1570–1555	m	m
Semicircle stretch	1480–1450	s	m	1480–1465	s	m	1500–1480	m	m
Semicircle stretch	1440–1415	s	mw	1430–1410	s	mw	1420–1410	s	mw
CH in-plane rock	1050–1040	m	ms	—			—		
Radial	1000–985	m	vs	1030–1010	m	vs	1000–985	m	vs
Adjacent H wag	780–740	s	—	820–770	s	—	850–790	s	—
Sextant out-plane bend	740–720	m	vw	730–690	s	vw	730–720	m	vw
Quadrant in-plane bend	630–615	w	m	630–615	w	m	670–660	—	m
Quadrant out-plane bend	410–395	—	vw	410–395	—	vw	400–385	—	vw
Quadrant in-plane bend[a]	850–800	wv	ms	805–750	wv	m	805–785	vw	ms

[a] Substituent-sensitive mode involves C—X stretch. Regions are for alkyl substituents.

carbon. Because of the adjacent electronegative nitrogen, this lone CH wag has very weak intensity compared with those of meta-substituted benzenes.

In the Raman there is a very strong band at 1000–985 cm^{-1} in 2- and 4-monosubstituted pyridines and 2,4- and 2,4,6-substituted pyridines—in other words, any isomer, as long as there are hydrogens on both the 3 and 5 carbons. In this substituent-insensitive mode, the 2,4,6 carbons must be nearly stationary. The nitrogen and carbons 3 and 5 (with their hydrogens) move radially in-phase. This band has medium intensity in the IR [30] (see Fig. 17-3, top row). When there is 3-substitution, there is a very strong Raman band and medium-intensity IR band at 1030–1010 cm^{-1}. Presumably, this involves carbons 2,4,6 moving radially in-phase.

17-8.2. Pyridinium Salts

Pyridinium salts have a strong broad $NH^+ X^-$ IR band whose frequency depends on the strength of the hydrogen bond. There can be multiple internal band structure so that some absorption may result in the broad region 3300–1900 cm^{-1}. The multiple internal band structure is explained as Fermi resonance between the broad NH stretch band and numerous overtones and combinations of lower-lying fundamentals [50, 51]. In pyridine salts, the NH stretch changes in frequency, depending on the anion and the hydration state. HCl salts of pyridines usually have bands in the 2800–2400 cm^{-1} region. The NH in-plane bending of salts of pyridines gives rise to a medium-intensity Raman band at 1250–1240 cm^{-1}, and the out-of-plane NH wag is found at 940–880 cm^{-1} with variable intensity [19]. The quadrant and semicircle ring stretch bands of pyridine salts tend to be 30–50 cm^{-1} higher than those of the free bases [19].

17-8.3. Pyridine *N*-Oxides

The *N*-oxides of pyridine [52, 53, 56], pyrimidines [52], and pyrazine [54, 55] have a strong IR band in the 1300–1200 cm^{-1} region involving NO stretch, and a weaker band at 880–845 cm^{-1} involving in-plane quadrant bend and NO stretch. In the Raman spectra [53], liquid pyridine *N*-oxide has bands at 1252 cm^{-1} (m), 835 cm^{-1} (s), and 541 cm^{-1} (w), all involving NO stretch mixed with ring vibrations.

17-8.4. Diazines

In diazines, two nitrogens replace two CH groups in benzene in the 1,2 positions (pyradazines), the 1,3 positions (pyrimidines), or the 1,4 positions (pyrazines).

Pyrimidine rings have quadrant stretch bands at 1590–1555 and 1565–1520 cm^{-1}, and semicircle stretch bands at 1480–1400 and 1410–1375 cm^{-1}, seen in both IR and Raman [19, 36, 37, 58]. Pyrimidines substituted in the positions 2, 4, 2,4, or 2,4,6—in other words, all isomers where a hydrogen remains in position 5—have a strong Raman band at 1005–980 cm^{-1} [19, 58] caused by the 1,3,5 ring atoms moving radially in-phase. This band is near 1050 cm^{-1} for 5-substitution [19, 58]. The substituent-insensitive quadrant in-plane bend is seen in the Raman at 645–625 cm^{-1} for 2-substitution, and 685–660 cm^{-1} for 4-substitution [19]. Amino-substituted pyrimidines have the NH_2 scissors bending band at 1680–1620 cm^{-1}, seen in the IR and Raman [19, 57]. As mentioned earlier, if pyrimidines have a hydroxy group ortho or para to a nitrogen, they usually exist in the keto form with a C=O band near 1700 cm^{-1} [57].

Pyrazine rings have characteristic Raman bands given in Table 17-9 [36, 59]. Quadrant stretch bands are seen in the 1586–1517 cm^{-1} region. The substituent-insensitive quadrant in-plane bend band is seen near 640 cm^{-1} for monosubstitution, and near 647 cm^{-1} for 2,5-disubstitution. Monosubstitution (in the 2 position) and 2,6-disubstitution give rise to a strong Raman band for the 1,35 in-phase radial mode near 1015 and 1023 cm^{-1}, respectively. Substituent-sensitive quadrant in-plane bending modes involve C—X stretch in mono- and in-plane C—X_2 stretch in disubstitution and are seen as strong Raman bands in the 865–706 cm^{-1} region for alkane substituents [36, 59].

Table 17-9.

Characteristic Raman Bands for Pyrazines [36, 59]

	cm^{-1}	R		cm^{-1}	R
Monosubstituted	1580–1560	m	Trisubstituted	1576–1570	m
	1530–1517	m		1540–1525	m
	1060–1050	s		955–915	m
	1024–1003	vs		748–725	m
	840–788	s		710–695	m
	660–617	w	Tetrasubstituted	1565–1580	w
2,3-Disubstituted	1580–1558	m		1550–1545	m
	1570–1525	m		720–710	s
	1292–1252	m			
	1100–1081	s			
	758–686	vs			
2,6-Disubstituted	1586–1572	w			
	1025–1021	m			
	709–706	vs			
2,5-Disubstituted	1586–1576	m			
	1540–1520	m			
	865–838	vs			
	650–642	m			

In the IR, most of the non-tautomeric, normal ring diazines have strong semicircle stretch bands in the 1480–1370 cm^{-1} region [30]. As in the benzene series, the IR intensities for the quadrant stretch bands for non-tautomeric diazines at 1600–1520 cm^{-1} are variable [30]. For non-tautomeric pyrimidines, these bands are usually very strong in the IR. For non-tautomeric pyrazines, these bands are IR-inactive when a center of symmetry is present, but may have medium intensity in non-symmetrical cases. For non-tautomeric pyridazines, these bands usually have medium intensity in the IR [30, 89].

17-8.5. Triazines [1, 3, 5]

In the 1,3,5 or *s*-triazines, three nitrogens replace the 1,3,5 CH groups in benzenes. The parent unsubstituted *s*-triazine has been vibrationally analyzed [60], and the results are given in Table 17-10. Note, in the IR, the strong ring-stretching bands at 1550 and 1410 cm^{-1}. These are doubly degenerate for *s*-triazine and also for substituted triazines that retain the three-fold axis of symmetry. Note the weak IR intensity of the lone CH wag at 930 cm^{-1} when the lone CH is between two ring nitrogens.

Table 17-10.

IR and Raman Bands of *s*-Triazine [60]

Symmetry	IR Liquid	R Liquid	Description
E'	3055 s	—	CH stretch
A_1'	—	3042 vs p	CH stretch
E'	1550 vs	1555 mw d	quandrant stretch
E'	1410 vs	1410 w d	semicircle stretch
E'	1172 s	1176 w d	CH rock
A_1'	1132 vw	1132 s p	ring radial
A_1'	—	992 s p	ring radial
A_2''	930 vw	—	CH wag
A_2''	730 vs	—	sextant out-of-plane deformation
E'	685 vs	676 ms d	quadrant in-plane deformation
E''	—	340 m d	quadrant out-of-plane deformation

Raman spectra of some trisubstituted *s*-triazines have a characteristic Raman band at 1000–980 cm^{-1} [19] (like the 2,4,5-substituted pyrimidines). This implies that the 2,4,6 ring carbons (with substituents) are nearly stationary and the 1,3,5 nitrogens move in-phase radially for this mode in trisubstituted *s*-triazines.

The IR spectra of many substituted *s*-triazines have been studied [23, 60–64] and have characteristic bands near 1550 cm^{-1} (vs) (quadrant stretch), 1420 cm^{-1} (s) (semicircle stretch), and 800 cm^{-1} (m) (out-of-plane sextant bend). The 1550 and 1420 cm^{-1} bands are doubly degenerate if a three-fold axis of symmetry is present, but each may become double if this symmetry is not present.

In alkyl- or aryl-substituted *s*-triazines, one or two strong IR bands are found at 1580–1525 cm^{-1}. There is also at least one band at 1450–1350 cm^{-1}, and a weak band in the 860–775 cm^{-1} region [23].

Melamines (triamino *s*-triazines) and quanamines (diamino triazines) have NH_2 groups that give rise to complex IR absorption at 3500–3100 cm^{-1} (NH_2 stretch). A band is also seen at 1680–1640 cm^{-1} (NH_2 deformation) that disappears if all NH_2 groups are replaced by NHR or NR_2 groups. The ring stretch modes give rise to strong IR bands at 1600–1500 cm^{-1} and usually multiple bands at 1450–1350 cm^{-1}. For these compound types, a medium-intensity, very characteristic IR band is seen at 825–800 cm^{-1}. In most melamines, the range is narrower at 812 ± 5 cm^{-1}. This mode involves out-of-plane ring bending by sextants where the ring nitrogens and carbons move in opposite directions out-of-plane. This band is characteristic for the normal ring form with all double bonds in the ring.

If the ring should be in the iso form with fewer than three double bonds in the ring and one or more double bonds external to the ring, then this band shifts to 795–750 cm^{-1} [23]. Examples of iso forms include ring *N*-alkyl isomelamines, or rings with OH or SH substituents that exist in the keto form with a ring NH. This latter group includes ammeline, ammelide, and cyanuric acid. In sodium salts of these, a ring NH is removed and an ONa group is formed. If all the ring NH units are removed, the ring is restored to the normal form [23]. In rings with NH_2 substituents, an HCl salt protonates a ring nitrogen and forms an external $C{=}NH_2^+$ group, generating the iso ring form.

In general, this out-of-plane sextant bend mode near 800 cm^{-1} can be used to tell normal triazine rings (825–795 cm^{-1}) from iso triazine rings (795–750 cm^{-1}) in rings with any combination of N or O substituents, or with two N atoms and any other substituent. It does not hold for trialkyl- or aryl-substituted triazines [23].

Trialkyl cyanurates have IR bands at 1600–1540, 1380–1320, 1160–1100 ($O{-}CH_2$ stretch), and 820–805 cm^{-1} (normal ring). Isocyanurates have C=O bands near 1700 cm^{-1} and an iso ring type band near 760 cm^{-1} [23].

17-9.

AROMATIC HETEROCYCLES, FIVE-MEMBERED RINGS

The five-membered ring aromatic heterocycles are five-membered rings containing two conjugated double bonds and a single-bonded atom that may be a nitrogen, oxygen, or sulfur, for example, as in pyrrole, furan, and thiophene. Many additional types are generated from these when one or more of the four CH units in the double bonds are replaced by nitrogens, as in imidazole, oxazole, and thiazole, all of which have a nitrogen in the 3-position, or pyrazole, isoxazole, or isothiazole, all of which have nitrogen in the 2-position. The double bonds are somewhat delocalized, although less so than in the six-membered ring aromatic heterocycles.

One characteristic of these five-membered rings is that the ring =CH stretch bands have somewhat higher frequencies than those in six-membered ring analogues, which generally appear at 3100–3000 cm^{-1}. When these five-membered rings contain oxygen or one or more nitrogens, as do furans or pyrroles, the CH stretch bands are seen at 3180–3090 cm^{-1}. Thiophenes have slightly lower CH stretch bands at 3120–3060 cm^{-1} [30]. These are seen in both the IR [23] and Raman spectra [19].

If at least one —CH=CH— double bond in the ring has two hydrogens on it, then a strong, somewhat broad IR band is seen at 800–690 cm^{-1}, which is caused by the in-phase, out-of-plane cis CH wag [23]. The CH wag intensity decreases as the number of ring CH units decreases. As a rough rule, the CH wag frequency increases in the substitution order 2 < 2,3 < 3 < 2,5 for a given ring type with similar substituents [30]. Thiophenes, for example, have a strong IR band for CH wag at 740–690 cm^{-1} for 2- and 2,3-substitution, near 755 cm^{-1} for 3-substitution, and at 800–770 cm^{-1} for 2,5-substitution [30]. Most common pyrroles and furans have a CH wag band in the 770–715 cm^{-1} region, and isoxazoles have the band at 835–770 cm^{-1} [30]. These bands are quite weak in the Raman.

As in the six-membered ring heterocycles, five-membered ring heterocycles with OH or SH groups on the carbon of a ring C=N usually exist in the keto form,

$$\mathrm{HO{-}C{=}N{-}} \longrightarrow \mathrm{O{=}C{-}NH{-}},$$

whereas NH_2 substituents usually do not tautomerize to the keto form. The NH_2 deformation band is usually seen at 1650–1615 cm^{-1}. Rings with exo C=O or C=S groups should be considered separately.

17-9.1. Five-Membered Ring Heterocycle Ring Stretch Bands

The fundamental modes have been assigned for pyrrole [65–69], furan [70–74], and thiophene [75, 76]. Valence force field calculations for these and some methyl derivatives have been reported [77, 78]. The spectra have been studied of substituted derivatives of pyrrole [69, 79–81], furan [81–84], and thiophene [81, 83, 85–87]. The three highest ring stretch frequencies in these are listed in Table 17-11 for non-tautomeric normal rings [19, 81]. The two highest ring stretching vibrations can be pictured as two quadrant stretch vibrations [23]. Since the bonds are not all identical in the ring, these

Table 17-11.

Ring Stretching Bands in cm^{-1} for Monosubstituted Pyrroles, Furans, and Thiophenes

Rings	Out-of-Phase C=C Stretch IR : var, R : var	In-Phase C=C Stretch IR : m, R : vs	C−X−C In-Phase Stretch + C−C Contract IR : ms, R : s
1-Substituted pyrroles	1560–1540	1510–1490	1390–1380
2-Substituted pyrroles	1570–1545	1475–1460	1420–1400
3-Substituted pyrroles	1570–1560	1490–1480	1430–1420
2-Substituted furans	1605–1560	1515–1460	1400–1370
2-Substituted thiophenes	1535–1514	1454–1430	1361–1347
3-Substituted thiophenes	1542–1492	1410–1380	1376–1362

consist primarily of C=C stretch, out-of-phase and in-phase (ca. 1550 and 1490 cm^{-1}). The third band (ca. 1380 cm^{-1}) in Table 17-11 is a semicircle stretch mode where the top half of the ring stretches while the bottom half contracts, and consists mainly of in-phase C—X—C stretch plus C—C contract. For more highly substituted rings, or rings with additional nitrogens, the overall regions for the above three ring bands are 1620–1490 cm^{-1}, 1515–1380 cm^{-1}, and 1430–1330 cm^{-1}, respectively [19, 30, 31]. The sulfur-containing rings tend to come at the lower end of the regions.

17-9.2. Additional Bands for Five-Membered Ring Heterocycles

The NH stretch of pyrrole has been studied in the IR and Raman [88]. In a dilute CCl_4 solution, a sharp band occurs at 3496 and 3498 cm^{-1}, respectively, and in the pure liquid a broader band is seen at 3403 and 3405 cm^{-1}, respectively. The bonded NH stretch band for substituted pyrroles is usually seen at 3450–3200 cm^{-1} [23, 30].

For imidazoles and other five-membered ring heteroaromatics with more than one nitrogen, such as triazoles and tetrazoles, there is a strong, broad IR band usually centering at 3000–2600 cm^{-1} for bonded NH···N=. In addition these have a weak broad IR band at 1880–1760 cm^{-1} due to a Fermi resonance-enhanced overtone of a weak IR NH···N wag band near 900 cm^{-1} [23, 30].

For 1-substituted pyrroles, where the nitrogen has an alkyl substituent, there is a band involving N—C(H_2) stretch at 1290–1280 cm^{-1} IR : s, R : vs [19]. The 1560–1540 cm^{-1} ring stretch band is weak in the IR and Raman [19, 30]. There are additional bands at 1095–1080 cm^{-1} IR : m, R : s and 1065–1055 cm^{-1} IR : mw, R : s, each involving in-plane CH deformation [19].

In 2-substituted furans, a distinction can be made between unconjugated and conjugated substituents. For a 2-CH_2 substituent, the two C=C stretch bands come at 1605–1590 cm^{-1} IR : mw, R : s and 1515–1490 cm^{-1} IR : m, R : vs. When a double bond or C=O is conjugated in the 2-position, these two bands come lower, at 1585–1560 cm^{-1} IR : mw, R : s, and 1480–1460 cm^{-1} IR : m, R : vs [30, 82, 83].

In most 2-substituted furans, medium-intensity IR bands are usually seen at 1163–1136 cm^{-1}, 1030–1010 cm^{-1}, and 815–795 cm^{-1}, with a weaker band at 1100–1072 cm^{-1} [30]. In the Raman, medium-intensity bands are seen at 1230–1220 cm^{-1}, 1160–1140 cm^{-1}, and 1080–1060 cm^{-1}, with a medium-strong band at 1020–992 cm^{-1} [19].

In 2-substituted thiophenes, a band is seen at 867–842 cm^{-1} IR : m, R : s [19, 30].

REFERENCES

1. J. R. Scherer, "Planar Vibrations of Chlorinated Benzenes," Dow Chemical Company, Midland, Michigan, 1963.
2. J. R. Scherer, *Spectrochim. Acta* **21,** 321, 1965.
3. J. R. Scherer, *Spectrochim. Acta* **24A,** 747, 1968.
4. A. R. Katritzky, *Quart. Rev. Chem. Soc.* **13,** 353, 1959.
5. K. W. F. Kohlrausch, "Ramanspectren," Akad. Verlagsges, Leipzig, 1943.
6. E. W. Schmid, J. Brandmuller, and G. Nonnenmacher, *Z. Elecrochem.* **64,** 726 and 940, 1960.
7. D. H. Whiffen, *J. Chem. Soc. London*, 1350, 1956.
8. K. S. Pitzer and D. W. Scott, *J. Am. Chem. Soc.* **65,** 803, 1943.
9. G. Varsanyi, "Vibrational Spectra of Benzene Derivatives," Academic Press, New York, 1969.
10. G. Varsanyi, "Assignments for Vibrational Spectra of Seven Hundred Benzene Derivatives," John Wiley & Sons, New York, 1975.
11. A. M. Bogomolov, *Opt. Spectrosc. (USSR)* **9,** 162, 1960.
12. A. M. Bogomolov, *Opt. Spectrosc. (USSR)* **10,** 162, 1961.
13. A. M. Bogomolov, *Opt. Spectrosc. (USSR)* **12,** 99, 1961.
14. A. M. Bogomolov, *Opt. Spectrosc. (USSR)* **13,** 90 and 183, 1962.
15. J. H. S. Green, *Spectrochim. Acta* **26A,** 1503, 1970.
16. J. H. S. Green, *Spectrochim. Acta* **26A,** 1523, 1970.
17. J. H. S. Green, *Spectrochim. Acta* **26A,** 1913, 1970.
18. J. H. S. Green and D. J. Harrison, *Spectrochim. Acta* **26A,** 1925, 1970.
19. F. R. Dollish, W. G. Fateley, and F. F. Bentley, "Characteristic Raman Frequencies of Organic Compounds," Wiley, New York, 1974.
20. S. E. Wiberley, S. C. Bunce, and W. H. Bauer, *Anal. Chem.* **32,** 217, 1960.
21. M. L. Josien and J. M. Lebas, *Bull. Soc. Chem. Fr.*, pp. 53, 57, and 62, 1952.
22. J. M. Lebas, C. Carrignou-Lagrange, and M. L. Josien, *Spectochim. Acta* **15,** 225, 1959.
23. N. B. Colthup, L. H. Daly, and S. E. Wiberley, "Introduction to Infrared and Raman Spectroscopy," 3rd Ed., Academic Press, Boston, 1990.
24. N. B. Colthup, *Appl. Spectrosc.* **30,** 589, 1976.
25. C. W. Young, R. B. DuVall, and N. Wright, *Anal. Chem.* **23,** 709, 1951.
26. D. H. Whiffen, *Spectrochim. Acta* **7,** 253, 1955.
27. F. F. Bentley, L. D. Smithson and A. L. Rozek, "Infrared Spectra of Characteristic Frequencies 700–300 cm^{-1}," Wiley Interscience, New York, 1968.
28. A. Stojiljkovic and D. H. Whiffen, *Spectrochim. Acta* **12,** 47, 1958.
29. A. R. Katritzky and J. M. Lagowski, *J. Chem. Soc. London*, 2421, 1960.
30. C. J. Pouchert, "The Aldrich Library of FT-IR Spectra," Aldrich Chemical Company, Inc., 1985.
31. K. G. R. Pachler, F. Matlock, and H. U. Gremlich, "Merck FT-IR Atlas," VCH Publishers, New York, 1988.
32. J. G. Hawkins, E. R. Ward, and D. H. Whiffen, *Spectrochim. Acta* **10,** 105, 1957.
33. C. G. Cannon and G. B. B. M. Sutherland, *Spectrochim. Acta* **4,** 373, 1951.
34. T. S. Wang and J. M. Sanders, *Spectrochim. Acta* **15,** 1118, 1959.
35. R. L. Werner, W. Kennard, and D. Rayson, *Aust. J. Chem.* **8,** 346, 1955.
36. S. K. Freeman, "Applications of Laser Raman Spectroscopy," Wiley Interscience, New York, 1974.
37. K. H. Michaelian and S. M. Ziegler, *Appl. Spectrosc.* **27,** 13, 1973.

38. P. Dizabo, H. E. Gatica, N. Le Calve, and G. Franck, *J. de Chim. Phys. et de Physiochemie Biol.* **66,** 1497, 1969.
39. H. Gockel, *Z. Phys. Chem.* **29B,** 79, 1935.
40. H. Luther, *Z. Electrochem.* **52,** 210, 1948.
41. H. Luther and C. Reichel, *Z. Phys. Chem.* **195,** 103, 1950.
42. R. Manzoni-Ansidei, *Gazz. Chim. Ital.* **67,** 790, 1937.
43. M. Brigodiot and J. M. Lebas, *Spectrochim. Acta* **27A,** 1315 and 1325, 1971.
44. R. Shunmugam and D. N. Sathyanarayana, *Spectrochim. Acta* **40A,** 757, 1984.
45. D. N. Sathyanarayana and S. V. Kasmir Raja, *Spectrochim. Acta* **41A,** 809, 1985.
46. H. E. Podal, *Anal. Chem.* **19,** 1423, 1957.
47. G. L. Cook and M. Church, *J. Phys. Chem.* **61,** 458, 1957.
48. M. P. Groenewege, *Spectrochim. Acta* **11,** 579, 1958.
49. J. H. S. Green, W. Kynaston, and H. M. Parsley, *Spectrochim. Acta* **19,** 549, 1963.
50. S. E. Odinokov, A. A. Mashkovsky, V. P. Glazunov, A. V. Iogansen, and B. V. Rassadin, *Spectrochim. Acta* **32A,** 1335, 1976.
51. V. P. Glazunov and S. E. Odinskov, *Spectrochim. Acta* **38A,** 399, 1982.
52. R. H. Wiley and S. C. Slaymaker, *J. Am. Chem. Soc.* **79,** 2233, 1957.
53. P. Mirone, *Atti Acad. Nazl. Lincei Rend. Cl. Sci. Fis. Mat. Nat.* **35,** 530, 1963.
54. B. Klein and J. Berkowitz, *J. Am. Chem. Soc.* **81,** 5160, 1959.
55. C. F. Koelsch and W. H. Grumprecht, *J. Org. Chem.* **23,** 1602, 1958.
56. V. L. Brezin, *Opt. Spectrosc.* **18,** 119, 1965.
57. L. N. Short and H. W. Thompson, *J. Chem. Soc. London*, 168, 1952.
58. S. B. Kartha, *Spectrochim. Acta* **38A,** 359, 1982.
59. R. P. Oertal and D. V. Myhre, *Anal. Chem.* **44,** 1589, 1972.
60. J. E. Lancaster, R. F. Stamm, and N. B. Colthup, *Spectrochim. Acta* **17,** 155, 1961.
61. A. Roosens, *Bull. Soc. Chim. Belg.* **59,** 377, 1950.
62. J. Goubean, E. L. Jahn, A. Kreutzberger and C. Grundmann, *J. Phys. Chem.* **58,** 1078, 1954.
63. W. M. Padgett II and W. F. Hammer, *J. Am. Chem. Soc.* **80,** 803, 1958.
64. H. Schroeder, *J. Am. Chem. Soc.* **81,** 5658, 1959.
65. R. C. Lord and F. A. Miller, *J. Chem. Phys.* **10,** 328, 1942.
66. P. Mirone, *Gas. Chim. Ital.* **86,** 165, 1956.
67. J. Morcillo and J. M. Orza, *Am. Real Soc. Esp. Fes. y Quim (Mad.)* **56B,** 231, 1960.
68. A. M. Prima, *Opt. Spektrosk., Akad. Nauk SSSR, Sb. Statei* **3,** 157, 1967.
69. R. A. Jones, in "Advances in Heterocyclic Chemistry," Vol. II (A. R. Katritzky and A. J. Boulton, Eds.), Academic Press, New York, 1970, pp. 443–459.
70. H. W. Thompson and R. B. Temple, *Trans. Faraday Soc.* **41,** 27, 1945.
71. B. Bak, S. Brodersen, and L. Hansen, *Acta Chem. Scand.* **9,** 749, 1955.
72. J. Lecomte, *Bull. Soc. Chem. Fr.*, 415, 1946.
73. G. B. Guthrie, D. W. Scott, W. N. Hubbard, C. Katz, J. P. McCullough, M. E. Gross, K. D. Williamson, and G. Waddington, *J. Am. Chem. Soc.* **74,** 4662, 1952.
74. M. Rico, M. Barrachina, and J. M. Orza, *J. Mol. Spectrosc.* **24,** 133, 1967.
75. N. Rico, J. M. Orza, and J. Morcillo, *Spectrochim. Acta* **21,** 689, 1965.
76. E. N. Bolotina and L. M. Sverdlov, *Opt. Spectrosc. Supp. 3, Mol. Spectrosc., Opt. Soc. Am.*, 75, 1968.
77. D. W. Scott, *J. Mol. Spectrosc.* **31,** 451, 1969.
78. D. W. Scott, *J. Mol. Spectrosc.* **37,** 77, 1971.
79. G. B. Bonino, R. Manzoni-Amsidei, and P. Pratesi, *Z. Phys. Chem.* **22B,** 21, 1933, and **25B,** 348, 1934.
80. A. Stern and K. Thalmayer, *Z. Phys. Chem.* **31B,** 403, 1936.

81. A. R. Katritzky, *Quart. Rev., Chem Soc.* **13,** 353, 1959.
82. K. Matsuno and K. Han, *Bull. Chem. Soc. Jpn.* **12,** 155, 1937.
83. G. B. Bonino and R. Manzoni-Ansidei, *Z. Phys. Chem.* **25B,** 327, 1934.
84. S. Hillers and A. Berzina, *Khim. Geterotikl. Soeden*, 487, 1966.
85. J. J. Peron, P. Saumagne, and J. M. Leas, *Spectrochim. Acta* **26A,** 1651, 1970.
86. M. Horak, I. J. Hyams, and E. R. Lippincott, *Spectrochim. Acta* **22,** 1355, 1966.
87. E. N. Bolotina and L. M. Sverdlov, *Zh. Prikl. Spectrosk.* **7,** 870, 1967.
88. A. Lautie and A. Novak, *J. Chem. Phys.* **56,** 2479, 1972.
89. J. F. Arenas, V. Hernandez, J. I. Marcos, and J. C. Otero, *J. Mol. Struct.* **189,** 307, 1988.

CHAPTER 18

Selected Infrared and Raman Spectra from the Sadtler Research Laboratories, Division of Bio-Rad Laboratories, Inc. for Compounds with Structures Discussed in Chapters 2–17

The following pairs of infrared and Raman spectra are presented to provide illustrative examples of compounds containing characteristic chemical groups discussed in Chapters 2–17. They are grouped in order of the book chapters. An alphabetical name index and a molecular formula index at the end of this book will be of assistance in locating compounds of interest.

The spectra were selected from the collection of Sadtler Standard Raman Spectra and they are reproduced with permission.*

The experimental conditions for the Raman data are noted on the spectra. In addition, the introduction to the Sadtler collection notes the following:

INSTRUMENTATION

All spectra contained in the publication were prepared on a Cary 83 laser Raman spectrophotometer. Solids were run in a single-pass solid sample holder and with a spike filter placed between the laser and the sample. Liquids were run using a multi-pass cell.

The Cary 83 instrument utilizes the 4880 Å argon line. The laser power is a nominal 100 mW at the laser head and 30–60 mW at the sample. The nominal power was seldom achieved during preparation of the spectra. The actual power for each sample is indicated in the heading data on each spectrum. Sample geometry is 90° in all cases.

SAMPLE PREPARATION

Every effort was made to obtain analytically pure samples for preparation of the spectra, however, many samples exhibited too much fluorescence in the initial rough Raman scan and further purification was necessary. Solid samples were recrystallized from a suitable solvent, and liquid samples were stored over activated decolorizing charcoal for a prolonged time and the charcoal was filtered off.

Solid samples are represented by one non-polarized spectrum in every case. There are no spectra of solids in solution included. Liquid samples are represented by two spectra prepared at identical instrument settings: the parallel polarized scan and the perpendicular polarized scan.

INFRARED SPECTRA

The infrared spectra were obtained on a Perkin-Elmer 521 grating spectrometer at 4 cm^{-1} resolution. Liquids were examined as films between CsBr plates, and solids were run as KBr pellets.

INDEX TO SPECTRA

362 R

OCTANE

RAMAN DATA

Spectrophotometer	Cary 83
Sensitivity	50000;
	10000
Suppression	52;
	0
Laser Power	24 mW
Spectral Slitwidth	6 cm^{-1}

C_8H_{18} Mol. Wt. 114.23

$CH_3CH_2(CH_2)_4CH_2CH_3$

B.P. 124-126°C

Source: Chem Service, Inc., West Chester, Pa.

WAVENUMBER / INTENSITY

2940/113.0	2900/107.7	2880/152.6
2860/140.0	280/49.5	

Capillary Cell: Neat

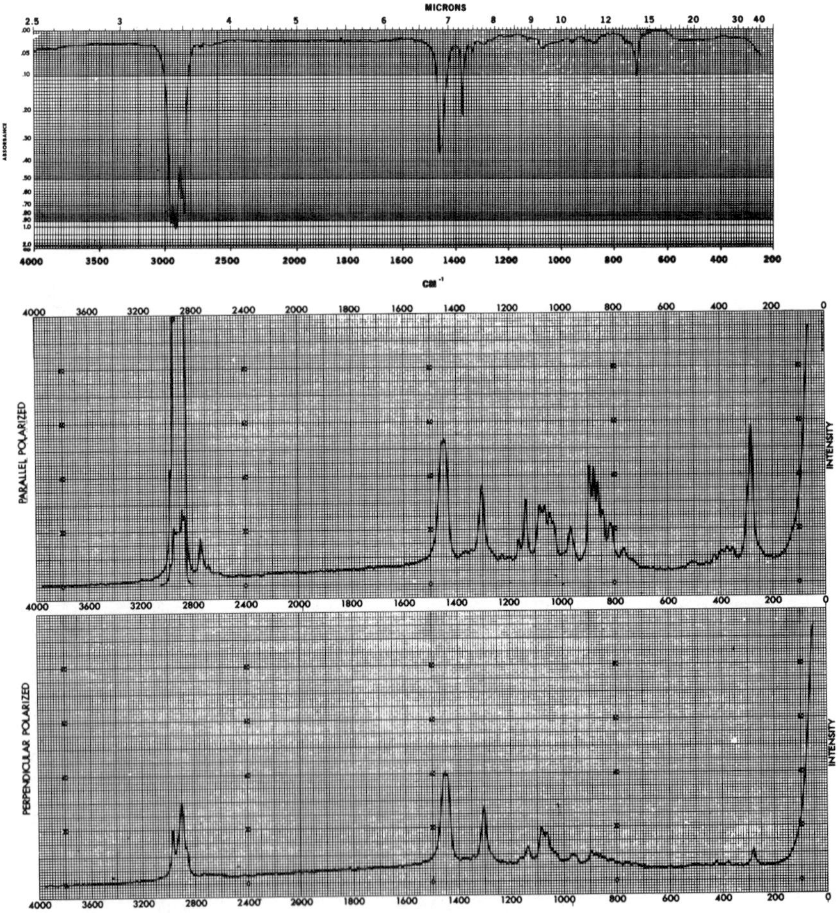

CYCLOHEXANE **355 R**

C_6H_{12} Mol. Wt. 84.16

Source: MCB Manufacturing Chemists, Norwood, Ohio

RAMAN DATA

Spectrophotometer	Cary 83
Sensitivity	50000
	-
Suppression	140
	-
Laser Power	23 mW
Spectral Slitwidth	2 cm^{-1}

WAVENUMBER / INTENSITY

2945/82.5 2860/110.2 809/366.6

1973 Sadtler

Capillary Cell: Neat

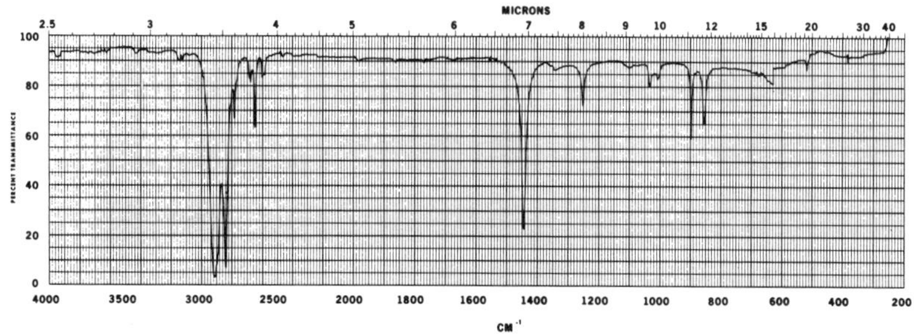

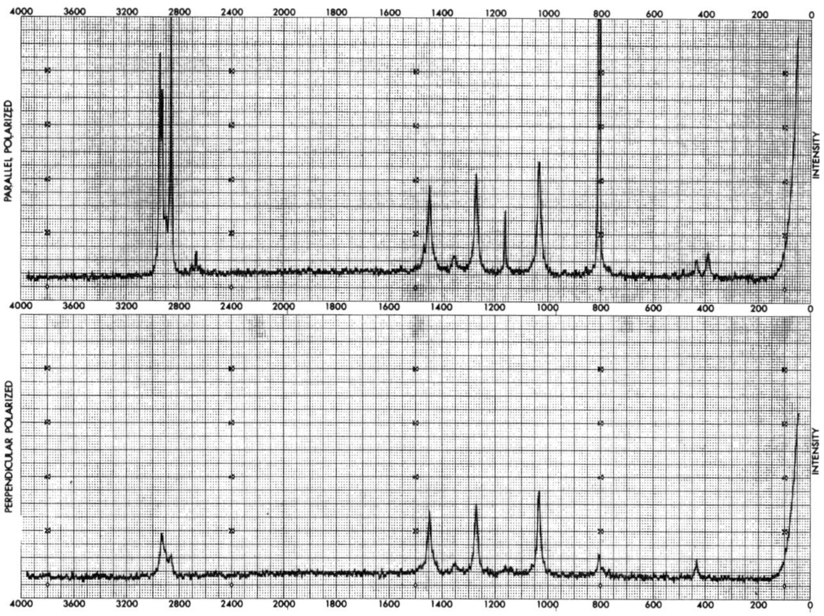

2,4-DIMETHYLPENTANE

793 R

RAMAN DATA

Spectrophotometer	Cary 83
Sensitivity	8750
	-
Suppression	0
	-
Laser Power	185 mW
Spectral Slitwidth	6 cm^{-1}

C_7H_{16} Mol. Wt. 100.21

B.P. 80.3°C

Source: Phillips Petroleum Company, Bartlesville, Oklahoma

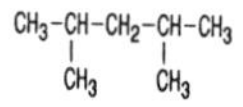

WAVENUMBER / INTENSITY

2920/89.6 2880/174.9 815/263.6

Capillary Cell: Neat

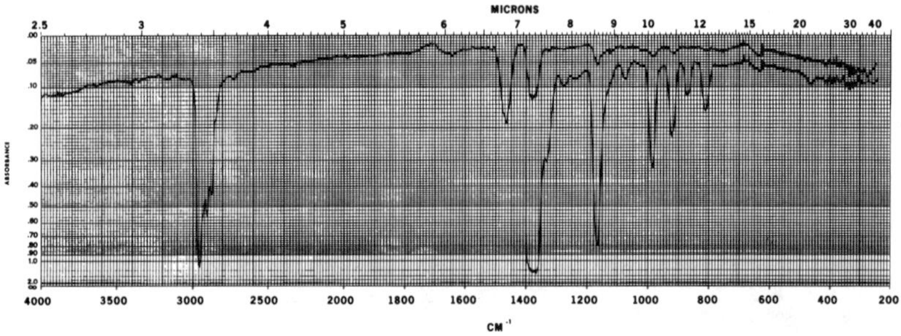

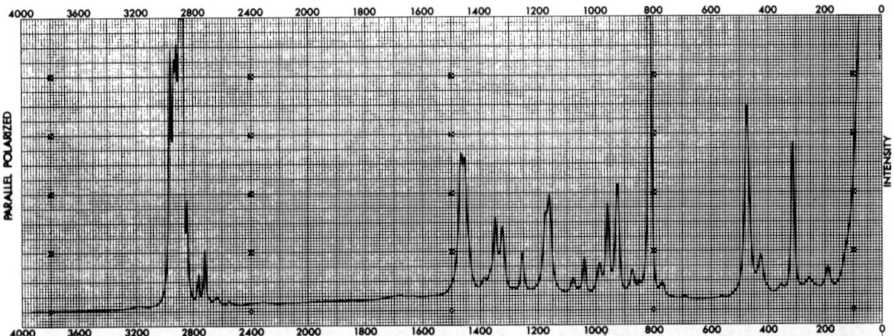

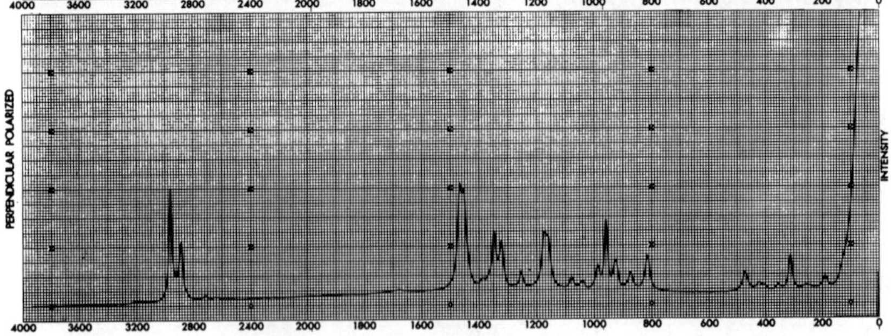

2205 R

2,2,3-TRIMETHYLPENTANE

C_8H_{18} Mol. Wt. 114.23

Source of Sample: E. Merck AG
Darmstadt, Germany

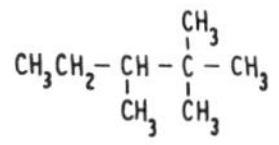

RAMAN DATA

Spectrophotometer	Cary 83
Sensitivity	20000; 2000
Suppression	0; 0
Laser Power	60 mW
Spectral Slitwidth	8 cm^{-1}

WAVENUMBER / INTENSITY

2940/88.3	2901/104.2	730/158.2

A=Capillary Cell: Neat
B=Capillary Cell: 0.015mm

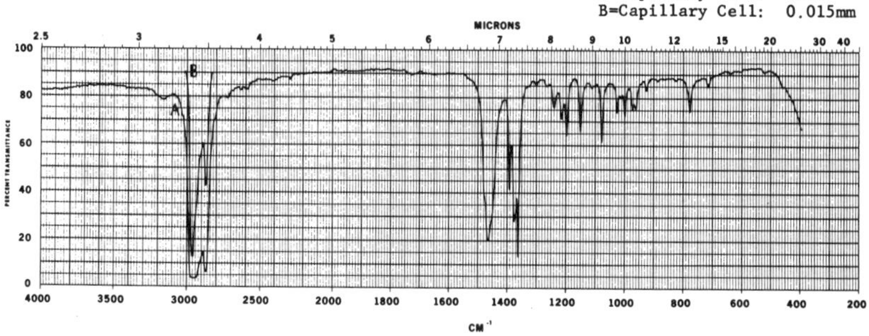

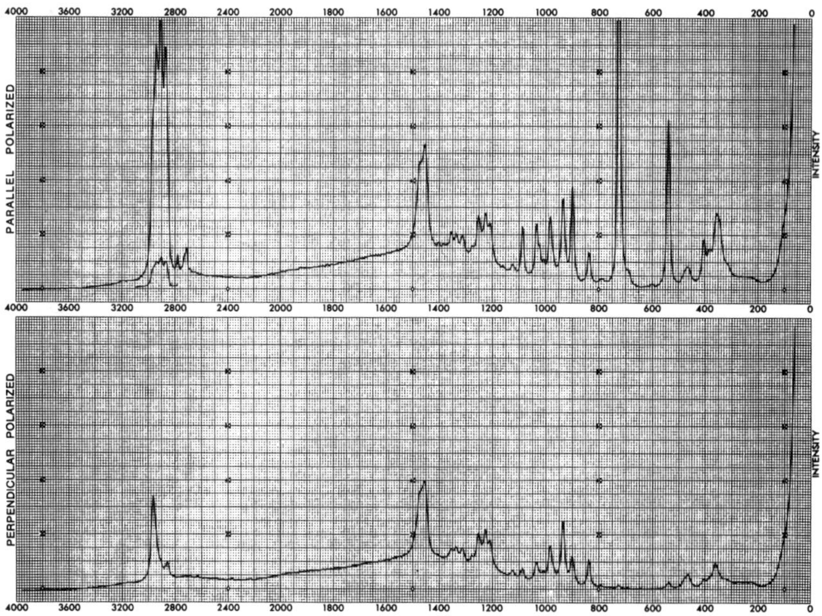

318 R

1-CHLOROHEXANE

$C_6H_{13}Cl$ Mol. Wt. 120.62

$CH_3CH_2(CH_2)_3CH_2-Cl$

B.P. 132-135°C

Source: Chem Service, Inc., West Chester, Pa.

RAMAN DATA

Spectrophotometer	Cary 83
Sensitivity	75000
	-
Suppression	40
	-
Laser Power	25 mW
Spectral Slitwidth	4 cm^{-1}

WAVENUMBER / INTENSITY

2870/94.9	1440/44.4	650/133.2

Capillary Cell: Neat

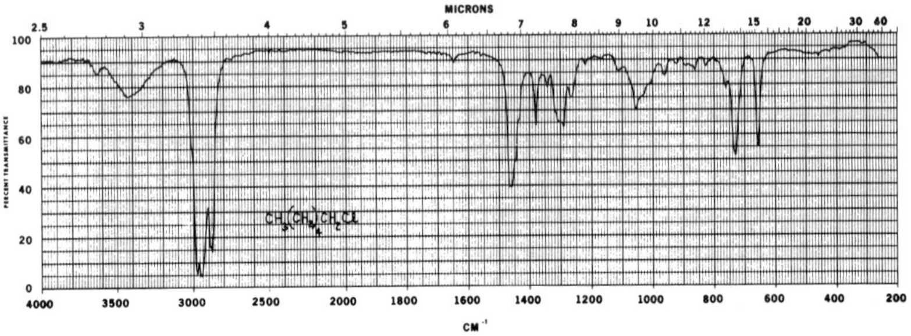

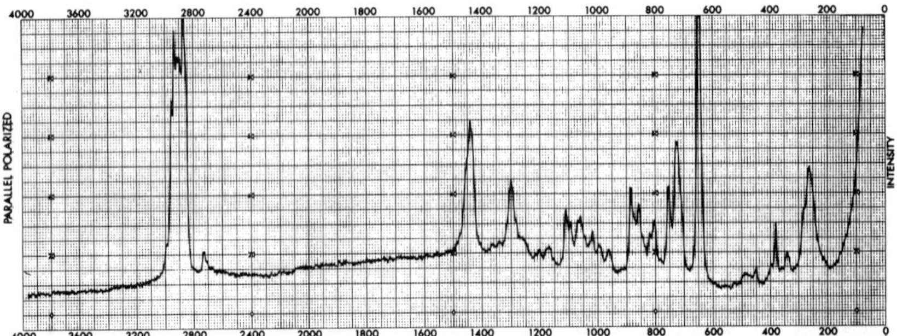

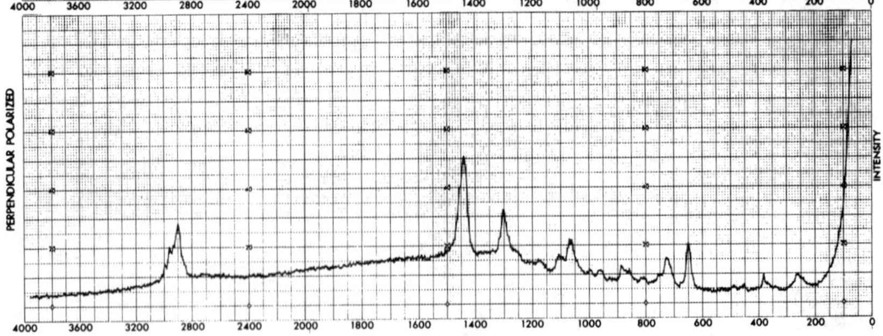

1-BROMOHEXANE

777 R

RAMAN DATA

Spectrophotometer	Cary 83
Sensitivity	75000
	-
Suppression	55
	-
Laser Power	25 mW
Spectral Slitwidth	6 cm^{-1}

$C_6H_{13}Br$ Mol. Wt. 165.08

$CH_3CH_2(CH_2)_3CH_2{-}Br$

B.P. 75-78°C/65 mm

Source: MCB Manufacturing Chemists, Norwood, Ohio

WAVENUMBER / INTENSITY		
2940/87.7	2870/93.7	1436/44.0
641/120.4	560/213.5	

Capillary Cell: Neat

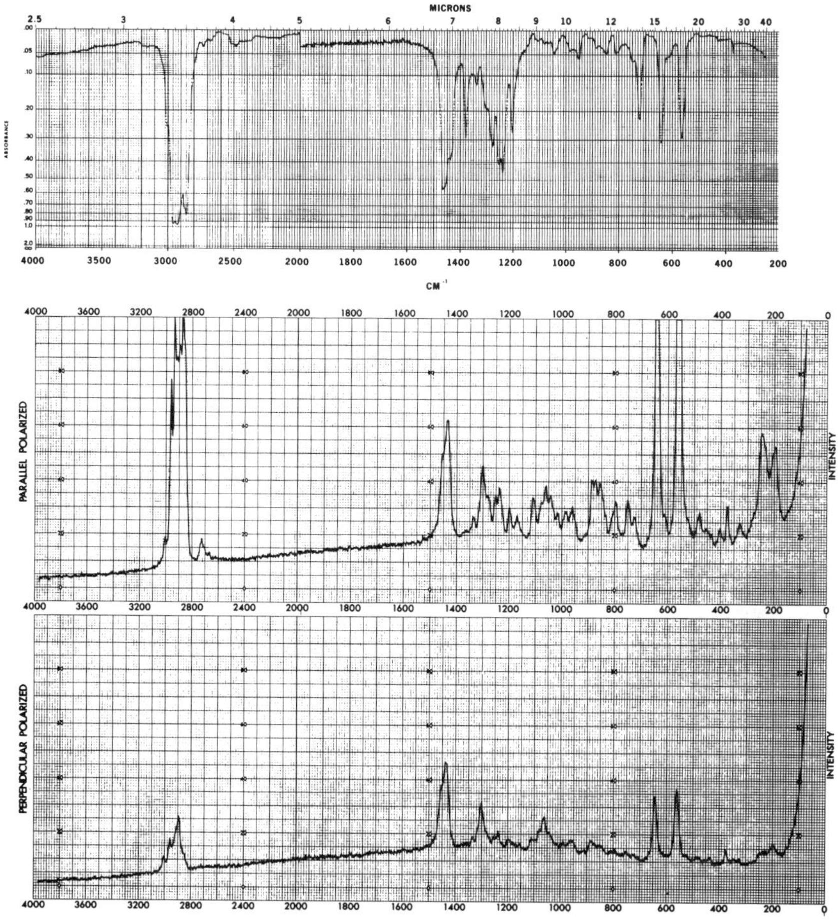

HEXACHLOROETHANE

309 R

RAMAN DATA

Spectrophotometer	Cary 83
Sensitivity	50000
	-
Suppression	0
	-
Laser Power	11 mW
Spectral Slitwidth	6 cm^{-1}

C_2Cl_6 Mol. Wt. 236.74

CCl_3-CCl_3

M.P. 185-186°C

Source: Chem Service, Inc., West Chester, Pa.

WAVENUMBER / INTENSITY

850/43.0 429/132.8 337/130.4
219/227.3

KBr Wafer

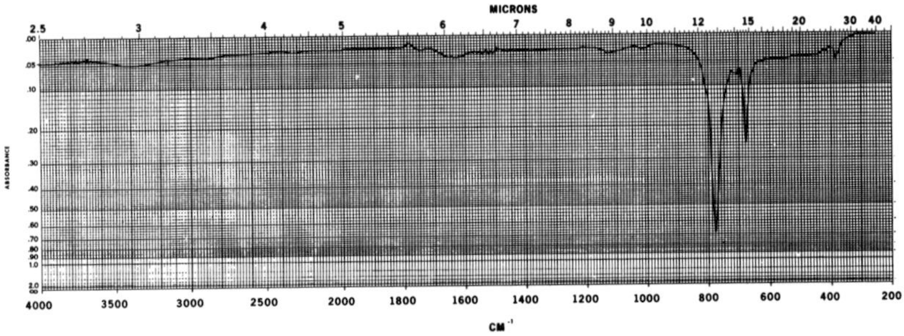

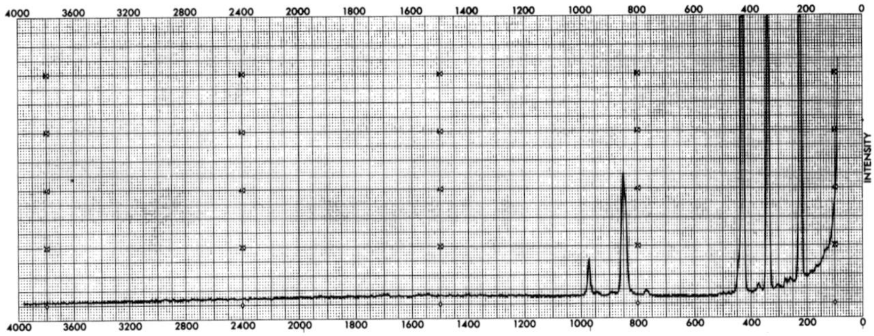

(TRIFLUOROMETHYL) UNDECAFLUOROCYCLOHEXANE

1150 R

RAMAN DATA

Spectrophotometer	Cary 83
Sensitivity	8500
	-
Suppression	0
	-
Laser Power	140 mW
Spectral Slitwidth	6 cm^{-1}

C_7H_{14} Mol. Wt. 350.06

M. P. 37°C B. P. 76°C

Source: Aldrich Chemical Company, Inc.
Milwaukee, Wisconsin

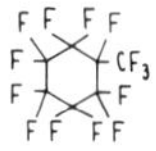

WAVENUMBER / INTENSITY

737/59.7 693/139.5 340/127.7

A=Capillary Cell: Neat
B=Capillary Cell: 0.015 mm

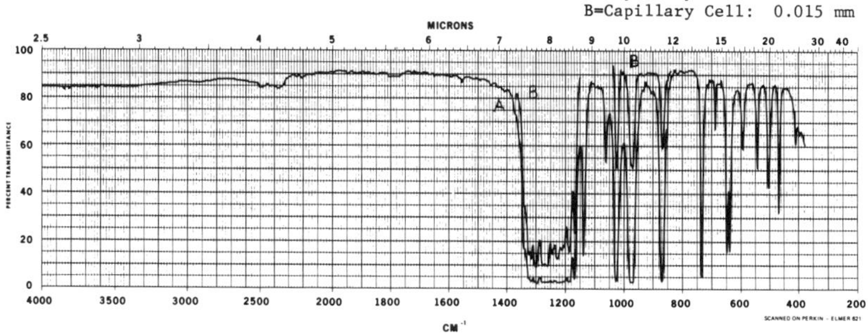

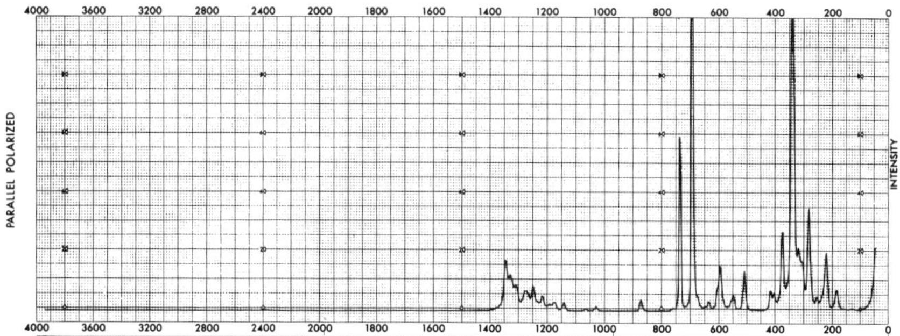

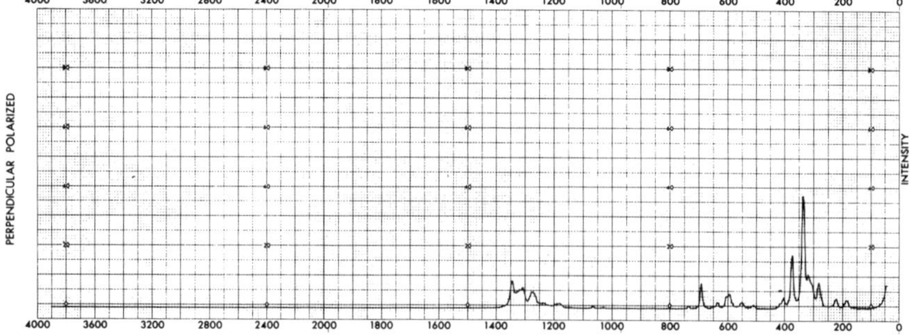

180 R

BUTYL ALCOHOL

RAMAN DATA

Spectrophotometer	Cary 83
Sensitivity	100000
	-
Suppression	0
	-
Laser Power	12 mW
Spectral Slitwidth	8 cm^{-1}

$C_4H_{10}O$ Mol. Wt. 74.12

$CH_3CH_2CH_2CH_2$-OH

B.P. 116-118°C

Source: Chem Service, Inc., West Chester, Pa.

WAVENUMBER / INTENSITY

2938/97.0 2905/101.2 2870/133.0
823/69.7

Capillary Cell: Neat

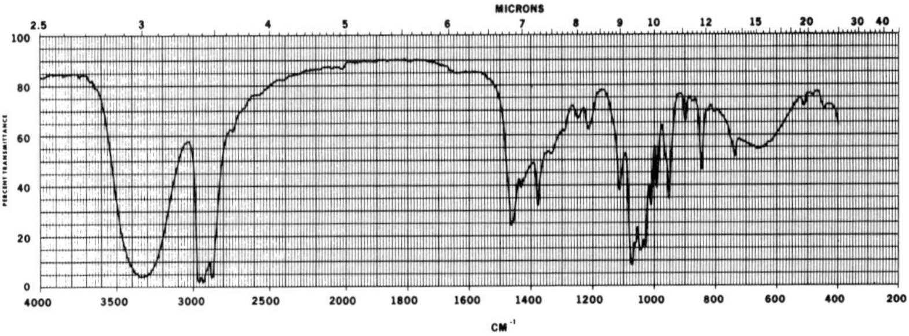

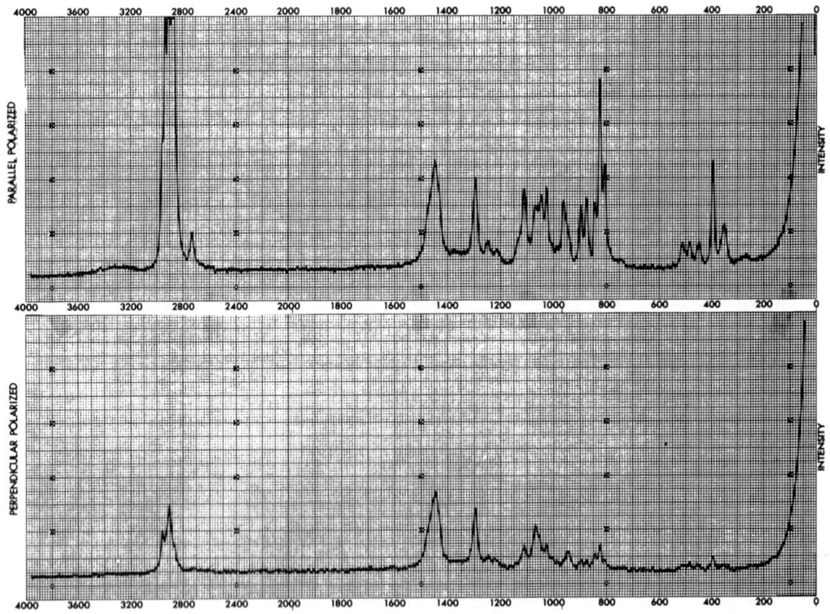

209 R

sec-BUTYL ALCOHOL

RAMAN DATA

Spectrophotometer	Cary 83
Sensitivity	100000
	-
Suppression	45
	-
Laser Power	12 mW
Spectral Slitwidth	8 cm-1

$C_4H_{10}O$ Mol. Wt. 74.12

B.P. 98-100°C

Source: Chem Service, Inc., West Chester, Pa.

$CH_3CH_2-CH-OH$ with CH_3 on the CH

WAVENUMBER / INTENSITY		
2925/110.8	2873/109.1	818/122.4
772/69.4		

Capillary Cell: Neat

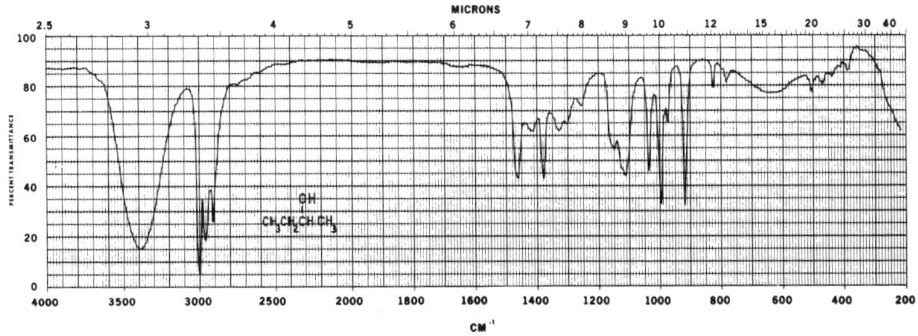

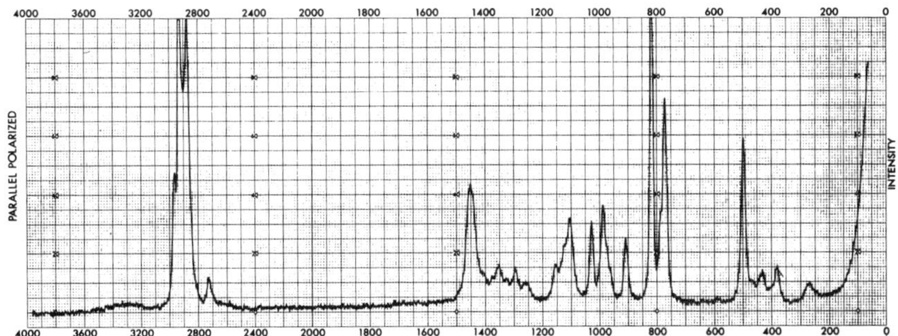

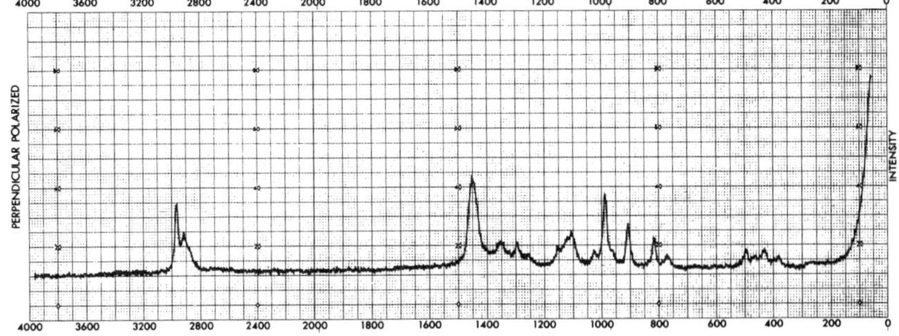

213 R

tert-BUTYL ALCOHOL

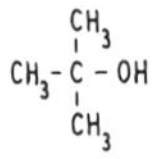

$C_4H_{10}O$ Mol. Wt. 74.12

M.P. 24-25.5°C

Source: Chem Service, Inc., West Chester, Pa.

RAMAN DATA

Spectrophotometer	Cary 83
Sensitivity	100000
	-
Suppression	50
	-
Laser Power	12 mW
Spectral Slitwidth	6 cm-1

WAVENUMBER / INTENSITY

2918/69.8 749/285.4

Capillary Cell: Neat

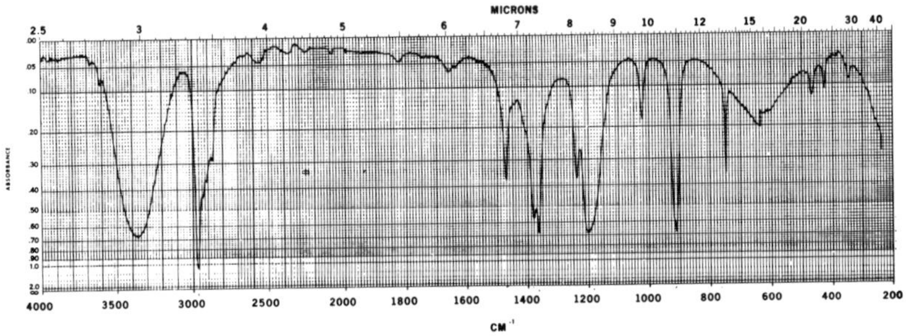

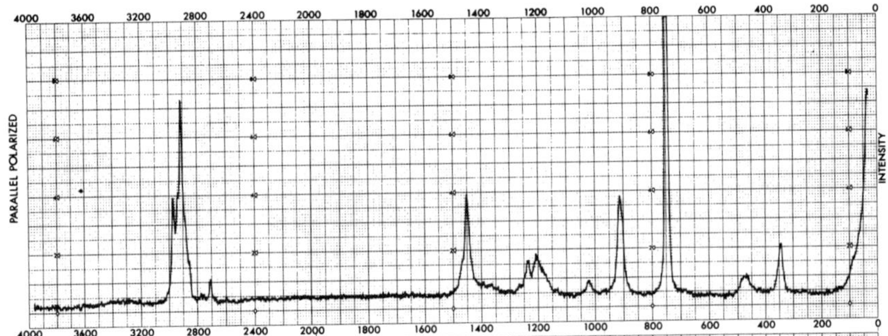

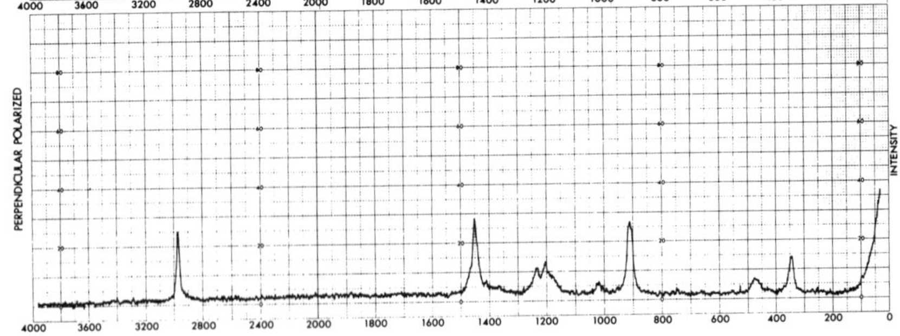

650 R

PHENOL

RAMAN DATA

Spectrophotometer	Cary 83
Sensitivity	11000;
	1100
Suppression	50;
	0
Laser Power	130 mW
Spectral Slitwidth	6 cm^{-1}

C_6H_6O Mol. Wt. 94.11

M.P. 42.5°C

Source: MCB Manufacturing Chemists, Norwood, Ohio

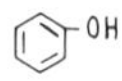

WAVENUMBER / INTENSITY

1005/237.1 820/90.4 280/366.3
250/130.2

Capillary Cell: Melt

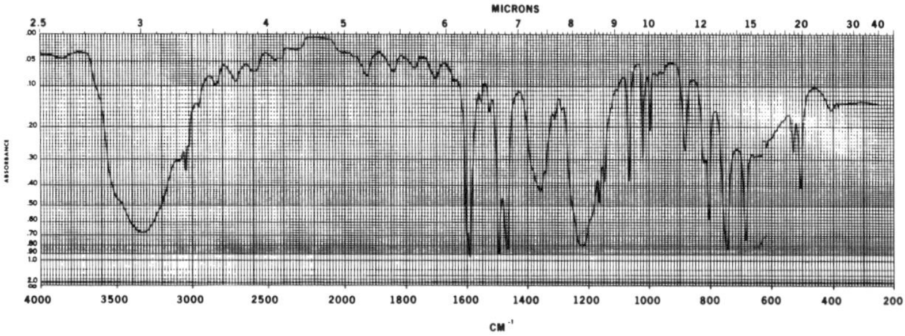

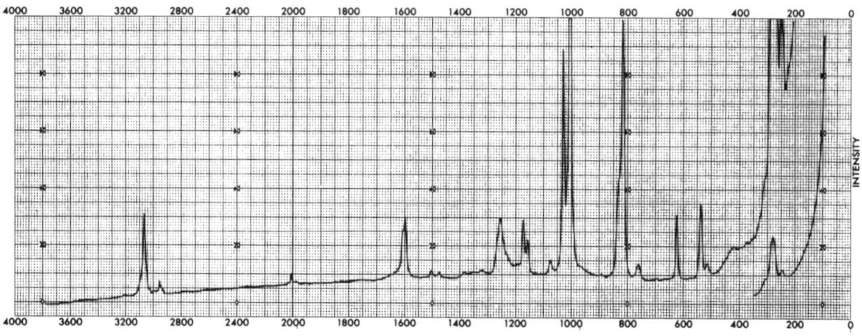

260 R BUTYL ETHER

RAMAN DATA

Spectrophotometer	Cary 83
Sensitivity	30000; 5000
Suppression	32; 0
Laser Power	24 mW
Spectral Slitwidth	6 cm^{-1}

$C_8H_{18}O$ Mol. Wt. 130.23

B.P. 138-141°C

Source: Chem Service, Inc., West Chester, Pa.

$CH_3CH_2CH_2CH_2-O-CH_2CH_2CH_2CH_3$

WAVENUMBER / INTENSITY		
2940/147.4	2915/153.6	2875/195.3
1450/58.9	838/100.9	

Capillary Cell: Neat

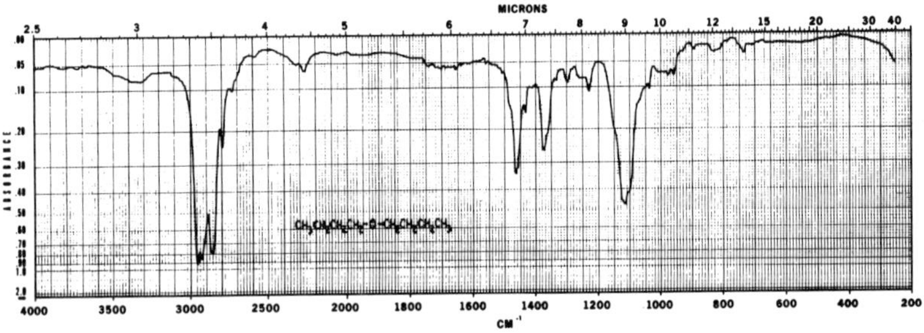

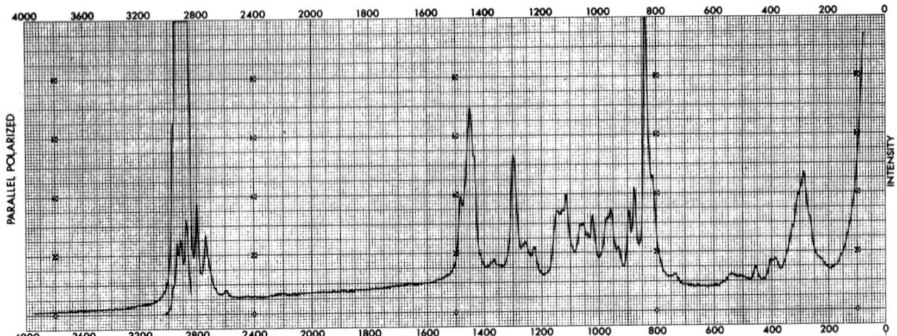

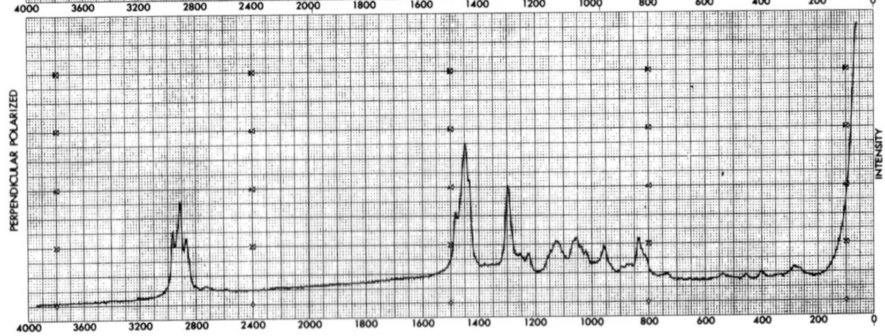

tert-BUTYL METHYL ETHER

1487 R

$C_5H_{12}O$ Mol. Wt. 88.15

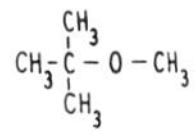

RAMAN DATA

Spectrophotometer	Cary 83
Sensitivity	15000
	-
Suppression	0
	-
Laser Power	140 mW
Spectral Slitwidth	6 cm^{-1}

Source: Aldrich Chemical Company, Inc.
Milwaukee, Wisconsin

WAVENUMBER / INTENSITY

2980/75.6 2932/132.0 728/545.9

Capillary Cell: .015 mm

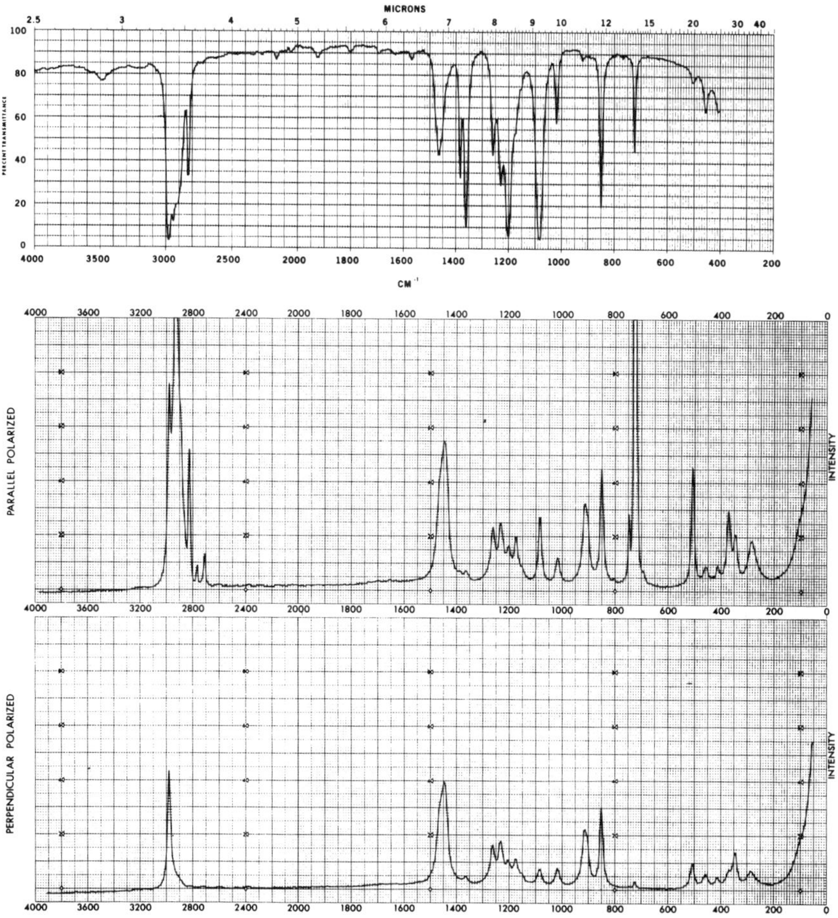

704 R

p-DIOXANE

RAMAN DATA

Spectrophotometer	Cary 83
Sensitivity	25000
	-
Suppression	35
	-
Laser Power	24 mW
Spectral Slitwidth	4 cm^{-1}

$C_4H_8O_2$ Mol. Wt. 88.11

B.P. 101-102°C

Source: Chem Service, Inc., West Chester, Pa.

WAVENUMBER / INTENSITY

2849/56.0 830/291.5

Capillary Cell: Neat

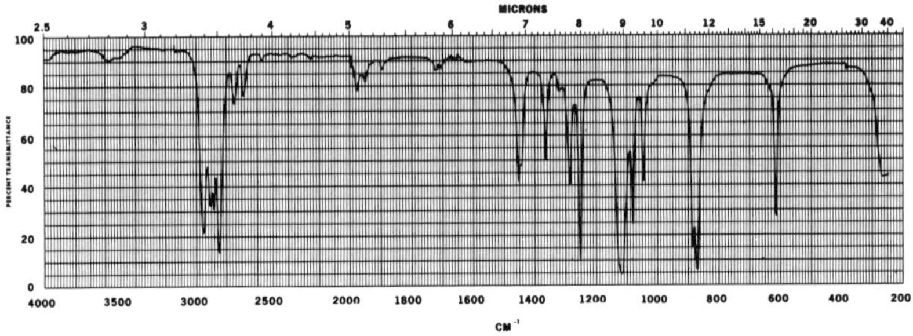

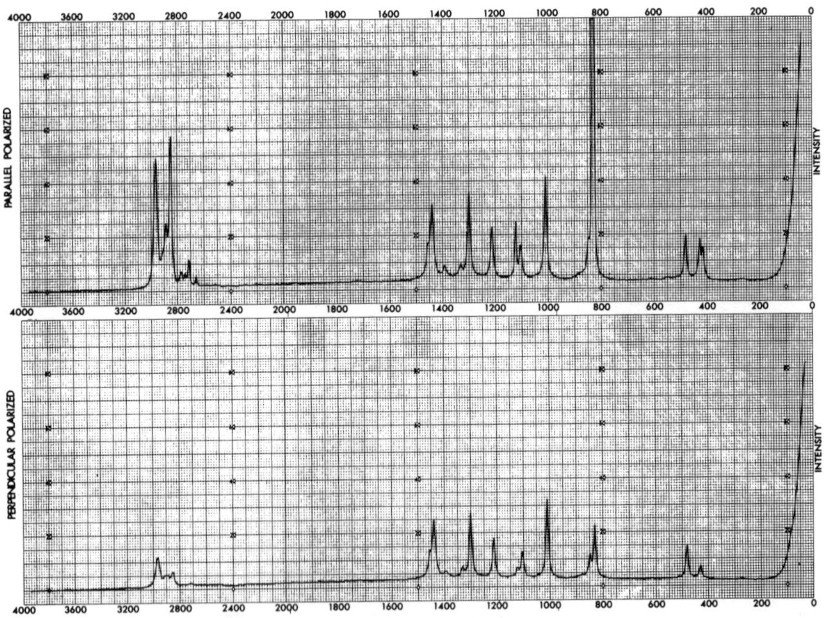

PROPYLENE OXIDE **287 R**

RAMAN DATA

Spectrophotometer	Cary 83
Sensitivity	62500
	-
Suppression	62
	-
Laser Power	24 mW
Spectral Slitwidth	5 cm^{-1}

C_3H_6O Mol. Wt. 58.08

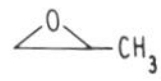

B.P. 34-35°C

Source: Chem Service, Inc.,
West Chester, Pa.

WAVENUMBER / INTENSITY

2930/106.9 1262/212.2 946/145.0
742/64.6

Gas Cell: 10 cm

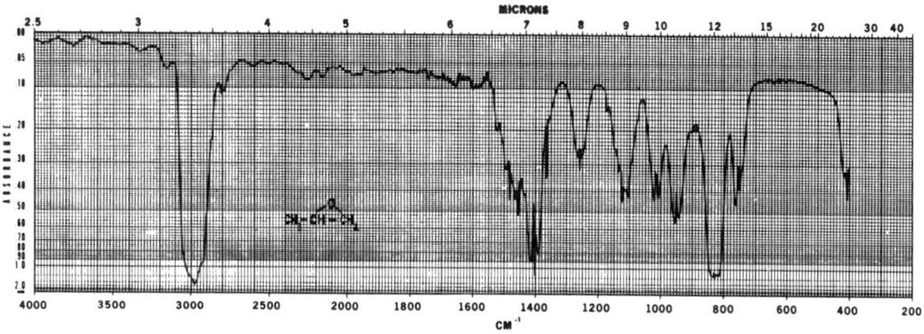

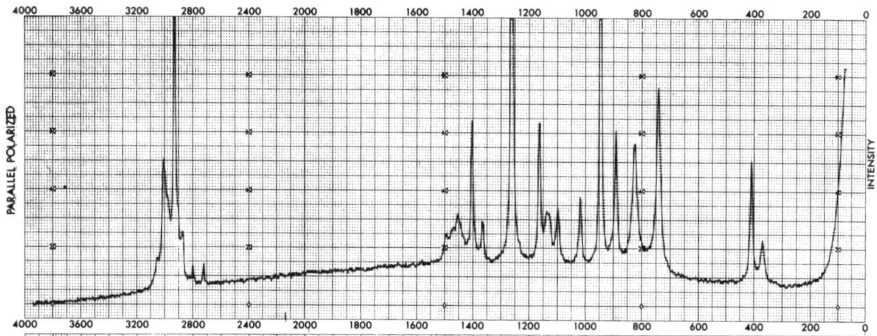

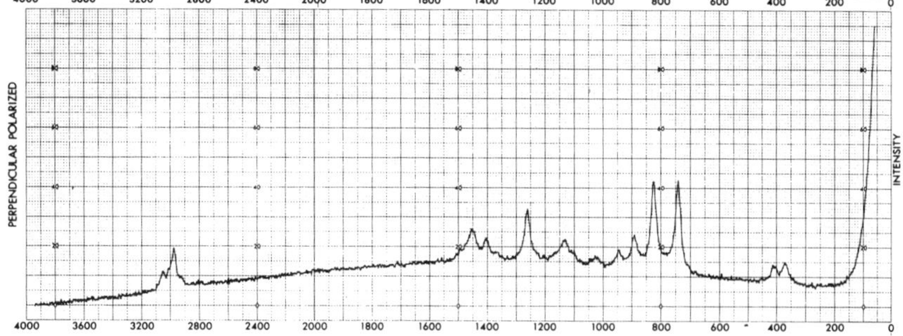

265 R

ANISOLE

C_7H_8O Mol. Wt. 108.14

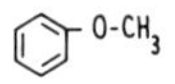

B.P. 152-154°C

Source: Chem Service, Inc., West Chester, Pa.

RAMAN DATA

Spectrophotometer	Cary 83
Sensitivity	30000
	-
Suppression	32
	-
Laser Power	24 mW
Spectral Slitwidth	4 cm^{-1}

WAVENUMBER / INTENSITY

1241/90.0	1017/109.8
990/608.5	777/291.5

Capillary Cell: Neat

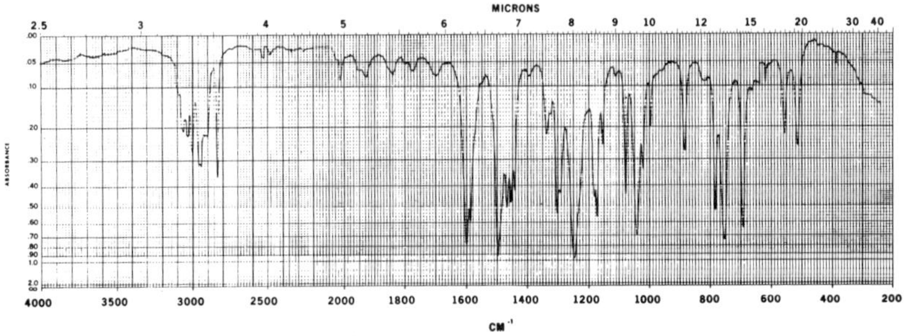

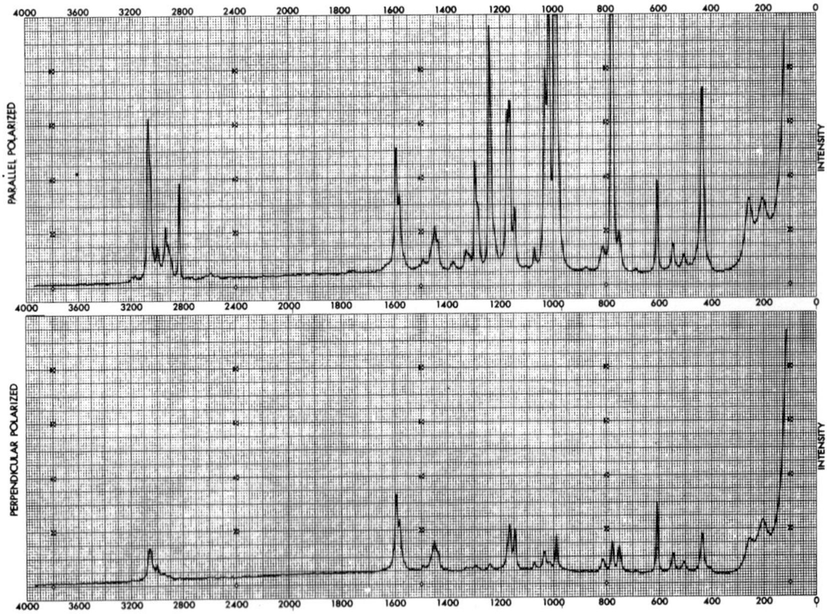

696 R

ACETALDEHYDE, DIETHYL ACETAL

RAMAN DATA

Spectrophotometer	Cary 83
Sensitivity	50000
	-
Suppression	160
	-
Laser Power	22 mW
Spectral Slitwidth	6 cm-1

$C_6H_{14}O_2$ Mol. Wt. 118.18

B.P. 102-104°C

Source: Chem Service, Inc., West Chester, Pa.

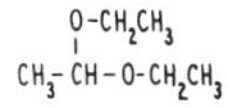

WAVENUMBER / INTENSITY	
2935/101.2	2870/55.1

A=Capillary Cell: Neat
B=Capillary Cell: 0.015 mm

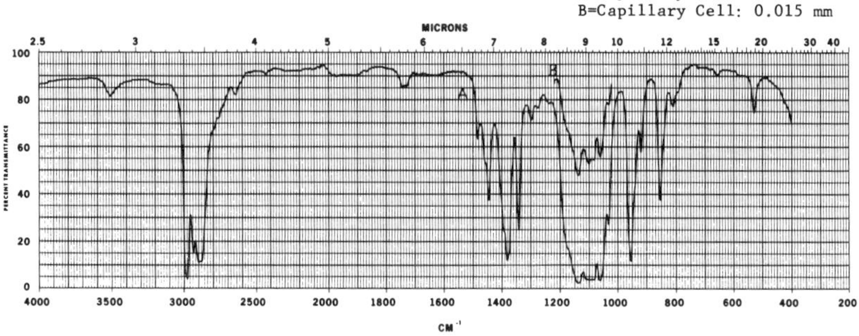

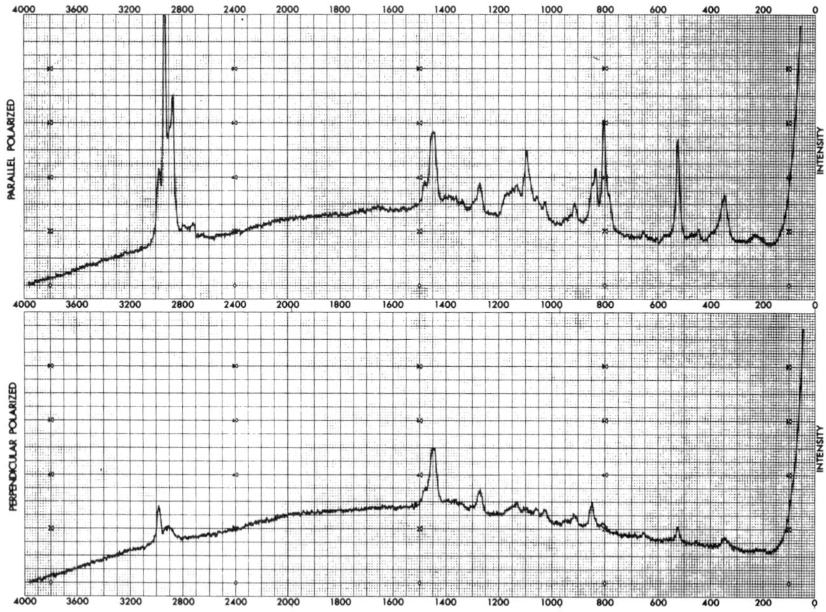

404 R

DI-tert-BUTYL PEROXIDE

$C_8H_{18}O_2$ Mol. Wt. 146.23

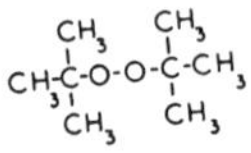

RAMAN DATA

Spectrophotometer	Cary 83
Sensitivity	5500
Suppression	0
Laser Power	109 mW
Spectral Slitwidth	6 cm^{-1}

Source: H. G. Beachell
University of Delaware

WAVENUMBER / INTENSITY		
2935/119.0	906/86.0	886/144.1
776/270.3	517/123.5	

Capillary Cell: 0.015 mm

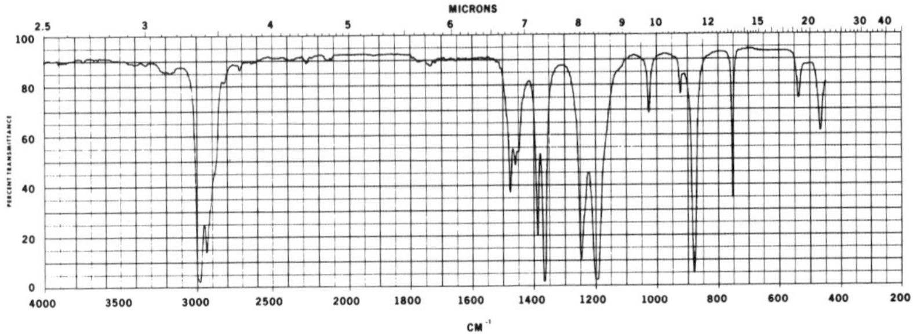

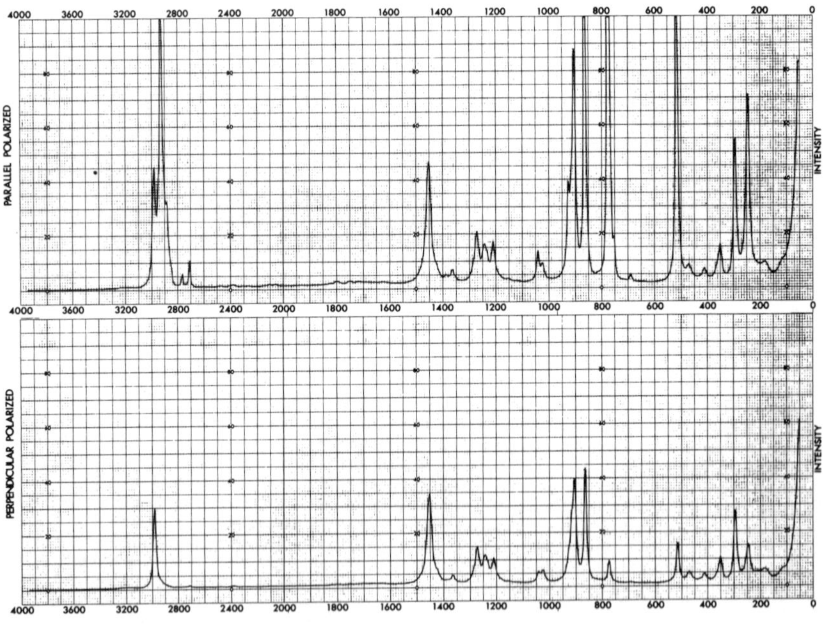

1-DODECENE **1511 R**

RAMAN DATA

Spectrophotometer	Cary 83
Sensitivity	15000;
	1500
Suppression	17;
	0
Laser Power	140 mW
Spectral Slitwidth	6 cm^{-1}

$C_{12}H_{24}$ Mol. Wt. 168.33

B.P. 95-96°C/15mm

Source: Eastman Organic Chemicals
Rochester, New York

$CH_3CH_2(CH_2)_7CH_2$-$CH{=}CH_2$

WAVENUMBER / INTENSITY

2900/95.1	2875/96.2	2853/131.1
1641/114.8	1300/71.0	

Capillary Cell: Neat

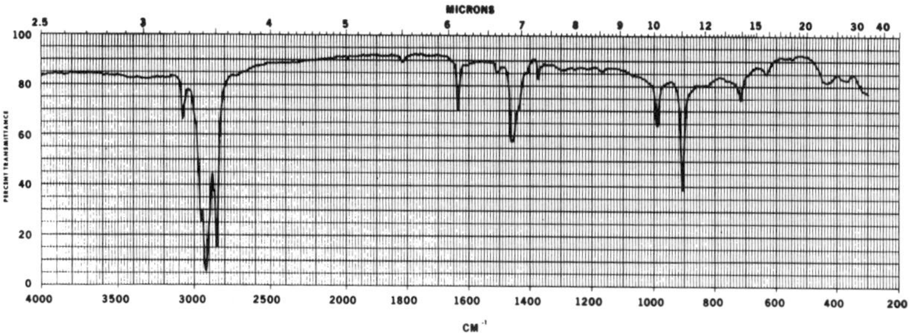

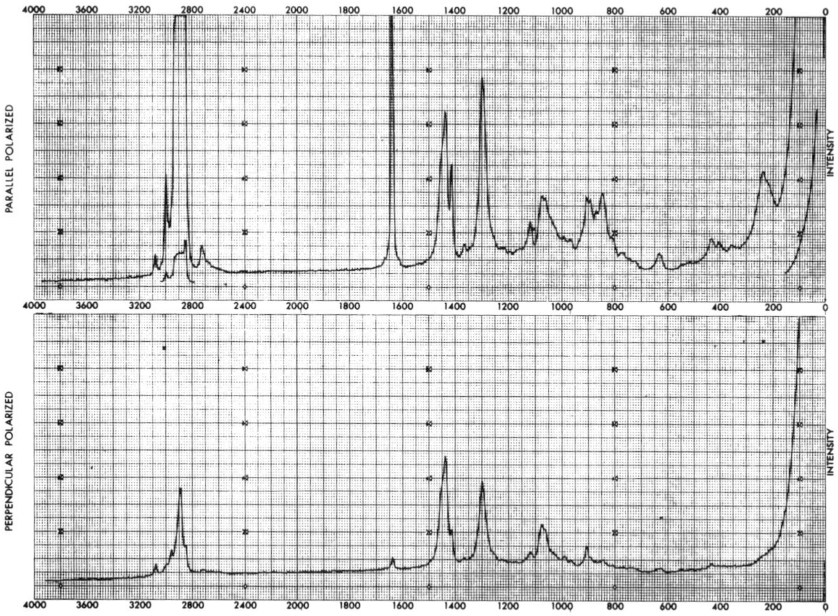

2-METHYL-1-HEPTENE

721 R

RAMAN DATA

Spectrophotometer	Cary 83
Sensitivity	16000;
	1600
Suppression	30;
	0
Laser Power	12 mW
Spectral Slitwidth	7 cm^{-1}

C_8H_{16} Mol. Wt. 112.22

$CH_3CH_2CH_2CH_2CH_2-C(CH_3)=CH_2$

Source: MCB Manufacturing Chemists, Norwood, Ohio

WAVENUMBER / INTENSITY

2990/54.0	2920/128.5	2880/132.7
1656/158.1		

A=Capillary Cell: Neat
B=Capillary Cell: 0.015 mm

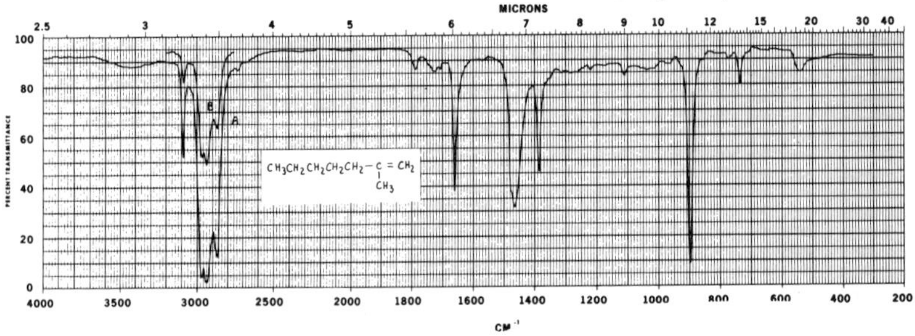

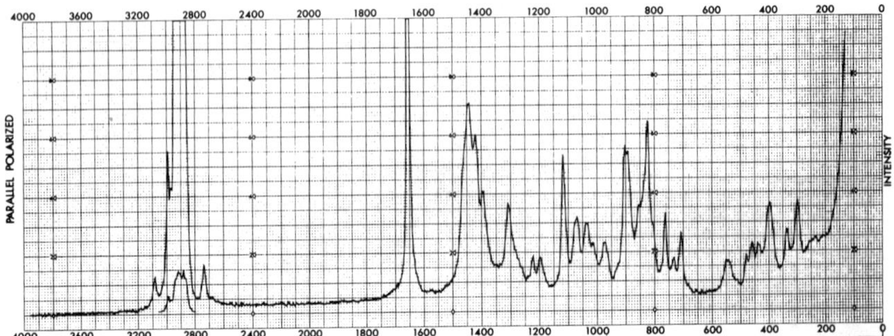

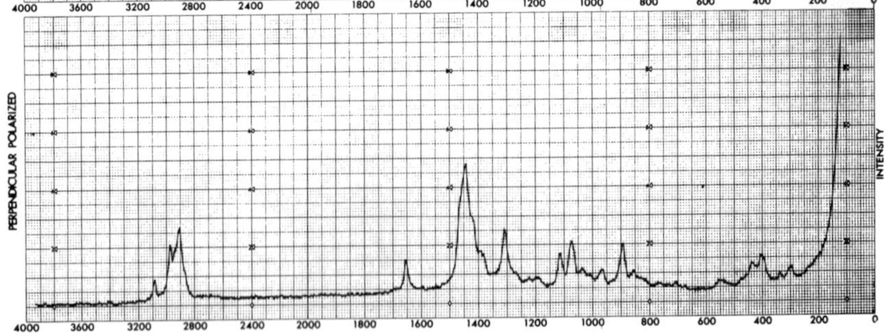

2798 R

cis-3-HEXENE

RAMAN DATA

Spectrophotometer	Cary 83
Sensitivity	6000;
	600
Suppression	18;
	-
Laser Power	150 mW
Spectral Slitwidth	6 cm^{-1}

C_6H_{12} Mol. Wt. 84.16

B.P. 67°C

Source of Sample: Tokyo Kasei Kogyo Co. Ltd.
Tokyo, Japan

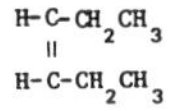

WAVENUMBER / INTENSITY

2935/116.5 2876/89.1 1662/181.6
853/129.6

Gas Cell: 10cm

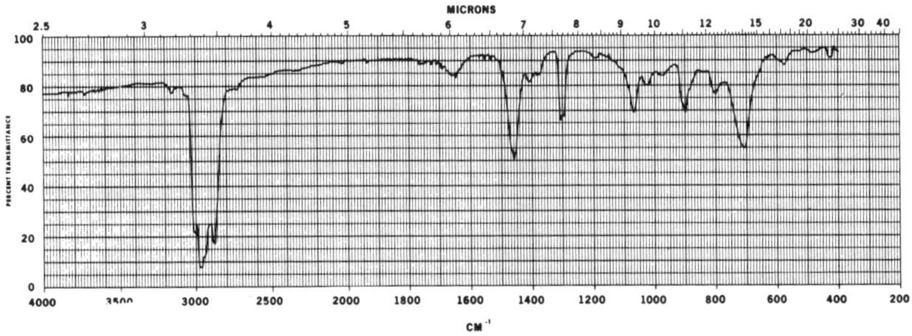

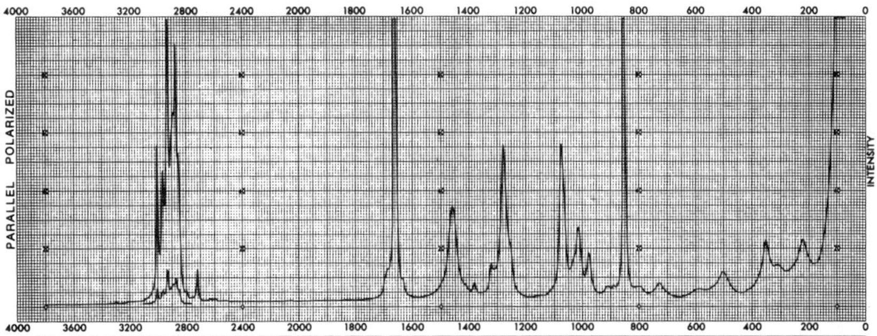

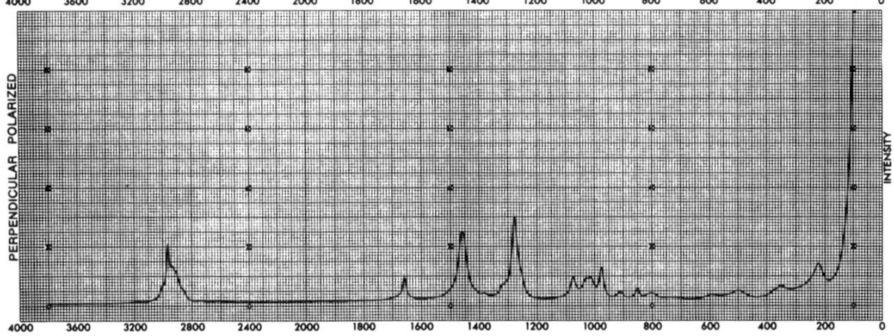

2790 R

trans-3-HEXENE

C_6H_{12} Mol. Wt. 84.16

B.P. 67°C

Source of Sample: Tokyo Kasei Kogyo Co. Ltd. Tokyo, Japan

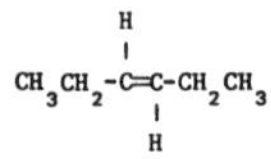

RAMAN DATA

Spectrophotometer	Cary 83
Sensitivity	5000; 500
Suppression	28; -
Laser Power	150 mW
Spectral Slitwidth	6 cm^{-1}

WAVENUMBER / INTENSITY

2937/95.3 2894/65.2 1679/130.4

Gas Cell: 10cm

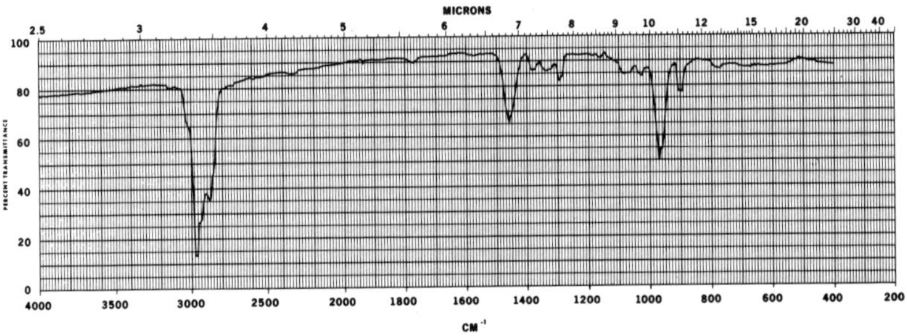

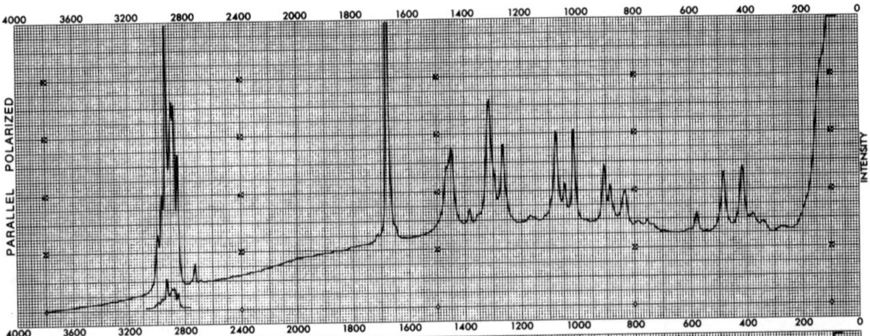

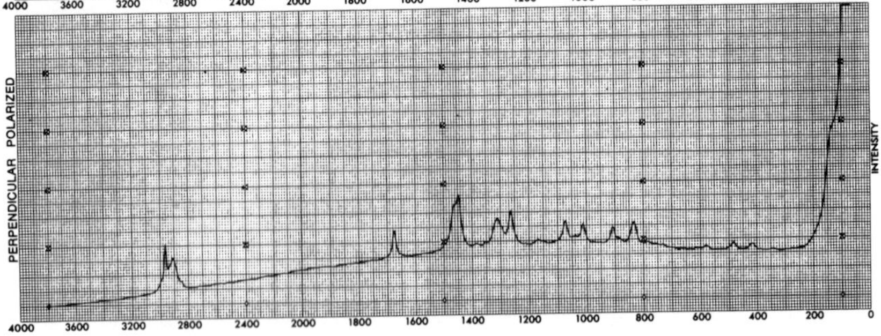

3092 R

2-METHYL-2-PENTENE

RAMAN DATA

Spectrophotometer	Cary 83
Sensitivity	12000; 1200
Suppression	0; -
Laser Power	78 mW
Spectral Slitwidth	8 cm^{-1}

C_6H_{12} Mol. Wt. 84.16

B.P. 152.3°F

Source of Sample: Phillips Petroleum Company
Bartlesville, Oklahoma

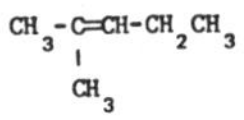

WAVENUMBER / INTENSITY

2900/107.8	2862/88.3	1673/158.8

Capillary Cell: 0.015mm

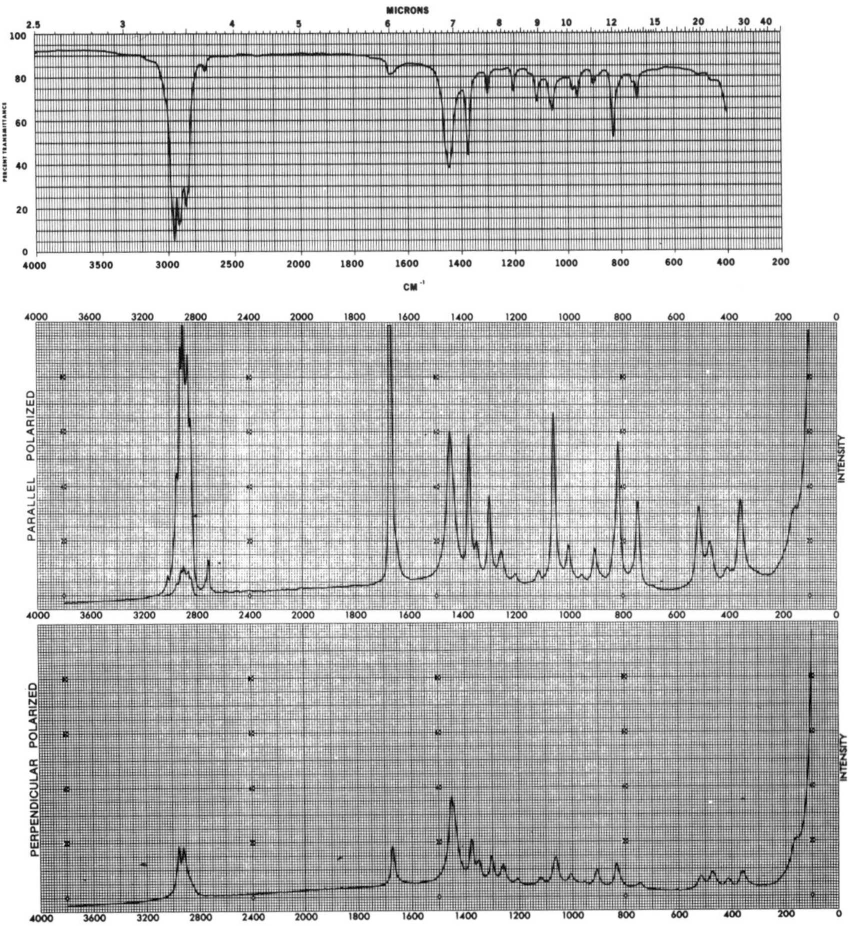

2775 R

1-PENTADECYNE

$C_{15}H_{28}$ Mol. Wt. 208.39

$CH_3CH_2(CH_2)_{10}CH_2-C\equiv CH$

Source of Sample: Farchan Research Labs. Willoughby, Ohio

RAMAN DATA

Spectrophotometer	Cary 83
Sensitivity	15000; 1500
Suppression	6; -
Laser Power	150 mW
Spectral Slitwidth	6 cm^{-1}

WAVENUMBER / INTENSITY

2902/125.1	2870/114.8	2852/157.4
2121/147.0	1448/69.5	

Capillary Cell: Neat

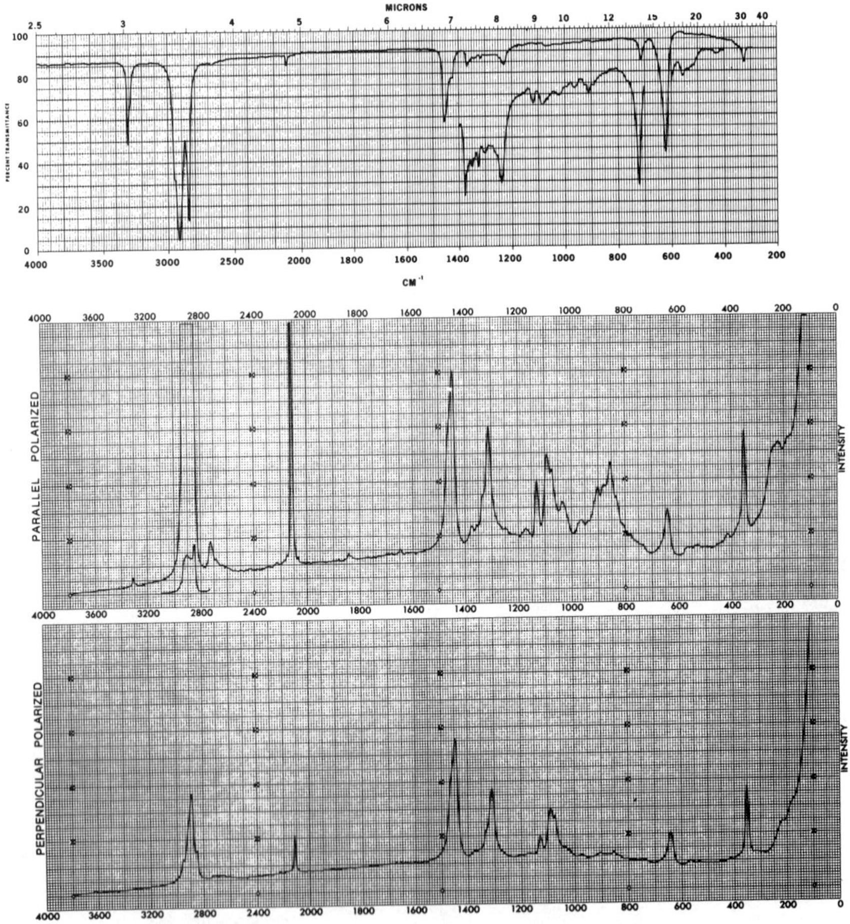

2779 R

6-DODECYNE

$C_{12}H_{22}$ Mol. Wt. 166.31

Source of Sample: Farchan Research Labs. Willoughby, Ohio

$CH_3CH_2(CH_2)_2CH_2-C\equiv C-CH_2(CH_2)_2CH_2CH_3$

RAMAN DATA

Spectrophotometer	Cary 83
Sensitivity	12000; 1200
Suppression	6; -
Laser Power	150 mW
Spectral Slitwidth	6 cm^{-1}

WAVENUMBER / INTENSITY

2915/157.6	2900/155.2	2875/141.4
2861/117.5	2234/102.5	1449/77.3

Capillary Cell: Neat

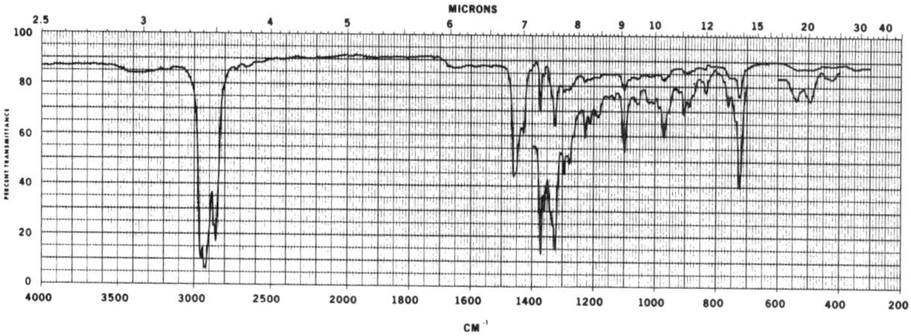

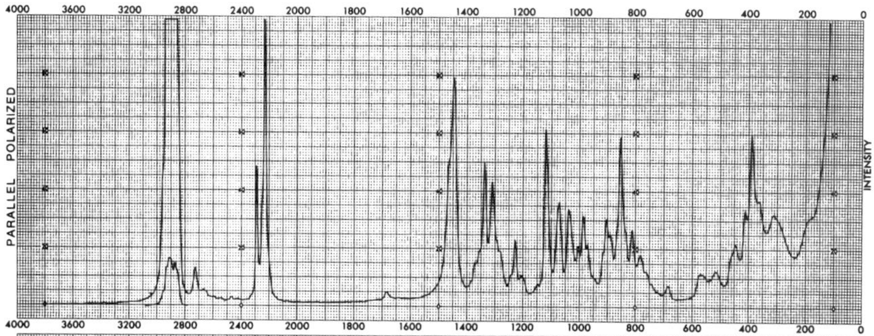

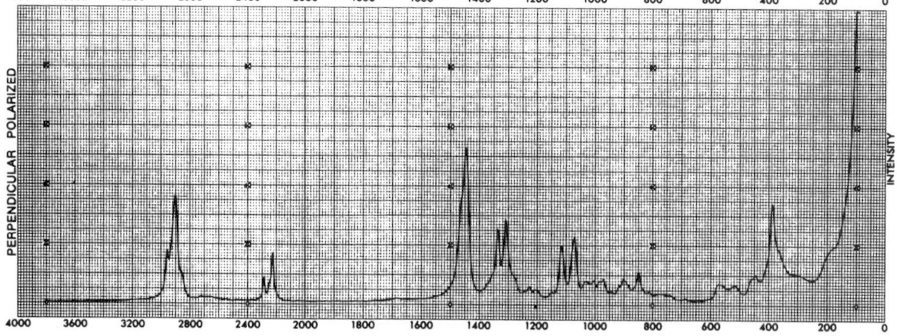

157 R

ACRYLONITRILE

C_3H_3N Mol. Wt. 53.06

$CH_2=CH-C\equiv N$

B.P. 73-77°C

Source: Chem Service, Inc., West Chester, Pa.

RAMAN DATA

Spectrophotometer	Cary 83
Sensitivity	37500; 25000
Suppression	90; 0
Laser Power	9 mW
Spectral Slitwidth	6 cm^{-1}

WAVENUMBER / INTENSITY

2230/91.2 1610/119.3 1415/34.1

Capillary Cell: Neat

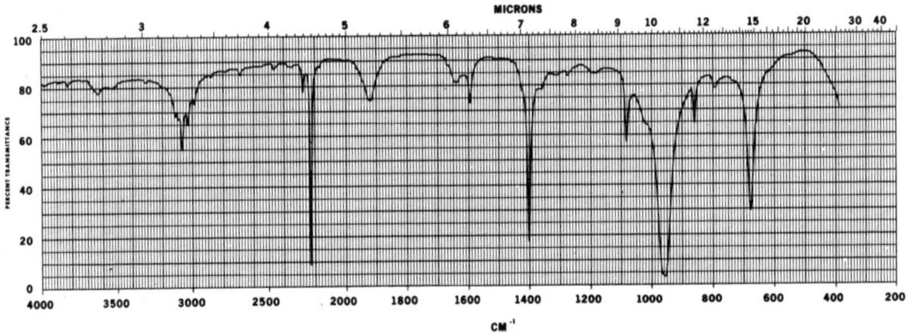

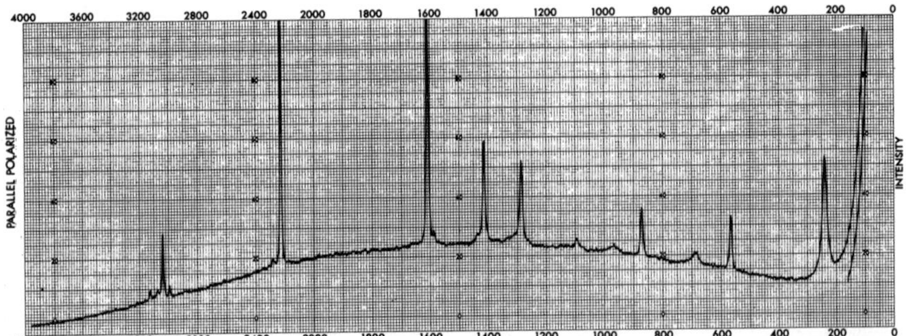

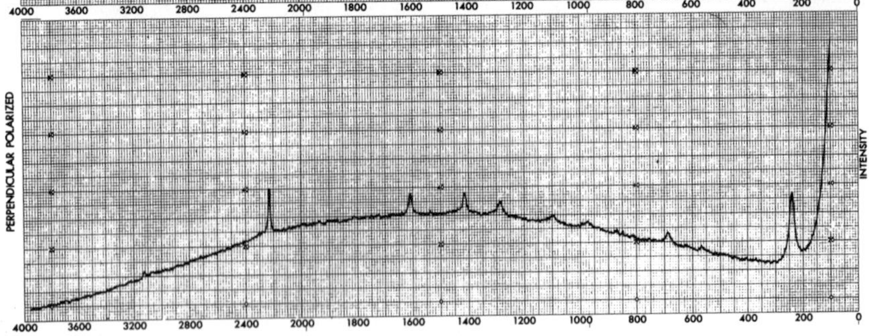

BUTYRONITRILE

575 R

RAMAN DATA

Spectrophotometer	Cary 83
Sensitivity	14000;
	1400
Suppression	35;
	0
Laser Power	210 mW
Spectral Slitwidth	6 cm^{-1}

C_4H_7N Mol. Wt. 69.11

B.P. 175.5°C

Source: Eastman Organic Chemicals, Rochester, New York

$CH_3CH_2CH_2-C\equiv N$

WAVENUMBER / INTENSITY		
2935/239.1	2880/144.6	2250/227.2
876/76.8	846/275.7	

Capillary Cell: 0.015 mm

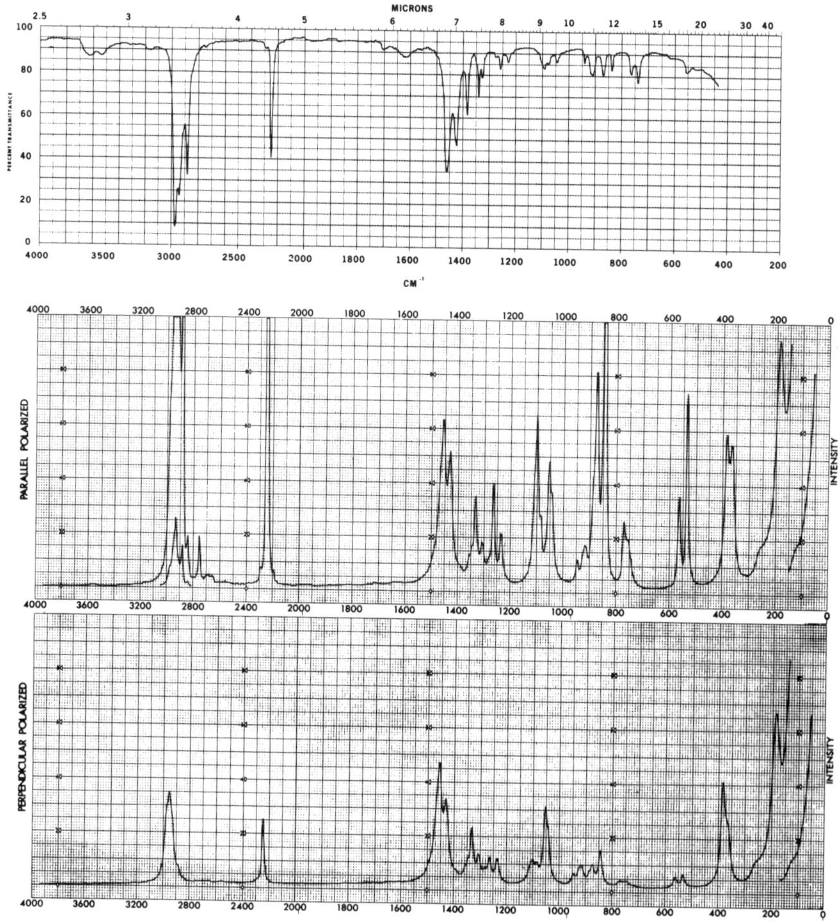

FUMARONITRILE

975 R

RAMAN DATA	
Spectrophotometer	Cary 83
Sensitivity	7000;
	700
Suppression	17;
	0
Laser Power	85 mW
Spectral Slitwidth	6 cm^{-1}

$C_4H_2N_2$ Mol. Wt. 78.07

M. P. 95-97°C B. P. 186°C

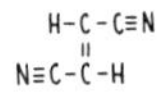

Source: Aldrich Chemical Company, Inc.
Milwaukee, Wisconsin

WAVENUMBER / INTENSITY		
2228/300.0	1612/273.7	1298/83.0
264/99.4	90/141.0	

KBr Wafer

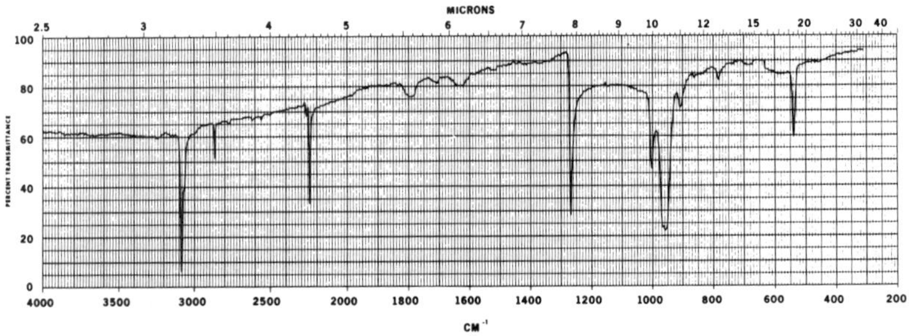

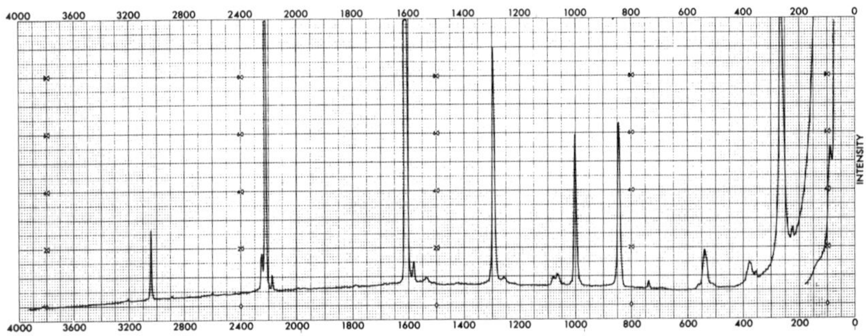

BENZONITRILE

1403 R

RAMAN DATA

Spectrophotometer	Cary 83
Sensitivity	2700
	-
Suppression	15
	-
Laser Power	135 mW
Spectral Slitwidth	6 cm^{-1}

C_7H_5N Mol. Wt. 103.12

B.P. 188-189°C

Source: MCB Manufacturing Chemists
Norwood, Ohio

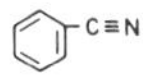

WAVENUMBER / INTENSITY

2226/158.5 2600/94.1 1180/78.7
1004/212.9

Capillary Cell: Neat

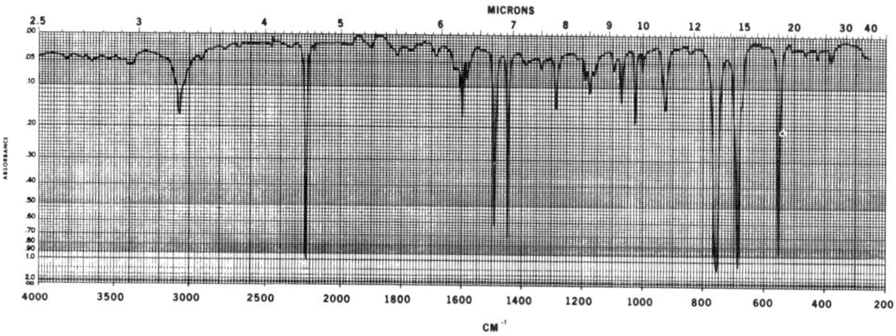

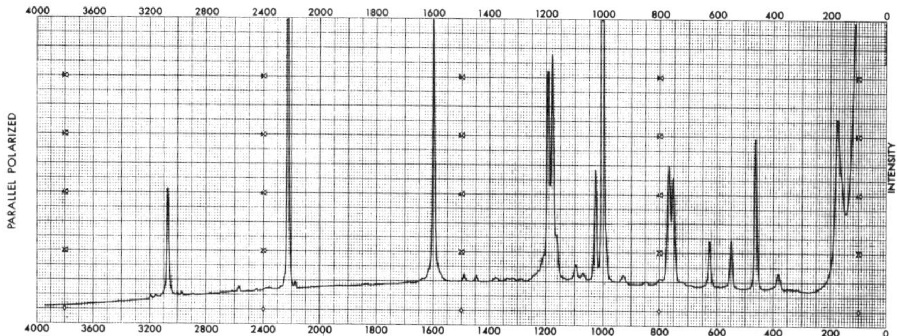

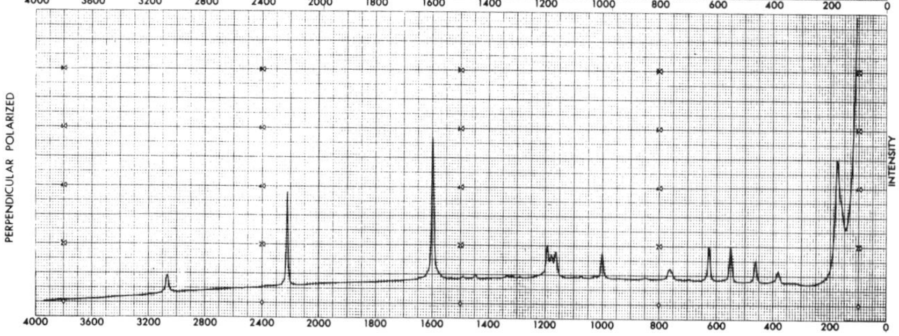

THIOCYANIC ACID, ETHYL ESTER

4137 R

C_3H_5NS Mol. Wt. 87.15

B.P. 143-145°C

Source of Sample: Fluka AG, Buchs, Switzerland

$CH_3CH_2-S-C{\equiv}N$

RAMAN DATA

Spectrophotometer	Cary 83
Sensitivity	26,000; 2600
Suppression	35; 4
Laser Power	140 mW
Spectral Slitwidth	6 cm^{-1}

WAVENUMBER / INTENSITY

2920/74.2	2140/188.0	620/233.8

Capillary Cell: 0.015 mm

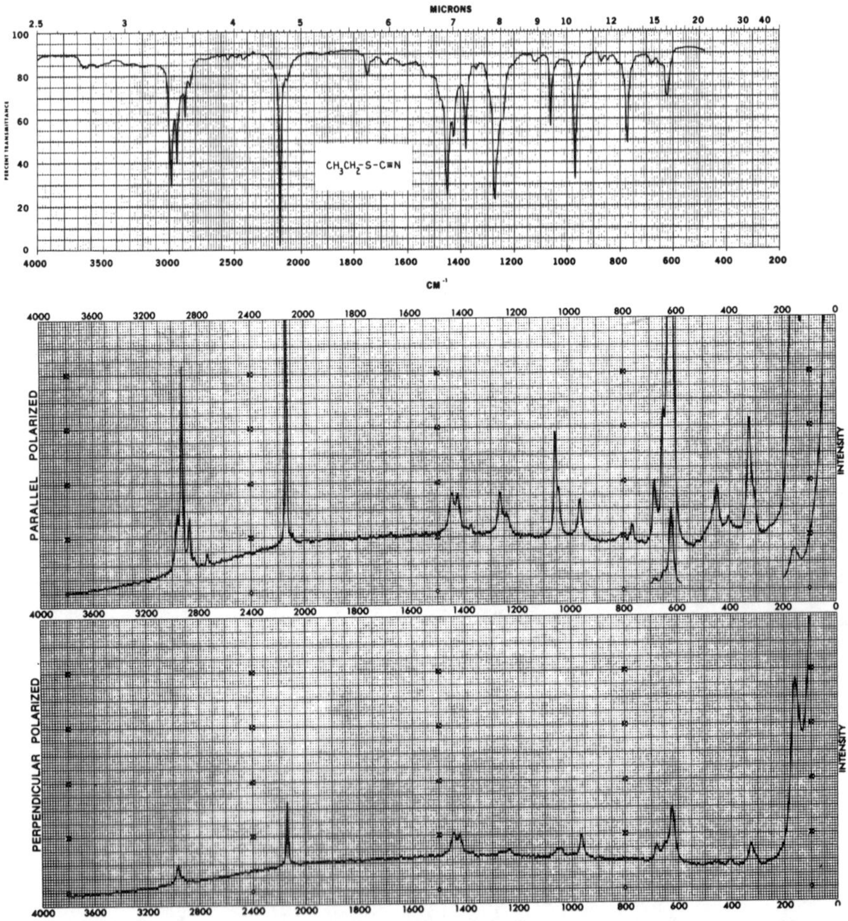

146 R

BUTYRALDEHYDE

RAMAN DATA

Spectrophotometer	Cary 83
Sensitivity	40000; 4000
Suppression	0; 0
Laser Power	40 mW
Spectral Slitwidth	6 cm^{-1}

C_4H_8O Mol. Wt. 72.11

B.P. 72-74°C

Source: Chem Service, Inc., West Chester, Pa.

$CH_3CH_2CH_2-\overset{O}{\overset{\|}{C}}H$

WAVENUMBER / INTENSITY

2935/192.3 2910/167.2 2872/234.8
1450/74.4

Capillary Cell: 0.015 mm

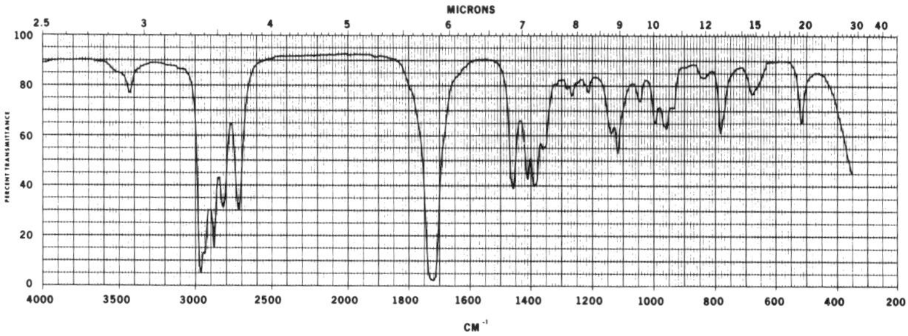

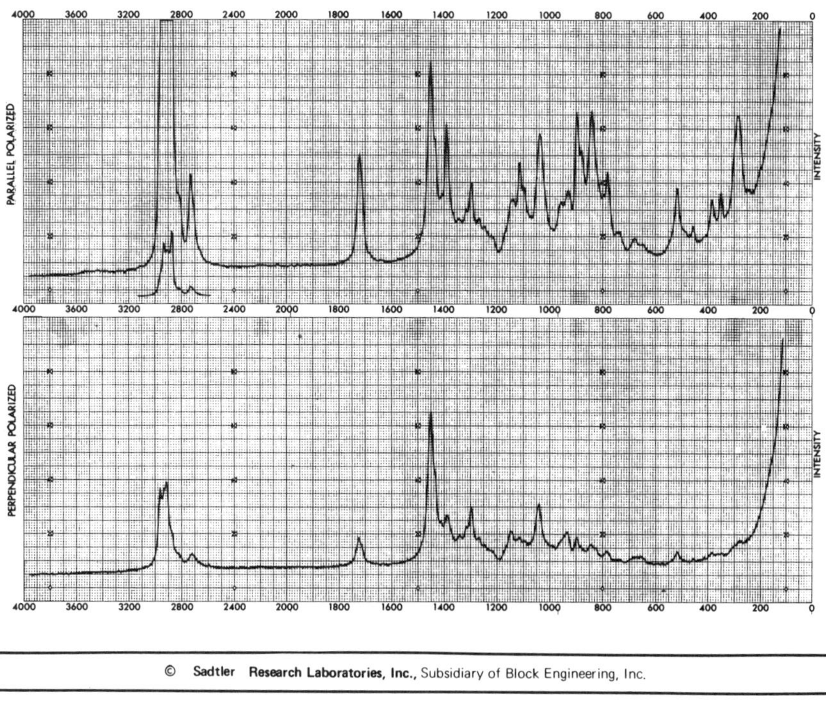

1393 R

BENZALDEHYDE

RAMAN DATA

Spectrophotometer	Cary 83
Sensitivity	2500
	-
Suppression	5
	-
Laser Power	140 mW
Spectral Slitwidth	6 cm^{-1}

C_7H_6O Mol. Wt. 100.13

B. P. 179°C (lit.)

Source: Eastman Organic Chemicals
Rochester, New York

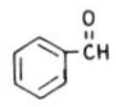

WAVENUMBER / INTENSITY

1695/133.5	1595/122.0	1201/115.5
1000/303.7	826/71.6	

Capillary Cell: Neat

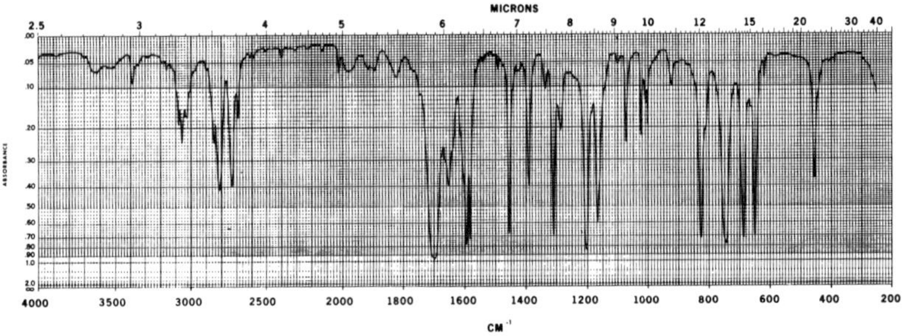

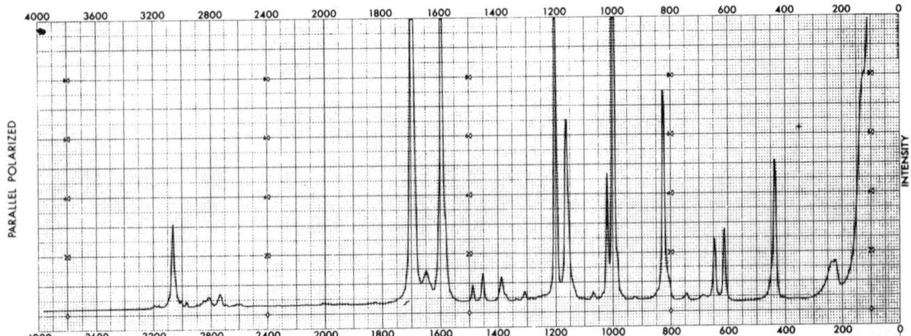

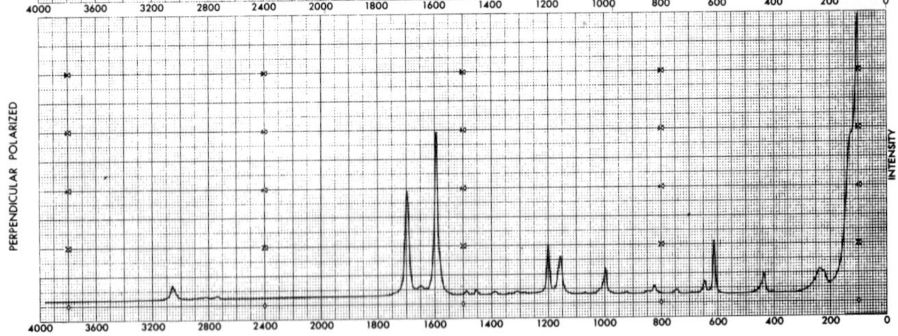

2-BUTANONE **163 R**

RAMAN DATA

Spectrophotometer	Cary 83
Sensitivity	62500;
	10000
Suppression	56;
	0
Laser Power	22 mW
Spectral Slitwidth	6 cm^{-1}

C_4H_8O Mol. Wt. 72.11

B.P. 79-80°C

Source: Chem Service, Inc., West Chester, Pa.

$CH_3CH_2-C(=O)-CH_3$

WAVENUMBER / INTENSITY

2945/133.1	2930/187.8	1710/59.2
1085/101.4	760/273.8	

Capillary Cell: Neat

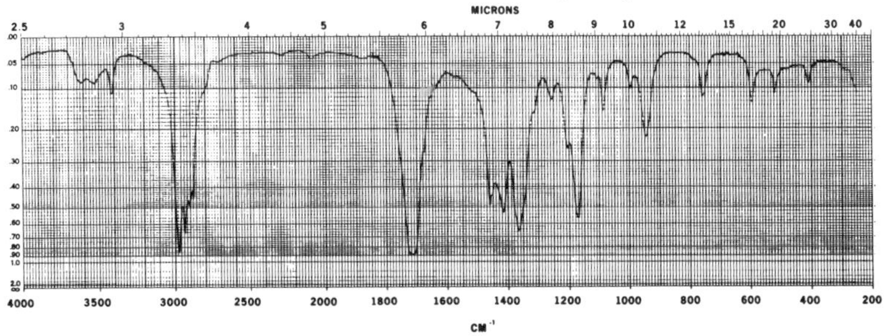

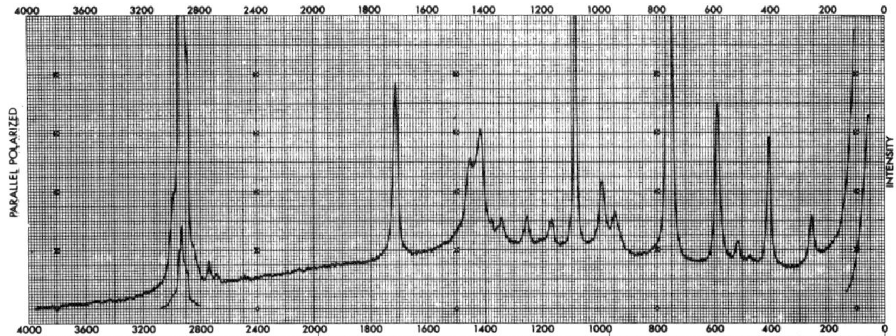

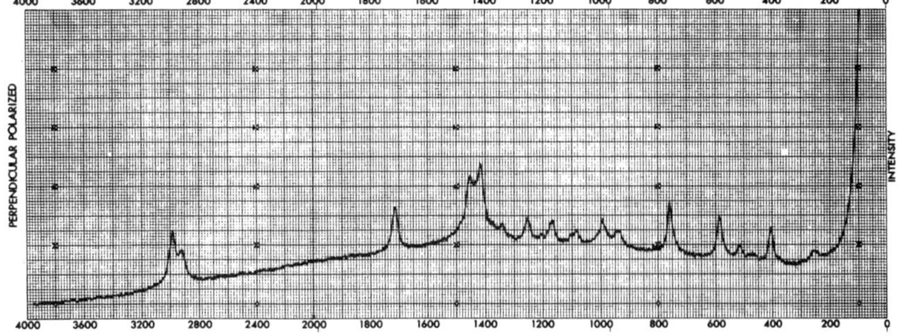

171 R

ACETOPHENONE

C_8H_8O Mol. Wt. 120.15

M.P. 19-20°C

Source: Chem Service, Inc., West Chester, Pa.

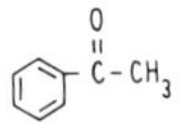

RAMAN DATA

Spectrophotometer	Cary 83
Sensitivity	15000
	-
Suppression	15
	-
Laser Power	22 mW
Spectral Slitwidth	4 cm^{-1}

WAVENUMBER / INTENSITY

1601/87.7 1003/204.7

Capillary Cell: Neat

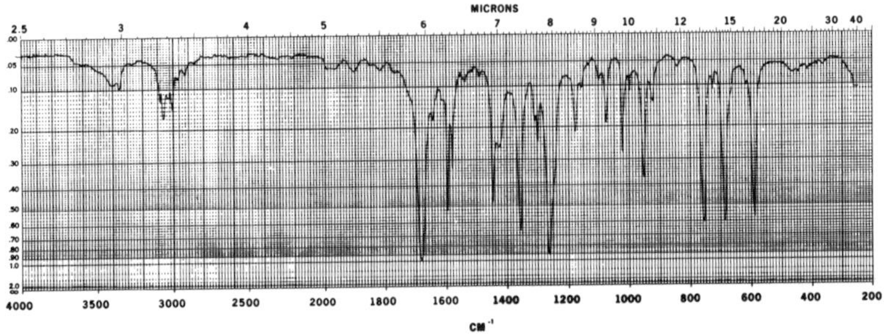

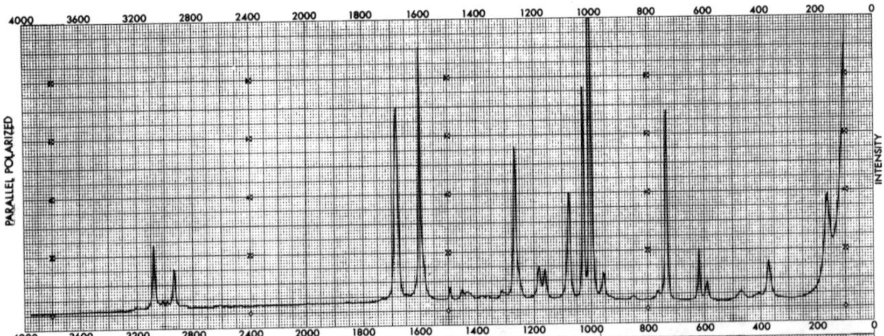

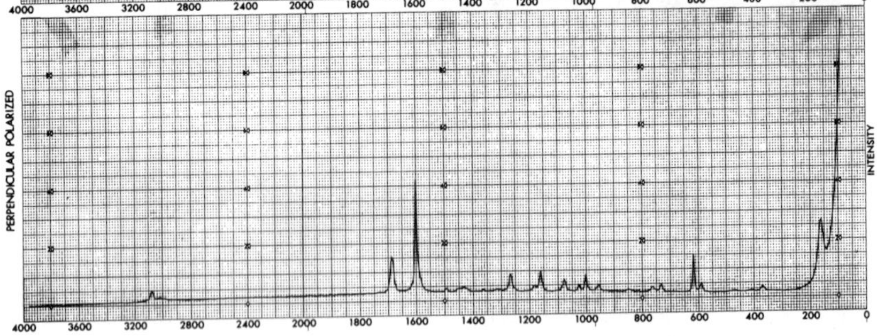

596 R

trans-4-PHENYL-3-BUTEN-2-ONE

$C_{10}H_{10}O$ Mol. Wt. 146.19

M.P. 37-39°C

Source: Chem Service, Inc., West Chester, Pa.

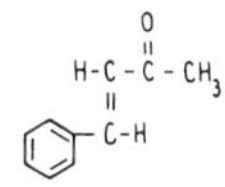

RAMAN DATA

Spectrophotometer	Cary 83
Sensitivity	31250;
	1000
Suppression	95;
	0
Laser Power	6 mW
Spectral Slitwidth	3 cm^{-1}

WAVENUMBER / INTENSITY	
1605/196.8	1008/34.4

Capillary Cell: Melt

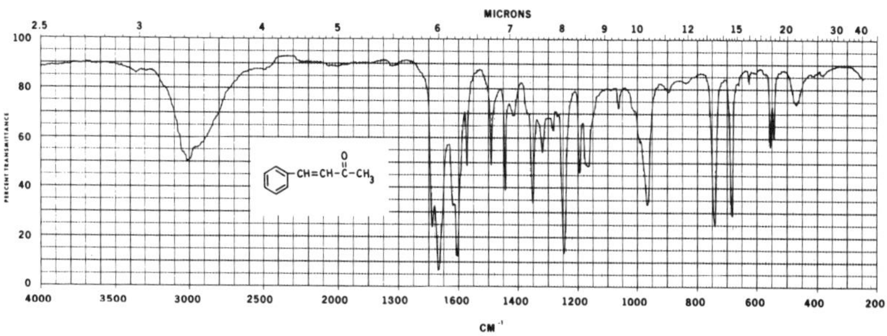

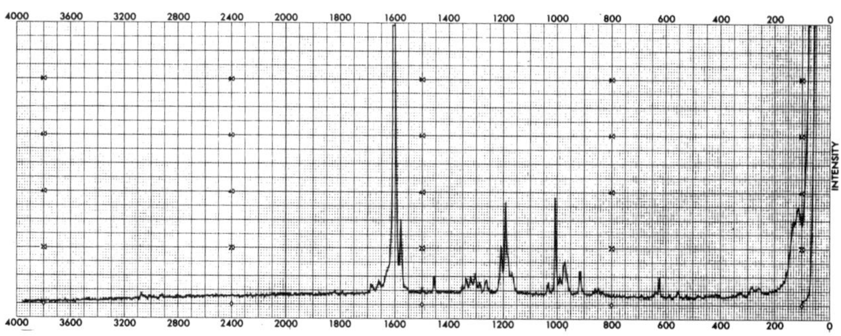

26 R

ACETIC ACID, METHYL ESTER

RAMAN DATA

Spectrophotometer	Cary 83
Sensitivity	25000
	-
Suppression	0
	-
Laser Power	25 mW
Spectral Slitwidth	6 cm^{-1}

$C_3H_6O_2$ Mol. Wt. 74.08

B.P. 57-59°C

Source: Chem Service, Inc., West Chester, Pa.

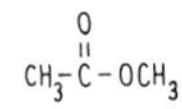

WAVENUMBER / INTENSITY

2940/114.9 1735/46.7 841/238.0
636/150.7

Gas Cell: 10 cm

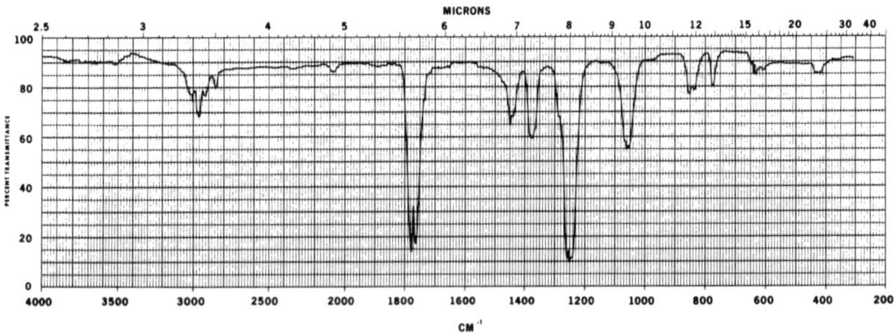

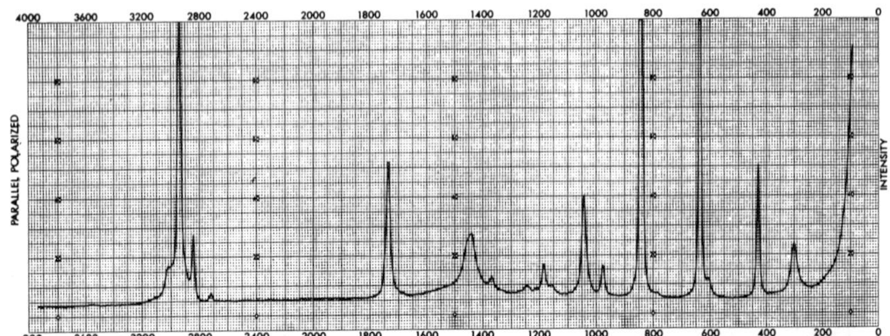

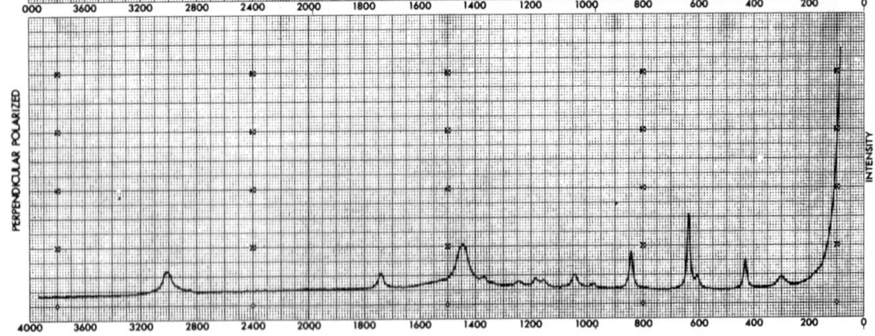

STEARIC ACID, METHYL ESTER

467 R

RAMAN DATA

Spectrophotometer	Cary 83
Sensitivity	125000; 25000
Suppression	245; 40
Laser Power	11 mW
Spectral Slitwidth	12 cm^{-1}

$C_{19}H_{38}O_2$ Mol. Wt. 298.51

M.P. 35-37°C

Source: Chem Service, Inc., West Chester, Pa.

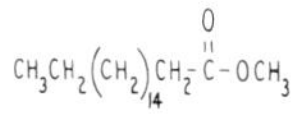

WAVENUMBER / INTENSITY	
2895/96.2	2860/60.3

Capillary Cell: Melt

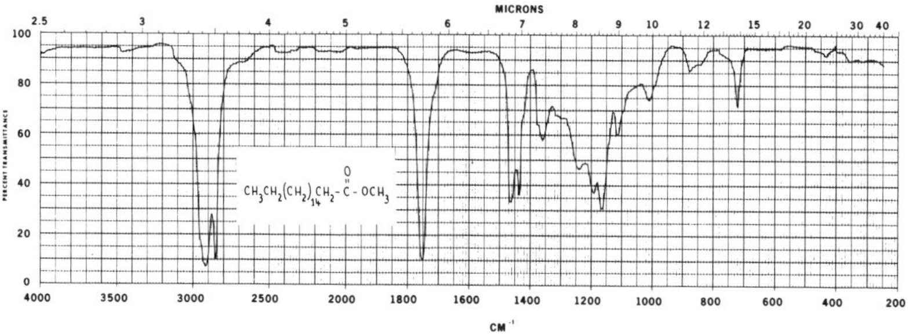

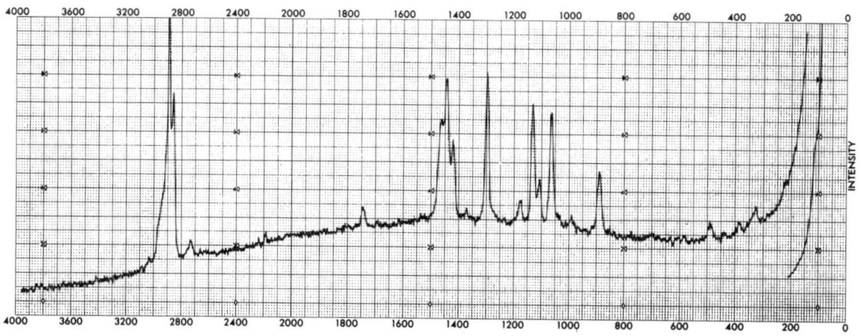

498 R

ACETIC ACID, VINYL ESTER

RAMAN DATA

Spectrophotometer	Cary 83
Sensitivity	20000;
	2000
Suppression	78;
	0
Laser Power	205 mW
Spectral Slitwidth	6 cm-1

$C_4H_6O_2$ Mol. Wt. 86.09

B.P. 72.8°C

Source: Chem Service, Inc., West Chester, Pa.

$$CH_3-\overset{O}{\overset{\|}{C}}-OCH=CH_2$$

WAVENUMBER / INTENSITY

1651/263.9 1301/143.5 414/85.0

0.025 mm CsBr Cell: Neat

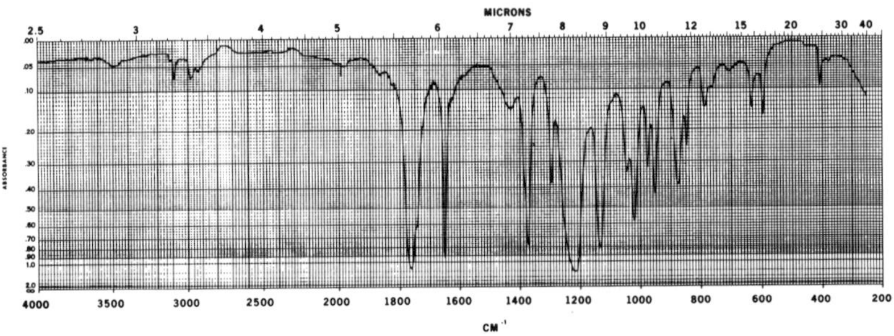

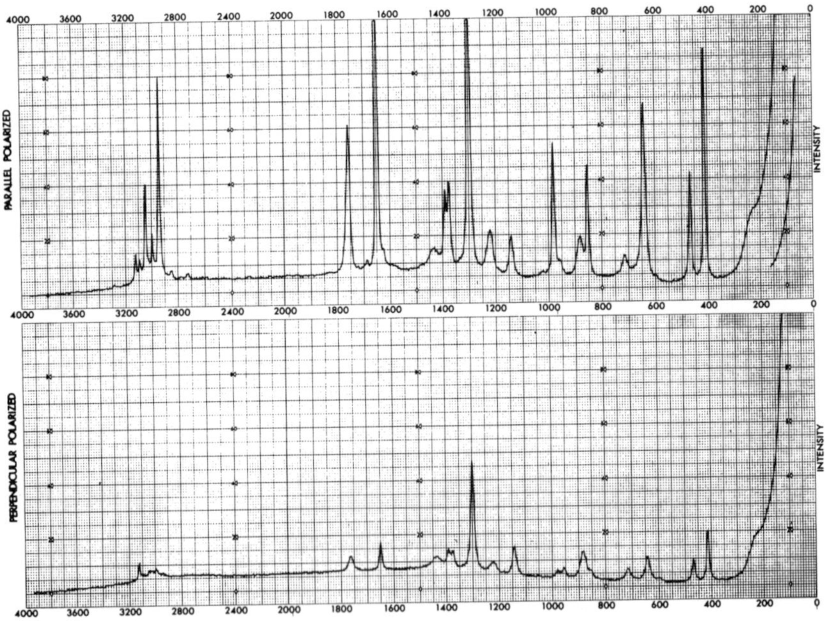

ACRYLIC ACID, ETHYL ESTER

857 R

RAMAN DATA

Spectrophotometer	Cary 83
Sensitivity	25000; 2500
Suppression	41; 0
Laser Power	195 mW
Spectral Slitwidth	6 cm^{-1}

$C_5H_8O_2$ Mol. Wt. 100.12

B. P. 98-99°C

Source: Chem Service, Inc.
West Chester, Pennsylvania

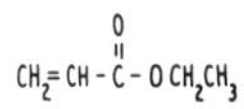

WAVENUMBER / INTENSITY

2940/79.7 1724/144.2 1640/374.7
864/147.7

A=Capillary Cell: Neat
B=Capillary Cell: 0.015 mm

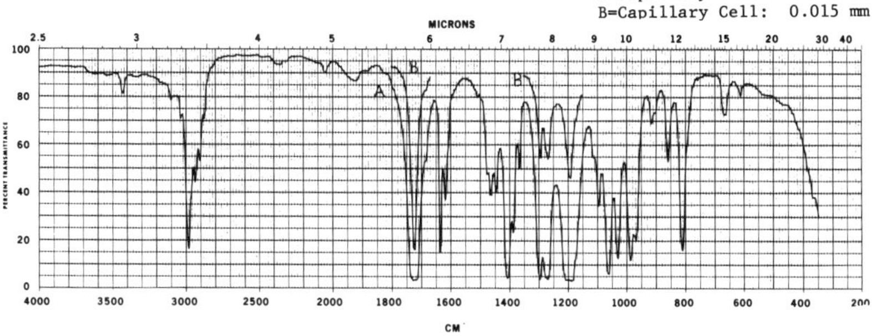

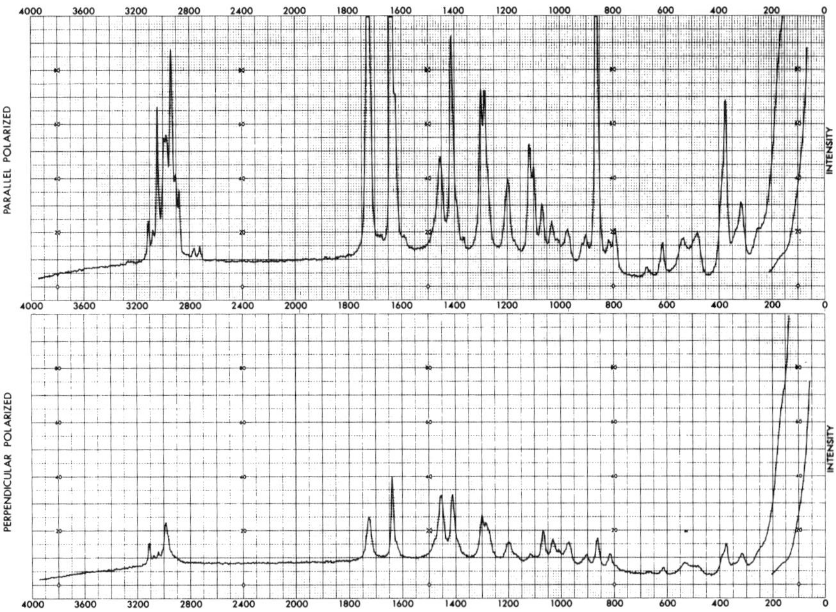

BENZOIC ACID, METHYL ESTER

63 R

RAMAN DATA

Spectrophotometer	Cary 83
Sensitivity	15000
	-
Suppression	0
	-
Laser Power	26 mW
Spectral Slitwidth	4 cm^{-1}

$C_8H_8O_2$ Mol. Wt. 136.15

B.P. 110-111°C/25 mm

Source: Chem Service, Inc.,
West Chester, Pa.

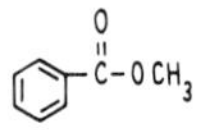

WAVENUMBER / INTENSITY	
1000/280.3	818/95.9

Capillary Cell: Neat

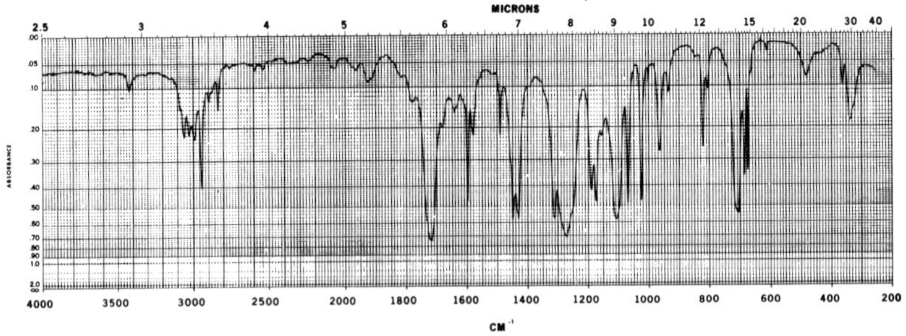

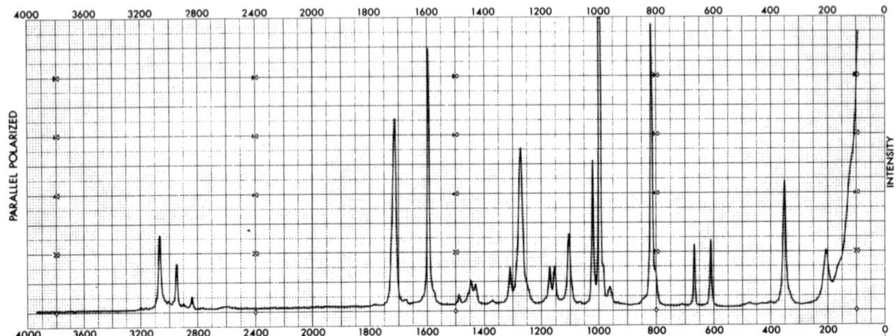

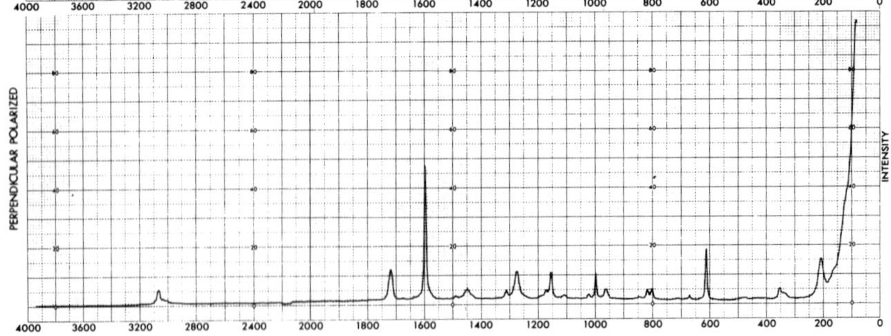

534 R

BUTYROLACTONE

RAMAN DATA

Spectrophotometer	Cary 83
Sensitivity	43750
	-
Suppression	100
	-
Laser Power	22 mW
Spectral Slitwidth	6 cm-1

$C_4H_6O_2$ Mol. Wt. 86.09

B.P. 90-92°C/17 mm

Source: Chem Service, Inc., West Chester, Pa.

WAVENUMBER / INTENSITY

2940/75.3	934/229.3	873/104.4
804/92.4	679/82.8	

Capillary Cell: Neat

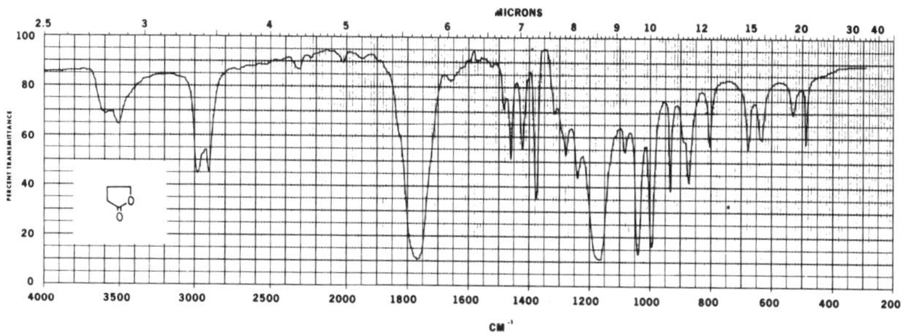

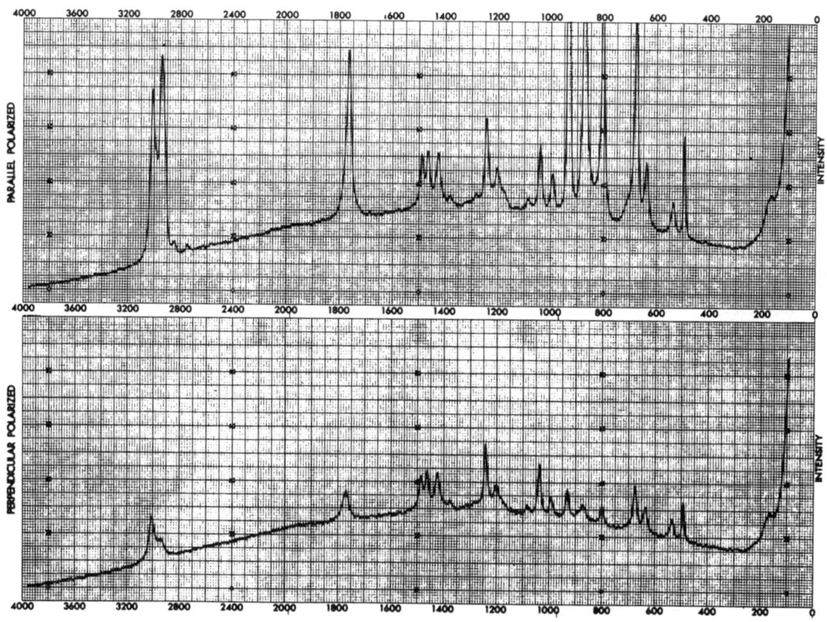

BENZOYL DISULFIDE

4375 R

RAMAN DATA

Spectrophotometer	Cary 83
Sensitivity	3700; 370
Suppression	21; 2
Laser Power	100 mW
Spectral Slitwidth	6 cm-1

$C_{14}H_{10}O_2S_2$ Mol. Wt. 274.36

Source of Sample: Aldrich Chemical Co., Inc., Milwaukee, Wisconsin

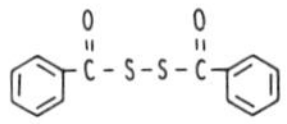

WAVENUMBER / INTENSITY		
1700/71.2	1592/109.6	1205/97.1
1002/146.4	120/120.0	

KBr Wafer

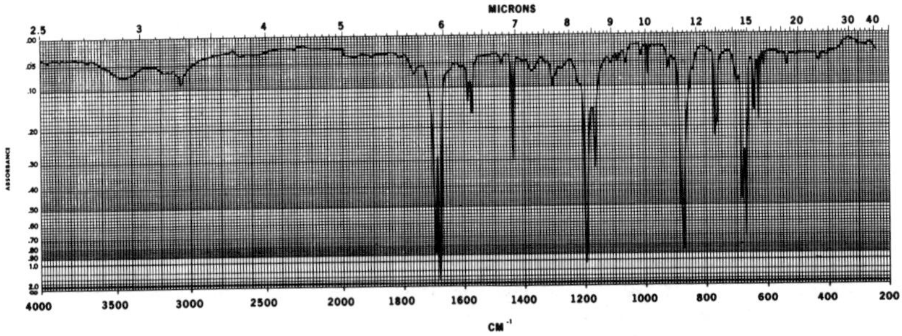

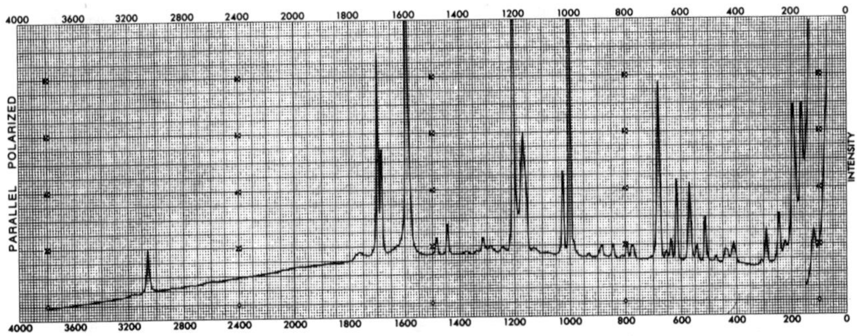

ACETIC ACID

407 R

RAMAN DATA

Spectrophotometer	Cary 83
Sensitivity	7000
	-
Suppression	150
	-
Laser Power	210 mW
Spectral Slitwidth	6 cm^{-1}

$C_2H_4O_2$ Mol. Wt. 60.05

Source: Merck & Co. Inc., Rahway, New Jersey

$CH_3-\overset{O}{\overset{\|}{C}}-OH$

WAVENUMBER / INTENSITY

2945/174.6 900/389.3 626/68.0

Capillary Cell: Neat

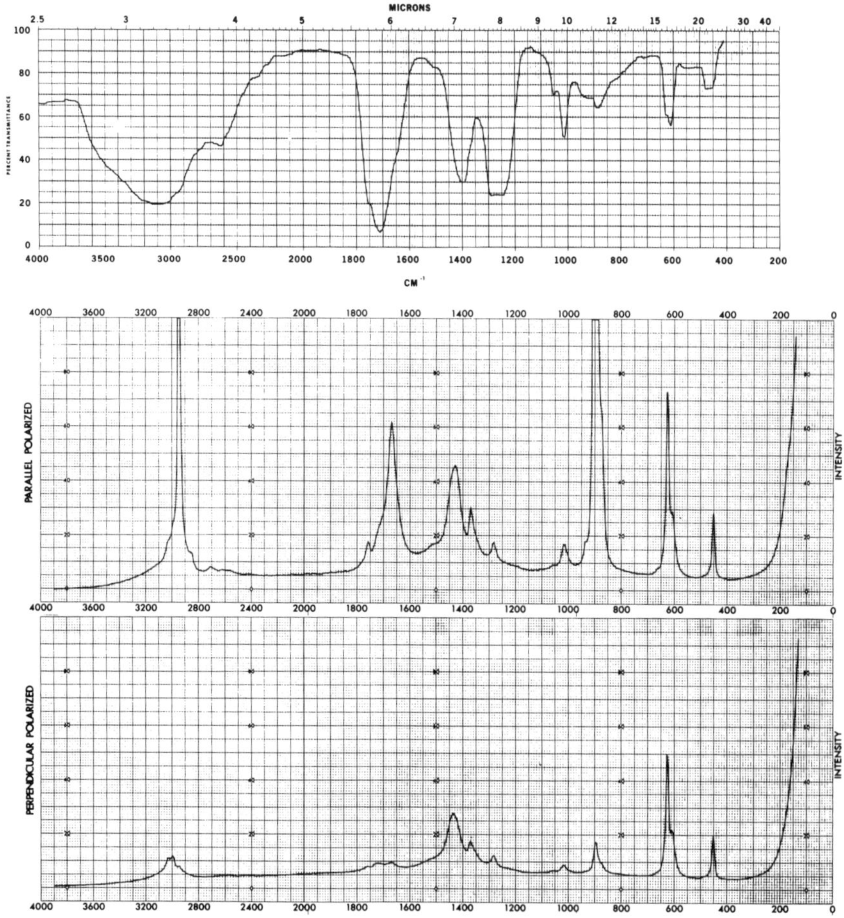

414 R

STEARIC ACID

RAMAN DATA

Spectrophotometer	Cary 83
Sensitivity	8000
	-
Suppression	170
	-
Laser Power	158 mW
Spectral Slitwidth	8 cm^{-1}

$C_{18}H_{36}O_2$ Mol. Wt. 284.49

M.P. 69.4°C

Source: Aldrich Chemical Co. Inc., Milwaukee, Wis.

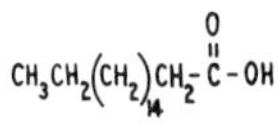

WAVENUMBER / INTENSITY

2885/151.6	2850/89.5	1303/94.1

Capillary Cell: Melt

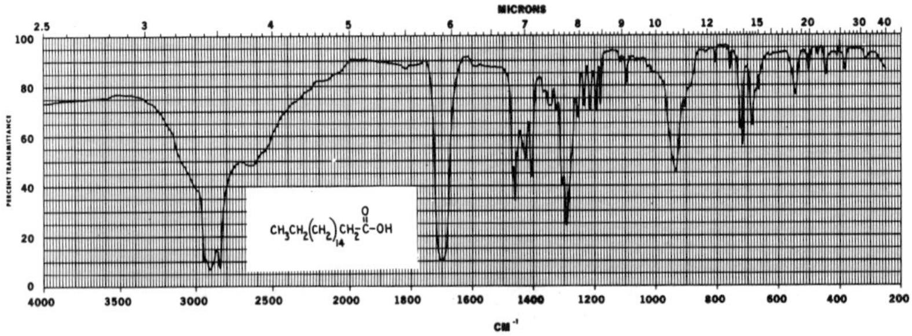

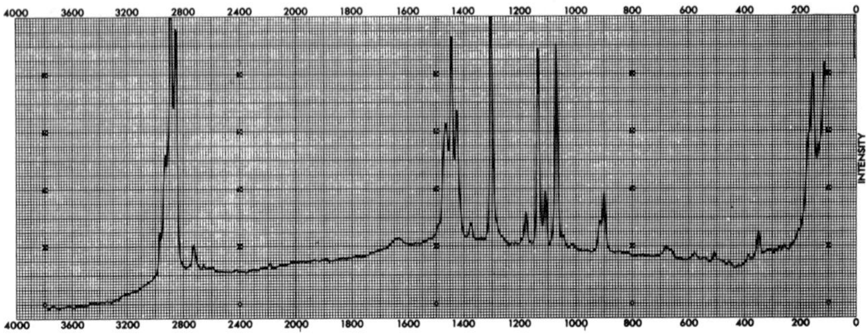

6 R

BENZOIC ACID

RAMAN DATA

Spectrophotometer	Cary 83
Sensitivity	50000; 8125
Suppression	43; 0
Laser Power	10 mW
Spectral Slitwidth	6 cm^{-1}

$C_7H_6O_2$ Mol. Wt. 122.12

M.P. 123-125°C

Source: Chem Service, Inc., West Chester, Pa.

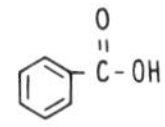

WAVENUMBER / INTENSITY		
1000/208.1	795/80.8	110/175.1

KBr Wafer

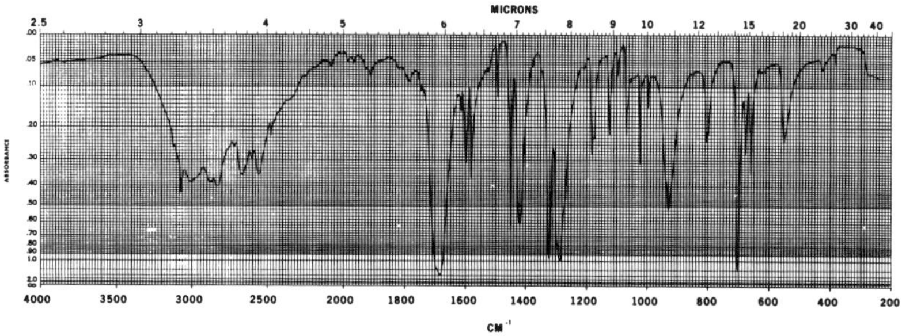

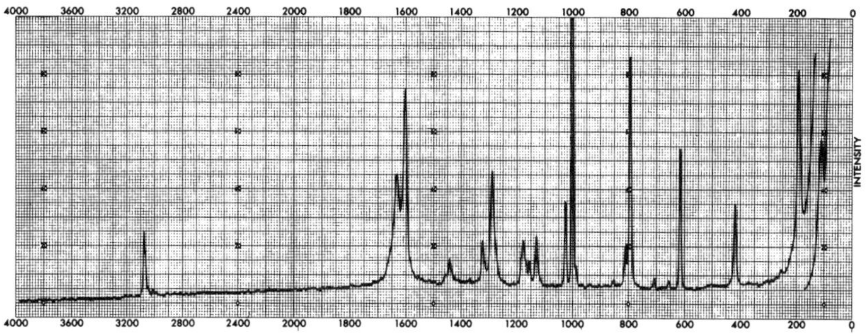

THIOLACETIC ACID

845 R

RAMAN DATA

Spectrophotometer	Cary 83
Sensitivity	10000
	-
Suppression	7
	-
Laser Power	80 mW
Spectral Slitwidth	6 cm^{-1}

C_2H_4OS Mol. Wt. 76.12

B. P. 93°C

Source: Aldrich Chemical Company, Inc.
Milwaukee, Wisconsin

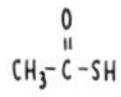

WAVENUMBER / INTENSITY

2925/67.3 634/190.5 455/119.0

Gas Cell: 10 cm

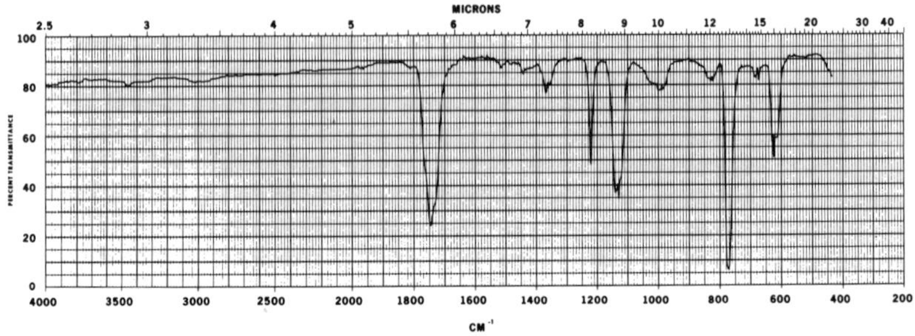

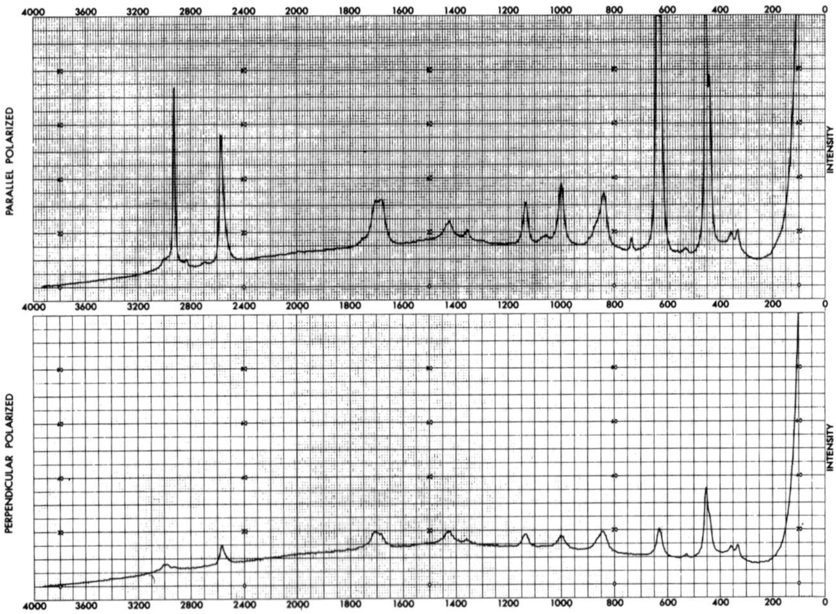

1640 R

RAMAN DATA

Spectrophotometer	Cary 83
Sensitivity	50000;
	5000
Suppression	0;
	0
Laser Power	35 mW
Spectral Slitwidth	10 cm-1

ACETIC ACID, SODIUM SALT

$C_2H_3NaO_2$ Mol. Wt. 82.03

M.P. 324°C (lit.)

Source: Fluka AG
Buchs, Switzerland

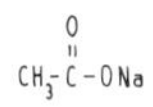

WAVENUMBER / INTENSITY

1422/58.2 939/113.0 230/17.5

KBr Wafer

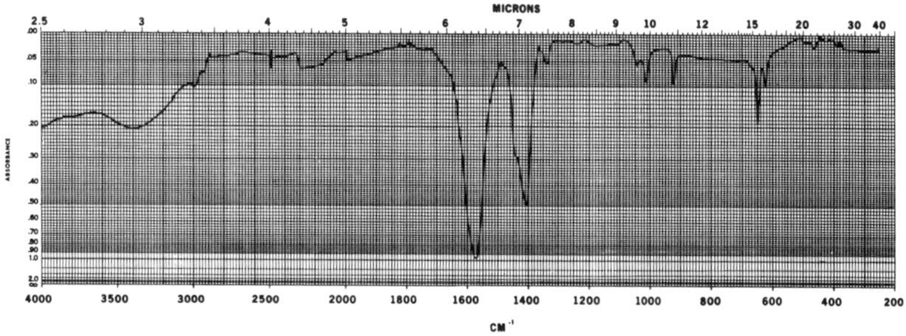

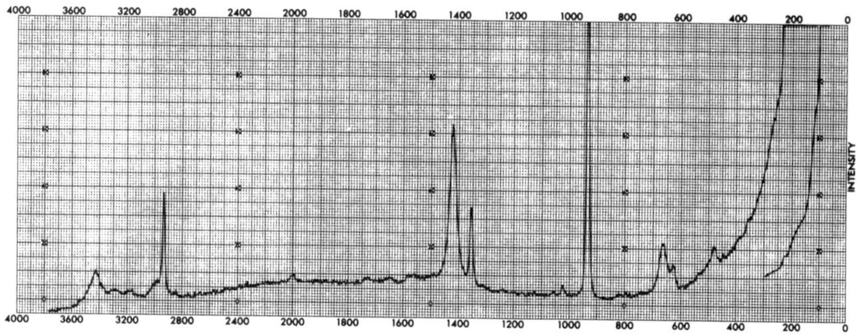

PROPIONIC ANHYDRIDE

445 R

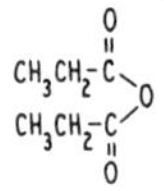

RAMAN DATA

Spectrophotometer	Cary 83
Sensitivity	5250
	-
Suppression	120
	-
Laser Power	210 mW
Spectral Slitwidth	6 cm^{-1}

$C_6H_{10}O_3$ Mol. Wt. 130.14

B.P. 165-169°C

Source: MCB Manufacturing Chemists, Norwood, Ohio

WAVENUMBER / INTENSITY		
2950/173.0	2895/76.0	1081/84.3

Capillary Cell: Neat

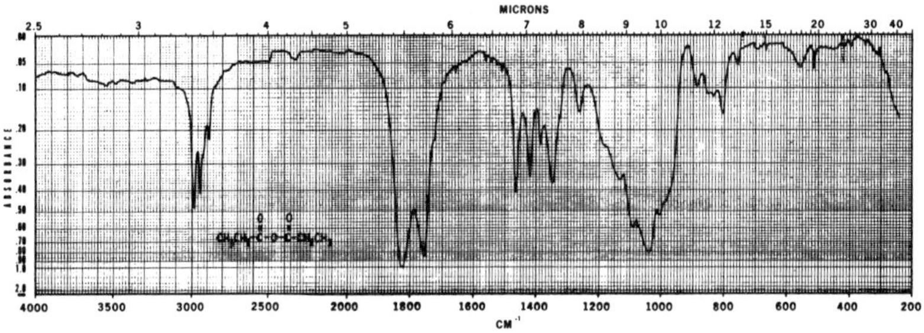

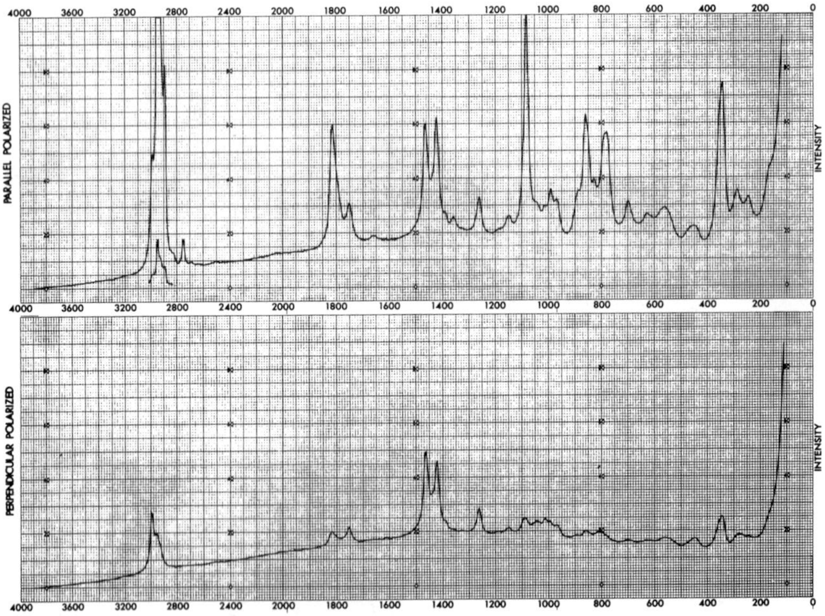

MALEIC ANHYDRIDE

19 R

RAMAN DATA

Spectrophotometer	Cary 83
Sensitivity	100000; 25000
Suppression	60; 0
Laser Power	13 mW
Spectral Slitwidth	6 cm^{-1}

$C_4H_2O_3$ Mol. Wt. 98.06

M.P. 52-54°C

Source: Chem Service, Inc., West Chester, Pa.

WAVENUMBER / INTENSITY

1846/111.7 1066/69.8

Capillary Cell: Melt

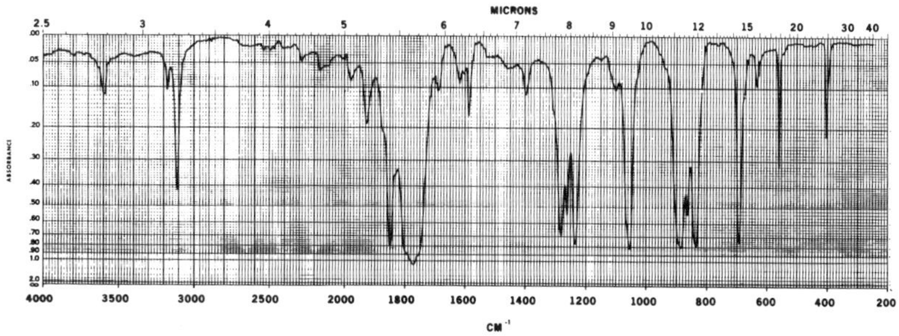

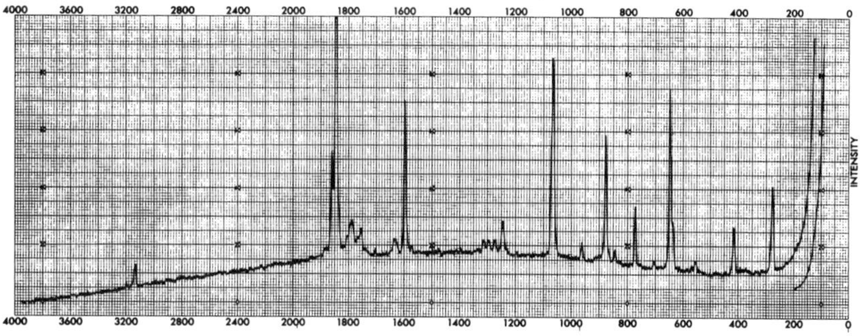

2791 R

HEXANOYL CHLORIDE

$C_6H_{11}ClO$ Mol. Wt. 134.61

B.P. 47-49°C/15mm

Source of Sample: Tokyo Kasei Kogyo Co. Ltd. Tokyo, Japan

$CH_3CH_2(CH_2)_2CH_2-C(=O)-Cl$

RAMAN DATA

Spectrophotometer	Cary 83
Sensitivity	6000
	-
Suppression	27
	-
Laser Power	150 mW
Spectral Slitwidth	6 cm-1

WAVENUMBER / INTENSITY

2935/68.0 449/172.5

Capillary Cell: 0.015mm

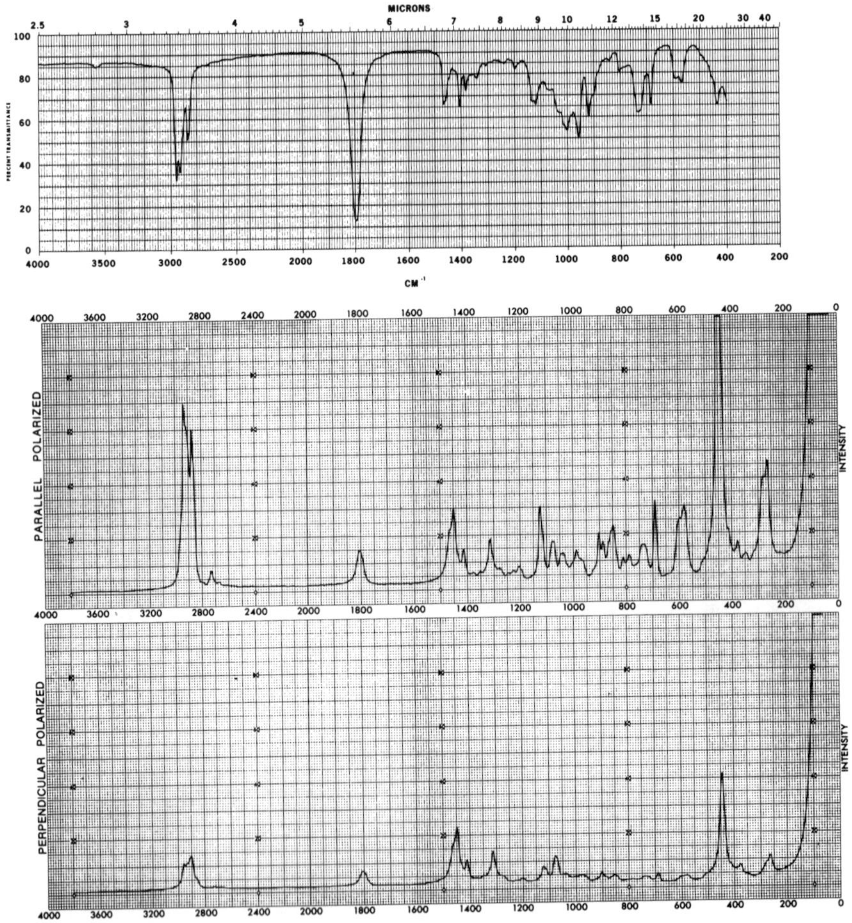

20 R

BENZOYL CHLORIDE

RAMAN DATA

Spectrophotometer	Cary 83
Sensitivity	24000
	-
Suppression	0
	-
Laser Power	12 mW
Spectral Slitwidth	6 cm-1

C_7H_5ClO Mol. Wt. 140.57

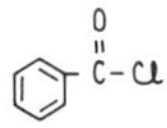

B.P. 195-197°C

Source: Chem Service, Inc., West Chester, Pa.

WAVENUMBER / INTENSITY		
1590/162.7	1200/110.1	1169/119.9
995/372.2	665/113.7	500/83.8

Capillary Cell: Neat

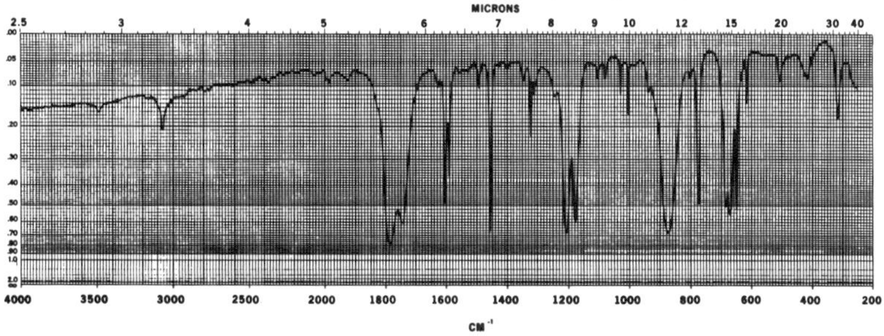

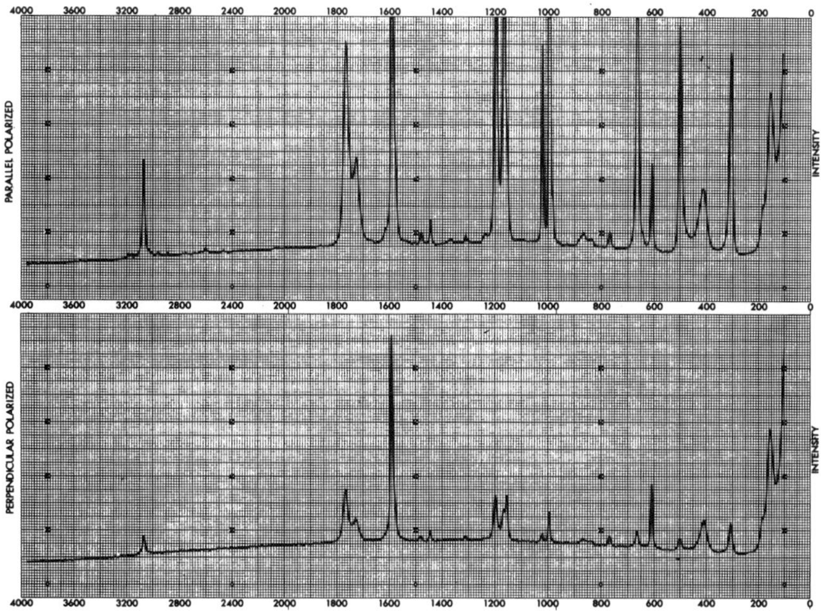

1706 R

RAMAN DATA

Spectrophotometer	Cary 83
Sensitivity	32000; 3200
Suppression	7; 0
Laser Power	35 mW
Spectral Slitwidth	10 cm^{-1}

BUTYRAMIDE

C_4H_9NO Mol. Wt. 87.12

M.P. 113-115°C

Source: Fisher Scientific Company
Fair Lawn, New Jersey

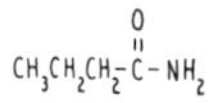

WAVENUMBER / INTENSITY		
1152/89.3	870/68.7	225/24.0

Film

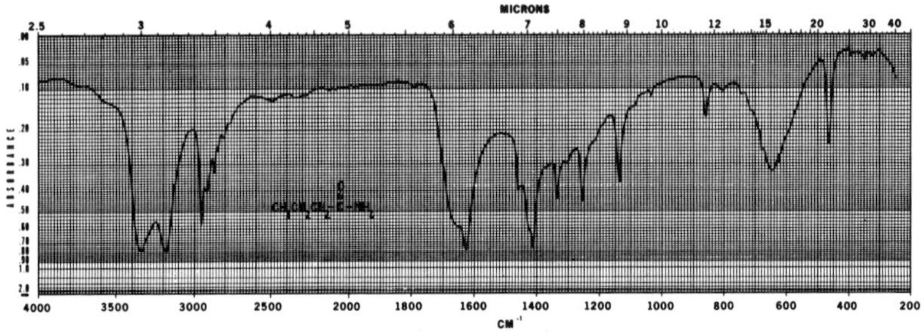

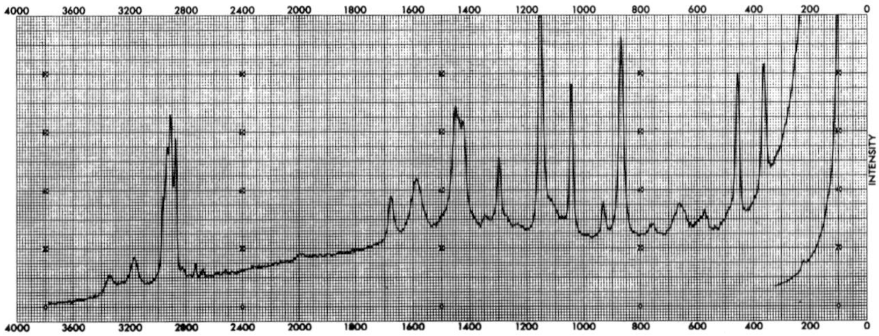

1294 R

N-ETHYLACETAMIDE

RAMAN DATA

Spectrophotometer	Cary 83
Sensitivity	10000; 1000
Suppression	3; 0
Laser Power	150 mW
Spectral Slitwidth	6 cm^{-1}

C_4H_9NO Mol. Wt. 87.12

B.P. 206-211°C

Source: Pennwalt Chemical Company
Philadelphia, Pennsylvania

$CH_3-\overset{O}{\overset{\|}{C}}-NH-CH_2CH_3$

WAVENUMBER / INTENSITY		
2911/165.2	890/90.9	621/86.1

Capillary Cell: Neat

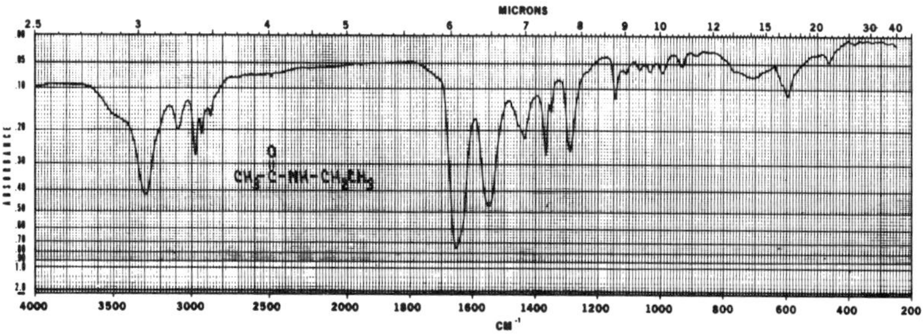

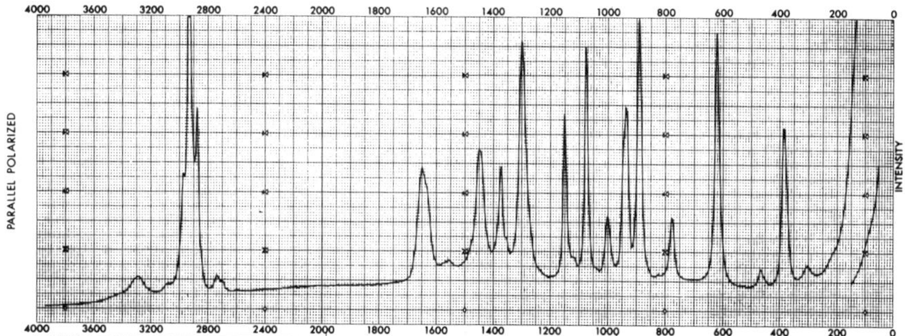

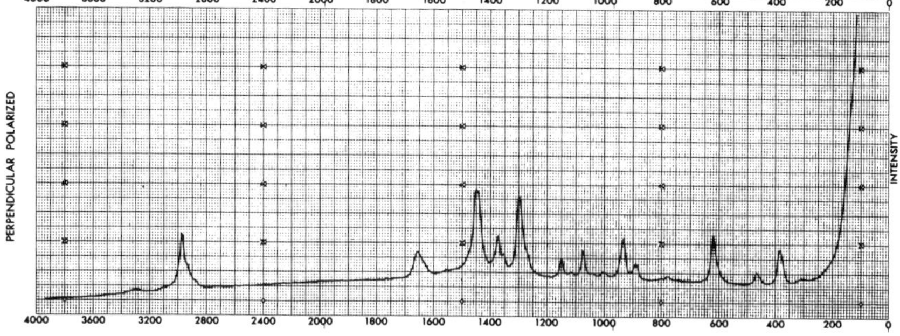

922 R

N,N-DIETHYLACETAMIDE

RAMAN DATA

Spectrophotometer	Cary 83
Sensitivity	7000
	-
Suppression	23
	-
Laser Power	80 mW
Spectral Slitwidth	6 cm^{-1}

$C_6H_{13}NO$ Mol. Wt. 115.18

B. P. 182-186°C

Source: Aldrich Chemical Company, Inc.
Milwaukee, Wisconsin

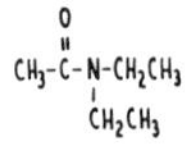

WAVENUMBER / INTENSITY

2940/44.5 714/92.9

Capillary Cell: Neat

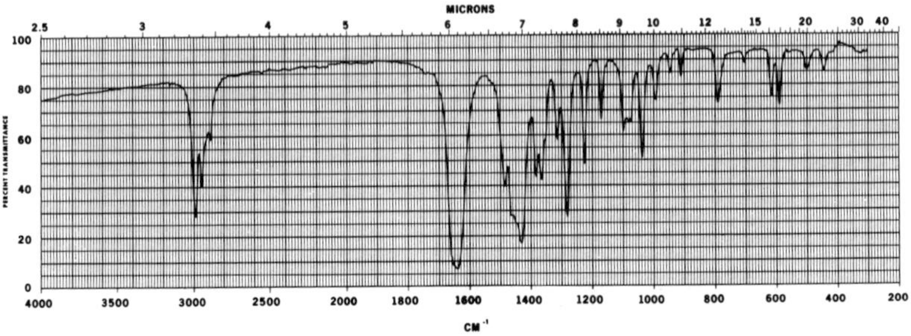

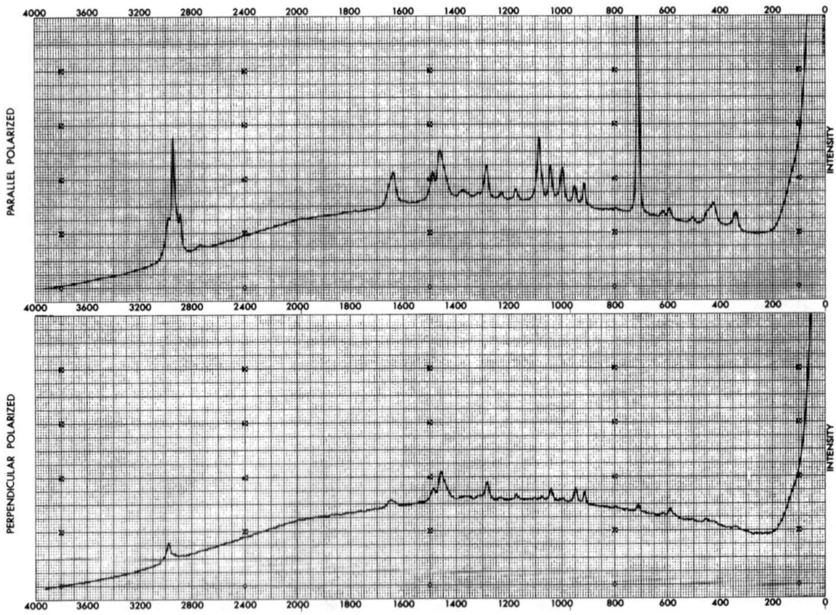

BENZAMIDE **123 R**

RAMAN DATA

Spectrophotometer	Cary 83
Sensitivity	45000;
	4500
Suppression	56;
	0
Laser Power	11 mW
Spectral Slitwidth	6 cm^{-1}

C_7H_7NO Mol. Wt. 121.14

M.P. 125-126°C

Source: Chem Service, Inc., West Chester, Pa.

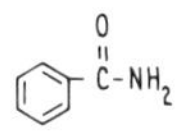

WAVENUMBER / INTENSITY		
1608/151.1	1148/55.0	1006/205.1
135/35.0	107/343.5	

KBr Wafer

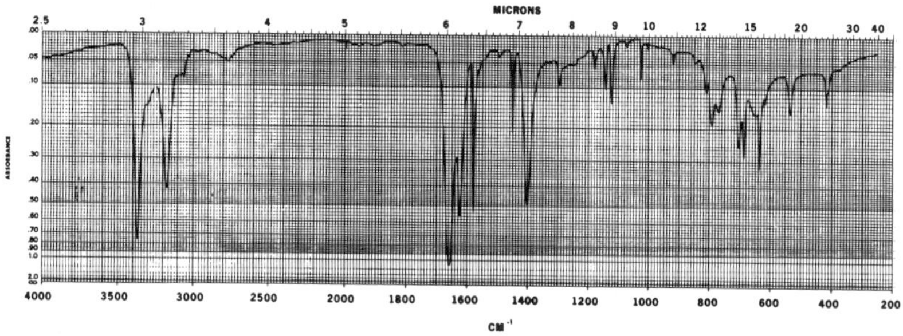

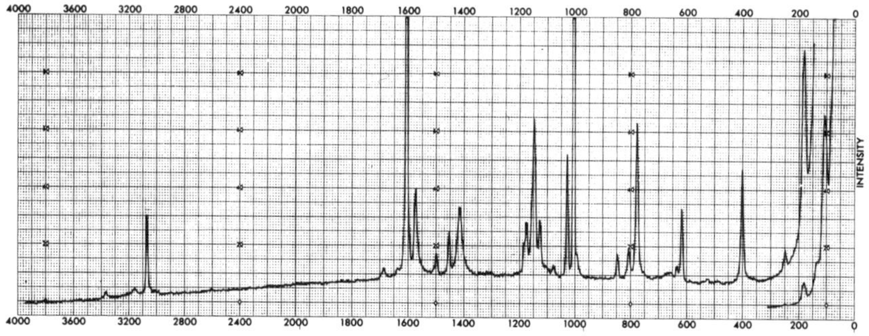

524 R

2-OXOHEXAMETHYLENIMINE

RAMAN DATA

Spectrophotometer	Cary 83
Sensitivity	10000;
	1000
Suppression	230;
	0
Laser Power	158 mW
Spectral Slitwidth	8 cm^{-1}

$C_6H_{11}NO$ Mol. Wt. 113.16

M.P. 68.8°C B.P. 236-38°C/10mm

Source: Fluka AG, Buchs, Switzerland

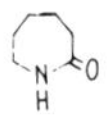

WAVENUMBER / INTENSITY

1029/77.9 751/111.6 707/82.0
140/130.0

Capillary Cell: Melt

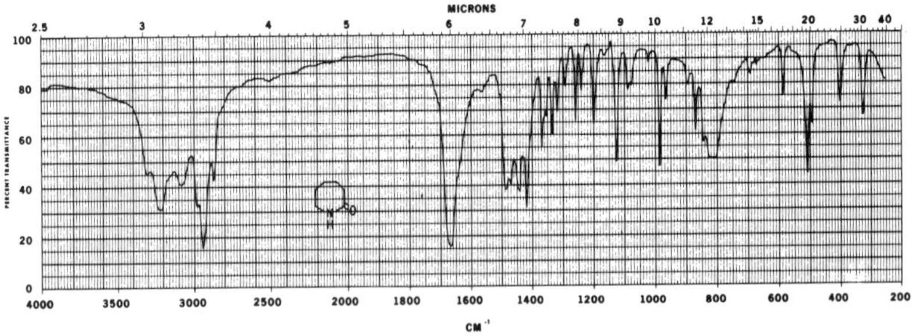

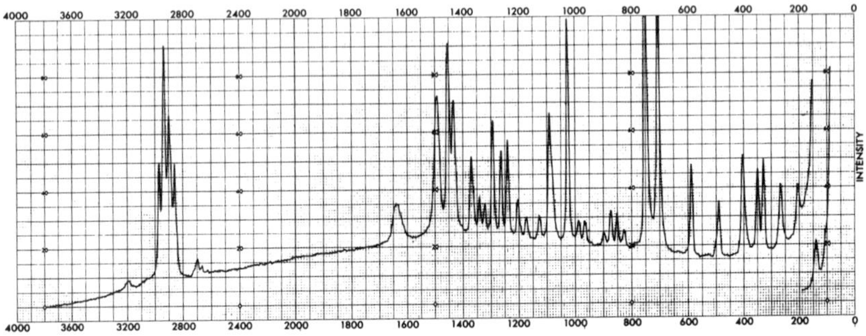

2626 R

RAMAN DATA

Spectrophotometer	Cary 83
Sensitivity	28000; 2800
Suppression	31; 0
Laser Power	35 mW
Spectral Slitwidth	8 cm^{-1}

BUTYLUREA

$C_5H_{12}N_2O$ Mol. Wt. 116.16

M.P. 94-95°C

Source of Sample: MCB Mfg. Chemists Norwood, Ohio

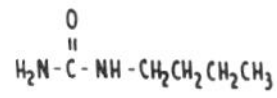

WAVENUMBER / INTENSITY

1451/49.1	1130/77.9	230/13.1

KBr Wafer

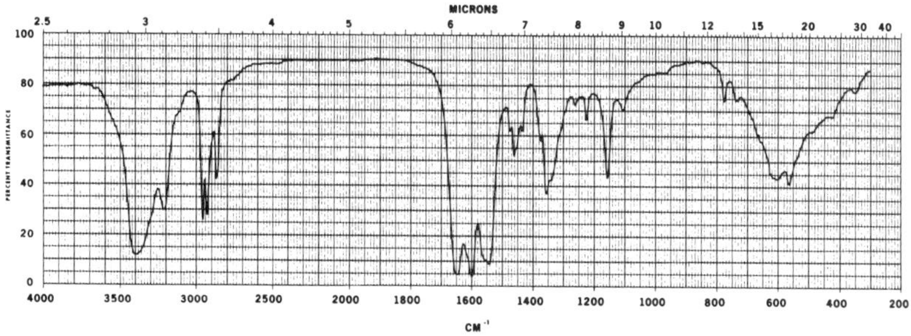

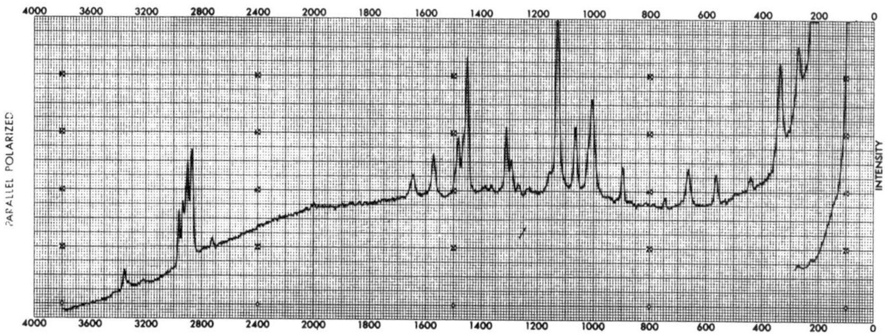

127 R

CARBANILIDE

RAMAN DATA

Spectrophotometer	Cary 83
Sensitivity	62500;
	10000
Suppression	115;
	0
Laser Power	11 mW
Spectral Slitwidth	6 cm^{-1}

$C_{13}H_{12}N_2O$ Mol. Wt. 212.25

M.P. 241-242°C

Source: Chem Service, Inc., West Chester, Pa.

WAVENUMBER / INTENSITY		
1605/110.6	1311/51.3	1003/92.0

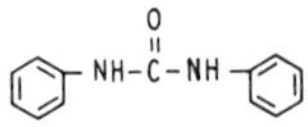

KBr Wafer

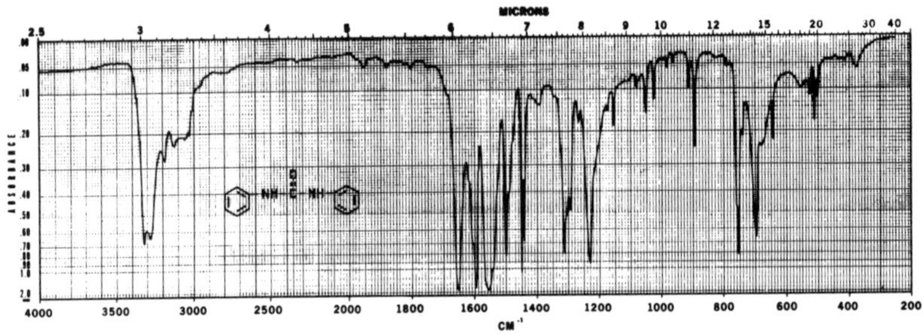

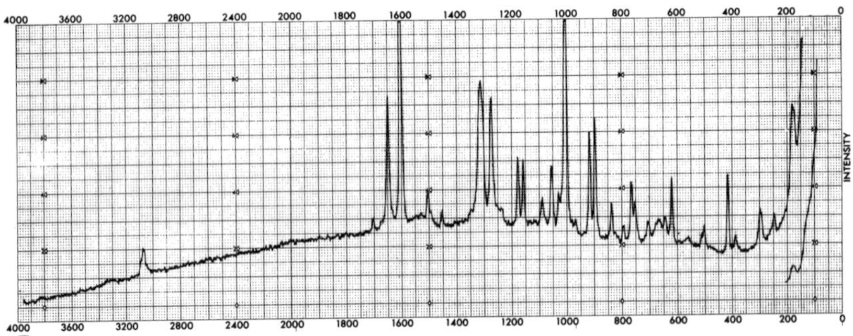

3096 R

THIOCARBANILIDE

RAMAN DATA

Spectrophotometer	Cary 83
Sensitivity	16000; 1600
Suppression	0; -
Laser Power	28 mW
Spectral Slitwidth	6 cm^{-1}

$C_{13}H_{12}N_2S$ Mol. Wt. 228.32

M.P. 153-154°C

Source of Sample: Philadelphia Col. of Pharm. & Sci. Philadelphia, Pennsylvania

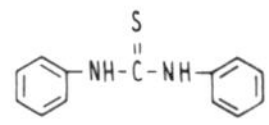

WAVENUMBER / INTENSITY

1000/95.6 130/86.3

KBr Wafer

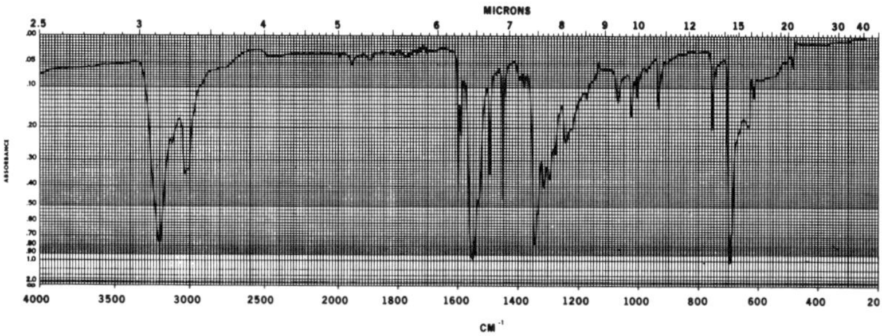

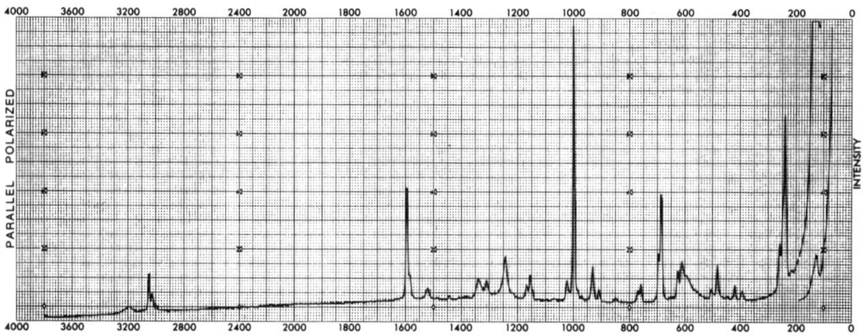

PROPYLAMINE **1461 R**

RAMAN DATA

Spectrophotometer	Cary 83
Sensitivity	15000; 1500
Suppression	5; 0
Laser Power	160 mW
Spectral Slitwidth	6 cm^{-1}

C_3H_9N Mol. Wt. 59.11

B.P. 46-51°C

Source: Merck & Company, Inc.
Rahway, New Jersey

$CH_3CH_2CH_2$-NH_2

WAVENUMBER / INTENSITY

2930/127.5	2892/101.2	2870/151.3
2860/125.4	457/71.3	

Capillary Cell: Neat

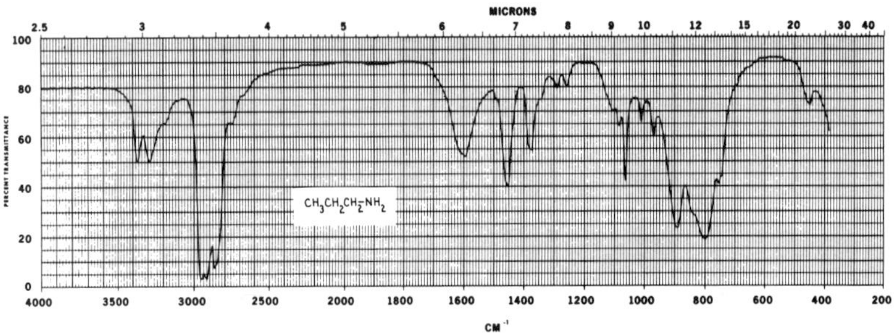

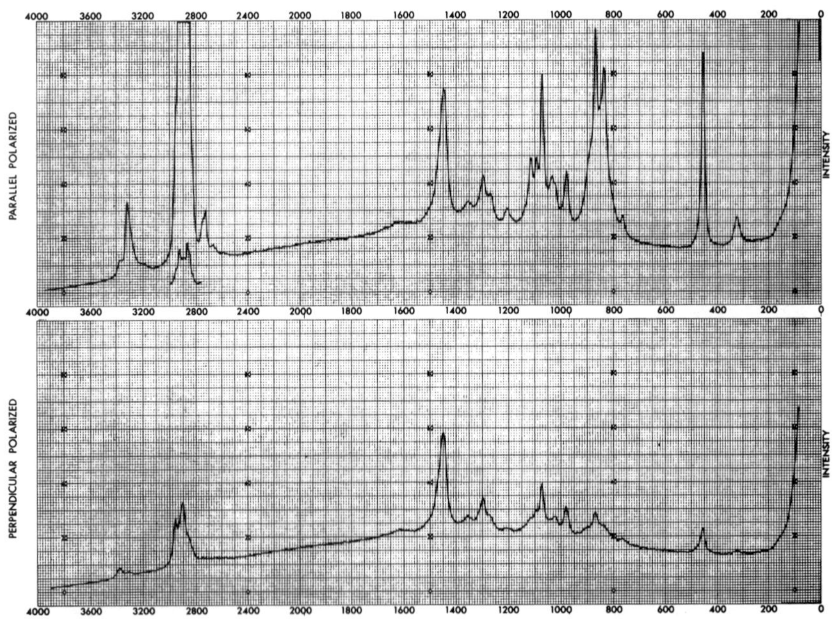

tert-BUTYLAMINE

673 R

RAMAN DATA

Spectrophotometer	Cary 83
Sensitivity	4000
	-
Suppression	77
	-
Laser Power	145 mW
Spectral Slitwidth	6 cm^{-1}

$C_4H_{11}N$ Mol. Wt. 73.14

B.P. 44-45°C

Source: Chem Service, Inc., West Chester, Pa.

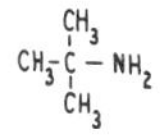

WAVENUMBER / INTENSITY

2910/48.8 752/277.6

Capillary Cell: 0.015 mm

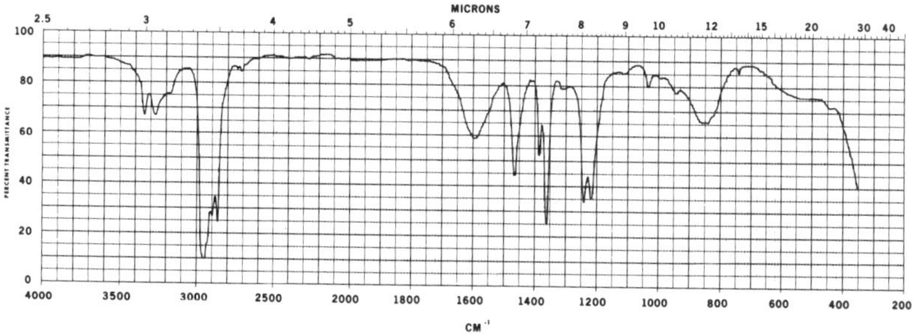

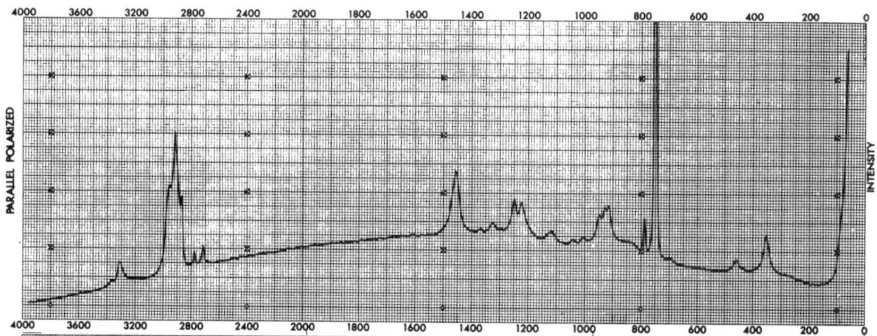

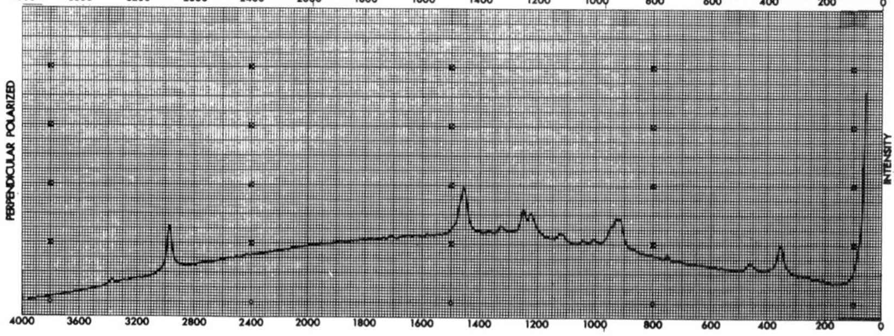

1905 R

DIBUTYLAMINE

$C_8H_{19}N$ Mol. Wt. 129.25

B. P. 158-161°C

Source: Eastman Organic Chemicals
Rochester, New York

$CH_3CH_2CH_2CH_2-NH-CH_2CH_2CH_2CH_3$

RAMAN DATA

Spectrophotometer	Cary 83
Sensitivity	18000; 1800
Suppression	1; 0
Laser Power	140 mW
Spectral Slitwidth	6 cm^{-1}

WAVENUMBER / INTENSITY

2935/160.6 2910/152.5 2874/203.4
1444/76.7

Capillary Cell: Neat

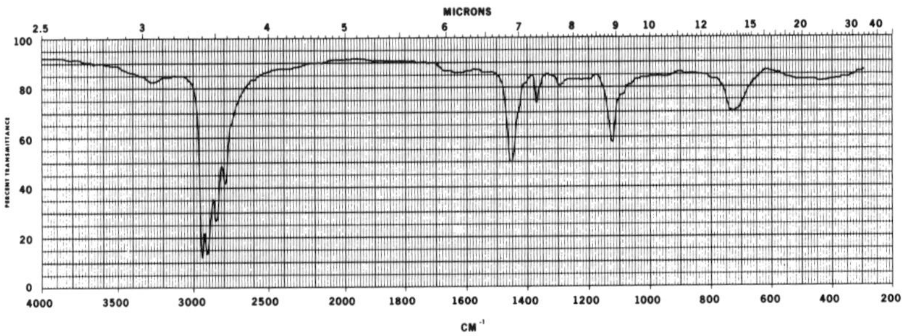

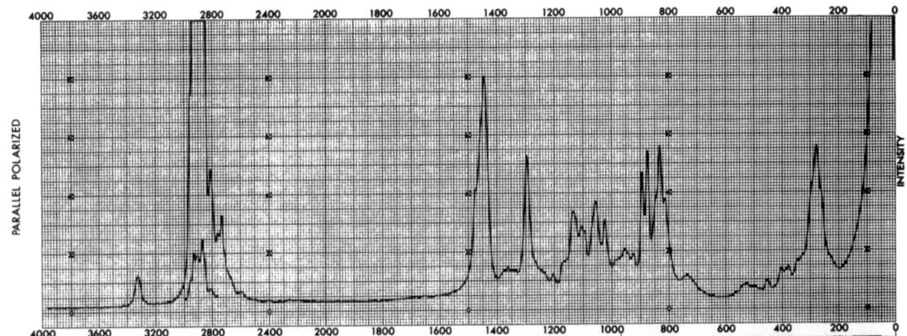

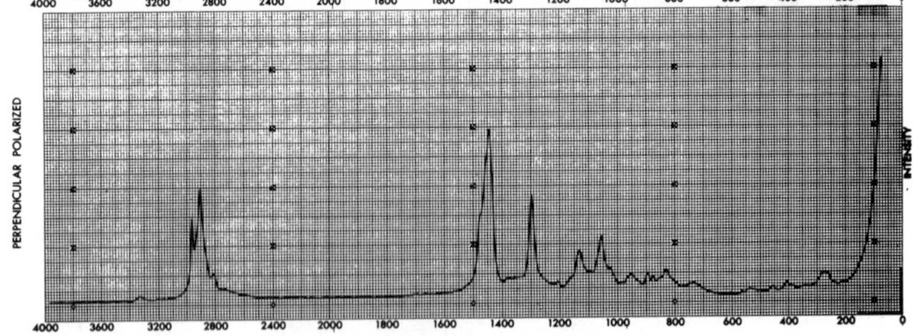

N,N,N',N'-TETRAMETHYL-1,3-PROPANEDIAMINE

1077 R

RAMAN DATA

Spectrophotometer	Cary 83
Sensitivity	13000; 1300
Suppression	9; 0
Laser Power	100 mW
Spectral Slitwidth	6 cm^{-1}

$C_7H_{18}N_2$ Mol. Wt. 130.23

B. P. 145-146°C

Source: Aldrich Chemical Company, Inc. Milwaukee, Wisconsin

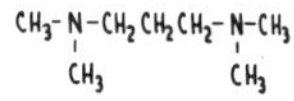

WAVENUMBER / INTENSITY

2949/129.1 2769/146.6 1469/75.9

Capillary Cell: Neat

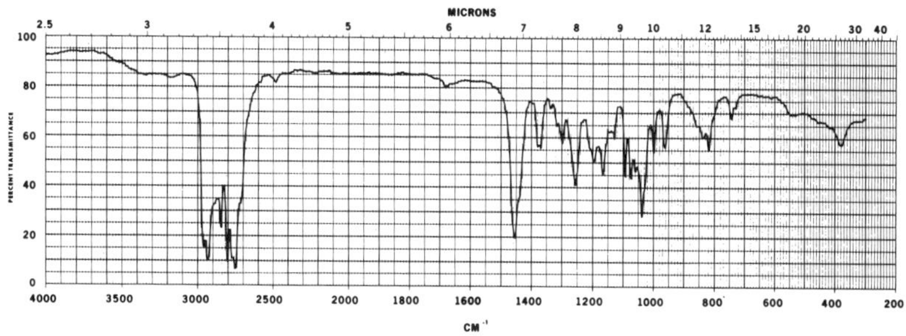

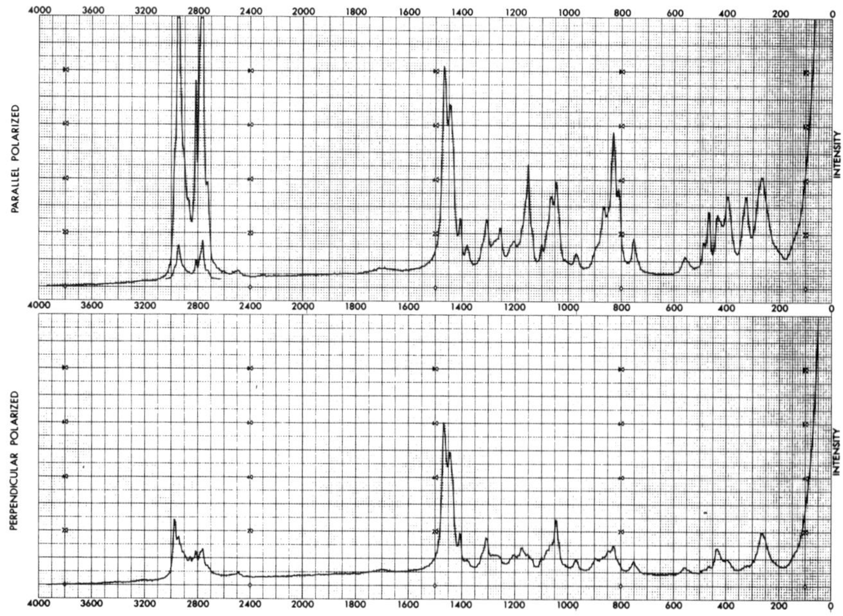

p-CHLOROANILINE **239 R**

RAMAN DATA

Spectrophotometer	Cary 83
Sensitivity	25000
	-
Suppression	25
	-
Laser Power	11 mW
Spectral Slitwidth	6 cm^{-1}

C_6H_6ClN Mol. Wt. 127.57

M.P. 70-72°C

Source: Chem Service, Inc., West Chester, Pa.

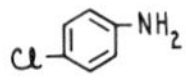

WAVENUMBER / INTENSITY	
1095/57.6	130/109.0

Capillary Cell: Melt

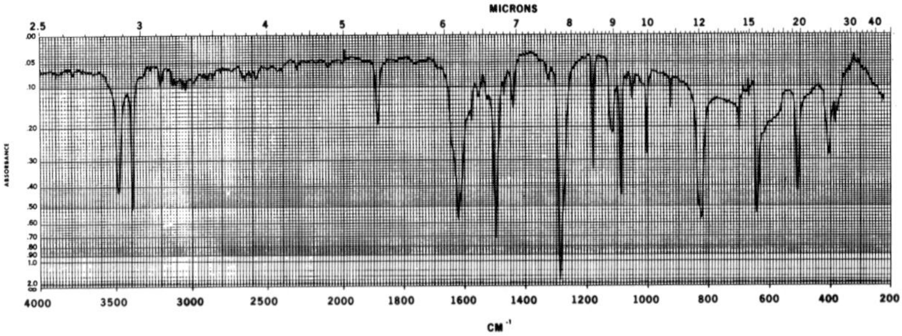

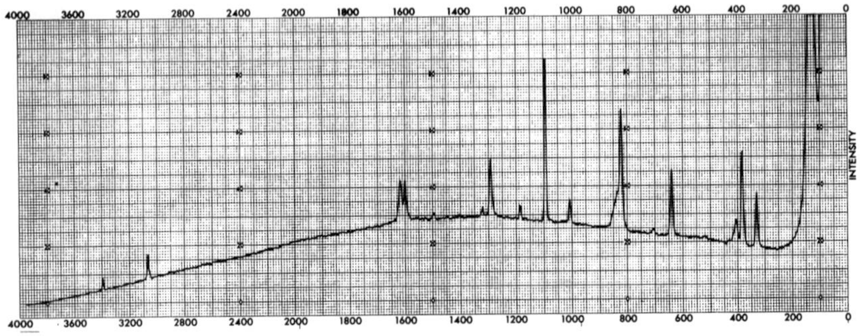

1906 R

N-METHYLANILINE

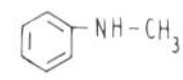

RAMAN DATA

Spectrophotometer	Cary 83
Sensitivity	40000
	-
Suppression	0
	-
Laser Power	60 mW
Spectral Slitwidth	8 cm^{-1}

C_7H_9N Mol. Wt. 107.16

B.P. 74-75°C/10mm

Source: Eastman Organic Chemicals
Rochester, New York

WAVENUMBER / INTENSITY

1030/63.4 997/187.0

Capillary Cell: Neat

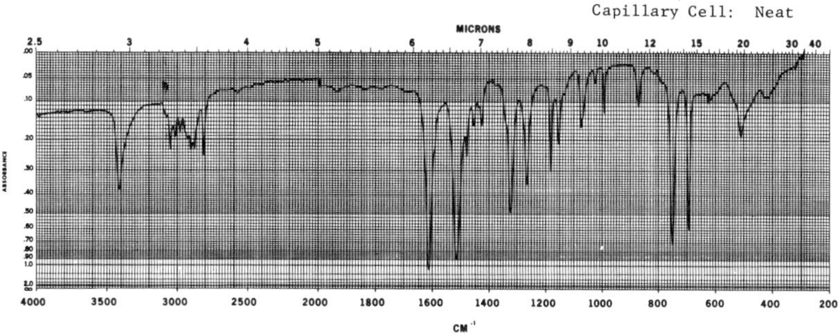

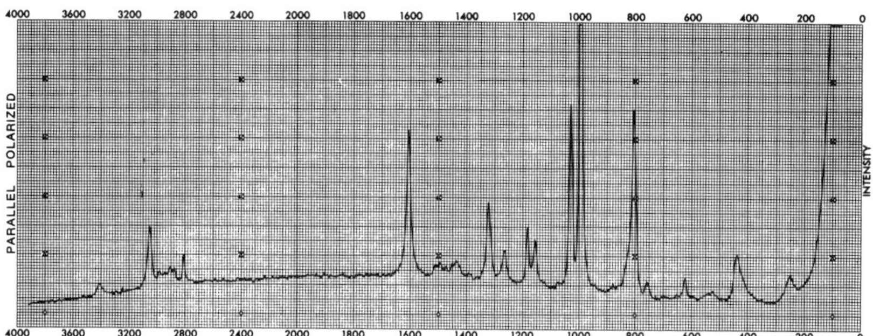

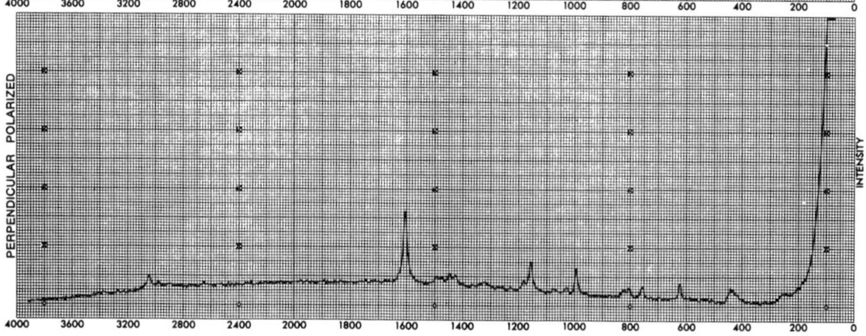

2987 R

tert-BUTYLAMINE, HYDROCHLORIDE

$C_4H_{11}N \cdot HCl$ Mol. Wt. 109.60

Source of Sample: Eastman Organic Chemicals
Rochester, New York

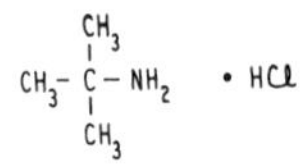

RAMAN DATA

Spectrophotometer	Cary 83
Sensitivity	30000; 3000
Suppression	0; -
Laser Power	80 mW
Spectral Slitwidth	8 cm^{-1}

WAVENUMBER / INTENSITY

898/58.9 750/170.9 467/98.2
187/81.8

KBr Wafer

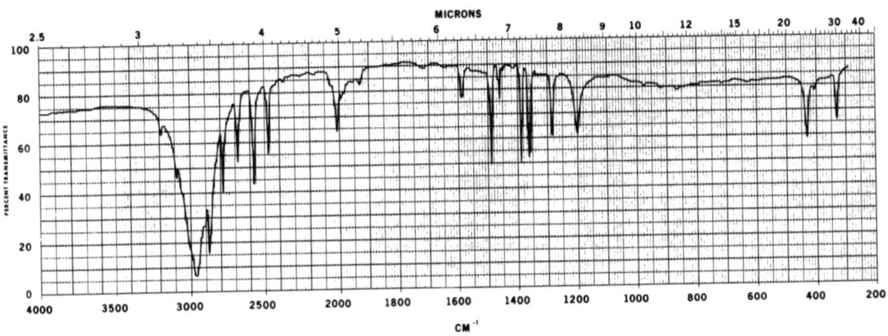

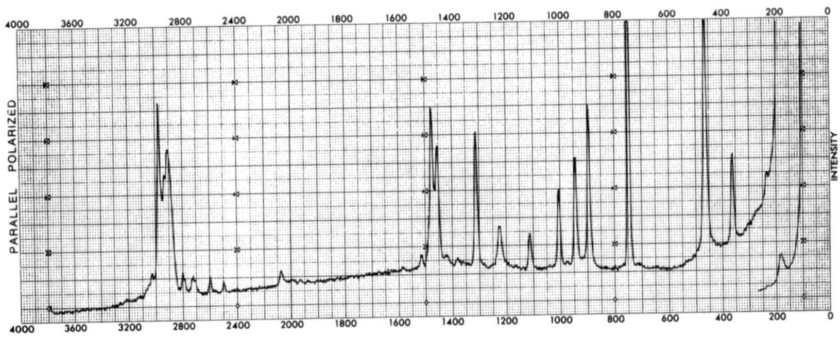

2578 R

ANILINE, SULFATE (2:1)

$(C_6H_7N)_2 \cdot H_2SO_4$ Mol. Wt. 284.34

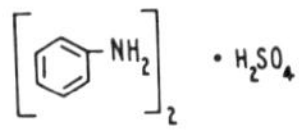

RAMAN DATA

Spectrophotometer	Cary 83
Sensitivity	15000; 1500
Suppression	47; 0
Laser Power	50 mW
Spectral Slitwidth	6 cm^{-1}

Source of Sample: MCB Mfg. Chemists
Norwood, Ohio

WAVENUMBER / INTENSITY

1007/137.8 980/62.9 174/31.7

KBr Wafer

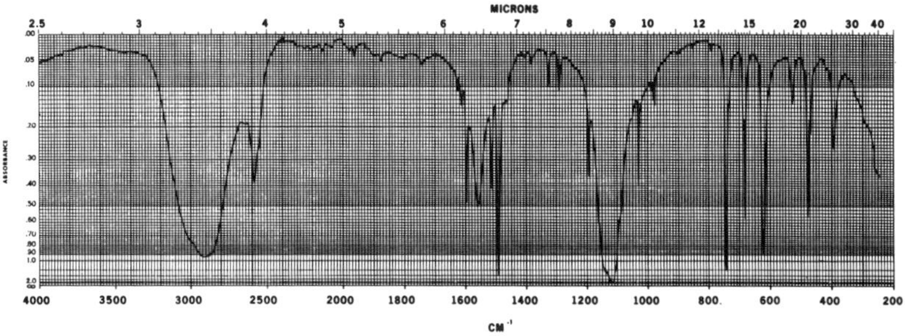

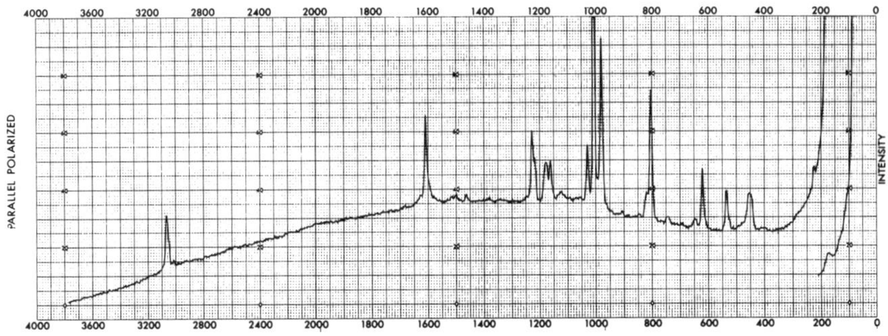

3572 R

DIPROPYLAMINE, HYDROCHLORIDE

$C_6H_{15}N \cdot HCl$ Mol. Wt. 137.65

M.P. 276-278°C

Source of Sample: Eastman Organic Chemicals
Rochester, New York

$CH_3CH_2CH_2-NH-CH_2CH_2CH_3 \cdot HCl$

RAMAN DATA

Spectrophotometer	Cary 83
Sensitivity	30000
	3000
Suppression	0
	-
Laser Power	80 mW
Spectral Slitwidth	8 cm-1

WAVENUMBER / INTENSITY

2928/88.0 332/85.2

KBr Wafer

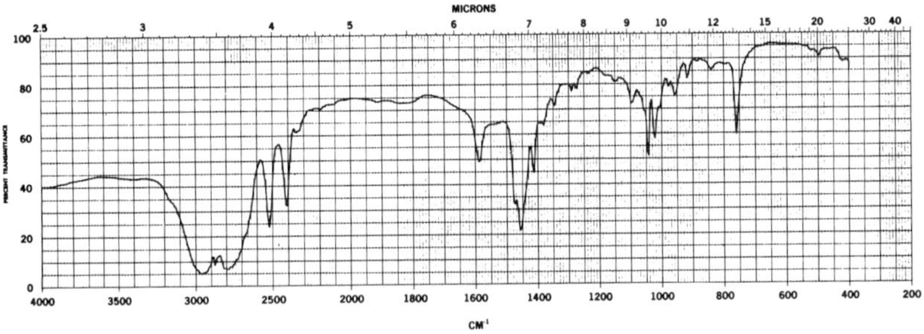

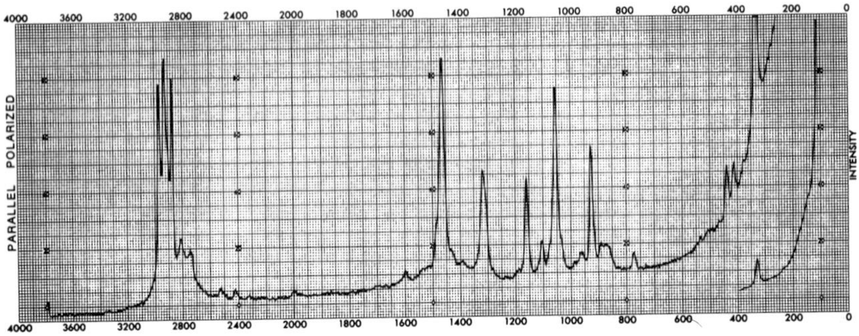

2623 R

TRIETHYLAMINE, HYDROCHLORIDE

$C_6H_{15}N \cdot HCl$ Mol. Wt. 137.65

M.P. 254°C (lit.)

Source of Sample: Fluka AG
Buchs, Switzerland

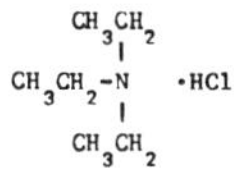

RAMAN DATA

Spectrophotometer	Cary 83
Sensitivity	36000;
	3600
Suppression	41;
	0
Laser Power	25 mW
Spectral Slitwidth	8 cm^{-1}

WAVENUMBER / INTENSITY

2982/81.2 466/108.0

KBr Wafer

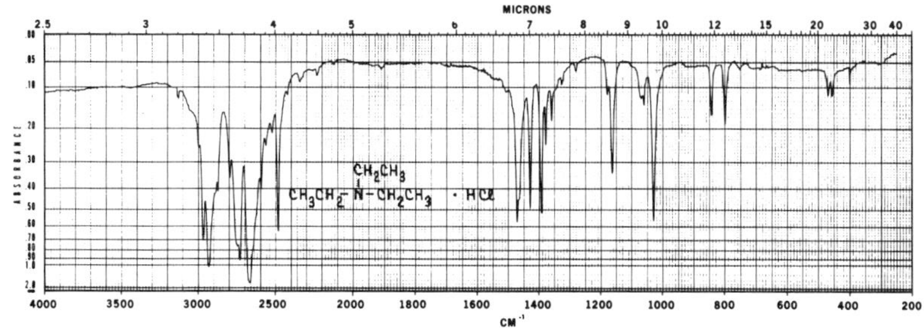

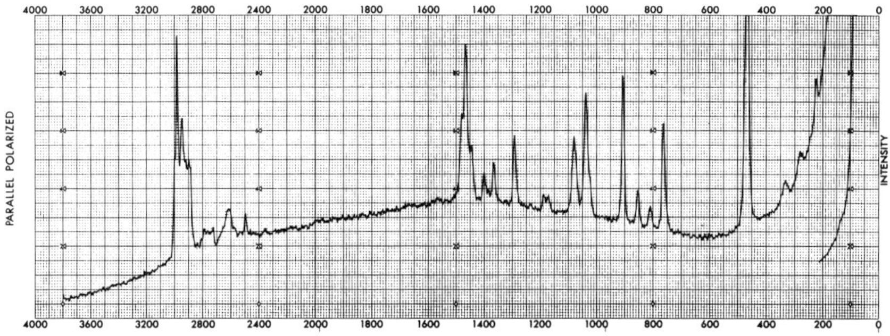

298 R

1-NITROPROPANE

RAMAN DATA

Spectrophotometer	Cary 83
Sensitivity	50000; 10000
Suppression	80; 0
Laser Power	22 mW
Spectral Slitwidth	6 cm^{-1}

$C_3H_7NO_2$ Mol. Wt. 89.09

B.P. 129-131°C

Source: Chem Service, Inc., West Chester, Pa.

$CH_3CH_2CH_2\text{-}NO_2$

WAVENUMBER / INTENSITY		
2940/93.7	2880/52.2	1380/124.2
895/138.0	868/149.9	

Capillary Cell: Neat

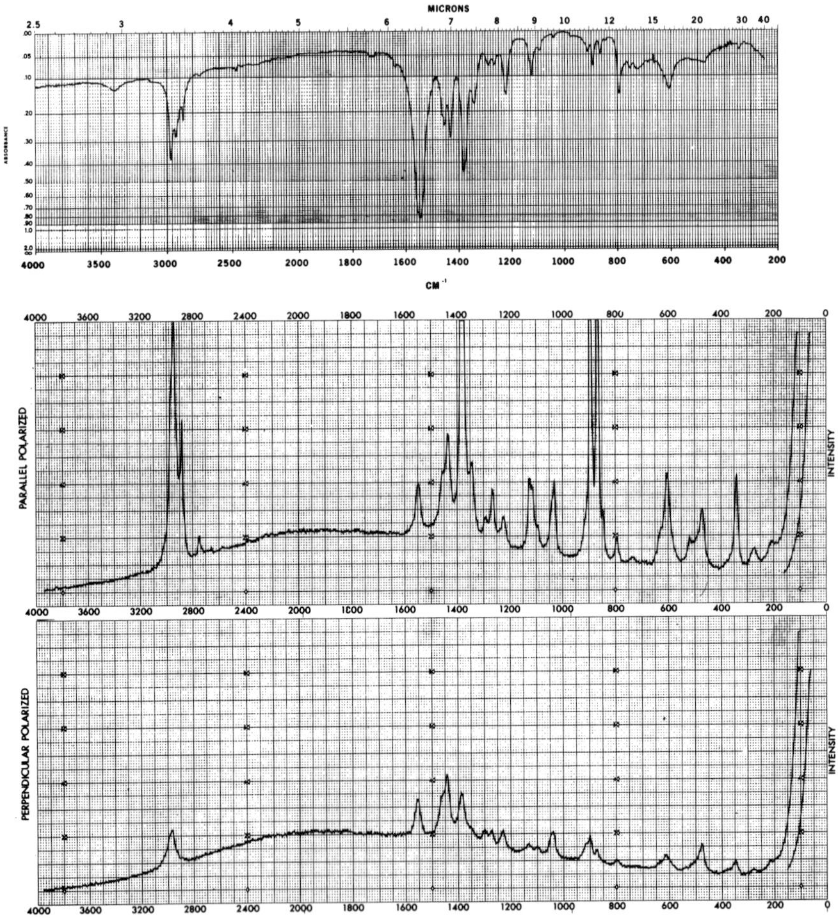

300 R

NITROBENZENE

RAMAN DATA

Spectrophotometer	Cary 83
Sensitivity	7500
	-
Suppression	0
	-
Laser Power	22 mW
Spectral Slitwidth	4 cm^{-1}

$C_6H_5NO_2$ Mol. Wt. 123.11

B.P. 211°C (lit.)

Source: Chem Service, Inc., West Chester, Pa.

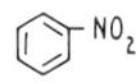

WAVENUMBER / INTENSITY		
1350/305.6	1010/112.4	856/48.8

Capillary Cell: Neat

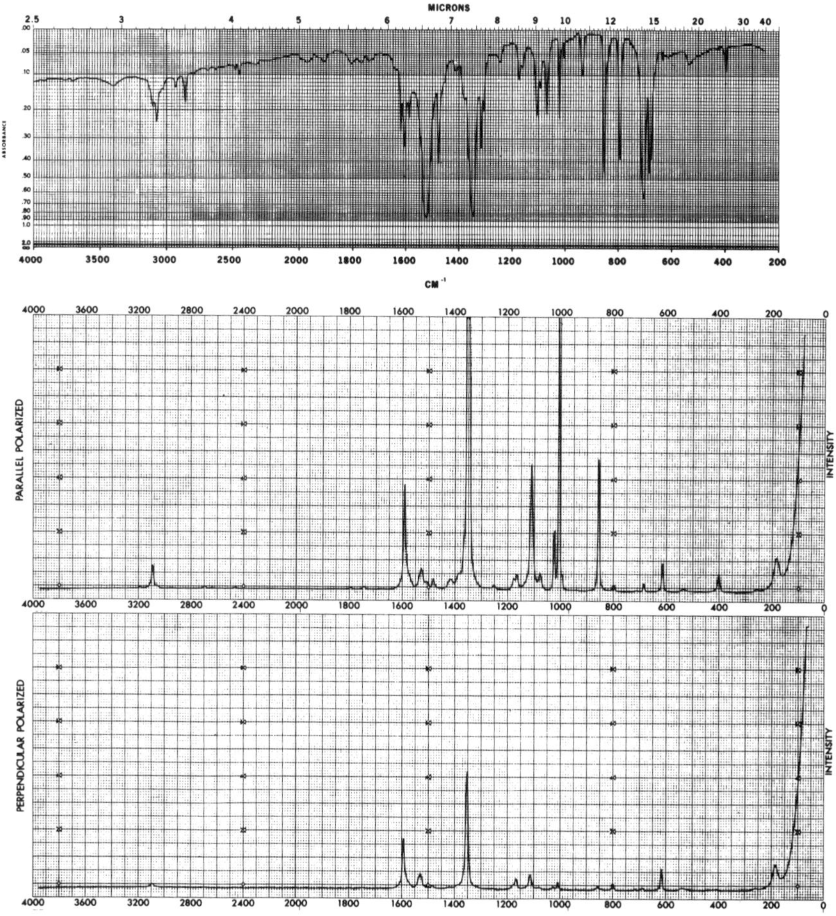

3455 R

2-METHYL-2-NITROSOPROPANE

C_4H_9NO Mol. Wt. 87.12

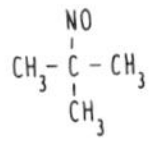

RAMAN DATA

Spectrophotometer	Cary 83
Sensitivity	20000; 2000
Suppression	0; 17
Laser Power	95 mW
Spectral Slitwidth	6 cm-1

Source of Sample: Aldrich Chemical Co., Inc.
Milwaukee, Wisconsin

WAVENUMBER / INTENSITY

1268/84.7 734/131.1 250/64.5
115/50.0

KBr Wafer

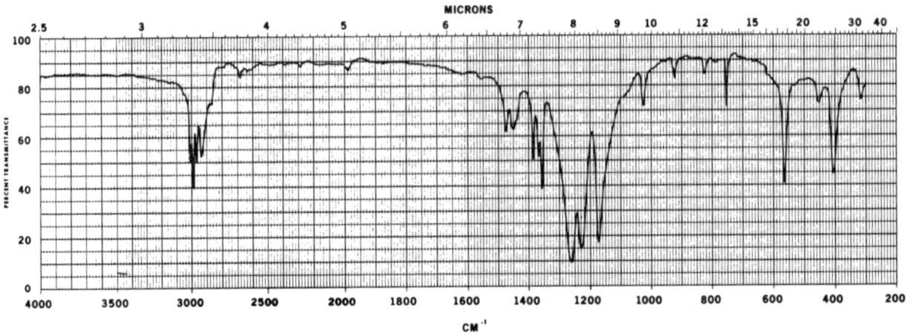

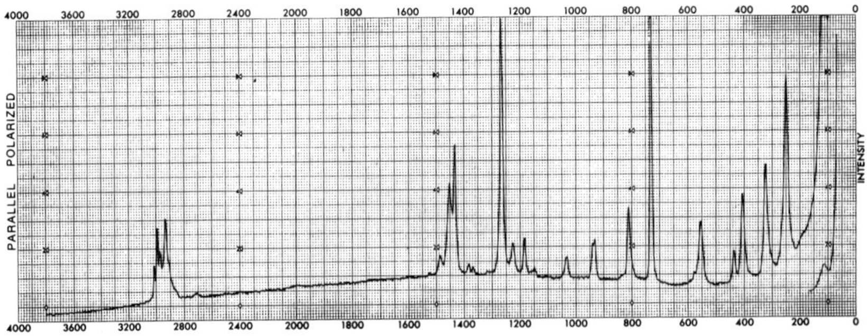

AZOXYBENZENE **295 R**

RAMAN DATA

Spectrophotometer	Cary 83
Sensitivity	43750
	-
Suppression	55
	-
Laser Power	11 mW
Spectral Slitwidth	12 cm^{-1}

$C_{12}H_{10}N_2O$ Mol. Wt. 198.23

M.P. 35-36°C

Source: Chem Service, Inc., West Chester, Pa.

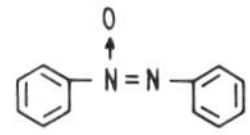

WAVENUMBER / INTENSITY		
1484/88.9	1335/138.2	1280/111.7

Capillary Cell: Melt

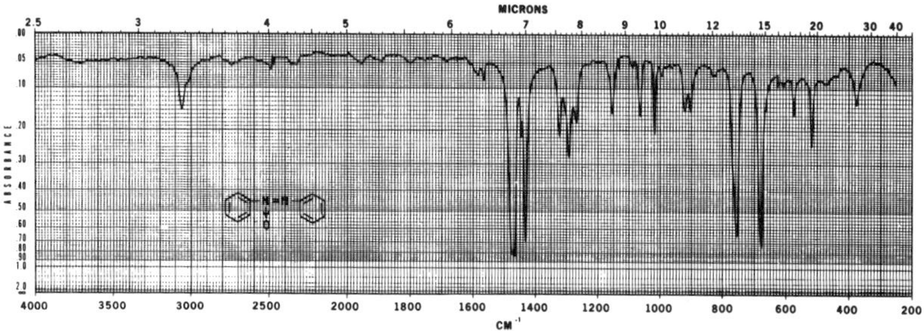

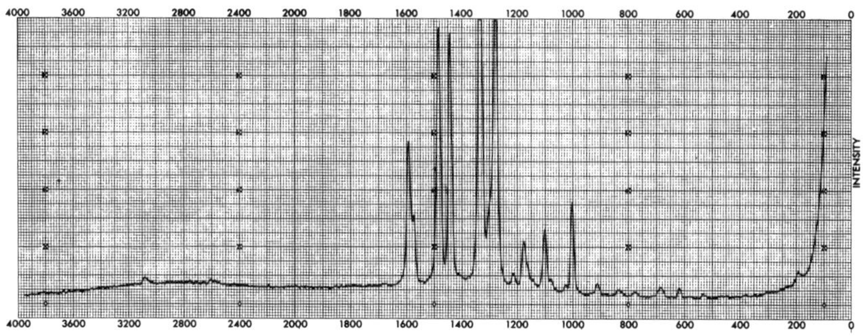

4238 R

PYRAZINE, 1,4-DIOXIDE

$C_4H_4N_2O_2$ Mol. Wt. 112.09

M.P. > 300°C

Source of Sample: B. Klein,
Veterans Administration Hospital,
East Orange, New Jersey

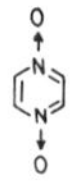

RAMAN DATA

Spectrophotometer	Cary 83
Sensitivity	1200;
	120
Suppression	5;
	0
Laser Power	100 mW
Spectral Slitwidth	6 cm^{-1}

WAVENUMBER / INTENSITY

1630/93.5	878/393.7	120/30.0
103/182.0	78/170.0	

KBr Wafer

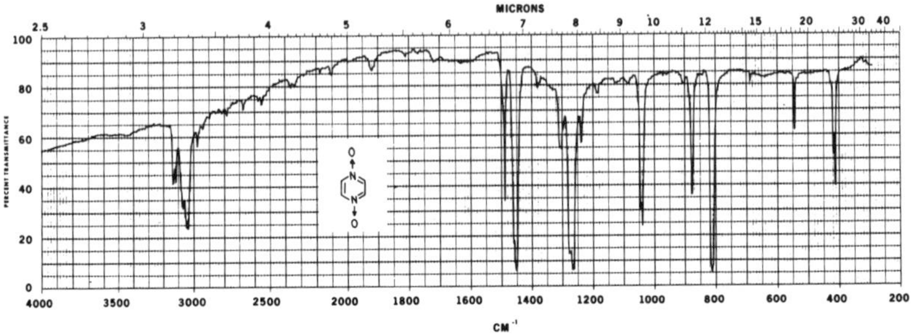

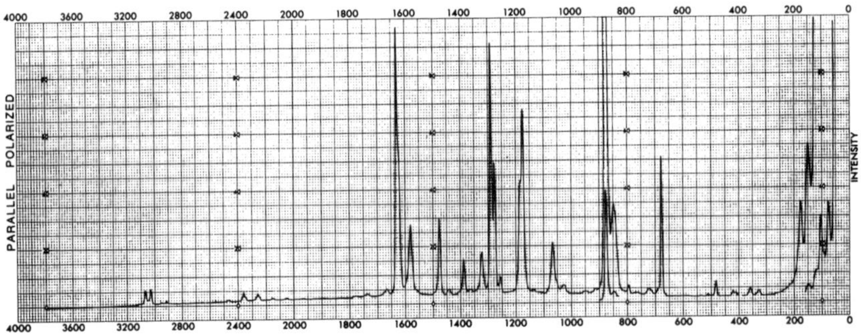

BUTYRALDEHYDE, OXIME **3675 R**

RAMAN DATA

Spectrophotometer	Cary 83
Sensitivity	19000; 1900
Suppression	27; 2
Laser Power	140 mW
Spectral Slitwidth	10 cm-1

C_4H_9NO Mol. Wt. 87.12

B.P. 152-153°C

Source of Sample: The Ames Laboratories, Inc. Milford, Connecticut

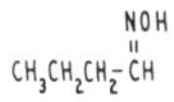

WAVENUMBER / INTENSITY

2925/110.0	2900/96.8	2860/106.8
1650/91.9	1450/47.0	882/62.7

Capillary Cell: Neat

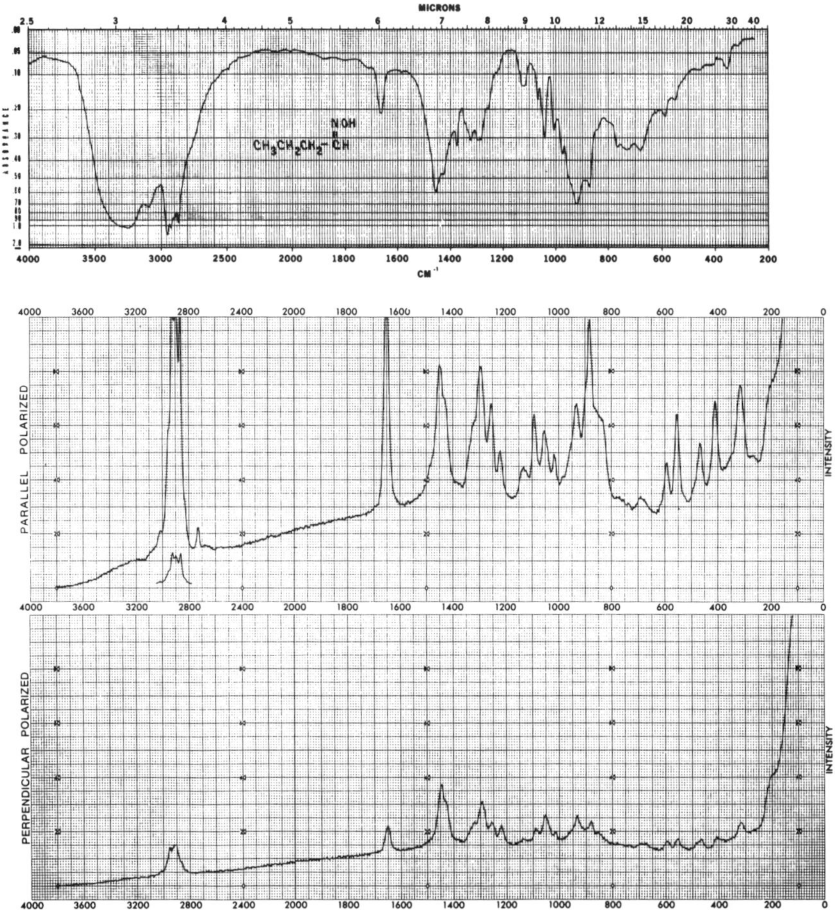

2207 R

OCTYL NITRITE

RAMAN DATA

Spectrophotometer	Cary 83
Sensitivity	20000; 2000
Suppression	0; 0
Laser Power	100 mW
Spectral Slitwidth	8 cm^{-1}

$C_8H_{17}NO_2$ Mol. Wt. 159.23

B.P. 52-53°C/6mm

Source of Sample: Eastman Organic Chemicals Rochester, New York

$CH_3CH_2(CH_2)_5CH_2$-O-N=O

WAVENUMBER / INTENSITY

2934/118.3	2916/115.3	2872/121.7
2858/126.6	1450/59.5	

Capillary Cell: Neat

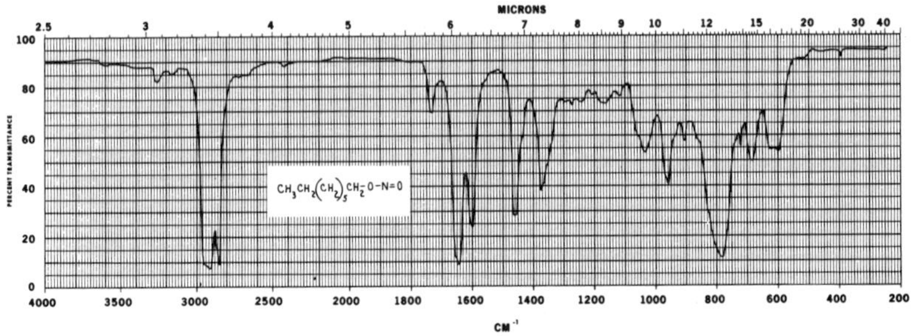

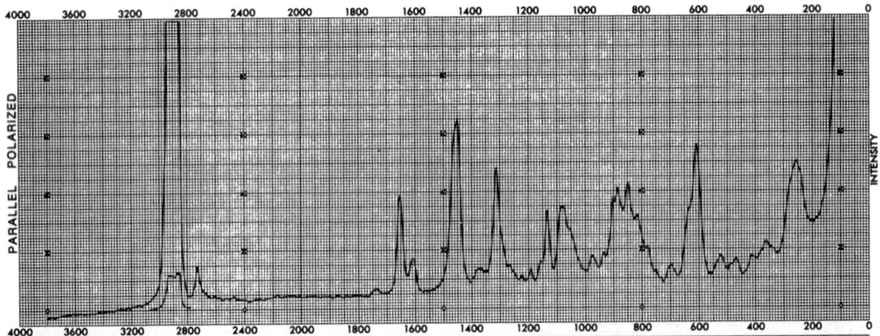

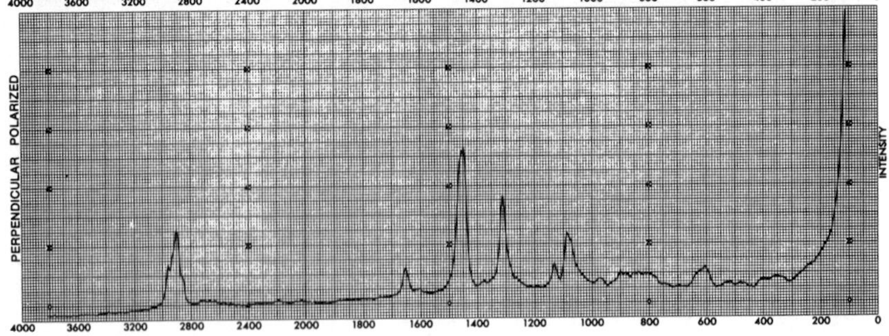

138 R

PROPYL NITRATE

RAMAN DATA

Spectrophotometer	Cary 83
Sensitivity	37500; 10000
Suppression	150; 40
Laser Power	18 mW
Spectral Slitwidth	6 cm^{-1}

$C_3H_7NO_3$ Mol. Wt. 105.09

B.P. 108-110°C d_4^{20} 1.0538 (lit.)

n_D^{20} 1.3973 (lit.)

Source: Eastman Organic Chemicals, Rochester, New York

$CH_3CH_2CH_2-ONO_2$

WAVENUMBER / INTENSITY		
2950/90.6	1282/194.1	865/162.4
574/76.4	354/98.6	

Capillary Cell: Neat

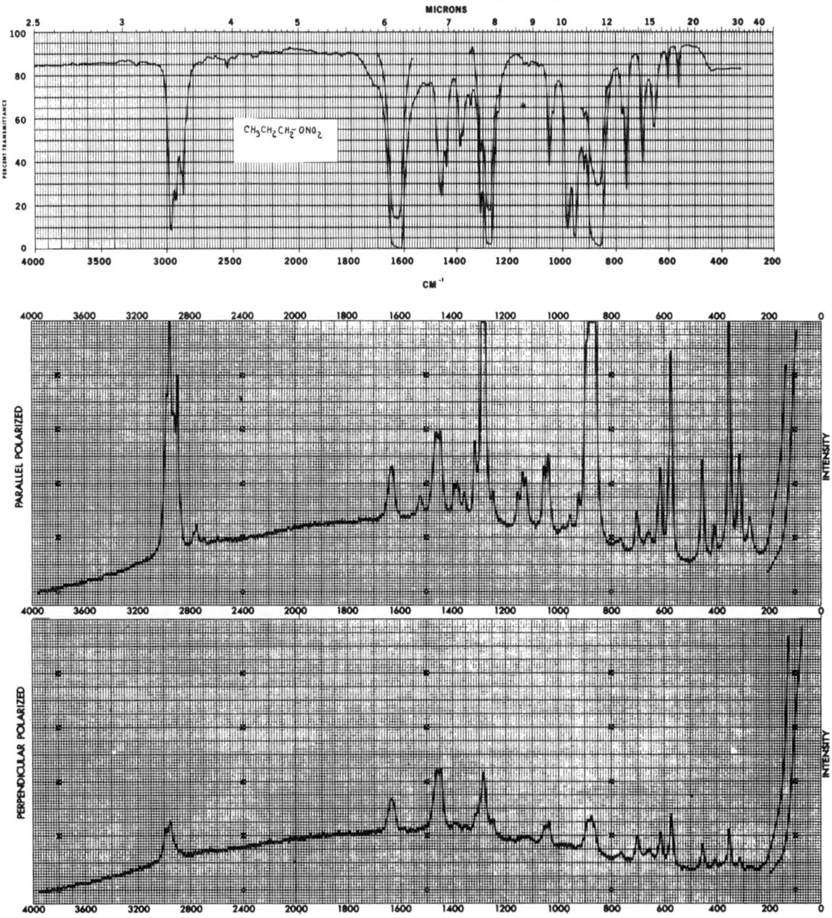

962 R

ISOCYANIC ACID, PHENYL ESTER

RAMAN DATA

Spectrophotometer	Cary 83
Sensitivity	1800; 180
Suppression	0; 0
Laser Power	195 mW
Spectral Slitwidth	6 cm^{-1}

C_7H_5NO Mol. Wt. 119.12

B. P. 163.7°C

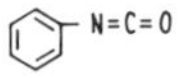

Source: Eastman Organic Chemicals
Rochester, New York

WAVENUMBER / INTENSITY

1602/116.4 1517/53.4 1447/94.5
1009/224.9

Capillary Cell: Neat

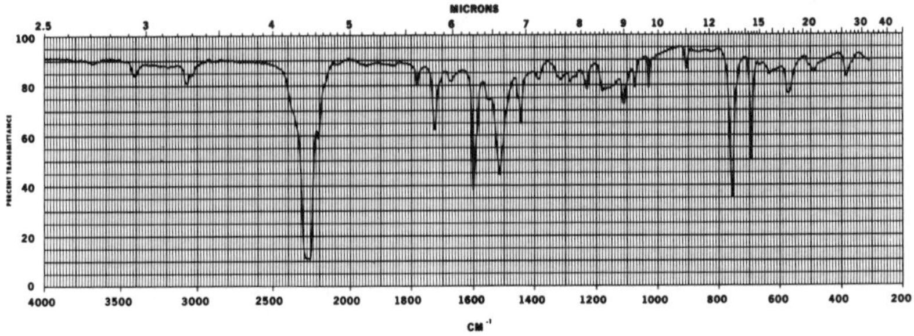

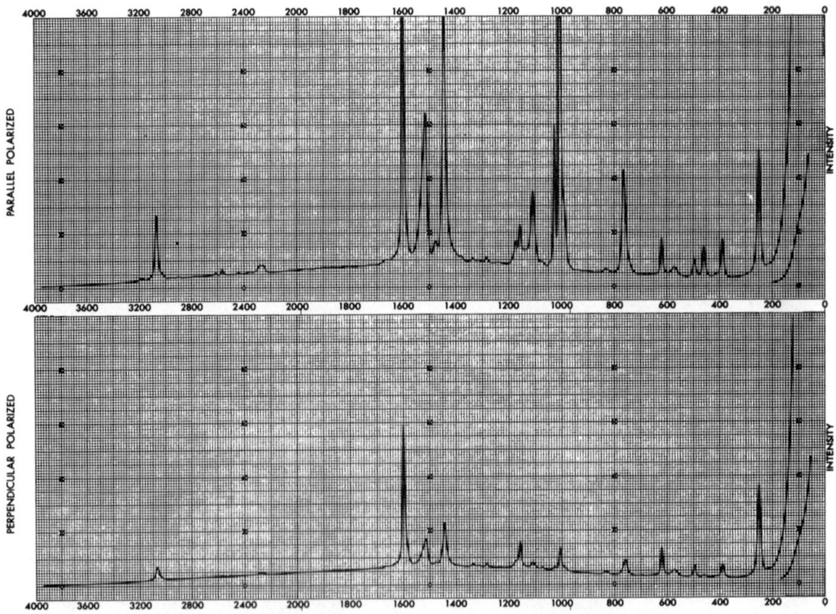

3152 R

ISOTHIOCYANIC ACID, PHENYL ESTER

C_7H_5NS Mol. Wt. 135.19

B.P. 89-90°C/10mm

Source of Sample: MCB Manufacturing Chemists
Norwood, Ohio

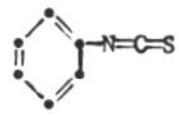

RAMAN DATA

Spectrophotometer	Cary 83
Sensitivity	5000;
	500
Suppression	14;
	-
Laser Power	90 mW
Spectral Slitwidth	6 cm^{-1}

WAVENUMBER / INTENSITY

1590/139.8	1245/144.6	1167/49.5
1030/130.4		

Capillary Cell: Neat

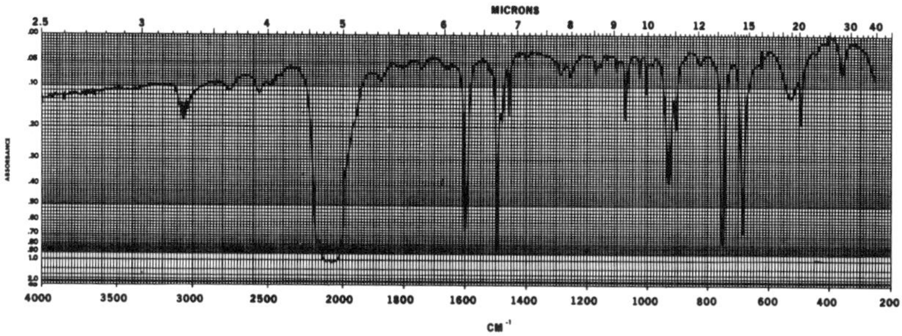

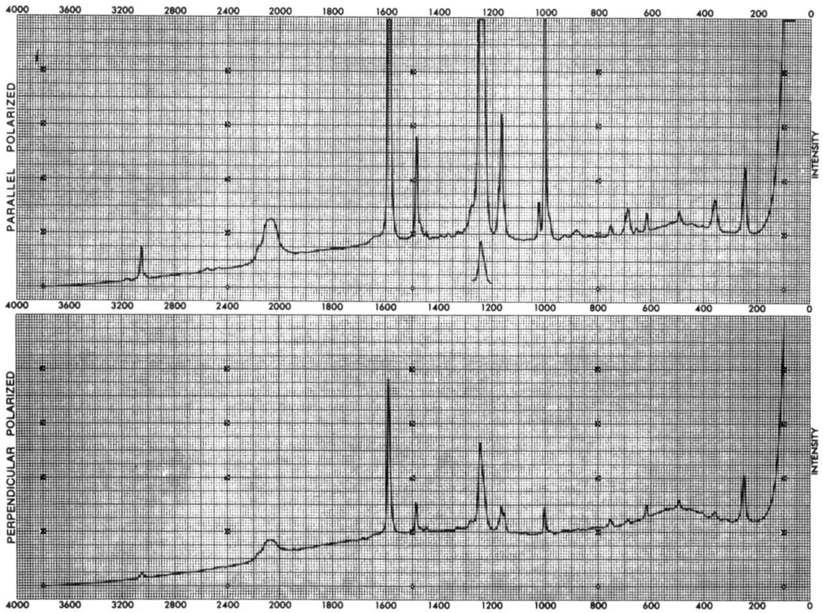

2785 R

1-BUTANETHIOL

$C_4H_{10}S$ Mol. Wt. 90.19

B.P. 98°C

Source of Sample: Tokyo Kasei Kogyo Co. Ltd. Tokyo, Japan

$CH_3CH_2CH_2CH_2$-SH

RAMAN DATA

Spectrophotometer	Cary 83
Sensitivity	7000
	-
Suppression	46
	-
Laser Power	150 mW
Spectral Slitwidth	6 cm^{-1}

WAVENUMBER / INTENSITY

2932/131.4 2872/96.3 2575/108.4
668/135.1

Capillary Cell: 0.015mm

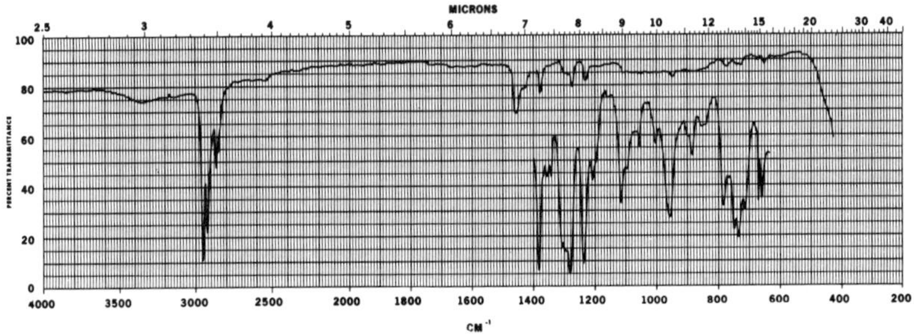

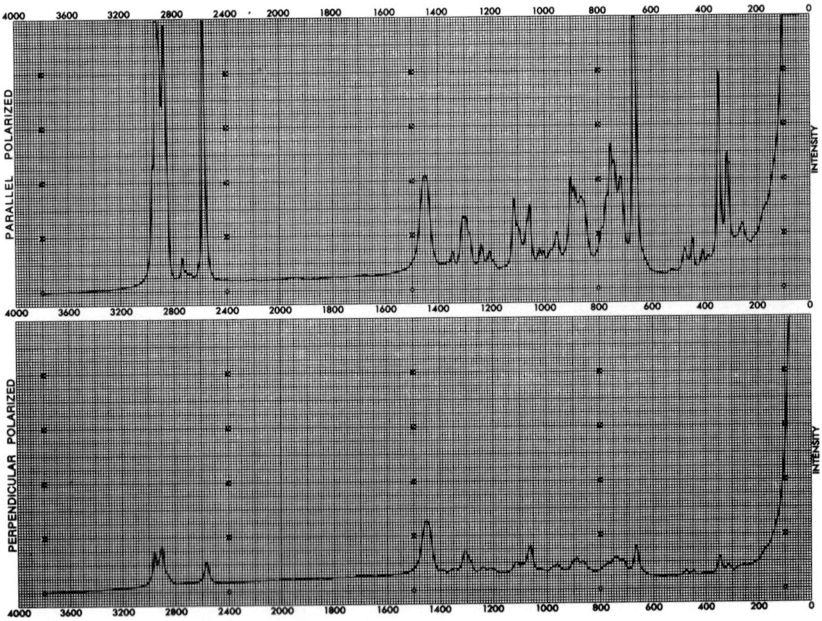

BENZENETHIOL **227 R**

RAMAN DATA	
Spectrophotometer	Cary 83
Sensitivity	25000;
	10000
Suppression	20;
	0
Laser Power	22 mW
Spectral Slitwidth	4 cm^{-1}

C_6H_6S Mol. Wt. 110.18

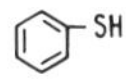

B.P. 62-63°C/15 mm

Source: Chem Service, Inc., West Chester, Pa.

WAVENUMBER / INTENSITY

1098/97.1 1030/200.9 1005/680.8
703/94.3

Capillary Cell: Neat

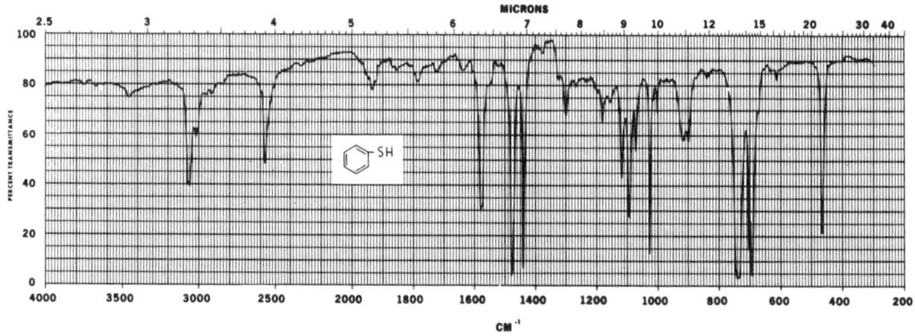

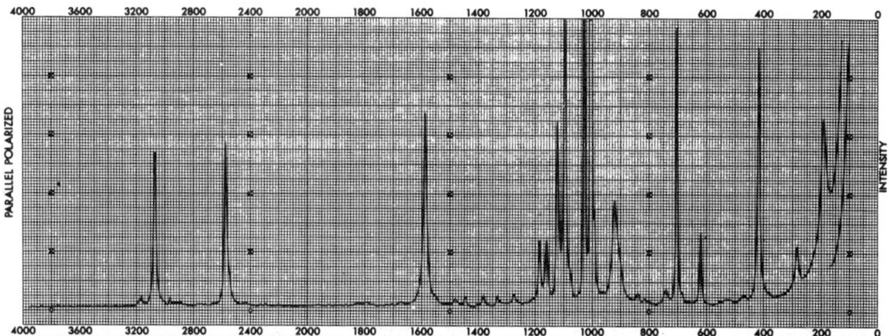

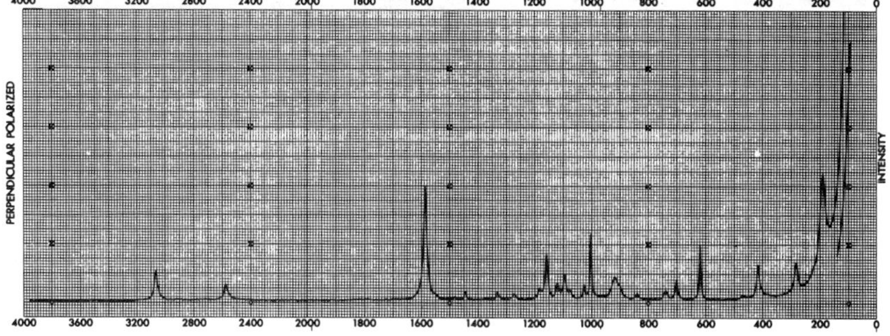

1947 R

BUTYL SULFIDE

$C_8H_{18}S$ Mol. Wt. 146.30

B.P. 182-186°C

Source: Eastman Organic Chemicals
Rochester, New York

$CH_3CH_2CH_2CH_2-S-CH_2CH_2CH_2CH_3$

RAMAN DATA

Spectrophotometer	Cary 83
Sensitivity	25000; 2500
Suppression	0; 0
Laser Power	76 mW
Spectral Slitwidth	6 cm^{-1}

WAVENUMBER / INTENSITY

2930/113.4 2902/141.4 2876/113.9
900/58.1

Capillary Cell: Neat

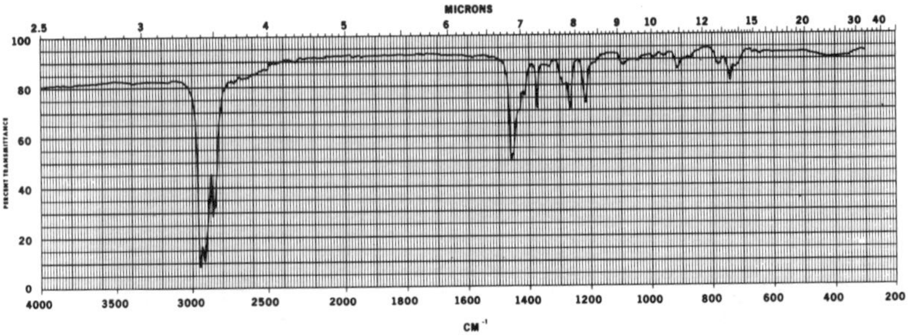

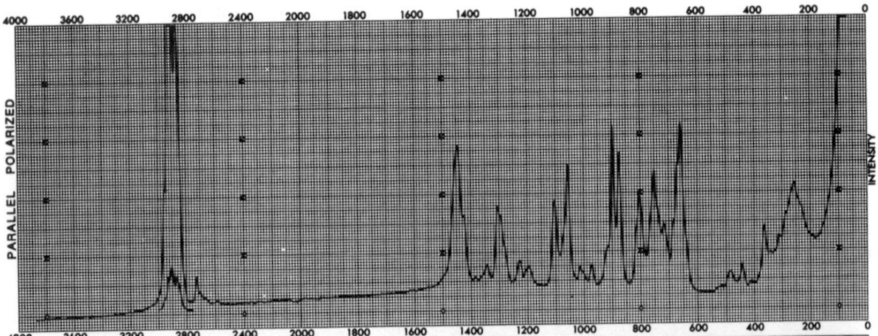

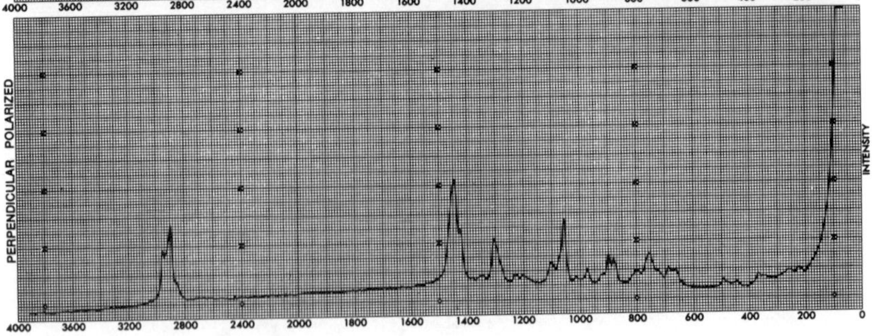

3204 R

BUTYL DISULFIDE

RAMAN DATA

Spectrophotometer	Cary 83
Sensitivity	20000; 2000
Suppression	0; -
Laser Power	90 mW
Spectral Slitwidth	10 cm^{-1}

$C_8H_{18}S_2$ Mol. Wt. 178.36

B.P. 110-112°C/13mm

Source of Sample: Pfaltz & Bauer Inc.
Stanford, Connecticut

$CH_3CH_2CH_2CH_2-S-S-CH_2CH_2CH_2CH_3$

WAVENUMBER / INTENSITY

2921/89.2	2898/118.9	512/118.3

Capillary Cell: Neat

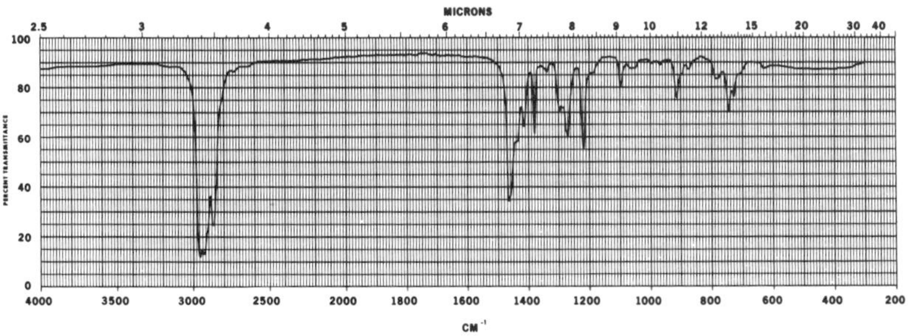

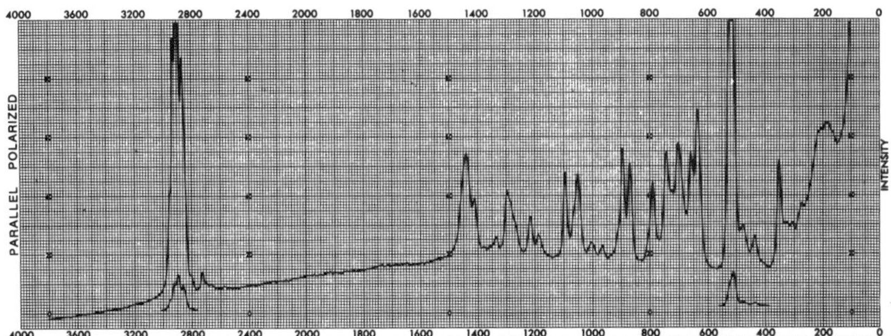

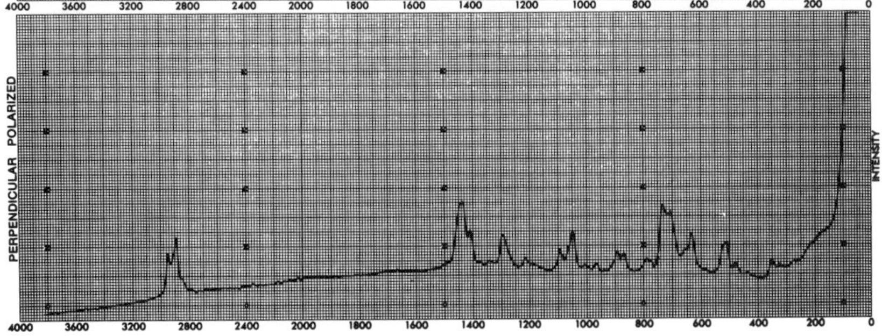

294 R

METHYL SULFOXIDE

RAMAN DATA

Spectrophotometer	Cary 83
Sensitivity	10000
	-
Suppression	0
	-
Laser Power	22 mW
Spectral Slitwidth	6 cm^{-1}

C_2H_6OS Mol. Wt. 78.13

M.P. 14-17°C

Source: Chem Service, Inc., West Chester, Pa.

$CH_3-S(=O)-CH_3$

WAVENUMBER / INTENSITY

2925/72.7 674/203.7

Capillary Cell: Neat

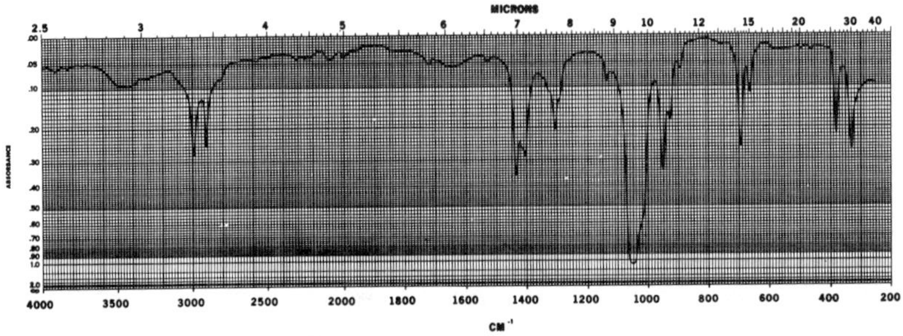

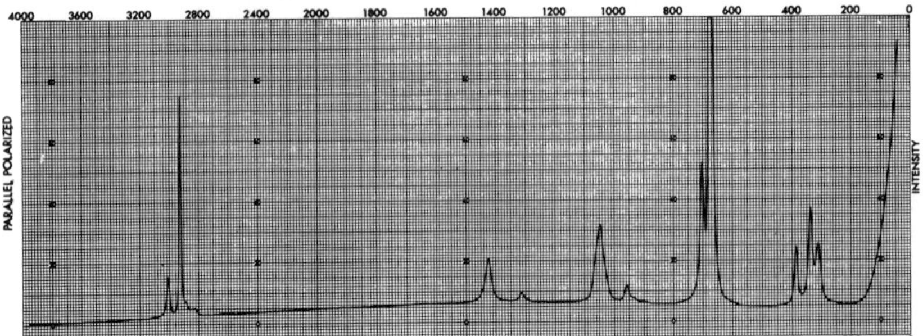

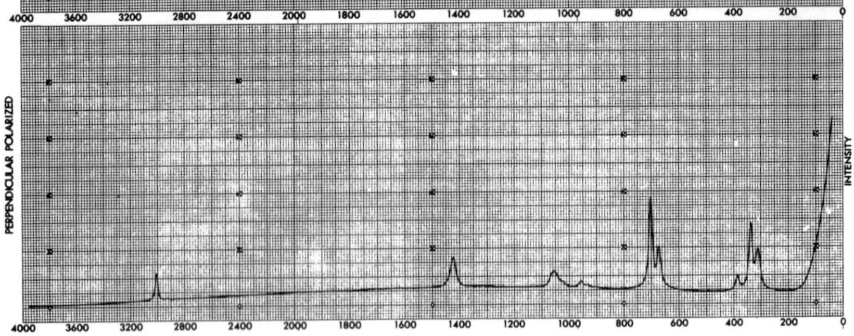

2914 R

SULFUROUS ACID, DIMETHYL ESTER

RAMAN DATA

Spectrophotometer	Cary 83
Sensitivity	16000; 1600
Suppression	28; -
Laser Power	100 mW
Spectral Slitwidth	6 cm^{-1}

$C_2H_6O_3S$ Mol. Wt. 110.13

B.P. 126-127°C

Source of Sample: Eastman Organic Chemicals
Rochester, New York

$CH_3O-S(=O)-OCH_3$

WAVENUMBER / INTENSITY

750/149.4 592/72.9

Capillary Cell: Neat

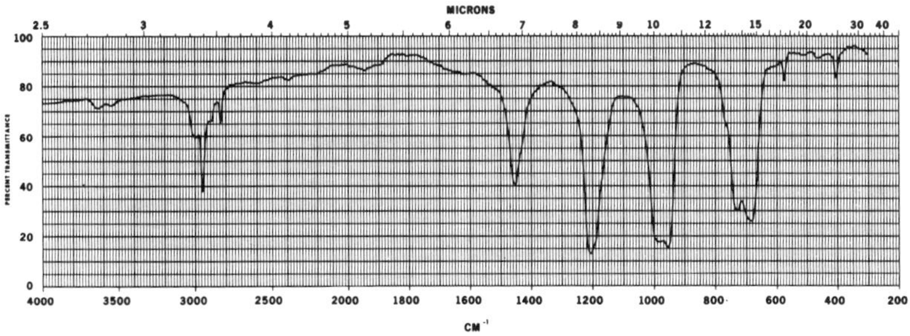

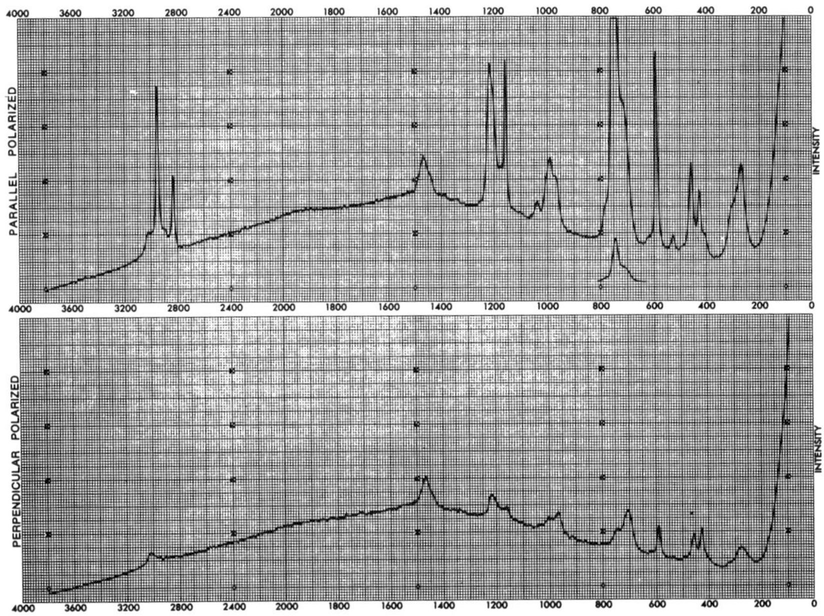

734 R

PHENYL SULFONE

RAMAN DATA

Spectrophotometer	Cary 83
Sensitivity	3300; 375
Suppression	52; 0
Laser Power	125 mW
Spectral Slitwidth	8 cm^{-1}

$C_{12}H_{10}O_2S$ Mol. Wt. 218.28

M.P. 128-129°C

Source: Eastman Organic Chemicals, Rochester, New York

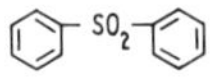

WAVENUMBER / INTENSITY

1587/122.4	1146/136.2	1005/263.7
706/66.5	260/83.1	112/166.3

KBr Wafer

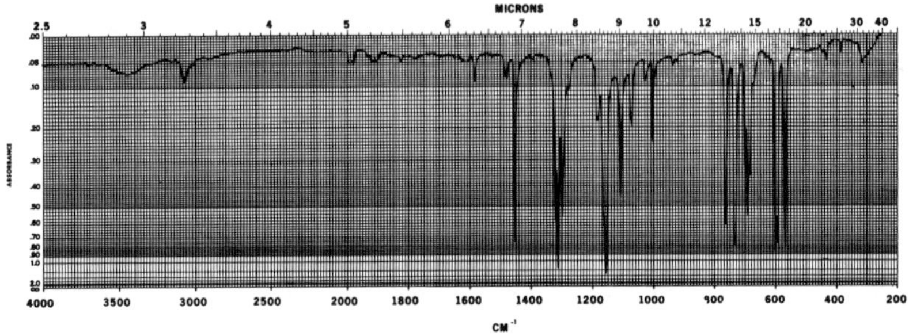

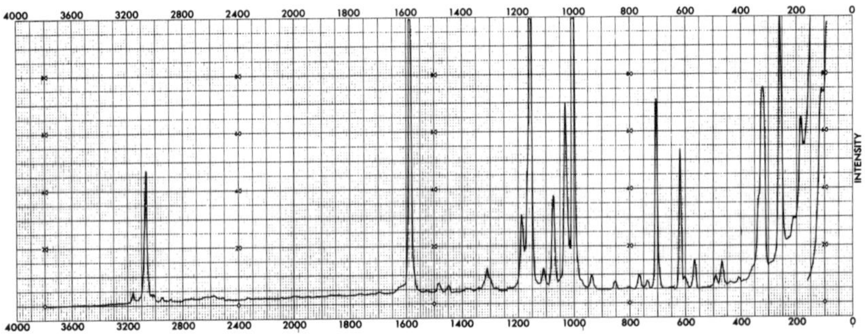

METHANESULFONIC ACID, BUTYL ESTER

2911 R

RAMAN DATA

Spectrophotometer	Cary 83
Sensitivity	15000; 1500
Suppression	7; -
Laser Power	100 mW
Spectral Slitwidth	6 cm^{-1}

$C_5H_{12}O_3S$ Mol. Wt. 152.21

B.P. 102-103°C/5mm

Source of Sample: Eastman Organic Chemicals Rochester, New York

$CH_3-SO_3CH_2CH_2CH_2CH_3$

WAVENUMBER / INTENSITY

2936/92.6 1179/120.7 766/94.5

540/57.0

Capillary Cell: Neat

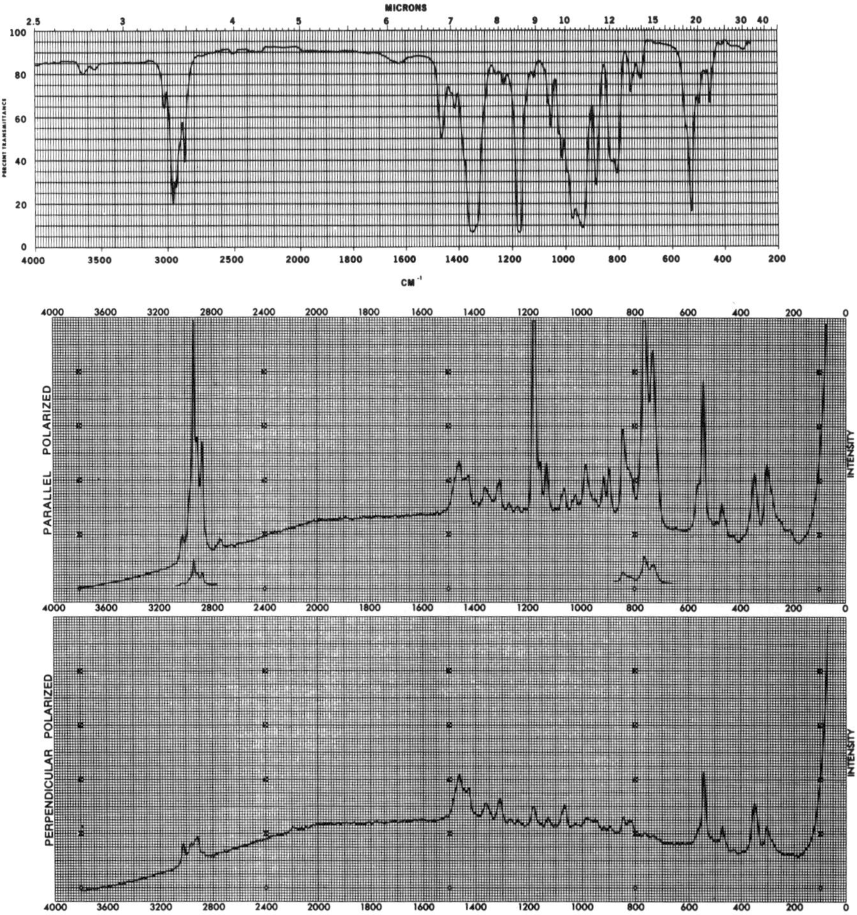

p-CHLOROBENZENESULFONAMIDE

4089 R

RAMAN DATA

Spectrophotometer	Cary 83
Sensitivity	11,000
	-
Suppression	40
	-
Laser Power	100 mW
Spectral Slitwidth	6 cm^{-1}

$C_6H_6ClNO_2S$ Mol. Wt. 191.58

M.P. 145-146°C

Source of Sample: MCB Manufacturing Chemists, Norwood, Ohio

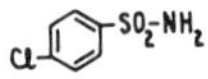

WAVENUMBER / INTENSITY		
1145/107.6	1085/101.9	752/61.1
260/101.5		

KBr Wafer

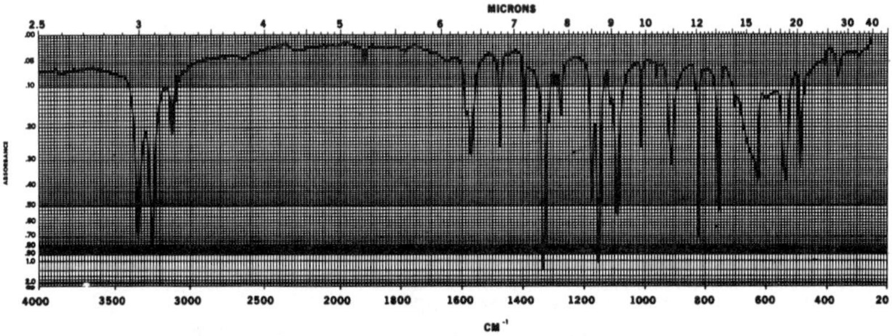

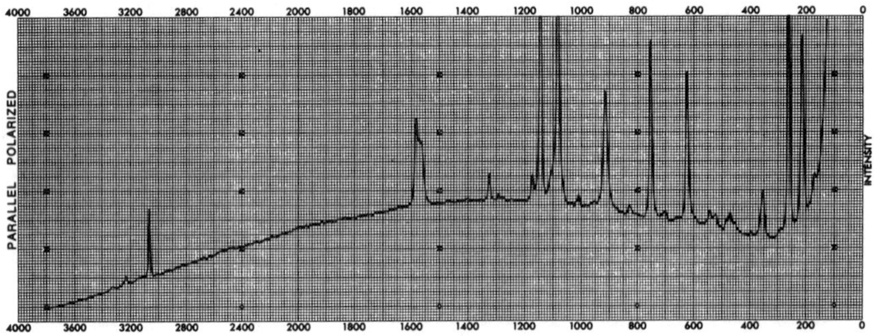

132 R

BENZENESULFONYL CHLORIDE

RAMAN DATA

Spectrophotometer	Cary 83
Sensitivity	5625
	-
Suppression	0
	-
Laser Power	22 mW
Spectral Slitwidth	6 cm^{-1}

$C_6H_5ClO_2S$ Mol. Wt. 176.62

M.P. 15-17°C

Source: Chem Service, Inc., West Chester, Pa.

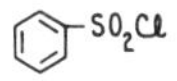

WAVENUMBER / INTENSITY

1175/49.6 1002/125.2 377/115.6
111/61.7

Capillary Cell: Neat

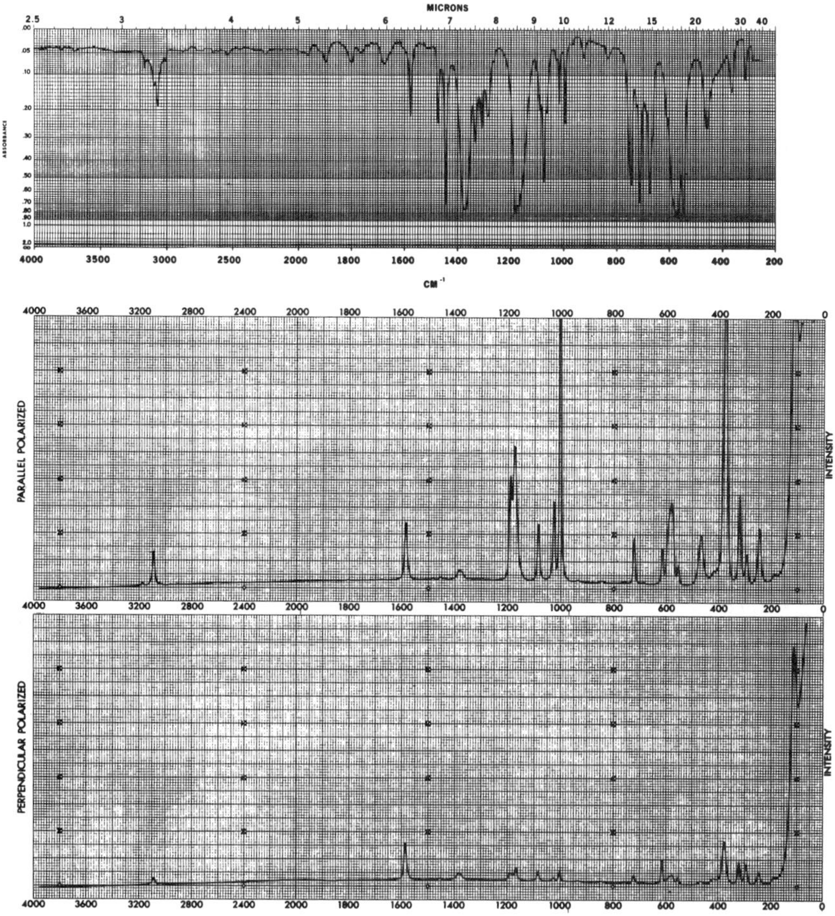

2591 R

1-BUTANESULFONIC ACID, SODIUM SALT

$C_4H_9NaO_3S$ Mol. Wt. 160.17

$CH_3CH_2CH_2CH_2-SO_3Na$

RAMAN DATA

Spectrophotometer	Cary 83
Sensitivity	25000; 2500
Suppression	17; 0
Laser Power	35 mW
Spectral Slitwidth	10 cm^{-1}

Source of Sample: Eastman Organic Chemicals Rochester, New York

WAVENUMBER / INTENSITY

1074/96.4 768/54.7 226/15.8

Split Mull

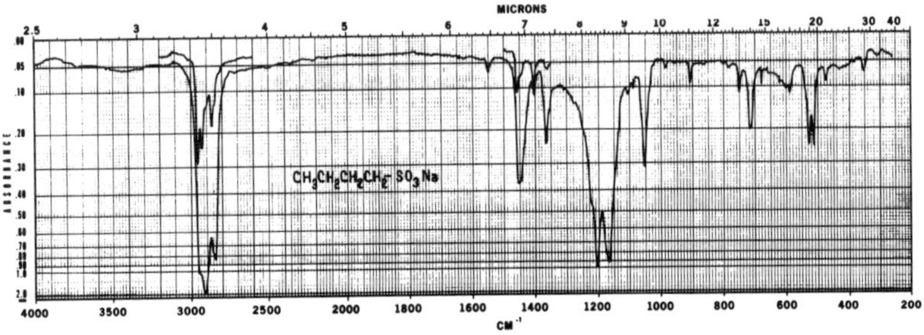

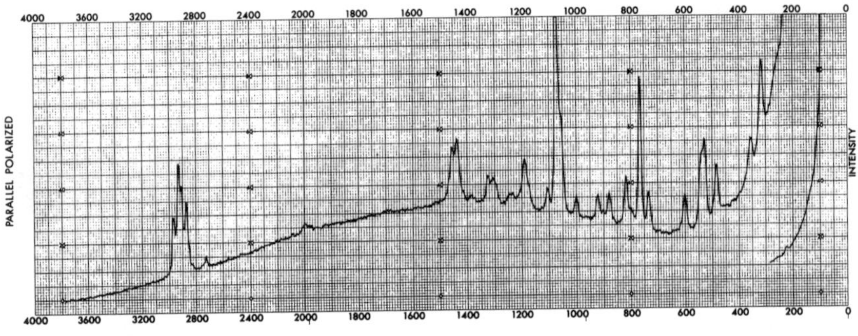

DIPHENYLSILANE

1195 R

$C_{12}H_{12}Si$ Mol. Wt. 184.32

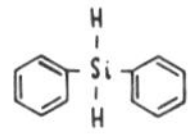

RAMAN DATA

Spectrophotometer	Cary 83
Sensitivity	2400
	-
Suppression	11
	-
Laser Power	75 mW
Spectral Slitwidth	6 cm^{-1}

Source: Aldrich Chemical Company, Inc.
Milwaukee, Wisconsin

WAVENUMBER / INTENSITY

1004/191.6 690/42.6

Capillary Cell: Neat

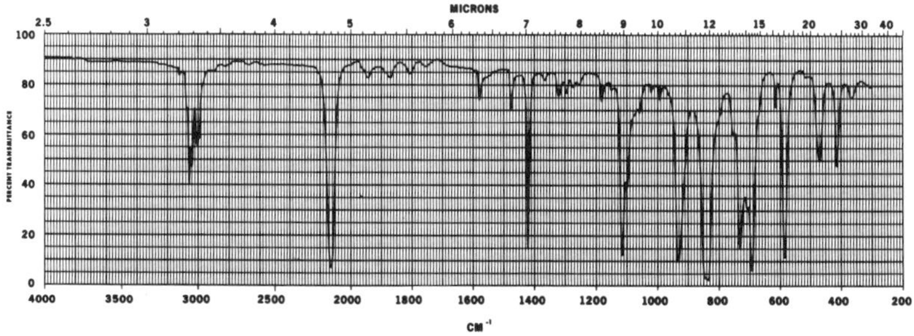

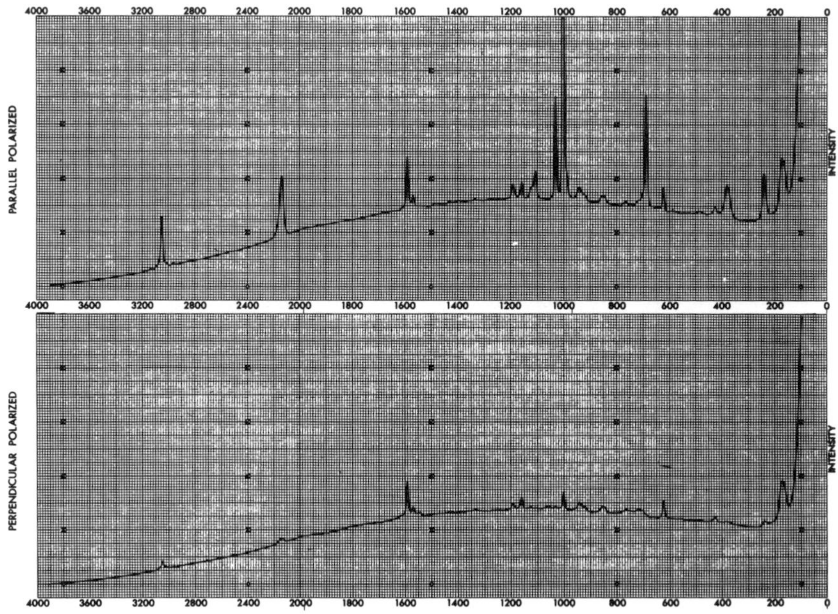

1194 R

HEXAMETHYLDISILOXANE

$C_6H_{18}OSi_2$ Mol. Wt. 162.38

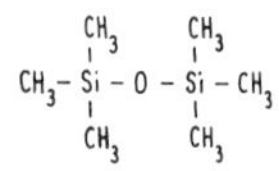

RAMAN DATA

Spectrophotometer	Cary 83
Sensitivity	5500
	-
Suppression	0
	-
Laser Power	155 mW
Spectral Slitwidth	6 cm^{-1}

Source: Harvard University
Converse Memorial Laboratory
Cambridge, Massachusetts

WAVENUMBER / INTENSITY

2900/94.7 524/122.4

Capillary Cell: Neat

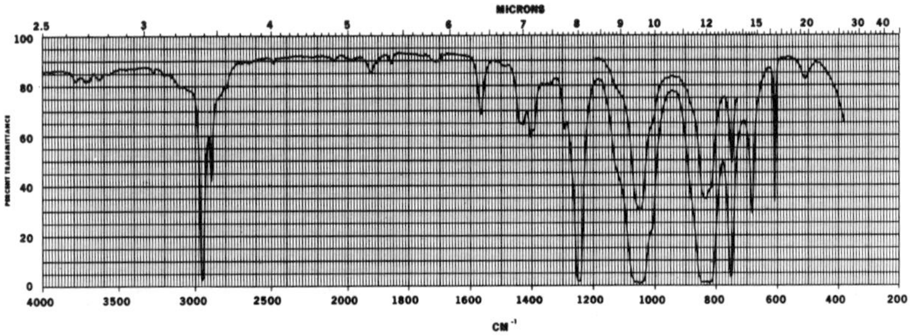

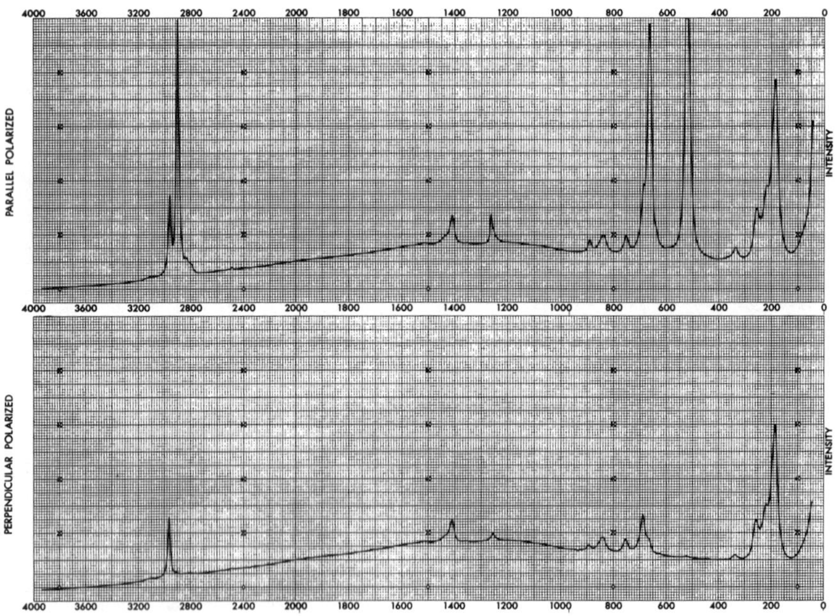

4160 R

TRIBENZYLPHOSPHINE OXIDE

$C_{21}H_{21}OP$ Mol. Wt. 320.38

M.P. 213-214°C

Source of Sample: V. Baliah, Annamalai University, Annamalainagar, India

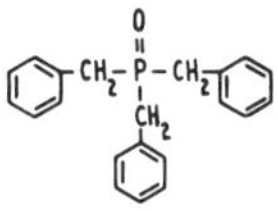

RAMAN DATA

Spectrophotometer	Cary 83
Sensitivity	16,000; 1600
Suppression	39; 4
Laser Power	100 mW
Spectral Slitwidth	6 cm^{-1}

WAVENUMBER / INTENSITY

1223/64.2	1000/155.6	650/68.2
165/91.0	112/161.0	84/50.0

KBr Wafer

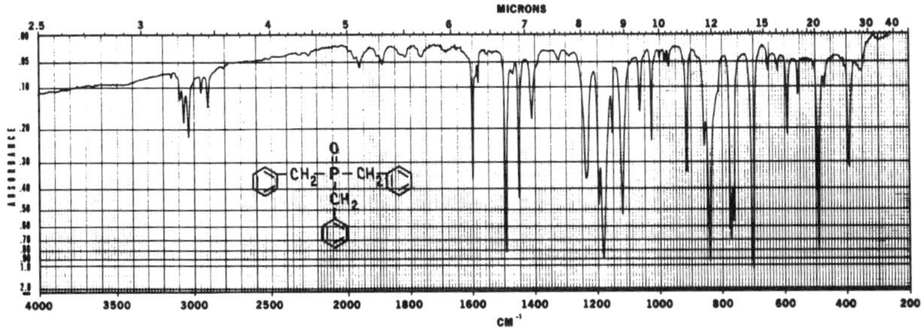

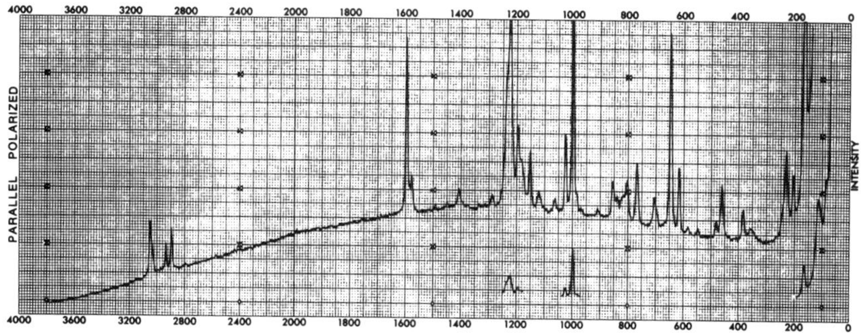

DIPHENYLPHOSPHINOUS CHLORIDE

953 R

RAMAN DATA

Spectrophotometer	Cary 83
Sensitivity	5500; 550
Suppression	3; 0
Laser Power	75 mW
Spectral Slitwidth	6 cm^{-1}

$C_{12}H_{10}ClP$ Mol. Wt. 220.64

B. P. 320°C

Source: Aldrich Chemical Company, Inc.
Milwaukee, Wisconsin

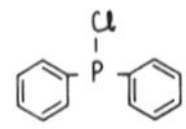

WAVENUMBER / INTENSITY		
1588/82.9	1031/121.4	1004/414.0

Capillary Cell: Neat

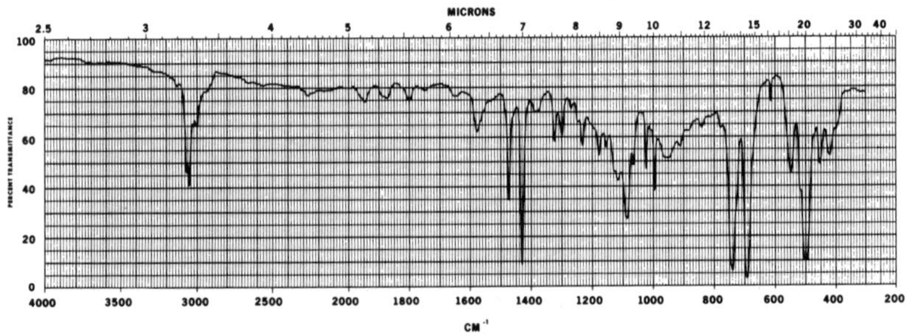

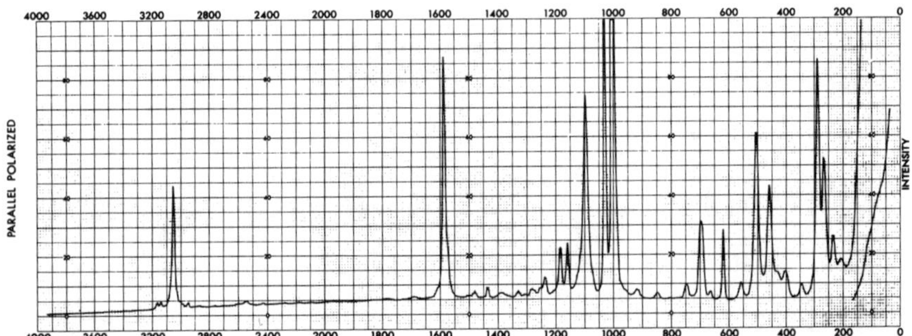

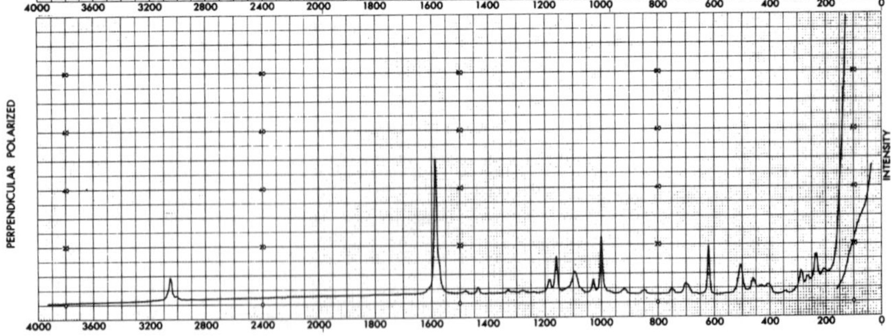

PHENYL PHOSPHITE

955 R

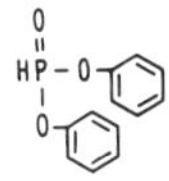

RAMAN DATA

Spectrophotometer: Cary 83
Sensitivity: 6500
-
Suppression: 28
-
Laser Power: 22 mW
Spectral Slitwidth: 6 cm^{-1}

$C_{12}H_{11}O_3P$ Mol. Wt. 234.19

Source: Chem Service, Inc.,
West Chester, Pa.

WAVENUMBER / INTENSITY	
1027/53.1	1007/194.7

Capillary Cell: Neat

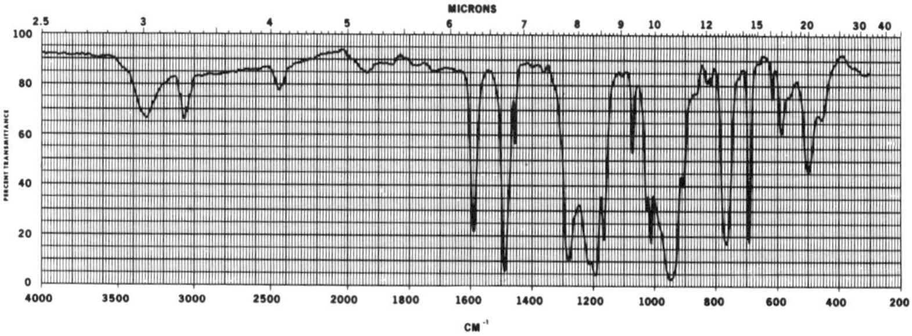

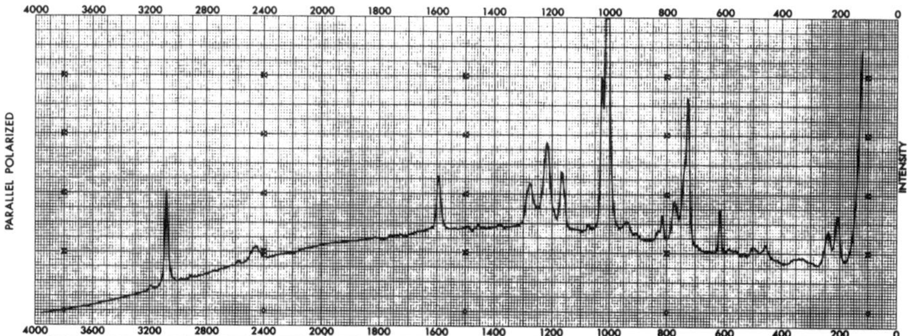

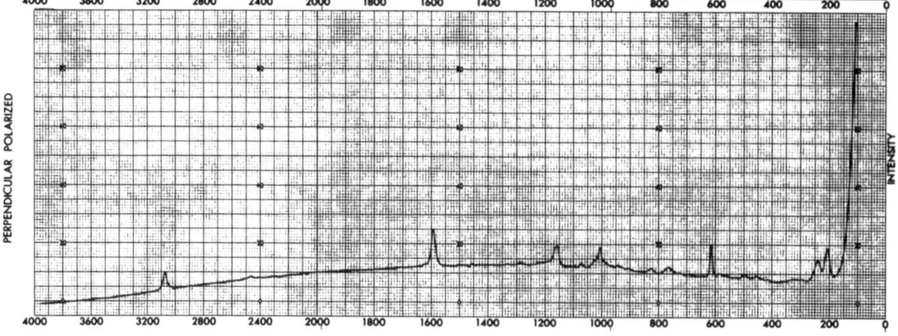

1366 R

DIPHENYLPHOSPHINIC ACID

RAMAN DATA

Spectrophotometer	Cary 83
Sensitivity	9000; 900
Suppression	49; 0
Laser Power	75 mW
Spectral Slitwidth	6 cm^{-1}

$C_{12}H_{11}O_2P$ Mol. Wt. 218.19

M.P. 195-197°C

Source: Doak Pharmacal Co., Inc.
Westbury, L. I., New York

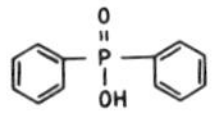

WAVENUMBER / INTENSITY

1590/141.6	1005/156.8	996/169.6
697/116.3	616/75.7	243/113.5
159/37.8	110/323.0	

KBr Wafer

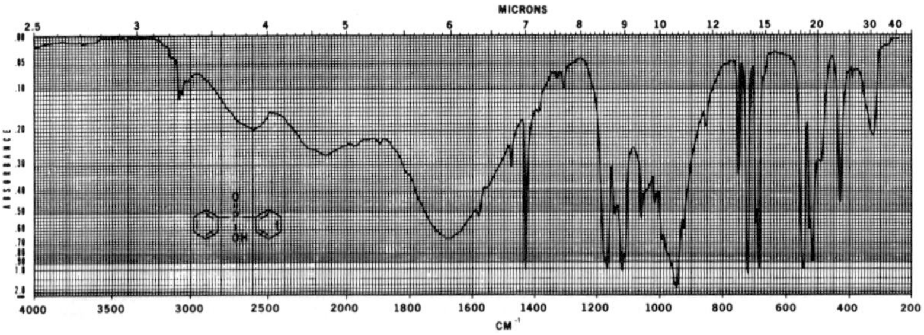

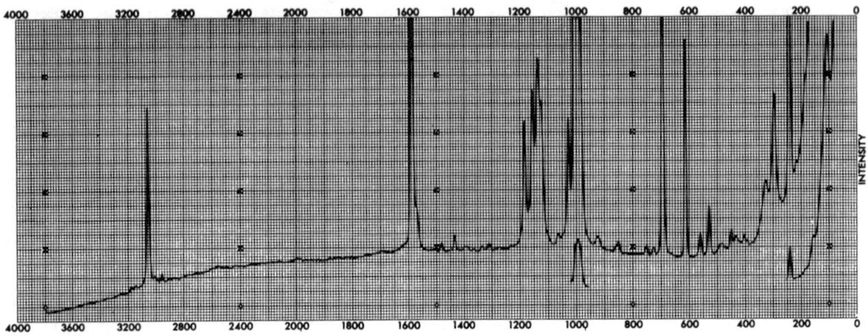

2575 R

1-BUTYLPHOSPHONIC ACID, DIBUTYL ESTER

RAMAN DATA

Spectrophotometer	Cary 83
Sensitivity	18000
	-
Suppression	47
	-
Laser Power	85 mW
Spectral Slitwidth	8 cm^{-1}

$C_{12}H_{27}O_3P$ Mol. Wt. 250.32

B.P. 127-128°C/2.5mm

Source of Sample: Virginia Chemicals, Inc. Portsmouth, Virginia

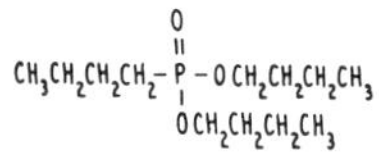

WAVENUMBER / INTENSITY		
2940/90.8	2918/108.4	2879/105.0

Capillary Cell: Neat

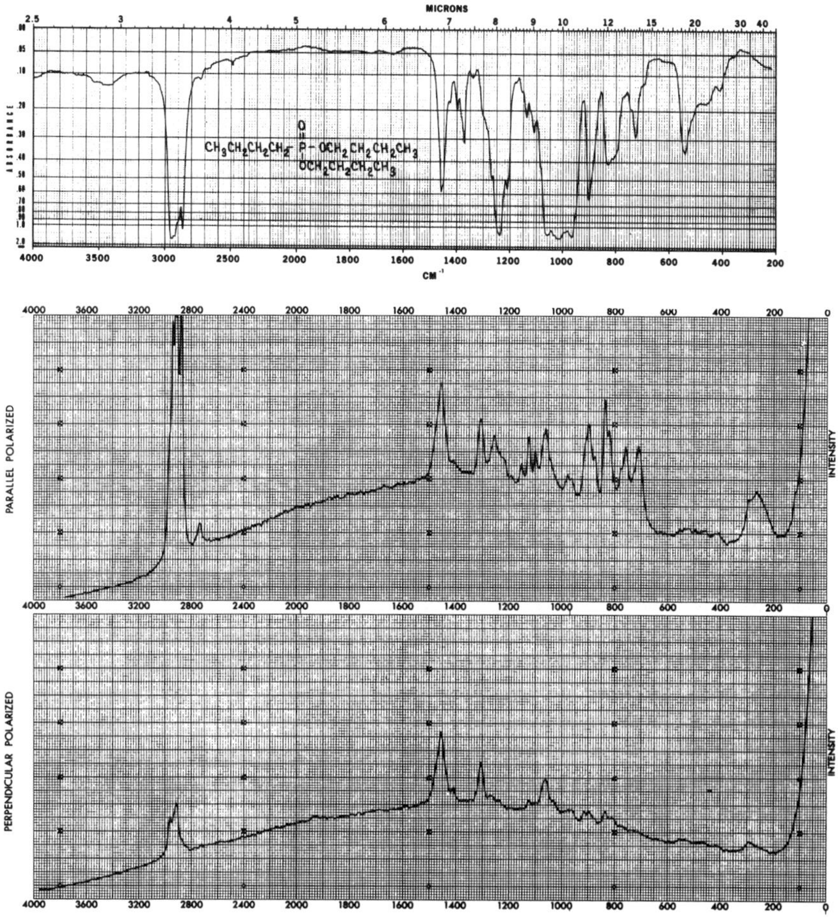

TRIPHENYLPHOSPHINE SULFIDE

4161 R

RAMAN DATA

Spectrophotometer	Cary 83
Sensitivity	19,000; 1900
Suppression	28; 3
Laser Power	100 mW
Spectral Slitwidth	6 cm^{-1}

$C_{18}H_{15}PS$ Mol. Wt. 294.36

M.P. 160-161°C

Source of Sample: V. Baliah, Annamalai University, Annamalainagar, India

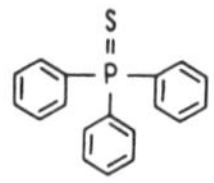

WAVENUMBER / INTENSITY

1579/106.4	1092/84.3	993/272.3
630/142.9	251/126.7	100/200.0
87/114.0		

KBr Wafer

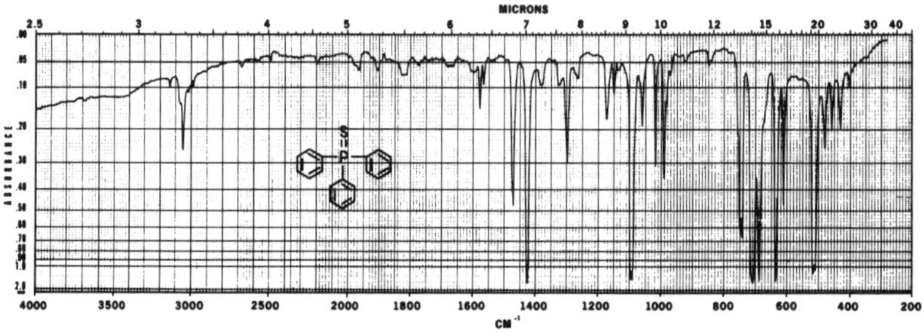

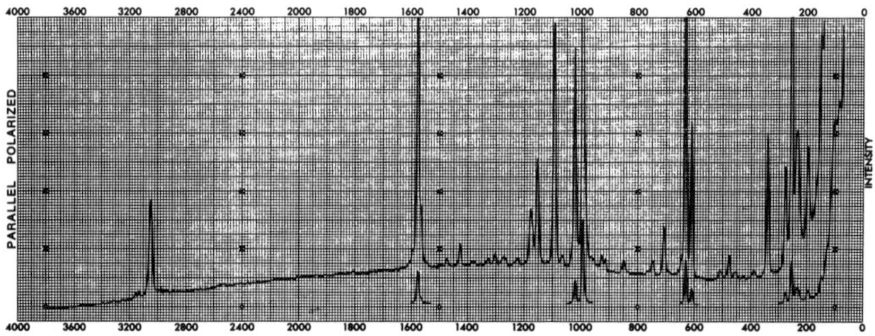

548 R

PHENYLPHOSPHONOTHIOIC ACID DICHLORIDE

RAMAN DATA

Spectrophotometer	Cary 83
Sensitivity	7800;
	780
Suppression	11;
	0
Laser Power	22 mW
Spectral Slitwidth	6 cm^{-1}

$C_6H_5Cl_2PS$ Mol. Wt. 211.05

Source: Chem Service, Inc., West Chester, Pa.

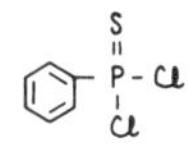

WAVENUMBER / INTENSITY

1099/97.2	1028/81.0	1000/212.2
500/229.6	377/369.8	215/119.7
100/85.0		

Capillary Cell: Neat

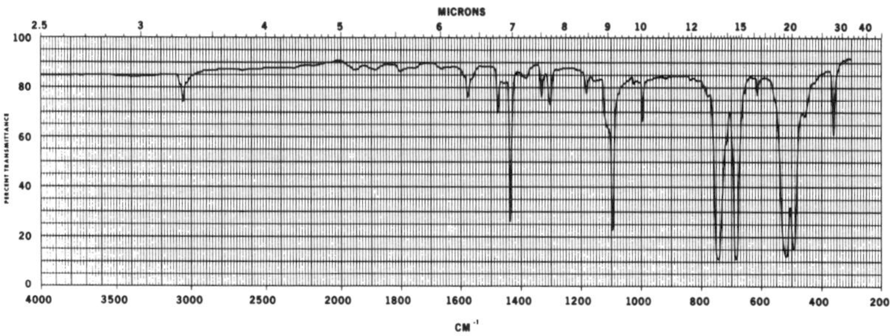

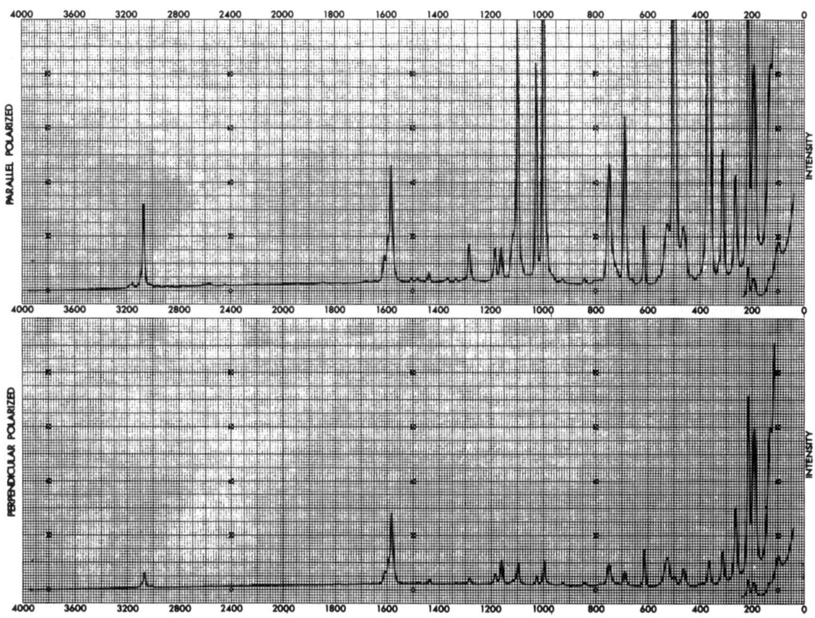

TOLUENE

383 R

RAMAN DATA

Spectrophotometer	Cary 83
Sensitivity	45000;
	4500
Suppression	100;
	0
Laser Power	22 mW
Spectral Slitwidth	4 cm^{-1}

C_7H_8 Mol. Wt. 92.14

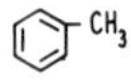

B.P. 110-111°C

Source: Chem Service, Inc., West Chester, Pa.

WAVENUMBER / INTENSITY

3060/100.6	2920/71.2	1207/226.0
1028/297.2	1000/1063.4	784/580.4
519/161.0		

Capillary Cell: Neat

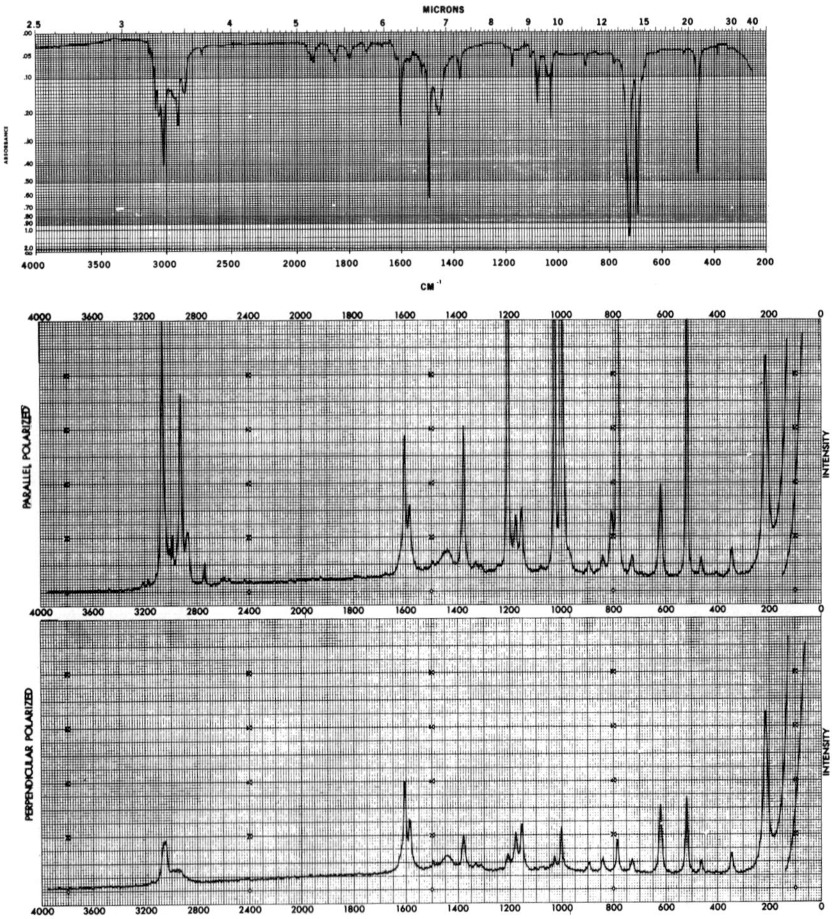

o-XYLENE

387 R

RAMAN DATA

Spectrophotometer	Cary 83
Sensitivity	50000; 10000
Suppression	105; 0
Laser Power	22 mW
Spectral Slitwidth	4 cm^{-1}

C_8H_{10} Mol. Wt. 106.17

M.P. -25° to -23°C

Source: Chem Service, Inc., West Chester, Pa.

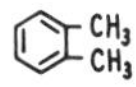

WAVENUMBER / INTENSITY

1385/74.3	1225/234.7	1055/335.5
735/621.6	583/202.3	

Capillary Cell: Neat

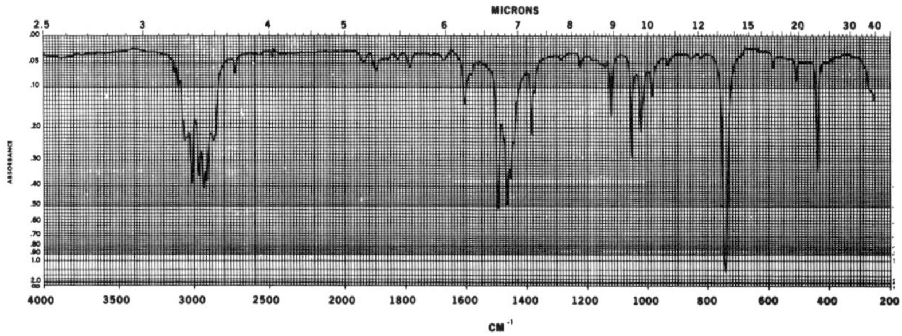

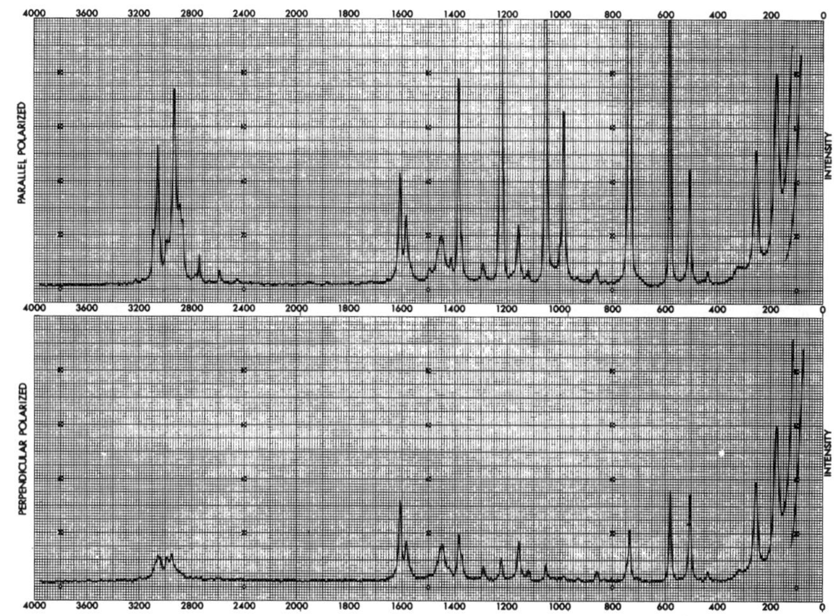

386 R m-XYLENE

RAMAN DATA

Spectrophotometer	Cary 83
Sensitivity	50000;
	6250
Suppression	31;
	0
Laser Power	22 mW
Spectral Slitwidth	4 cm^{-1}

C_8H_{10} Mol. Wt. 106.17

B.P. 138-139°C

Source: Chem Service, Inc.,
West Chester, Pa.

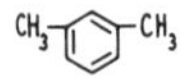

WAVENUMBER / INTENSITY

2930/78.8	1252/129.6	1003/499.5
728/404.7	539/207.7	

Capillary Cell: Neat

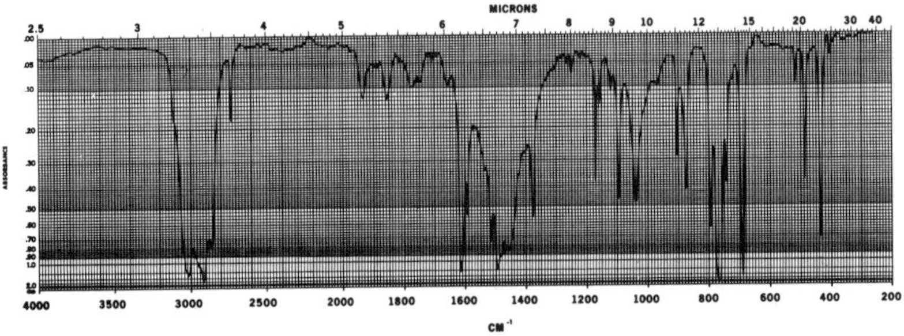

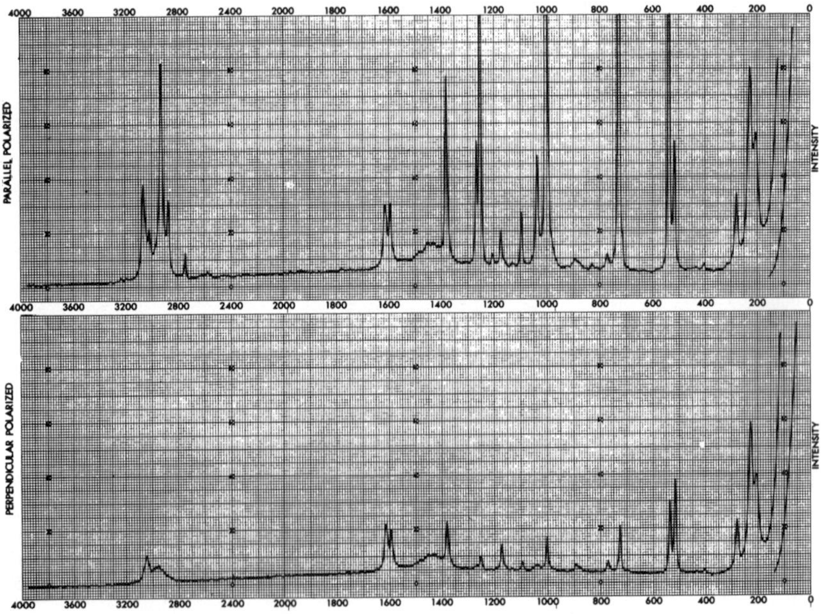

p-XYLENE **385 R**

RAMAN DATA

Spectrophotometer	Cary 83
Sensitivity	25000; 5000
Suppression	18; 0
Laser Power	22 mW
Spectral Slitwidth	4 cm^{-1}

C_8H_{10} Mol. Wt. 106.17

M.P. 12-13°C

Source: Chem Service, Inc., West Chester, Pa.

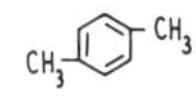

WAVENUMBER / INTENSITY

2925/70.0	1205/293.3	828/367.3
810/109.4	458/153.7	

Capillary Cell: Neat

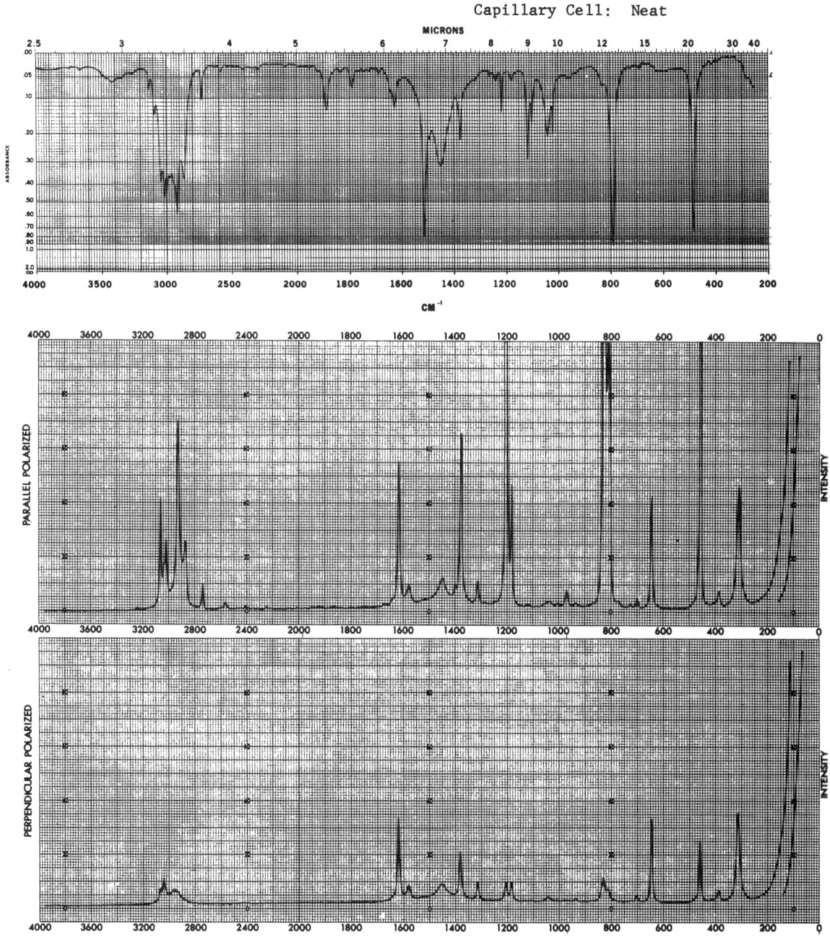

762 R

CHLOROBENZENE

RAMAN DATA

Spectrophotometer	Cary 83
Sensitivity	87500
	-
Suppression	500
	-
Laser Power	22 mW
Spectral Slitwidth	4 cm^{-1}

C_6H_5Cl Mol. Wt. 112.56

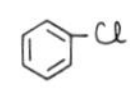

B.P. 130-132°C

Source: Chem Service, Inc., West Chester, Pa.

WAVENUMBER / INTENSITY		
1015/116.2	996/326.8	695/74.9

Capillary Cell: Neat

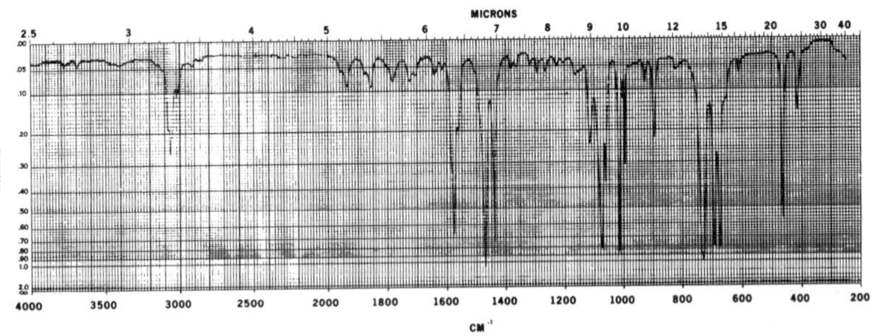

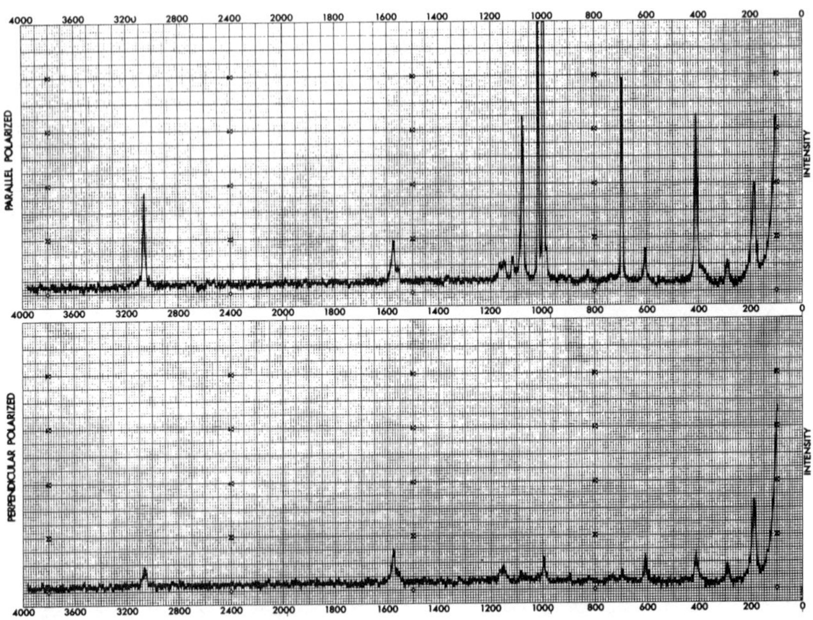

770 R

o-CHLOROTOLUENE

RAMAN DATA

Spectrophotometer	Cary 83
Sensitivity	2400
	-
Suppression	55
	-
Laser Power	210 mW
Spectral Slitwidth	6 cm-1

C_7H_7Cl Mol. Wt. 126.59

B.P. 159.5°C

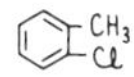

Source: MCB Manufacturing Chemists, Norwood, Ohio

WAVENUMBER / INTENSITY		
1203/132.4	1049/247.7	810/126.7
685/157.7	557/174.7	451/71.7

Capillary Cell: Neat

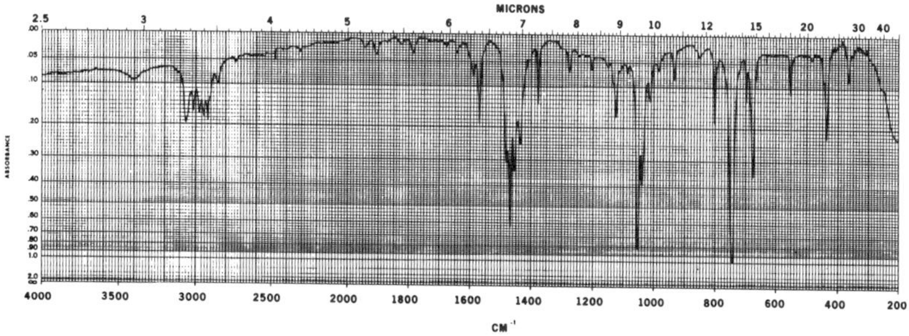

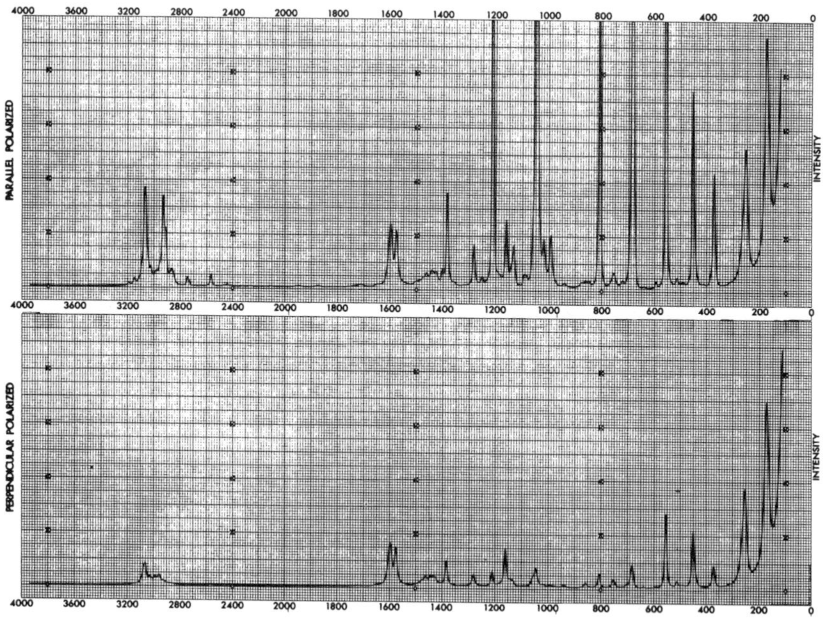

m-CHLOROTOLUENE

1973 R

RAMAN DATA

Spectrophotometer	Cary 83
Sensitivity	7500; 750
Suppression	11; 0
Laser Power	110 mW
Spectral Slitwidth	6 cm-1

C_7H_7Cl Mol. Wt. 126.59

B.P. 160.5-161.5°C

Source: MCB Manufacturing Chemists
Norwood, Ohio

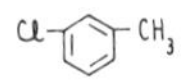

WAVENUMBER / INTENSITY

1216/68.7	1000/424.6	684/160.3
521/97.8	417/111.2	

Capillary Cell: Neat

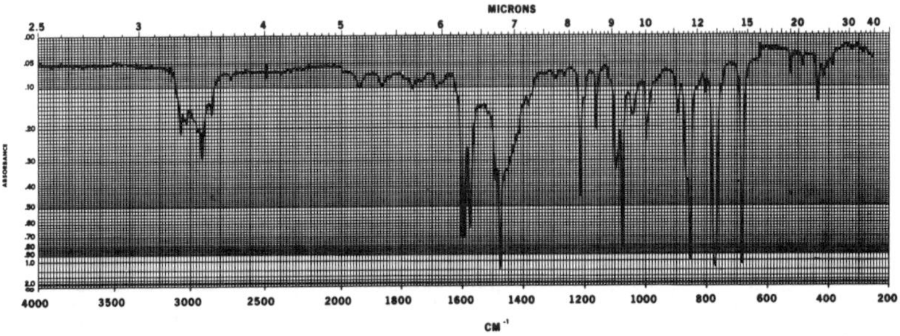

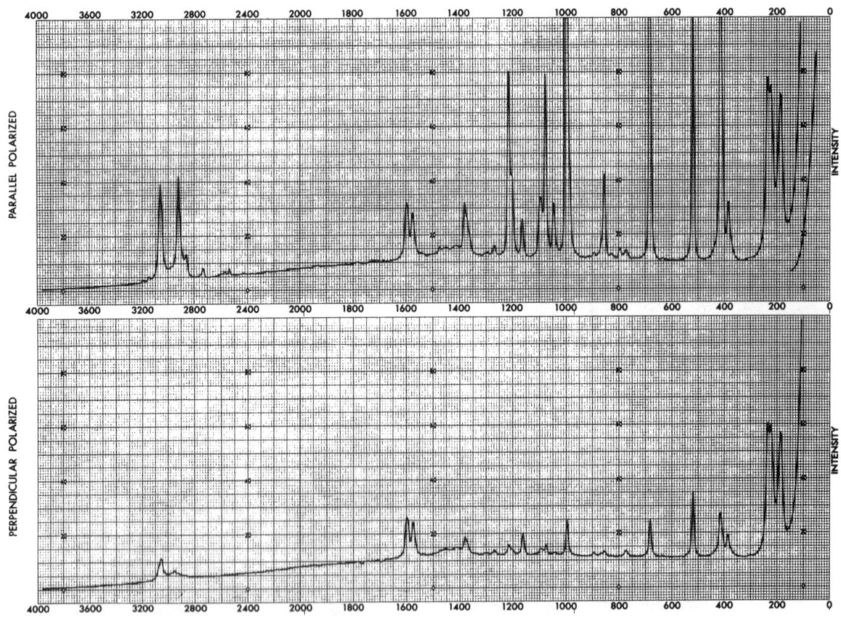

p-CHLOROTOLUENE **327 R**

RAMAN DATA		
Spectrophotometer	Cary 83	C_7H_7Cl Mol. Wt. 126.59
Sensitivity	62500	M.P. 6-8°C
	-	Source: Chem Service, Inc., West Chester, Pa.
Suppression	115	
	-	
Laser Power	25 mW	
Spectral Slitwidth	4 cm^{-1}	

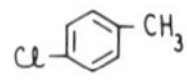

WAVENUMBER / INTENSITY

1205/136.8	1088/297.0	792/308.8
631/46.9	375/215.0	

Capillary Cell: Neat

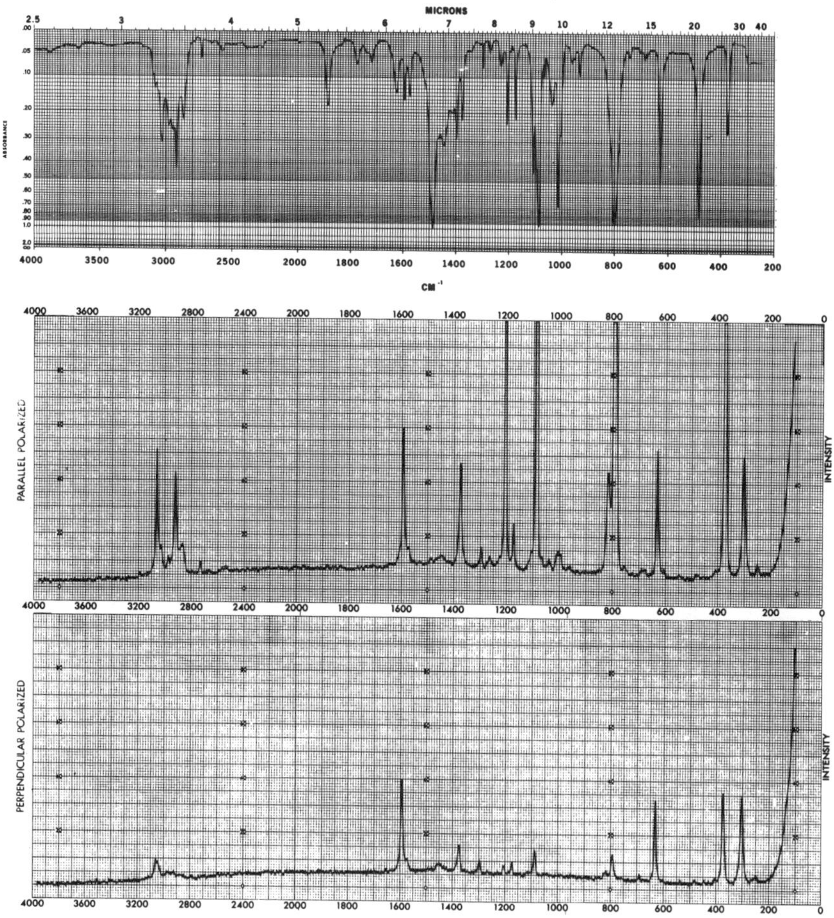

4-tert-BUTYLPYRIDINE

1083 R

RAMAN DATA

Spectrophotometer	Cary 83
Sensitivity	4000; 400
Suppression	21; 0
Laser Power	185 mW
Spectral Slitwidth	6 cm^{-1}

$C_9H_{13}N$ Mol. Wt. 135.21

B. P. 196.3°C/749mm

Source: Purdue University
Lafayette, Indiana

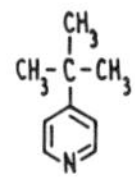

WAVENUMBER / INTENSITY		
1125/87.7	1000/350.4	718/213.4
536/54.3	140/109.0	

Capillary Cell: Neat

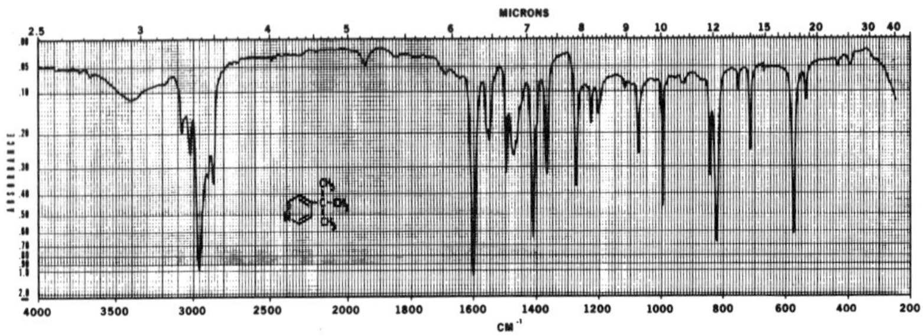

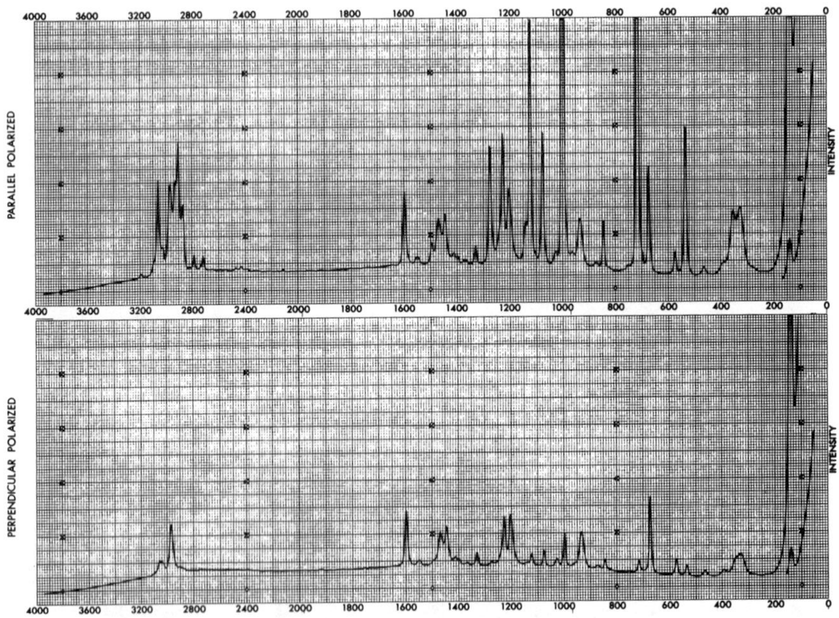

1497 R

FURAN

C_4H_4O Mol. Wt. 68.07

B.P. 31-32°C

Source: Fluka AG,
Buchs, Switzerland

RAMAN DATA

Spectrophotometer	Cary 83
Sensitivity	10000;
	1000
Suppression	5;
	0
Laser Power	100 mW
Spectral Slitwidth	6 cm^{-1}

WAVENUMBER / INTENSITY		
3159/79.0	1485/446.8	1380/186.3
1140/431.5	1060/105.2	

Gas Cell: 10 cm

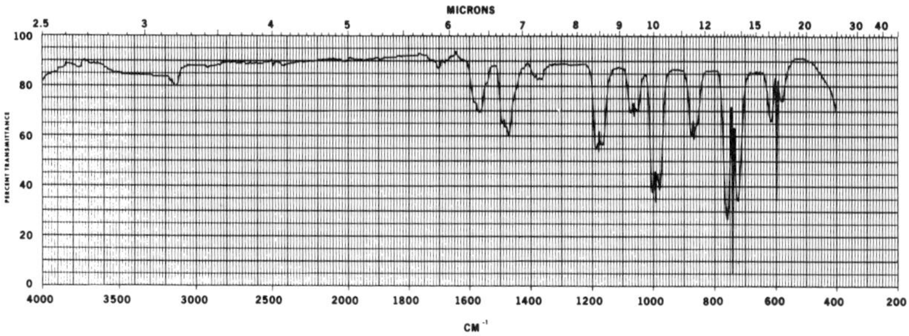

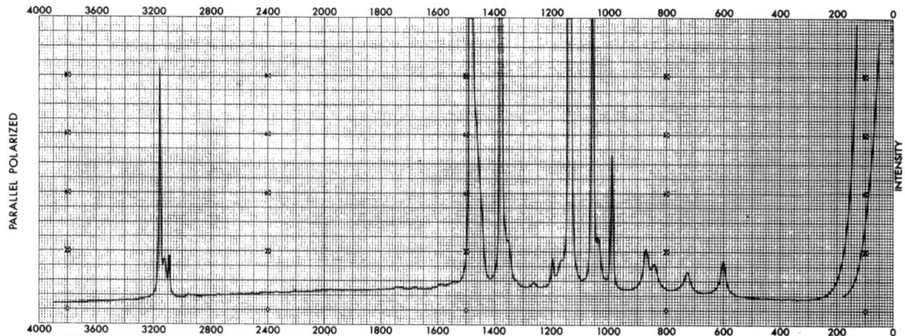

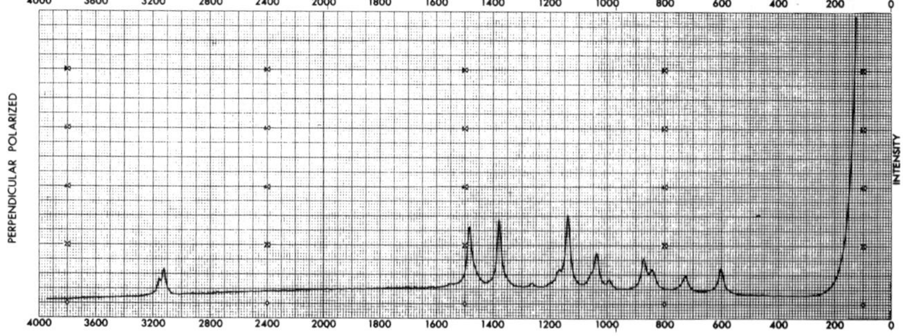

PYRAZOLE

1071 R

RAMAN DATA

Spectrophotometer	Cary 83
Sensitivity	9000; 900
Suppression	21; 0
Laser Power	66 mW
Spectral Slitwidth	6 cm^{-1}

$C_3H_4N_2$ Mol. Wt. 68.08

M. P. 67-70°C B. P. 186-188°C

Source: Aldrich Chemical Company, Inc.
Milwaukee, Wisconsin

WAVENUMBER / INTENSITY

1263/70.9 1151/256.8 160/34.5
140/8.9 109/108.0

Capillary Cell: Melt

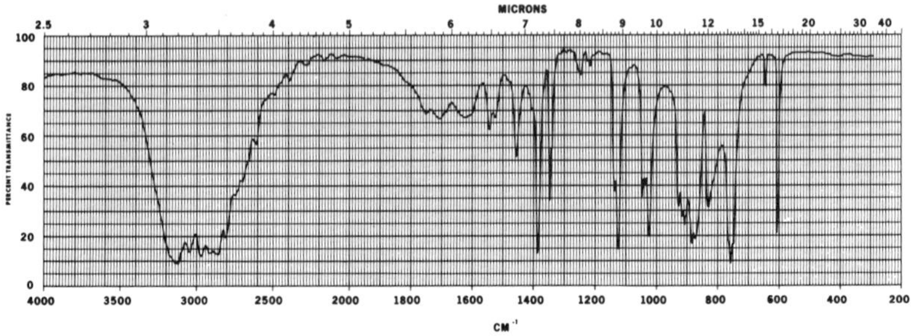

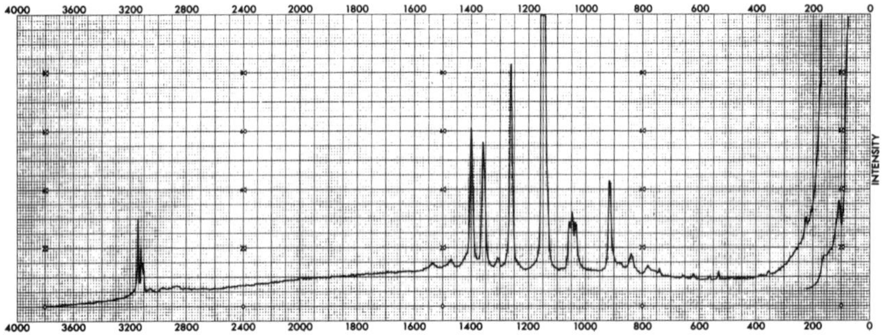

APPENDIX 1

Infrared and Raman Spectra of Common Organic Compounds—From the Aldrich Library of FT-IR Spectra* and the Dollish, Fateley and Bentley Collection†

In order to provide some additional reference spectra of common organic compounds, 30 Raman spectra are reproduced from the 1974 text by Dollish, Fateley and Bentley. The corresponding infrared spectra for these 30 compounds were obtained from the Aldrich FT-IR library and they are reproduced with permission. The compounds are also listed in the composite name and molecular formula indexes at the back of this book.

The Raman spectra were obtained with a Cary Model 81 Raman spectrometer modified for laser excitation. An argon ion laser operating on the 4880 Å line was used. The high-purity samples were contained in a Pyrex capillary and were illuminated coaxially to the laser beam.

The infrared spectra were run on a Nicolet 20SX FT-IR spectrometer at 2 cm^{-1} resolution. Liquid samples were run neat between two KBr pellets. Solid samples were run as Nujol mulls. Additional sample information is presented on the spectrum.

* *The Aldrich Library of FT-IR Spectra*, by C. J. Pouchert, and reproduced with permission from the Aldrich Chemical Company, Inc.

† F. R. Dollish, W. G. Fateley, and F. F. Bentley, *Characteristic Raman Frequencies of Organic Compounds*, John Wiley and Sons, Inc., New York, 1974, with permission.

INDEX TO SPECTRA

B5770-2 *CAS [108-86-1]*
Bromobenzene, 99%

FW 157.02
mp -31 °C
bp 156 °C

d 1.491
Fp 124 °F
n_D^{20} 1.5580

IR III, 588C
NMR II, 1,773C
Merck 10,1376

3064.8	1068.7	733.9
1578.4	999.5	683.8
1474.3	902.6	457.1

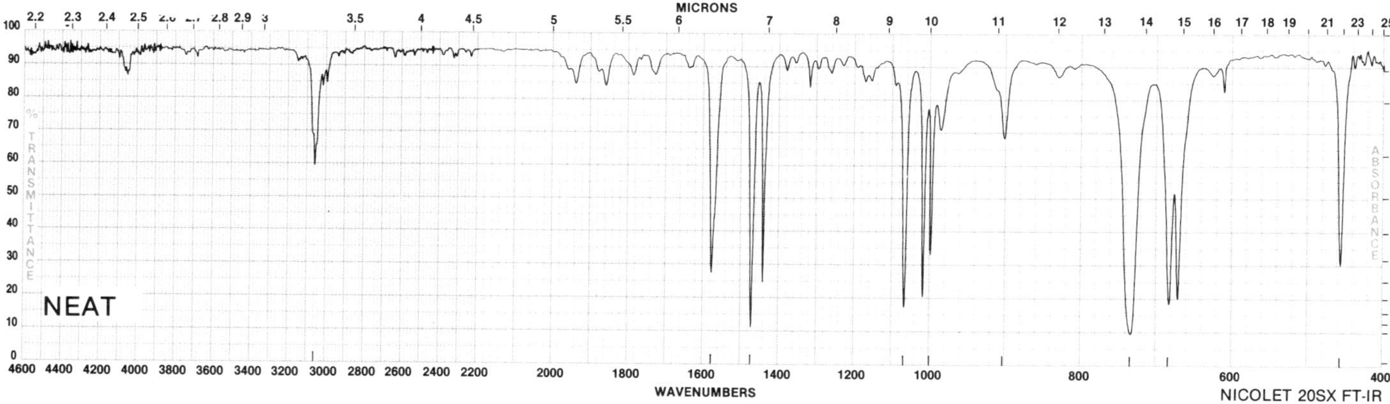

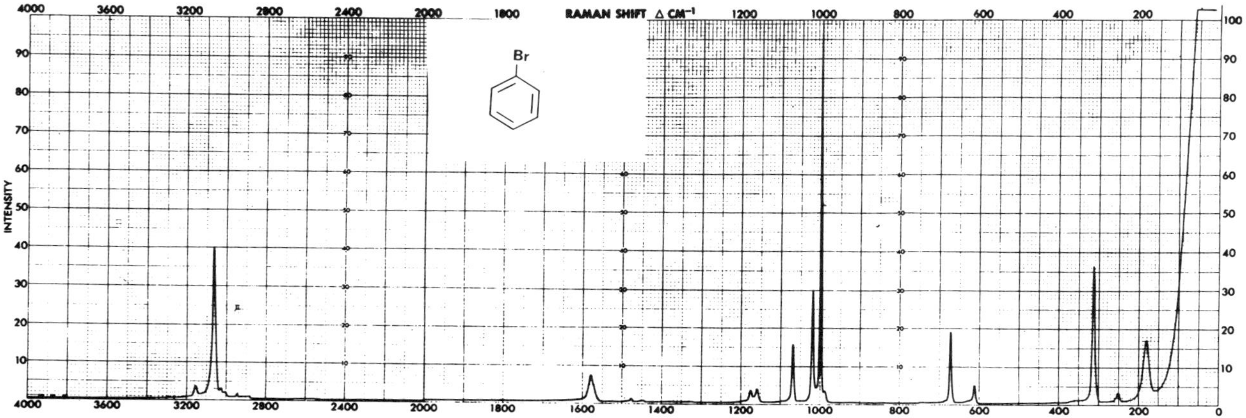

24276-4 *CAS [95-50-1]*
1,2-Dichlorobenzene, 99 + %

FW 147.00
mp -15°C
bp 179-180°C

d 1.306
Fp 150°F
n_D^{20} 1.5501

Merck 10,3040

1573.5	1127.6	748.0
1456.5	1035.2	659.4
1251.6	941.2	485.7

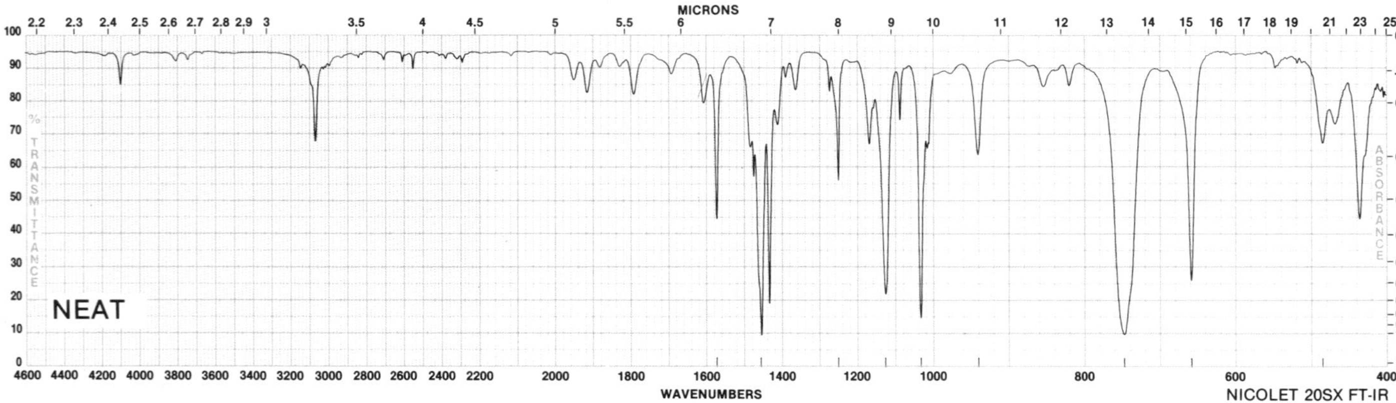

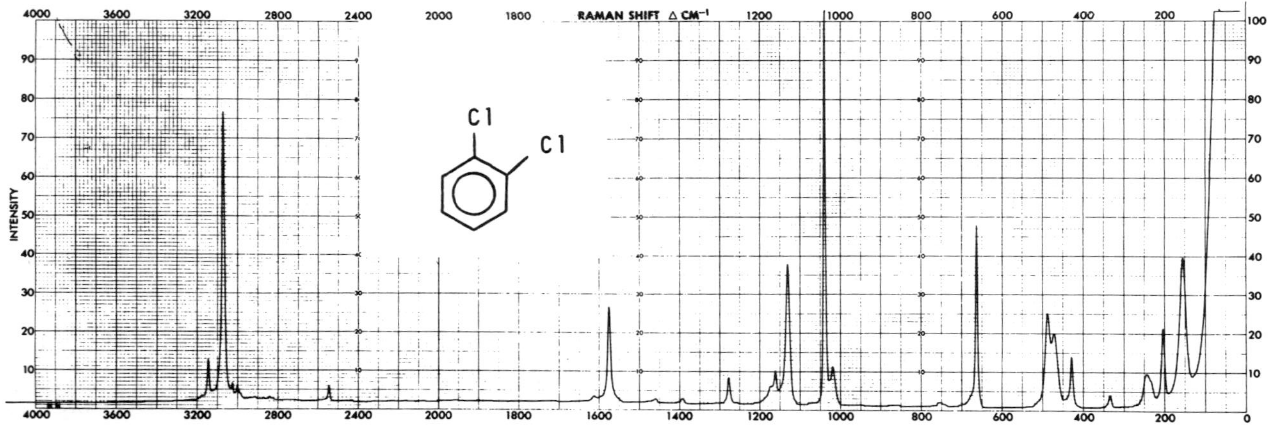

11380-8 *CAS [541-73-1]*
1,3-Dichlorobenzene, 98%

FW 147.00
mp -24°C
bp 172-173°C

d 1.288
Fp 146°F
n_D^{20} 1.5457

IR III, 593C
NMR II, 1,785D
Merck 10,3039

1577.8 1160.7 868.1
1461.0 1079.9 782.4
1400.2 997.8 672.7

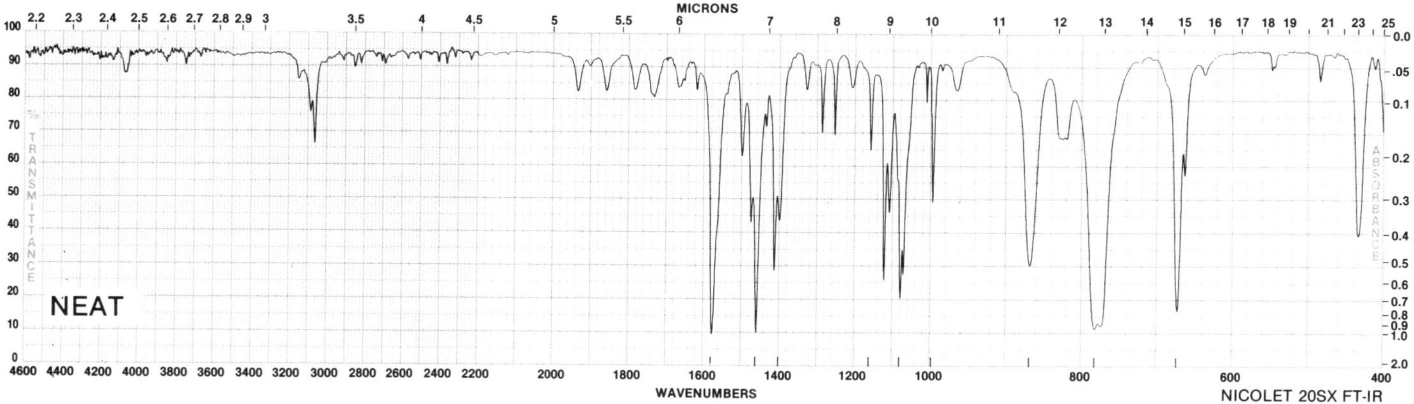

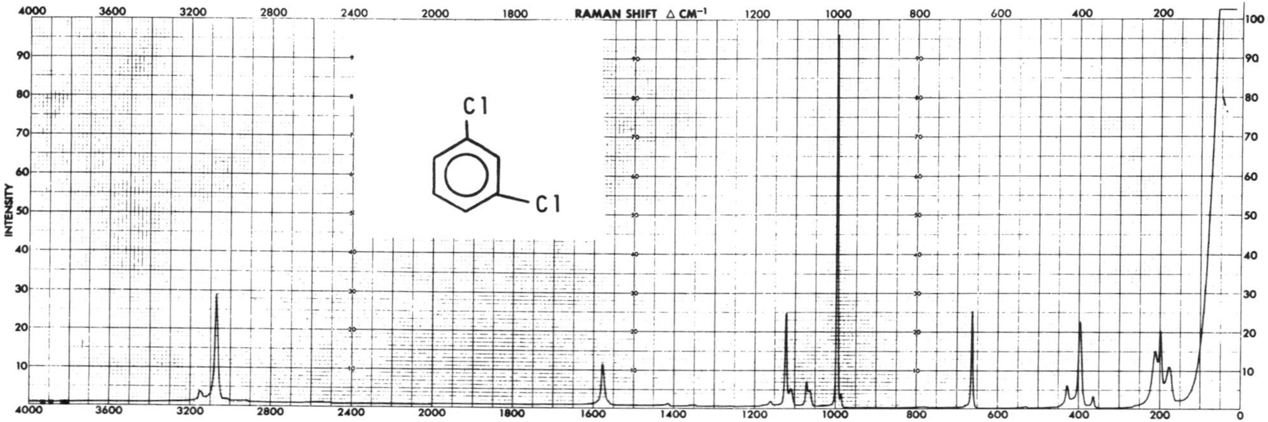

D5682-9 *CAS [106-46-7]*
1,4-Dichlorobenzene, 99 + %

FW 147.00
mp 54-56 °C
bp 173 °C

d 1.241
Fp 150 °F

IR III, 596A
Merck 10,3041

1886.5 1013.4 545.7
1390.6 951.2 483.7
1088.6 817.4

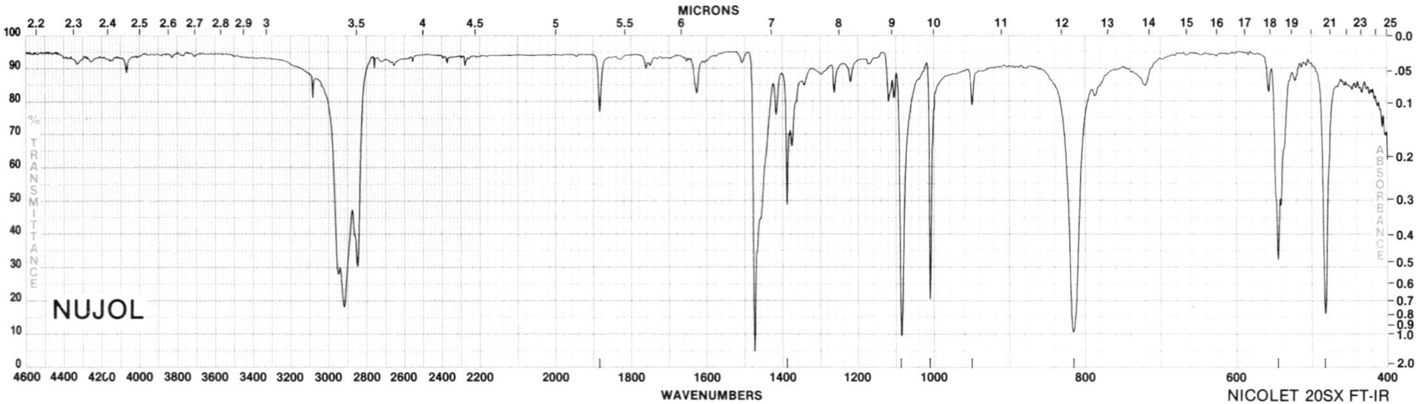

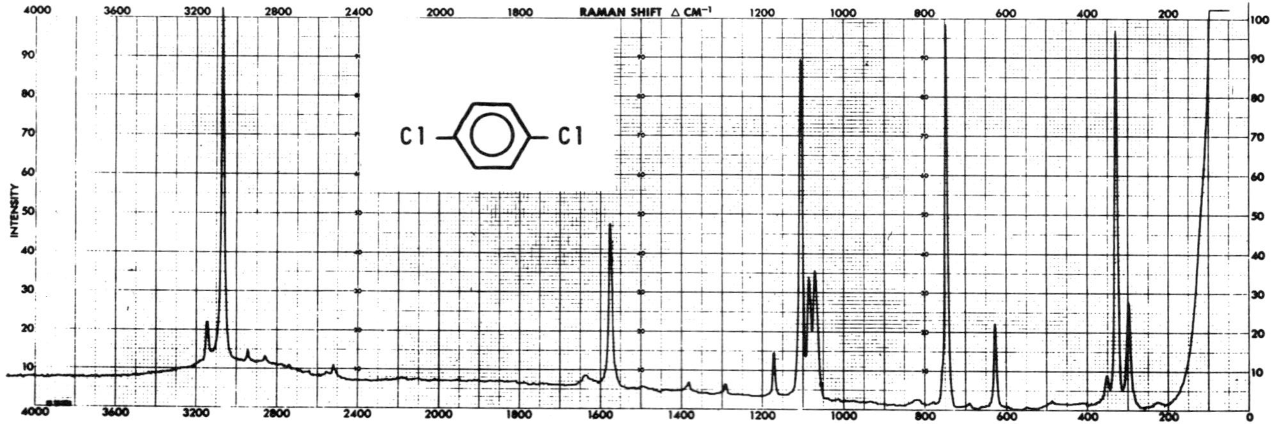

F600-1 *CAS [462-06-6]*
Fluorobenzene, 99%

FW 96.10	d 1.024	IR III, 588A	1595.7 1155.1 752.4
mp -42°C	Fp 9°F	NMR II, 1,773A	1494.8 1065.8 685.1
bp 85°C	n_D^{20} 1.4653	Merck 10,4075	1218.9 805.5 499.0

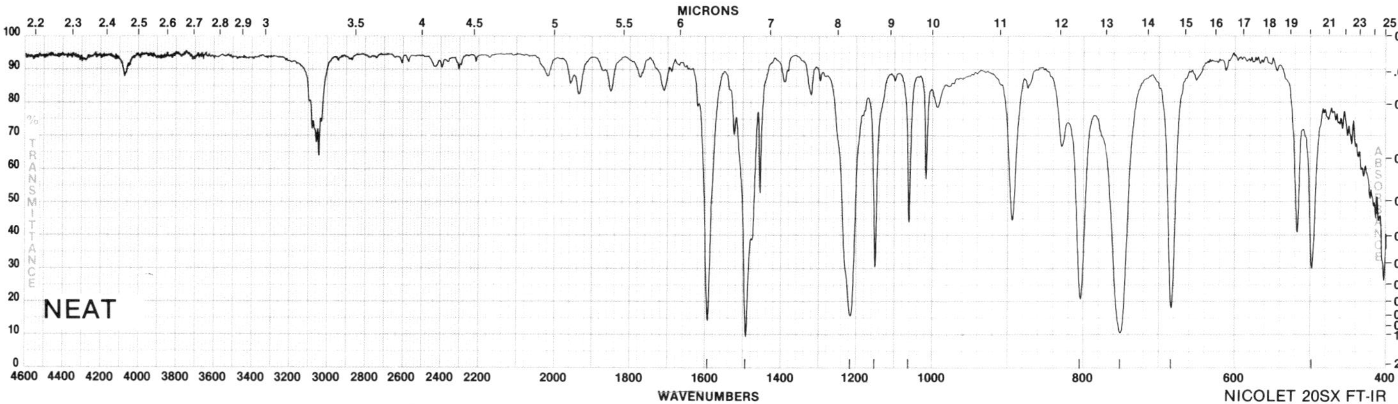

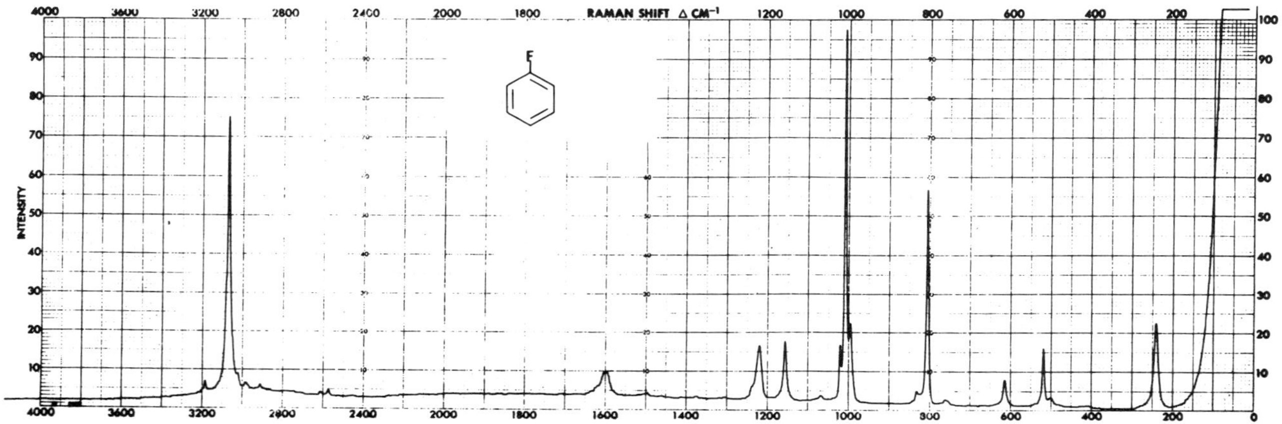

18579-5 CAS [98-82-8]
Cumene, 99%

FW 120.20
mp -96°C
bp 152-154°C

d 0.864
Fp 115°F
n_D^{20} 1.4917

IR III, 561E
Merck 10,2605

2961.3	1383.7	698.4
1604.0	1027.9	534.7
1493.7	760.6	404.5

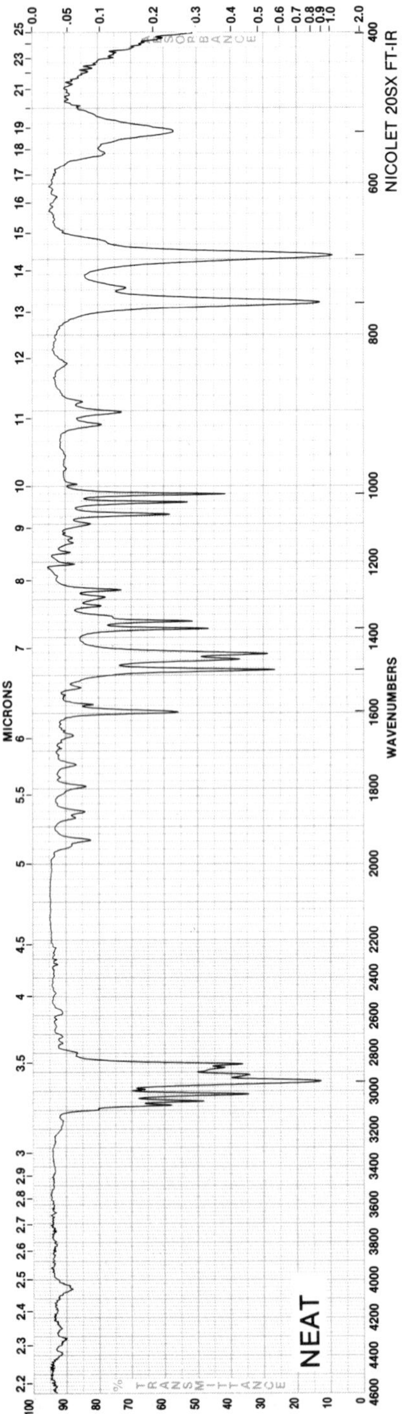

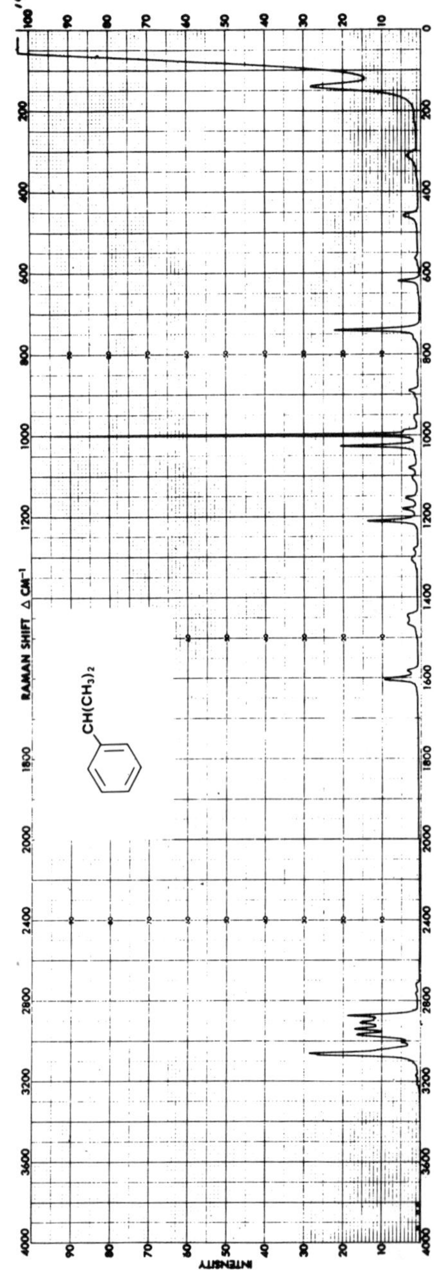

B9020-3 *CAS [104-51-8]*
Butylbenzene, 99 + %

FW 134.22
mp -88°C
bp 183°C

d 0.860
Fp 139°F
n_D^{20} 1.4895

IR III, 561F
NMR II, 1,734A
Merck 10,1522

2930.6	1378.2	697.5
1603.9	1030.2	570.0
1453.4	744.9	508.0

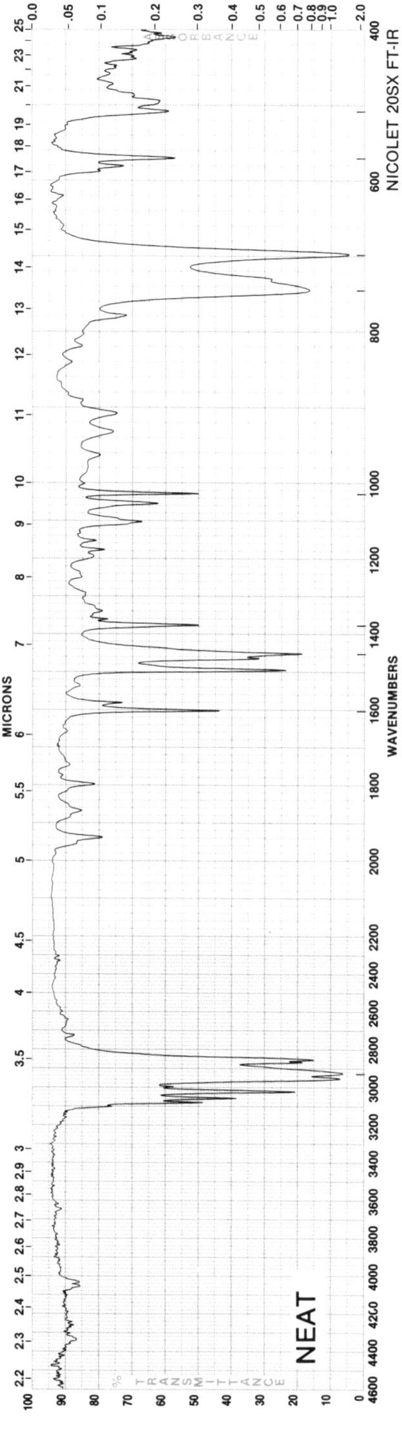

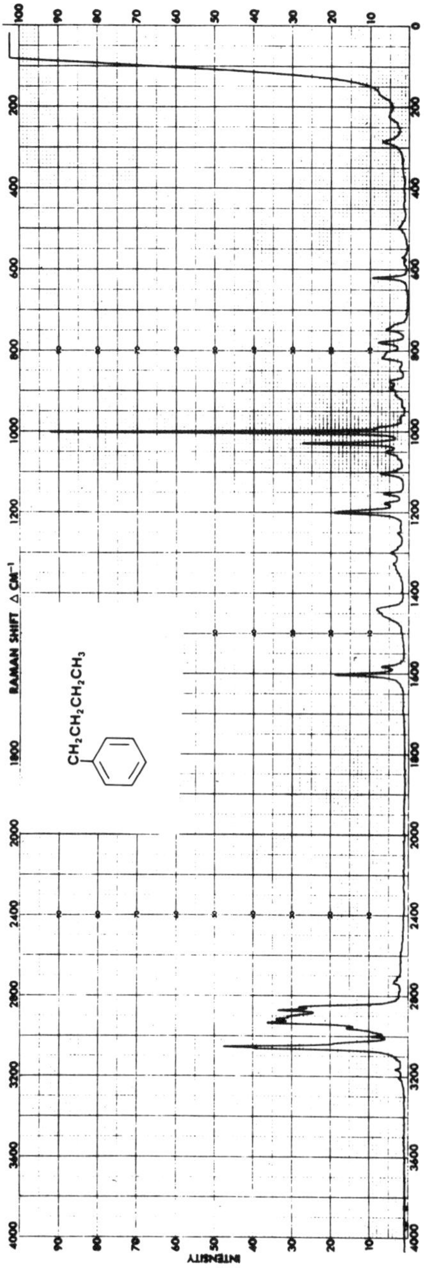

B9060-2 *CAS [98-06-6]*
***tert*-Butylbenzene**, 99%

FW 134.22	d 0.867	IR III, 562A	2964.2 1365.5 762.6
mp -58°C	Fp 94°F	NMR II, 1,737B	1600.0 1268.5 698.0
bp 169°C	n_D^{20} 1.4920	Merck 10,1524	1496.5 1031.6 546.5

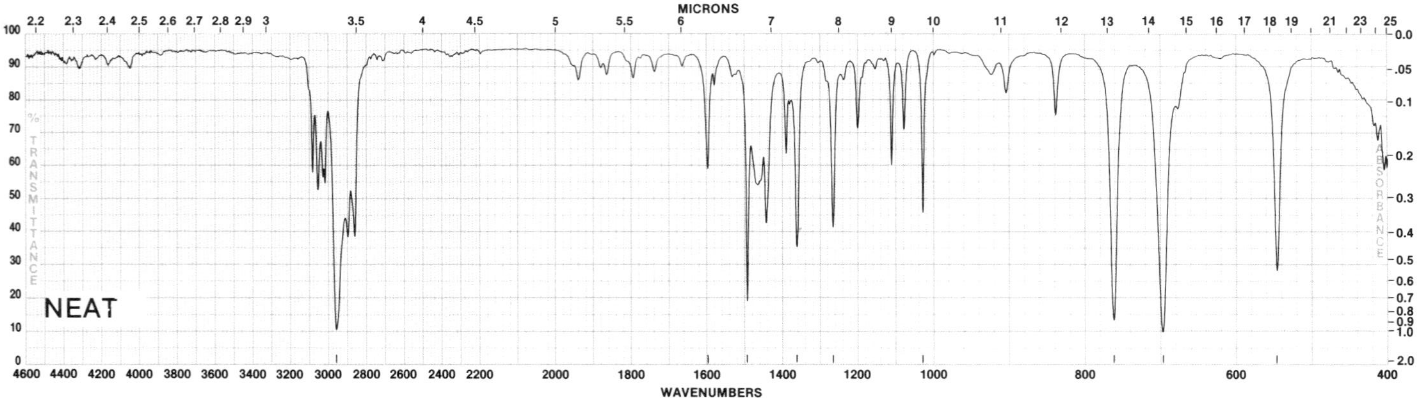

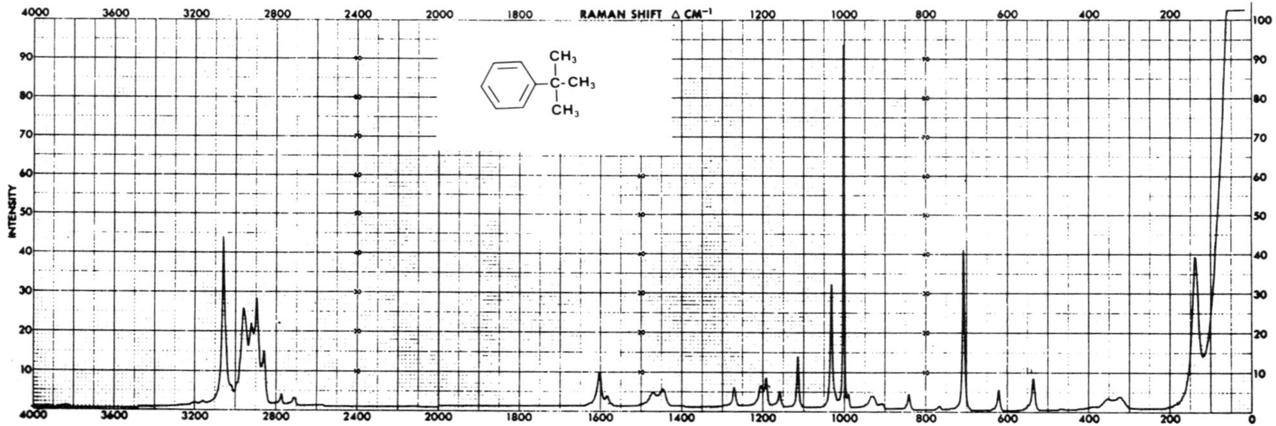

T7320-2 *CAS [526-73-8]*
1,2,3,-Trimethylbenzene, tech., 90%

FW 120.20
mp -25°C
bp 175-176°C

d 0.894
Fp 119°F
n_D^{20} 1.5149

IR III, 567B
NMR II, 1,744C

2941.8	1378.0	766.1
1587.7	1094.3	739.2
1475.5	1009.4	708.6

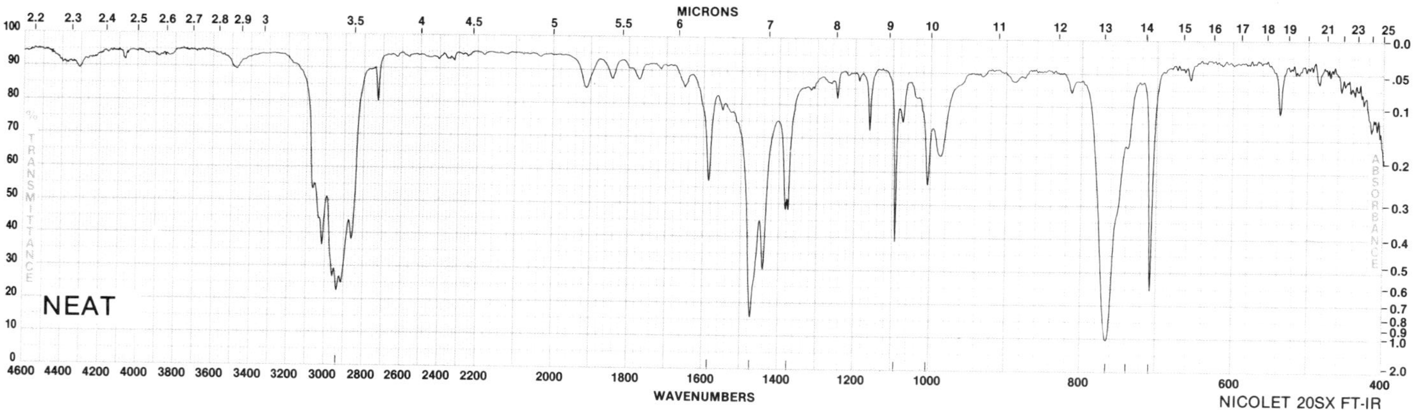

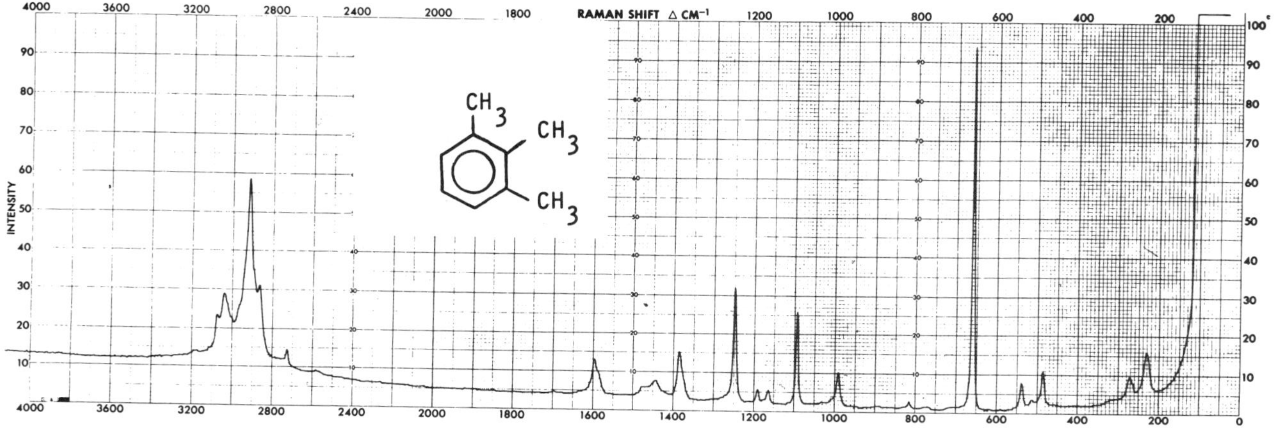

T7360-1 *CAS [95-63-6]*
1,2,4-Trimethylbenzene, 99%

FW 120.20	d 0.889	IR III, 567C	2921.0 1451.2 873.3
mp -44°C	Fp 120°F	NMR II, 1,744D	1618.2 1382.7 805.7
bp 168°C	n_D^{20} 1.5040	Merck 10,7816	1505.8 1021.1 539.1

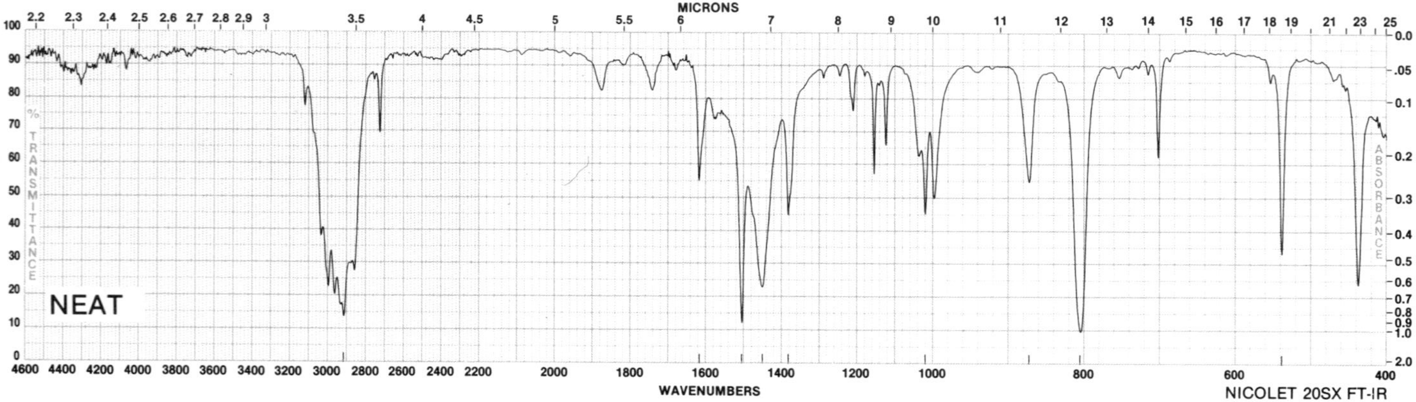

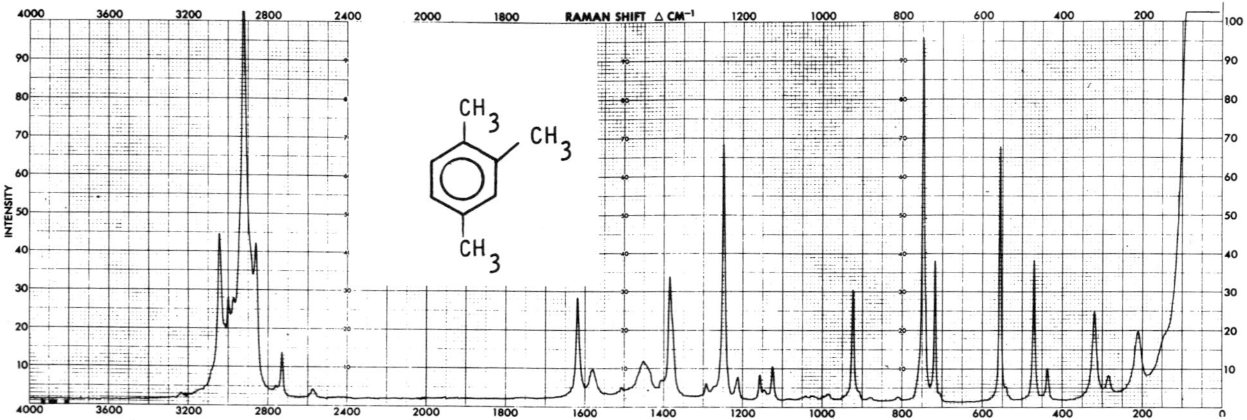

M720-0 *CAS [108-67-8]*
Mesitylene, 99%

FW 120.20
mp -45°C
bp 162-164°C

d 0.864
Fp 112°F
n_D^{20} 1.4978

IR III, 567D
Merck 10,5752

2917.7	1549.3	1037.2
1758.6	1472.6	834.9
1608.1	1374.7	687.2

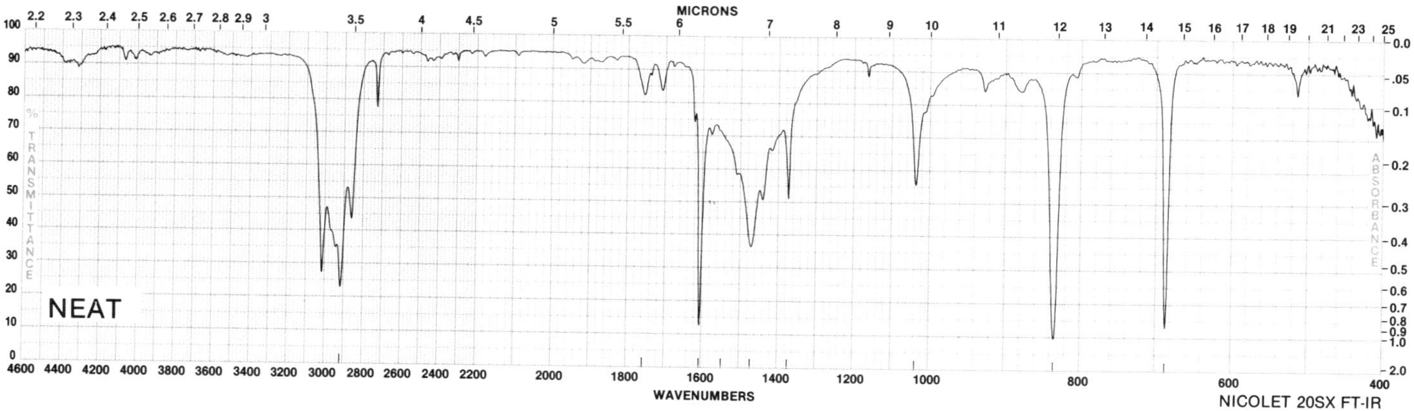

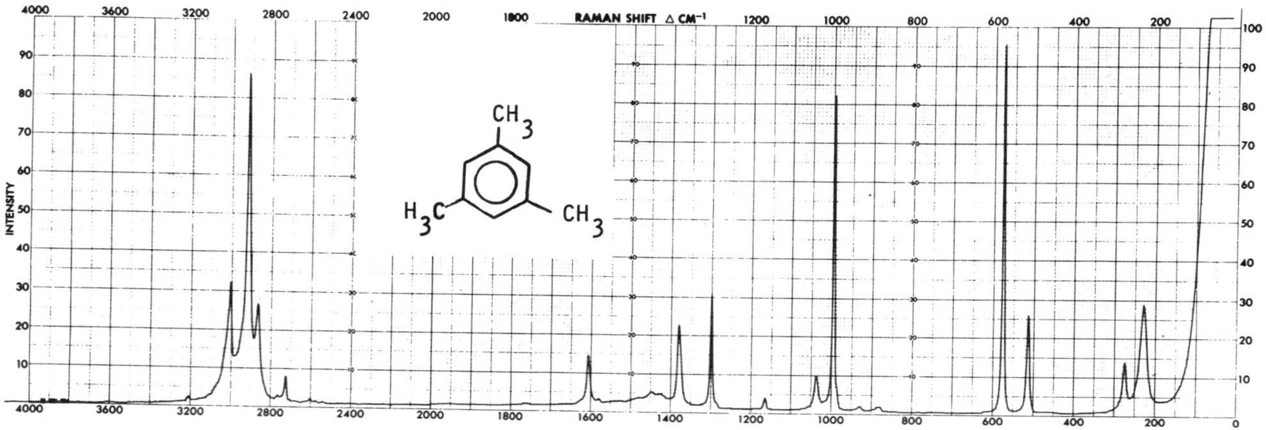

B3465-6 *CAS [92-52-4]*
Biphenyl, 99%

FW 154.21
mp 69-72°C
bp 255°C

d 0.992

IR III, 571E
NMR II, 1,755A
Merck 10,3326

3033.0	1344.6	728.5
1569.2	1170.1	698.6
1481.1	903.3	610.5

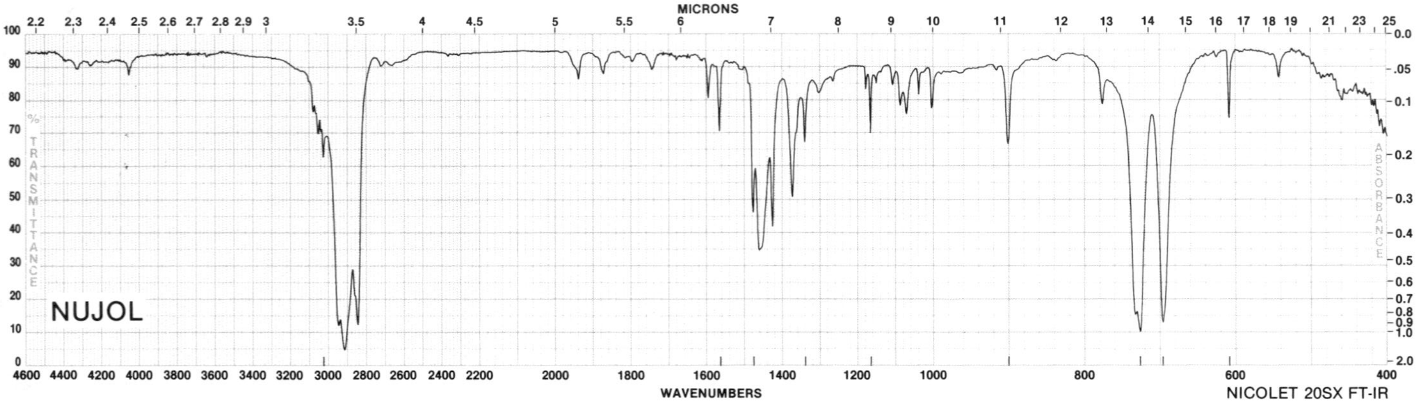

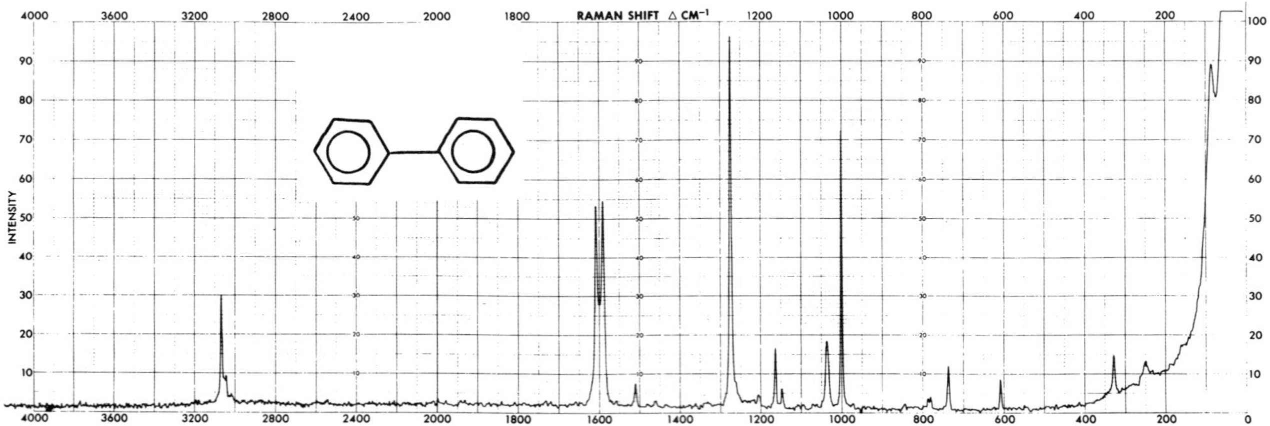

1-1955-1 Isoprene, puriss.

$H_2C{:}CHC(CH_3){:}CH_2$ M.W. 68.12 n_D^{20} 1.4217

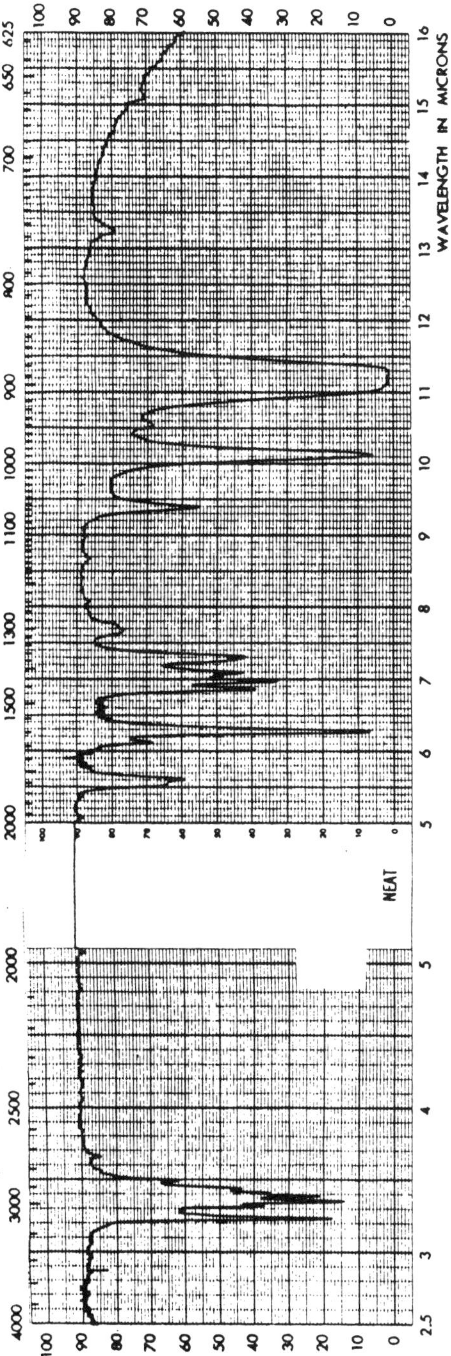

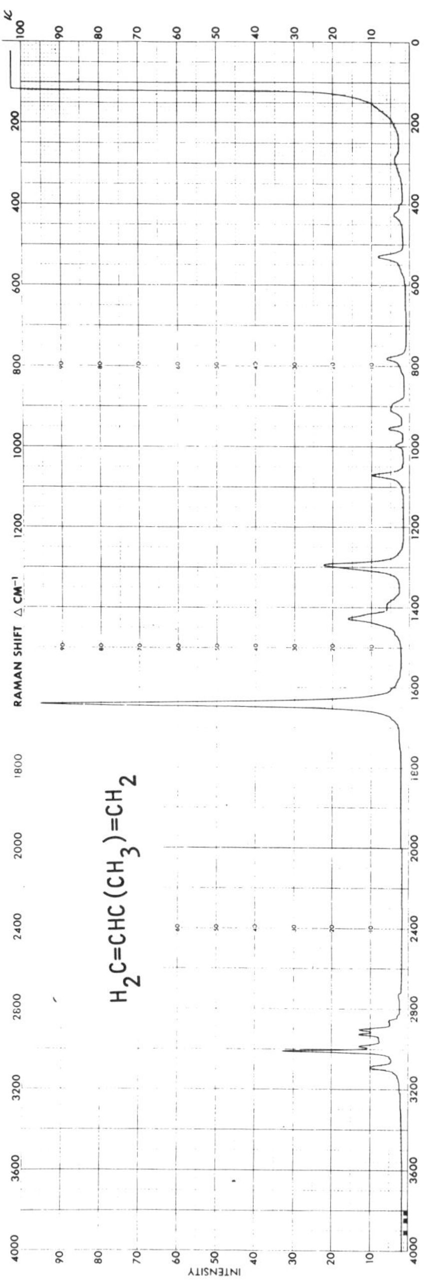

12543-1 *CAS* [*110-83-8*]
Cyclohexene, 99%

FW 82.15
mp -104°C
bp 83°C

d 0.811
Fp 10°*F*
n_D^{20} 1.4460

IR III, 30F
NMR II, 1,45A
Merck 10,2721

2926.2	1137.0	876.4
1652.8	1036.6	718.5
1437.4	916.4	639.6

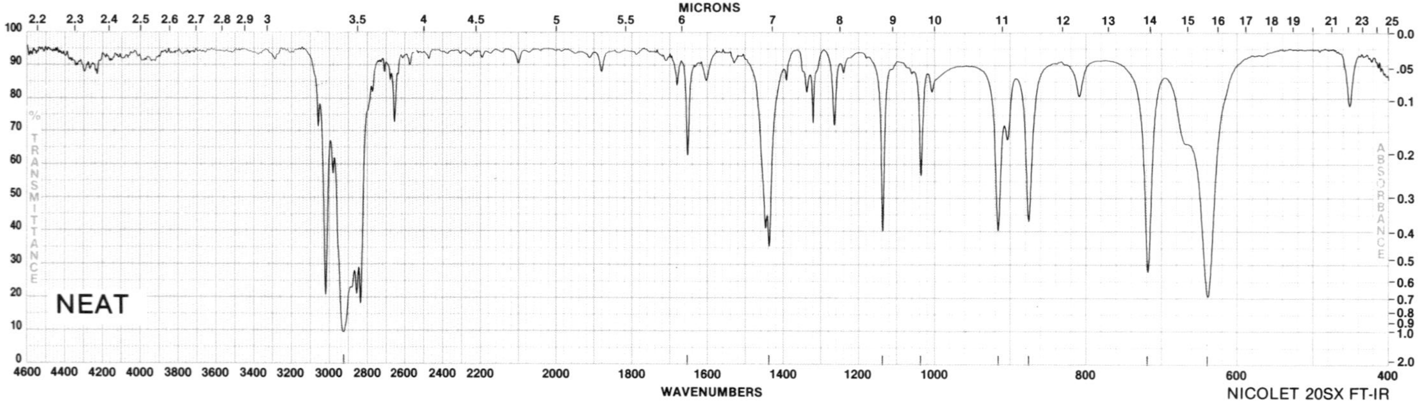

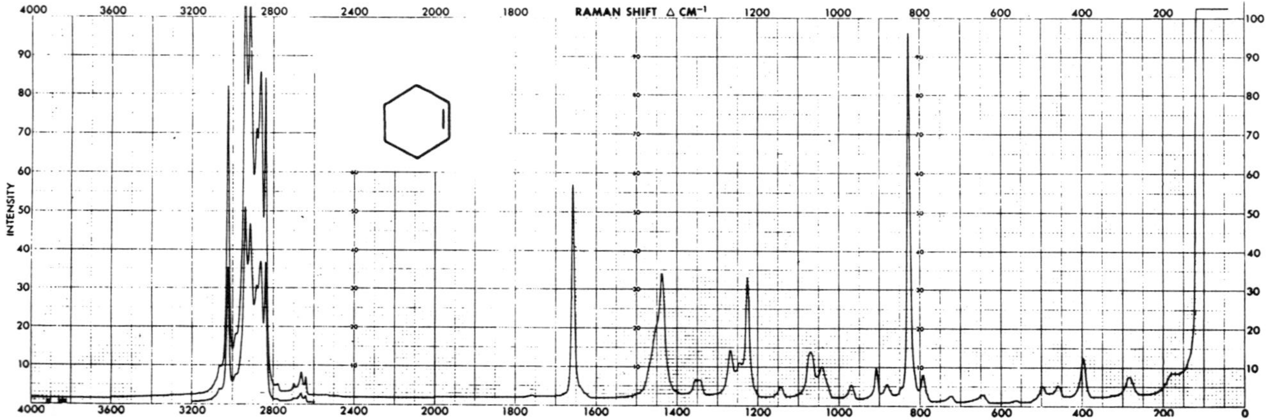

22119-8 CAS [75-12-7]
Formamide, 99%

FW 45.04	d 1.130	IR III, 437A	3314.3	1603.6	630.7
mp 2-3°C	Fp 310°F	NMR II, 1,629A	2881.8	1391.2	600.4
bp 210°C	n_D^{20} 1.4475	Merck 10,4121	1682.0	1308.3	564.6

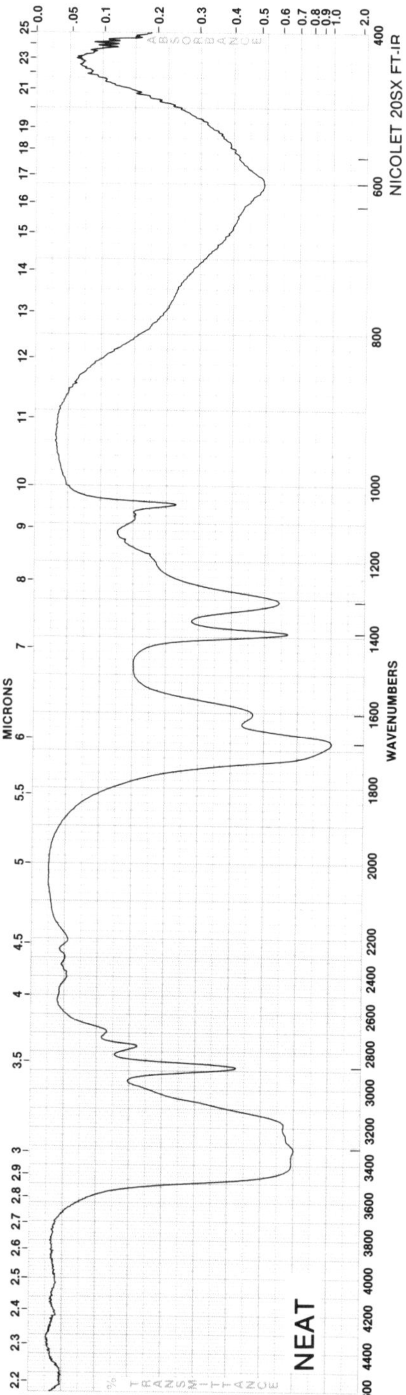

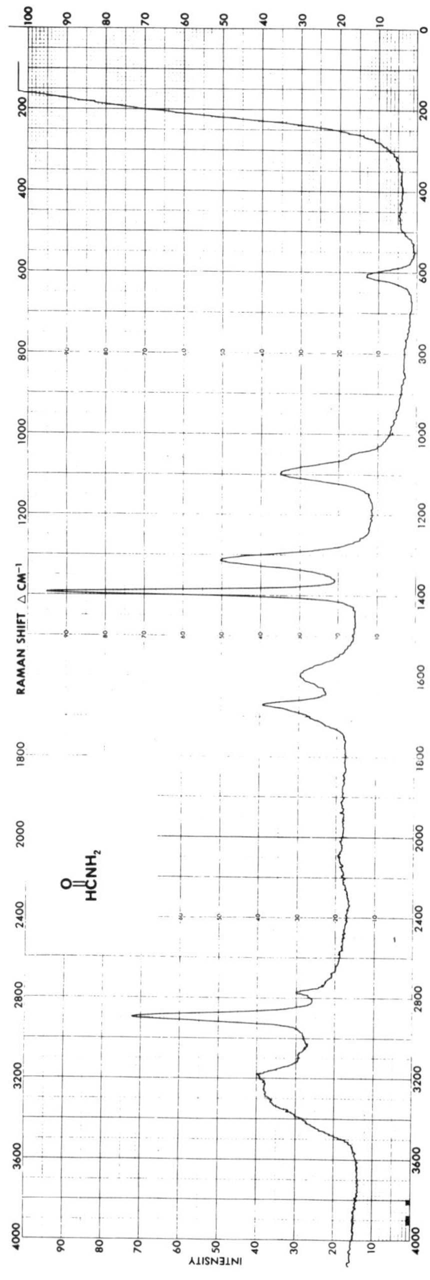

M4670-5 *CAS [123-39-7]*
***N*-Methylformamide, 99%**

FW 59.07
mp -40°C
bp 180-185°C

d 1.011
Fp 210°F
n_D^{20} 1.4302

IR III, 440G
NMR II, 1,634D

3297.4	1541.4	1244.0
2878.6	1453.7	1150.4
1665.8	1384.4	770.3

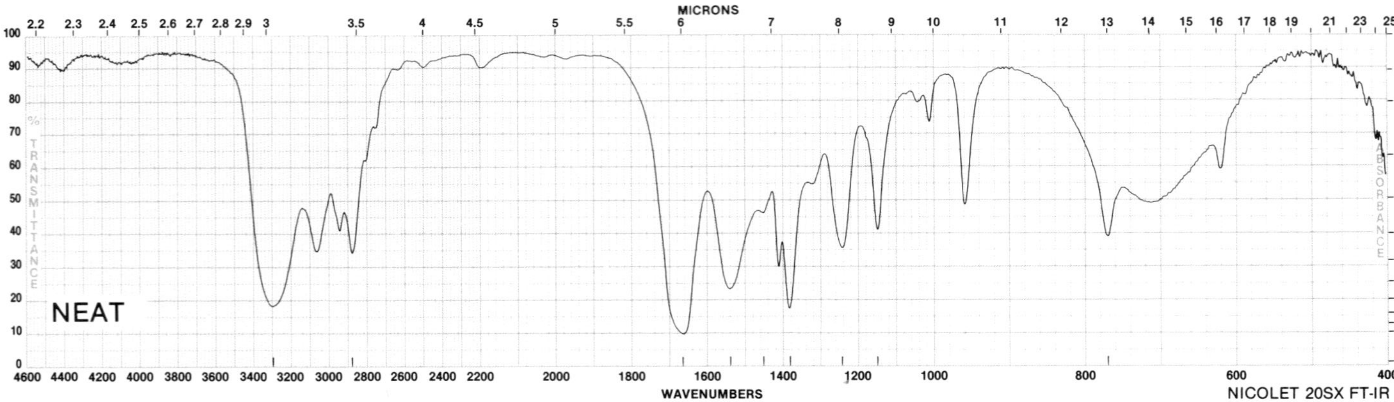

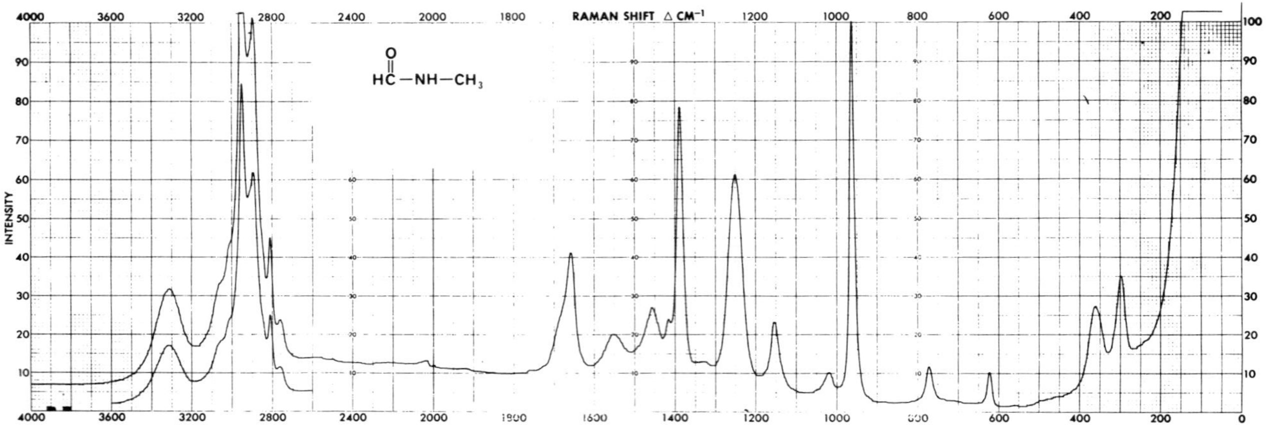

15481-4 *CAS [68-12-2]*
***N,N*-Dimethylformamide**, 99 + %

FW 73.10
mp -61 °C
bp 153 °C

d 0.944
Fp 136 °F
n_D^{20} 1.4305

IR III, 443D
Merck 10,3237

2928.4 1438.4 1092.8
1676.3 1387.7 658.3
1505.6 1257.0

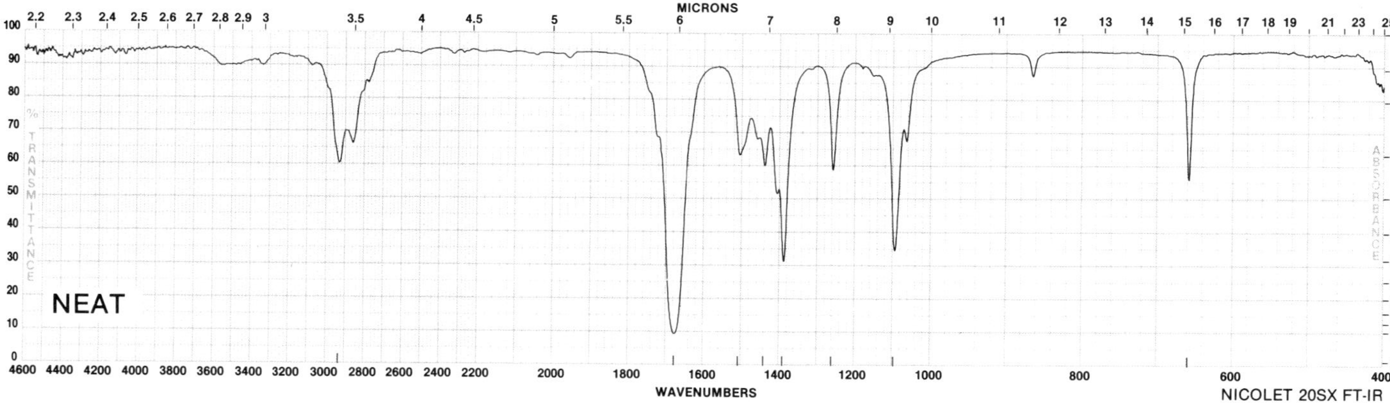

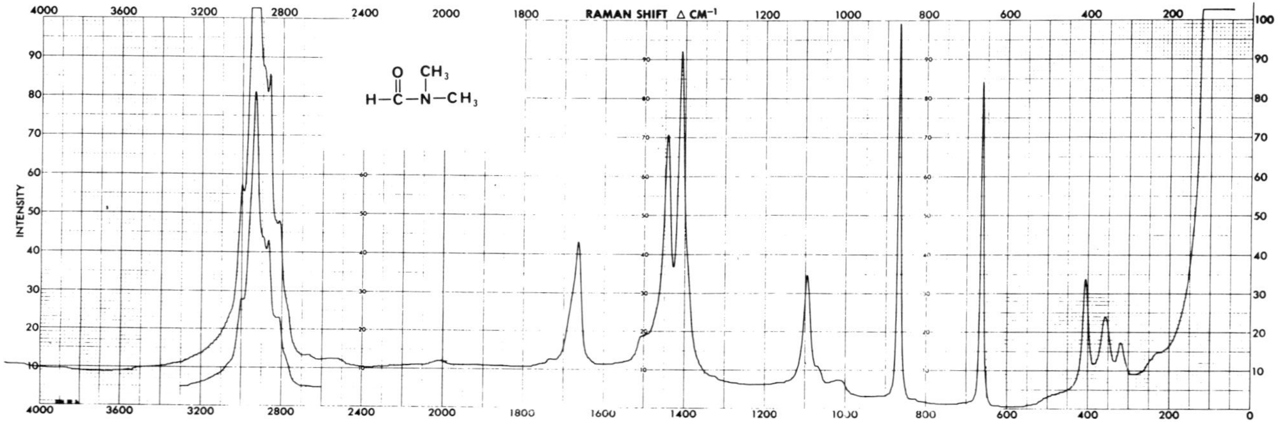

M4683-7 *CAS [107-31-3]*
Methyl formate, 99%

FW 60.05
mp -100.4°C
bp 34°C

d 0.974
Fp -26°*F*
n_D^{20} 1.3430

IR III, 359A
NMR II, 1,507A
Merck 10,5947

2957.7 1381.8 911.3
1728.1 1209.2 768.0
1434.1 1030.1

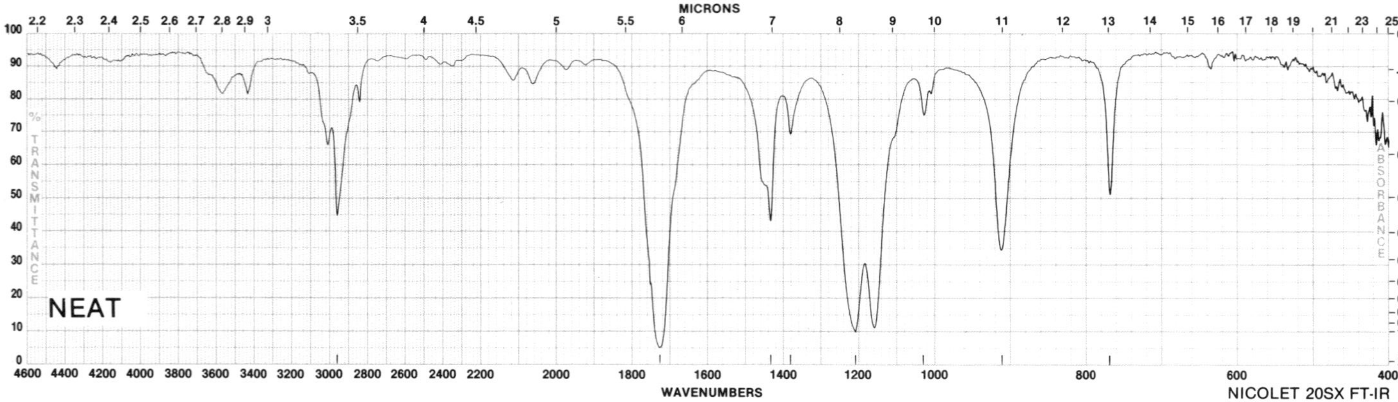

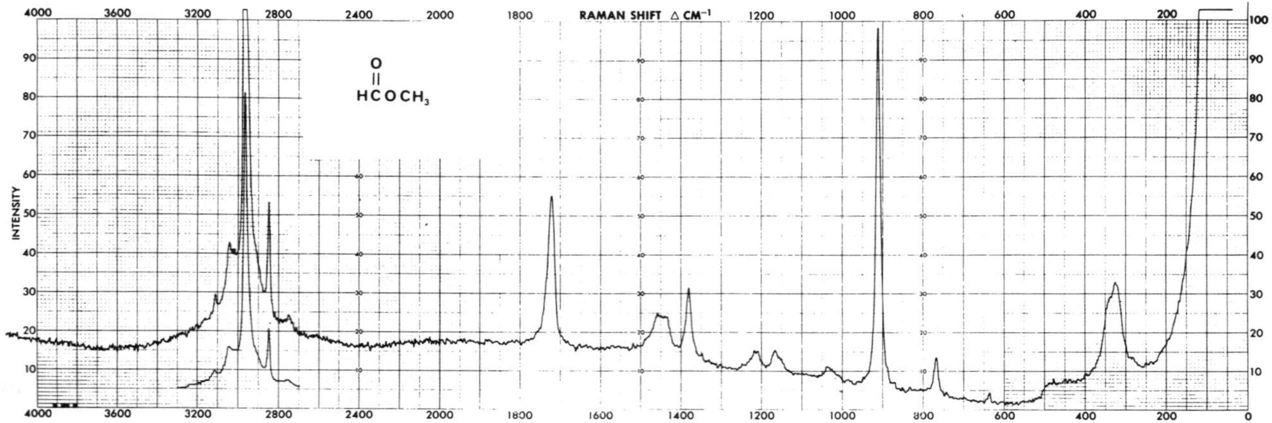

12855-4 *CAS [592-42-7]*
1,5-Hexadiene, 98%

FW 82.15	d 0.692	IR III, 22B	2927.1	1439.3	911.4
mp -141°C	Fp -17°*F*	NMR II, 1,33C	1828.2	993.8	648.0
bp 60°C	n_D^{20} 1.4045		1642.3	958.8	621.6

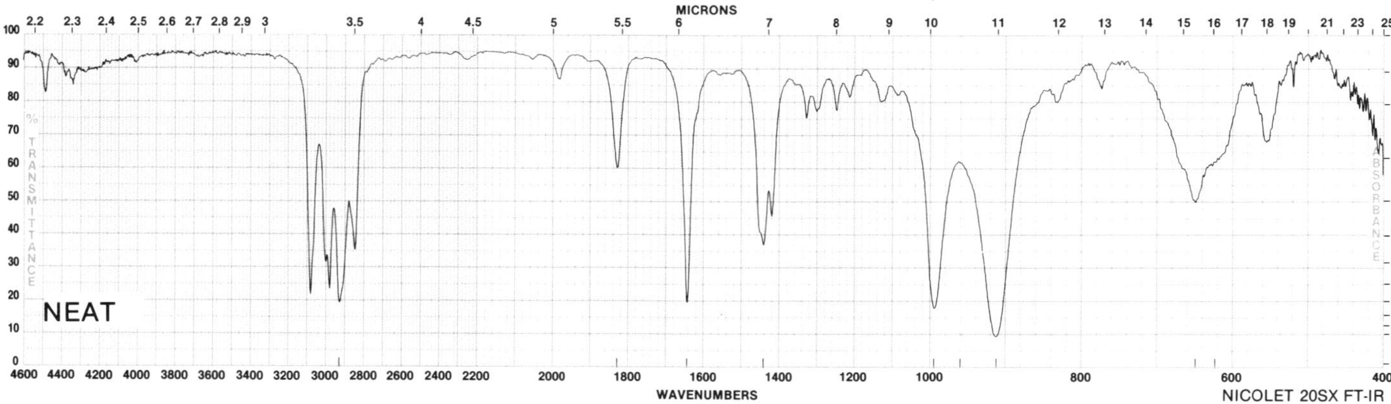

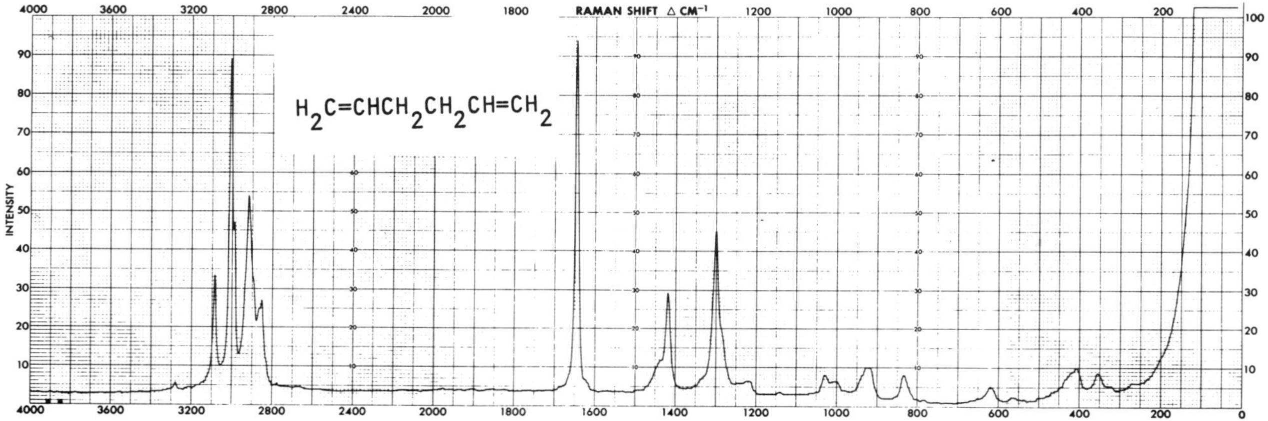

M4970-4 *CAS [591-76-4]*
2-Methylhexane, 99%

FW 100.21
mp -118°C
bp 90°C

d 0.679
Fp 25°*F*
n_D^{20} 1.3840

IR III, 6D
NMR II, 1,15A

4334.3	1467.9	1074.2
2957.4	1384.1	918.0
2730.7	1171.3	728.1

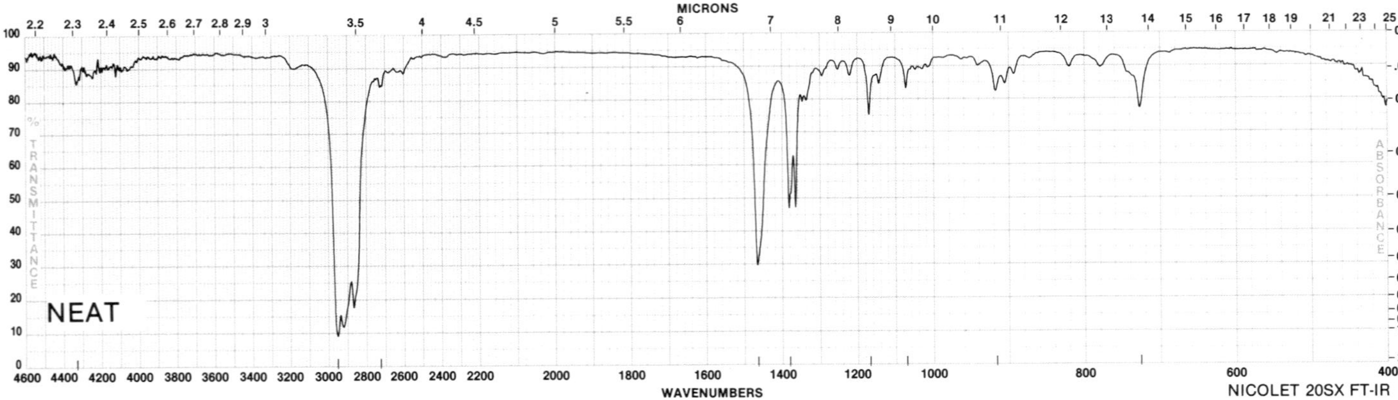

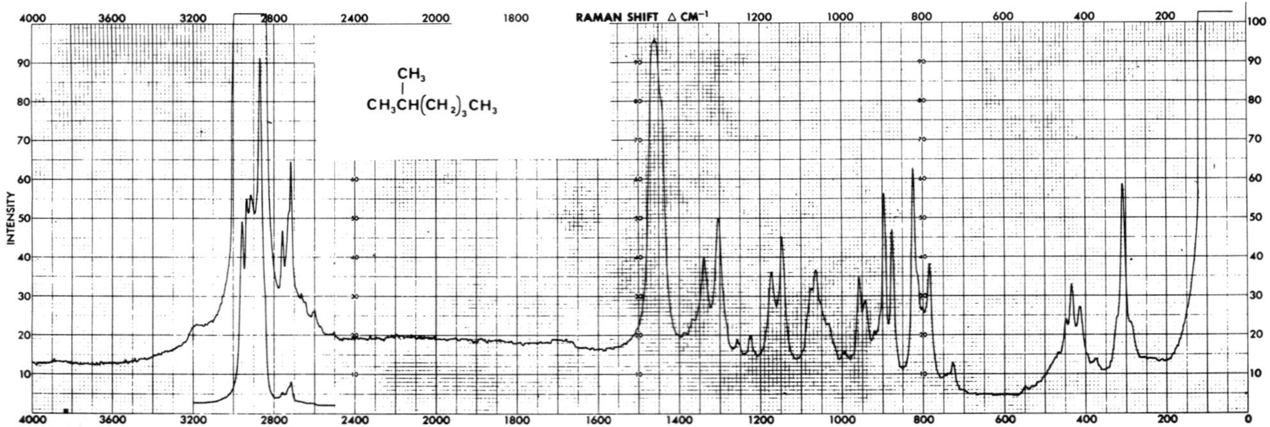

25090-2 *CAS [1127-76-0]*
1-Ethylnaphthalene, 99+%

FW 156.23
mp 14°C
bp 258-260°C

d 1.008
Fp 233°F
n^{20}_{D} 1.6054

2966.3 1453.3 860.0
1596.8 1394.0 776.3
1510.6 1165.3 731.2

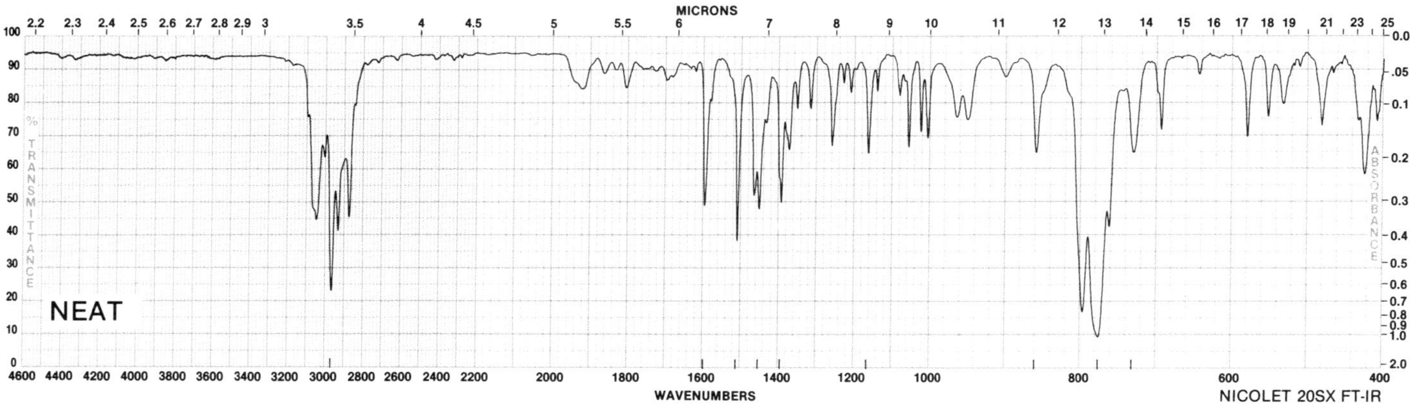

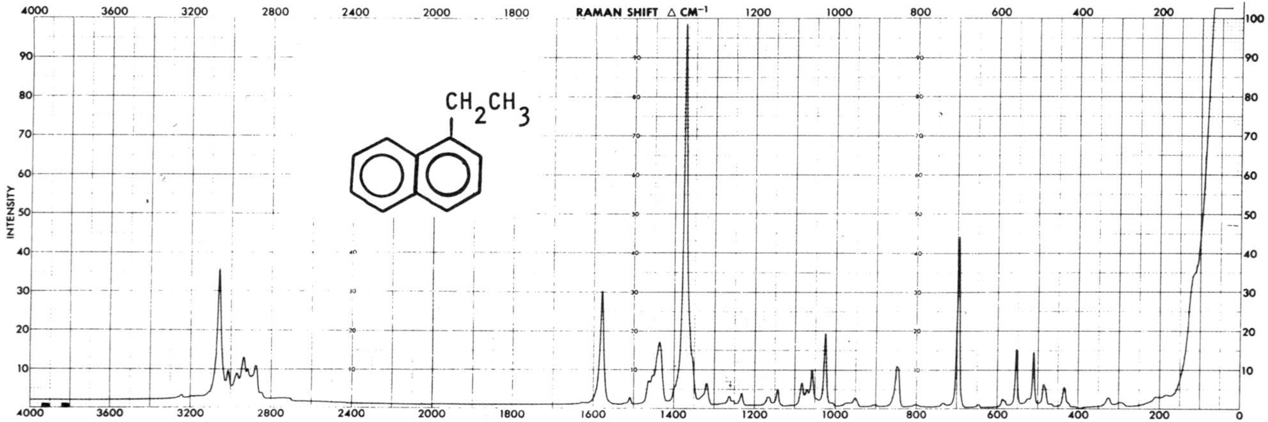

11067-1 *CAS [590-35-2]*
2,2-Dimethylpentane, 99 + %

FW 100.21
bp 78 °C
d 0.674

Fp 60 °*F*
n_D^{20} 1.3824

IR III, 8D
NMR II, 1,16B

2956.8	1207.7
1477.1	945.7
1364.4	740.1

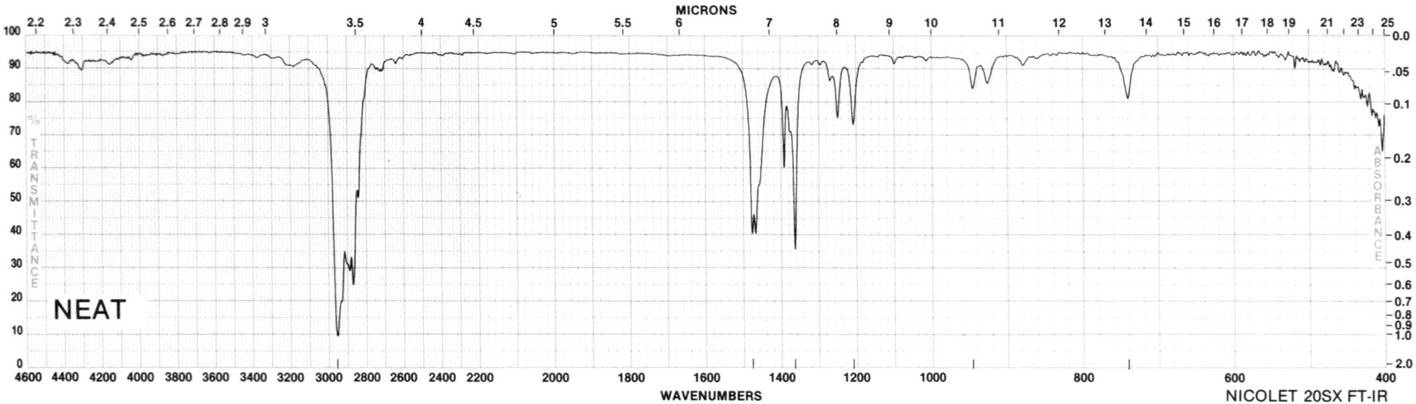

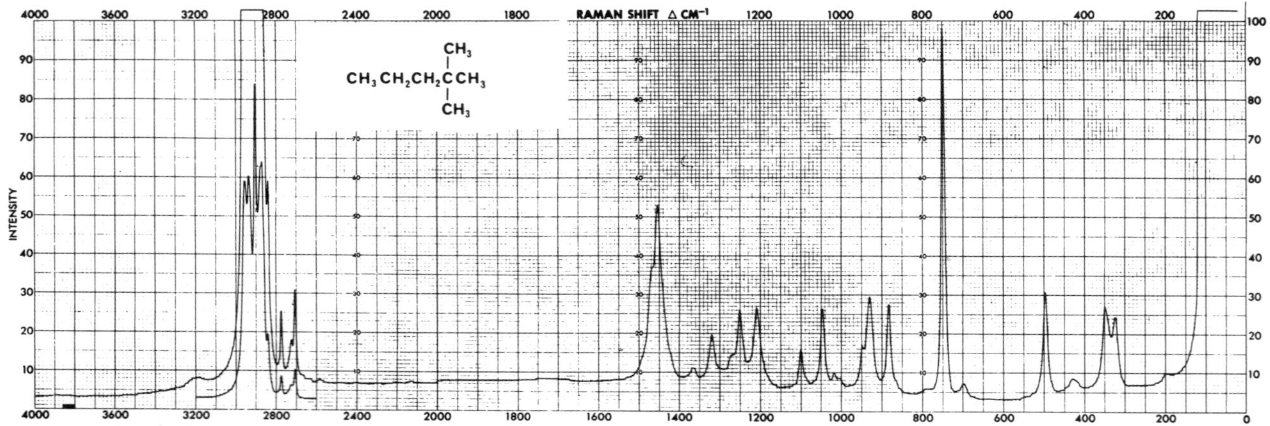

P810-6 *CAS [107-87-9]*
2-Pentanone, 97%

FW 86.13
mp -78 °C
bp 100-101 °C

d 0.812
Fp 45 °*F*
n_D^{20} 1.3897

IR III, 240D
NMR II, 1,369C
Merck 10,5988

2963.9	1366.0	1170.7
1717.4	1295.5	727.0
1422.9	1235.5	591.9

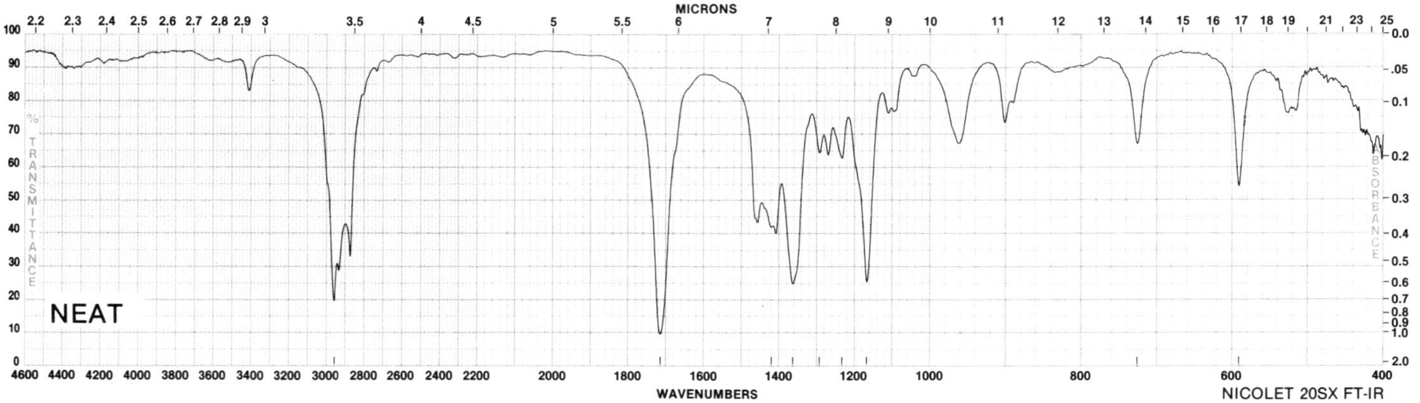

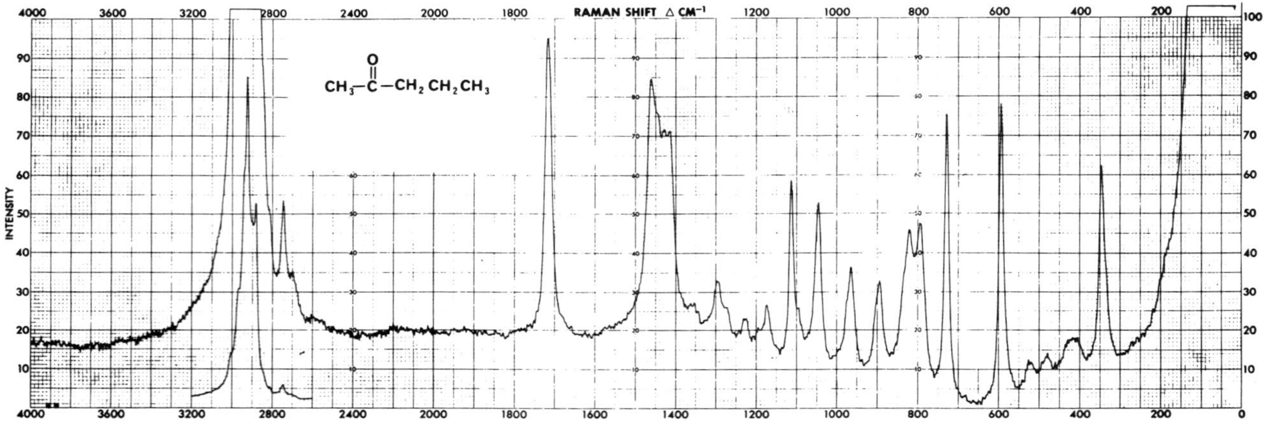

18452-7 Pyridine, 99 + % CAS [110-86-1]

FW 79.10
mp -42°C
bp 115°C

d 0.978
Fp 68°F
n_D^{20} 1.5102

IR III, 1306A
Merck 10,7869

3024.8	1146.5	747.6
1580.9	1030.3	703.6
1437.7	990.5	602.9

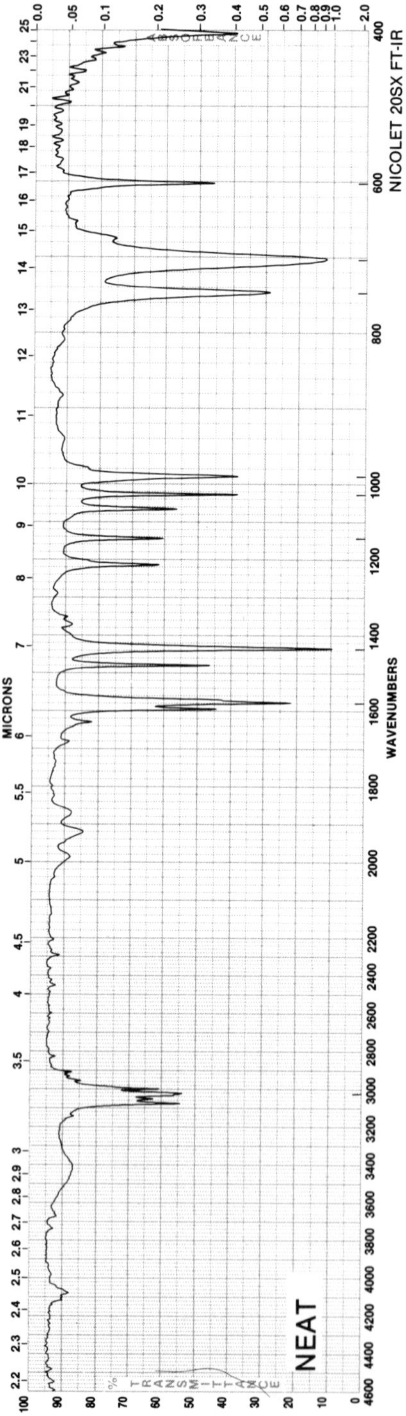

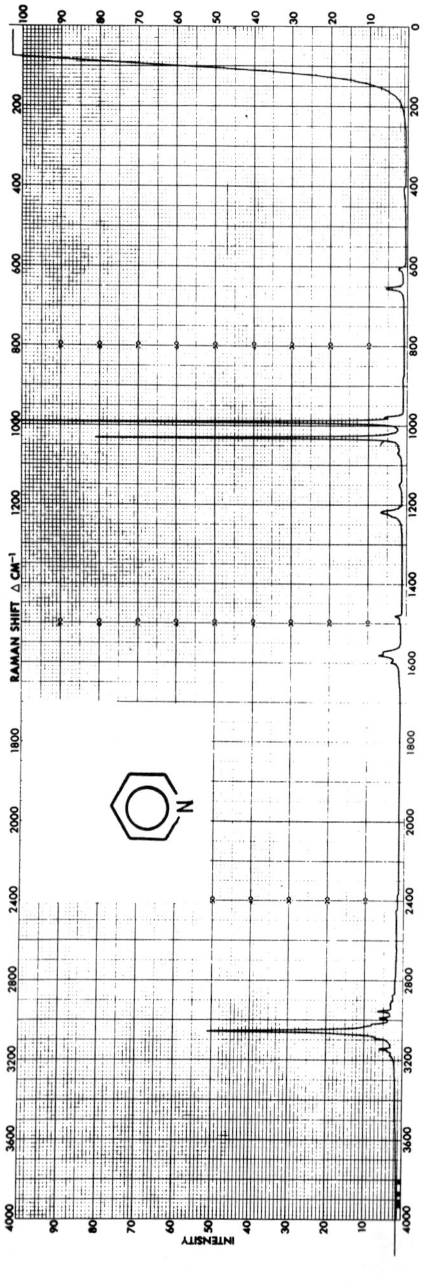

23951-8 CAS *[109-06-8]*
2-Picoline, 99+%

FW 93.13
mp -70°C
bp 128-129°C

d 0.943
n_D^{20} 1.5010

IR III, 1307E

3011.7 1294.9 997.7
1592.4 1149.0 753.0
1477.7 1049.7 471.0

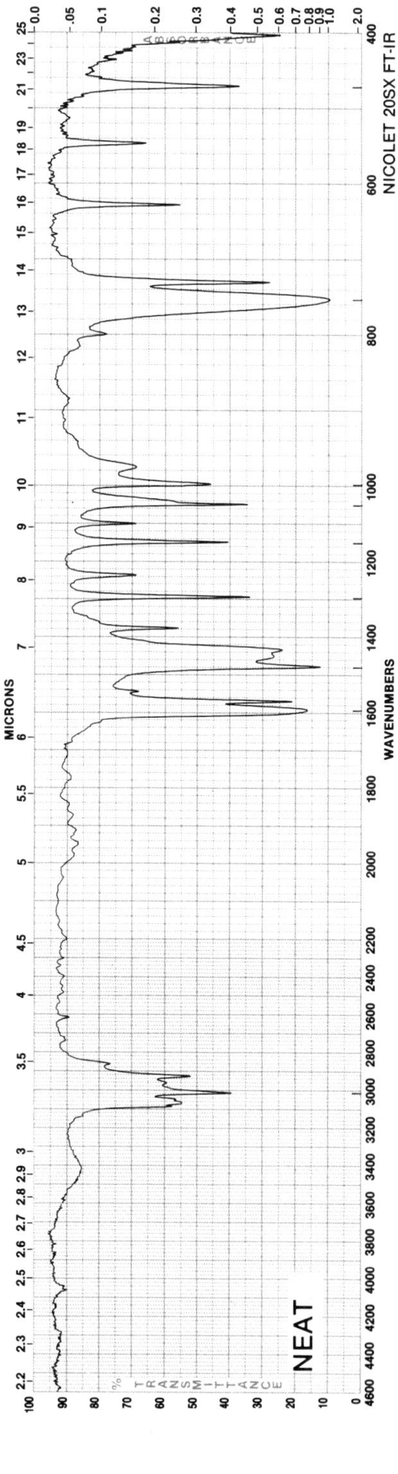

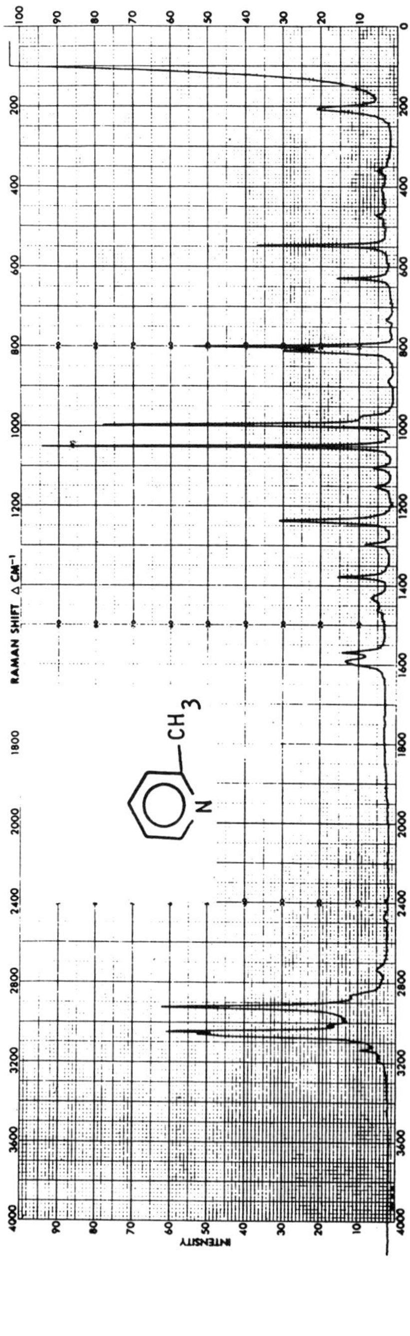

P4205-3 CAS [*108-99-6*]
3-Picoline, 99%

FW 93.13
mp -19°C
bp 143-144°C

d 0.957
Fp 97°F
n_D^{20} 1.5060

IR III, 1308G
NMR II, 2,615C
Merck 10,7282

3027.8	1412.6	787.0
1577.8	1190.9	712.0
1479.0	1029.4	630.2

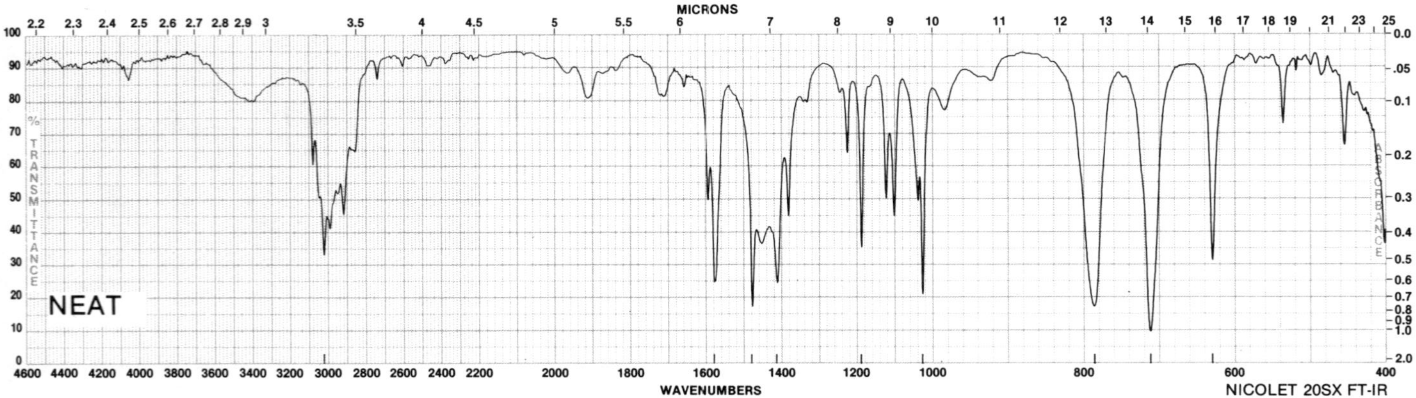

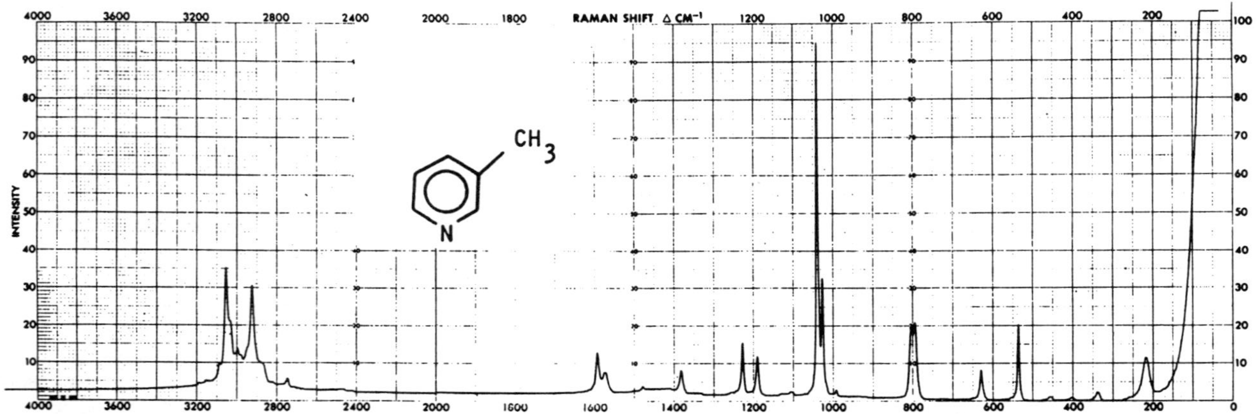

13149-0 *CAS [108-89-4]*
4-Picoline, 98%

FW 93.13	d 0.957	IR III, 1309B	3027.9 1412.3 798.5
mp 2.4 °C	Fp 134 °F	NMR II, 2,616A	1605.0 1223.6 515.4
bp 145 °C	n_D^{20} 1.5045	Merck 10,7283	1496.2 994.9 485.2

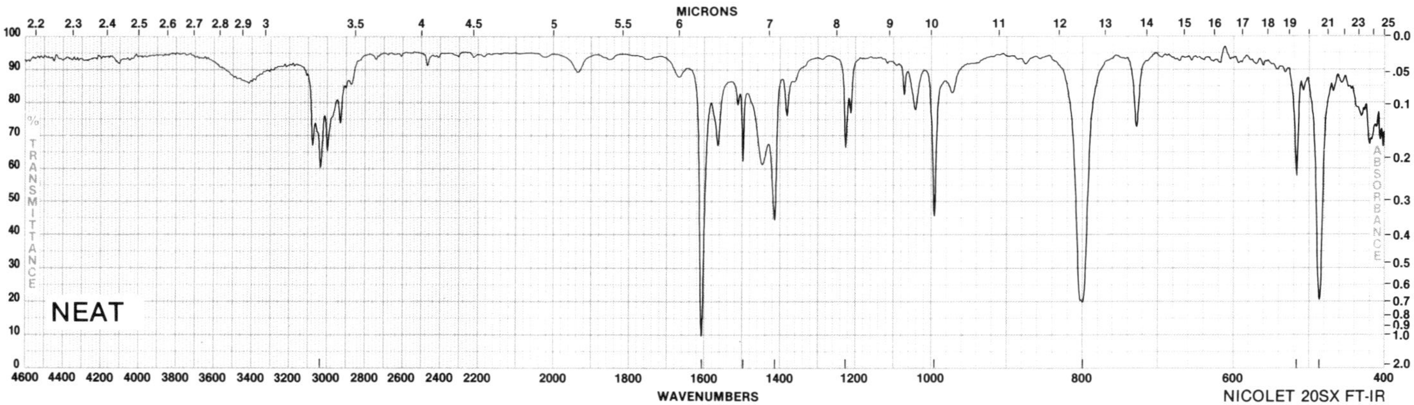

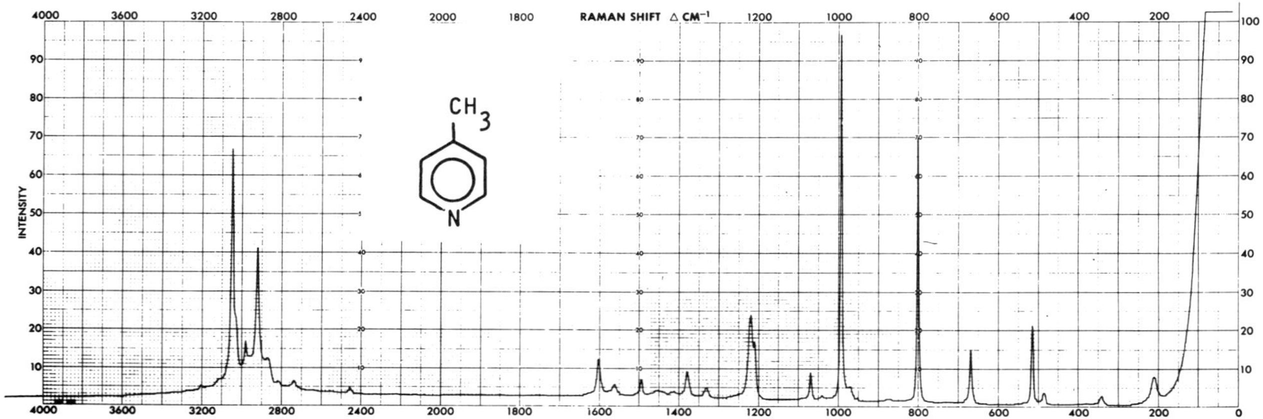

M8090-3 *CAS [98-83-9]*
α-Methylstyrene, 99%

FW 118.18
mp -24°C
bp 165-169°C

d 0.909
Fp 114°F
n_D^{20} 1.5375

IR III, 569D
NMR II, 1,748D

3084.0 1443.0 891.1
1629.4 1374.9 776.4
1495.4 1027.9 704.1

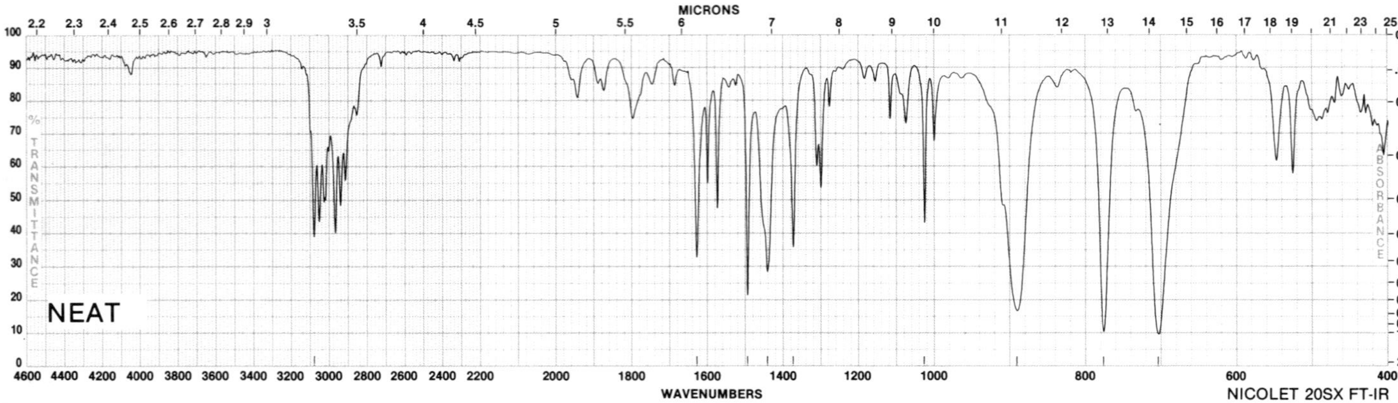

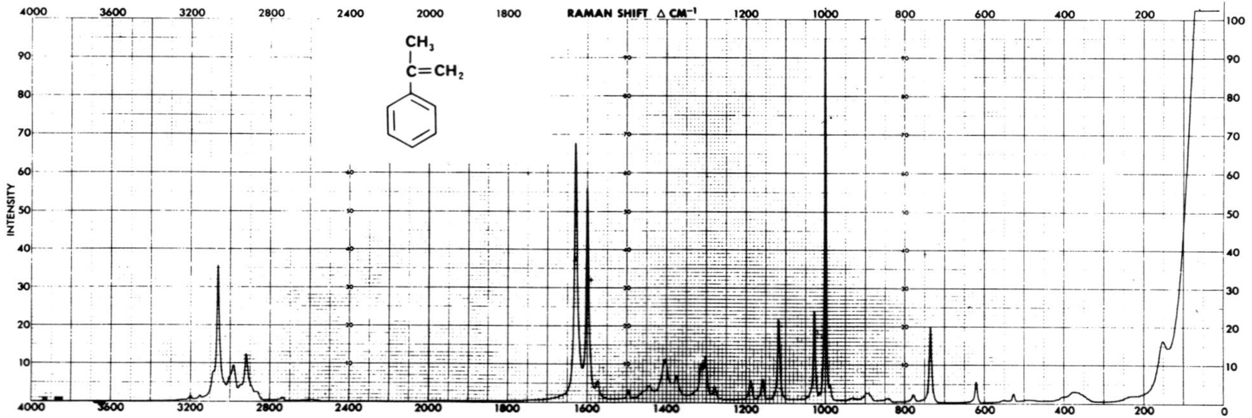

T3180-1 *CAS [110-02-1]*
Thiophene, 99 + %

FW 84.14	d 1.051	IR III, 1206E	3107.5 1251.1 712.7
mp -38 °C	Fp 30 °F	NMR II, 2,466B	1587.5 1081.0 452.7
bp 84 °C	n_D^{20} 1.5278	Merck 10,9195	1407.2 833.7

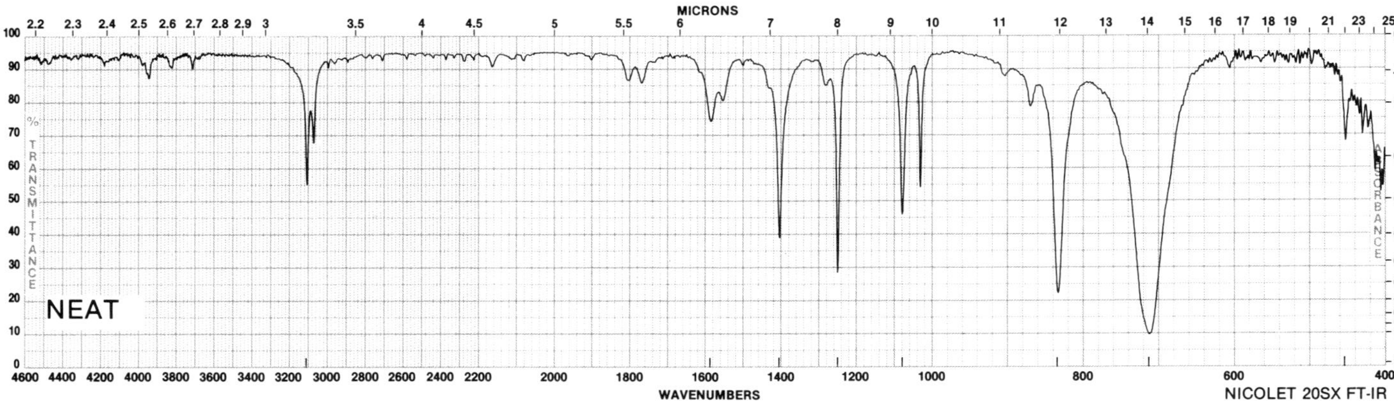

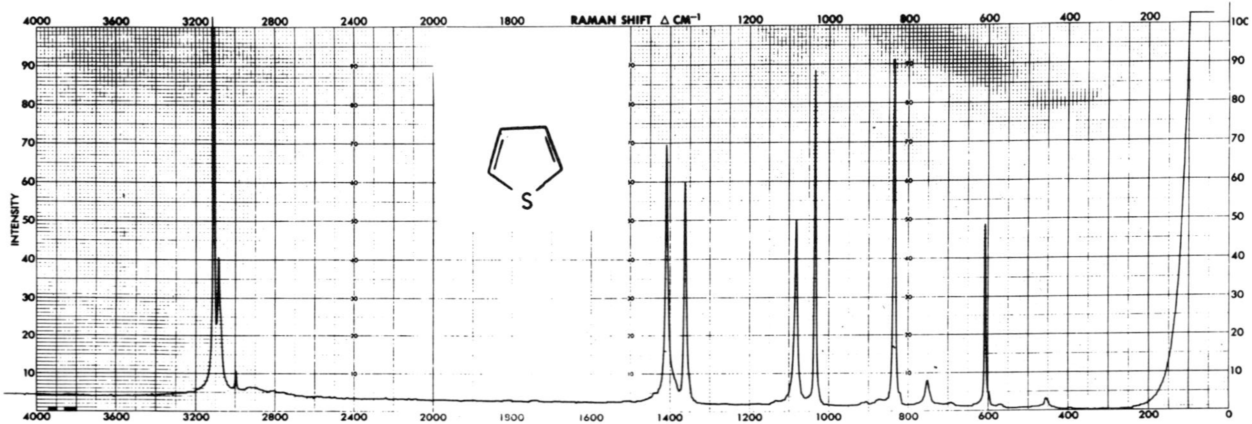

13206-3 *CAS [121-44-8]*
Triethylamine, 99%

FW 101.19
mp -7°C
bp 88.8°C

d 0.726
Fp 20°*F*
n_D^{20} 1.4000

IR III, 175G
NMR II, 1,258B
Merck 10,9477

2970.4	1382.2	1137.1
2754.1	1293.0	1070.9
1467.6	1204.7	736.4

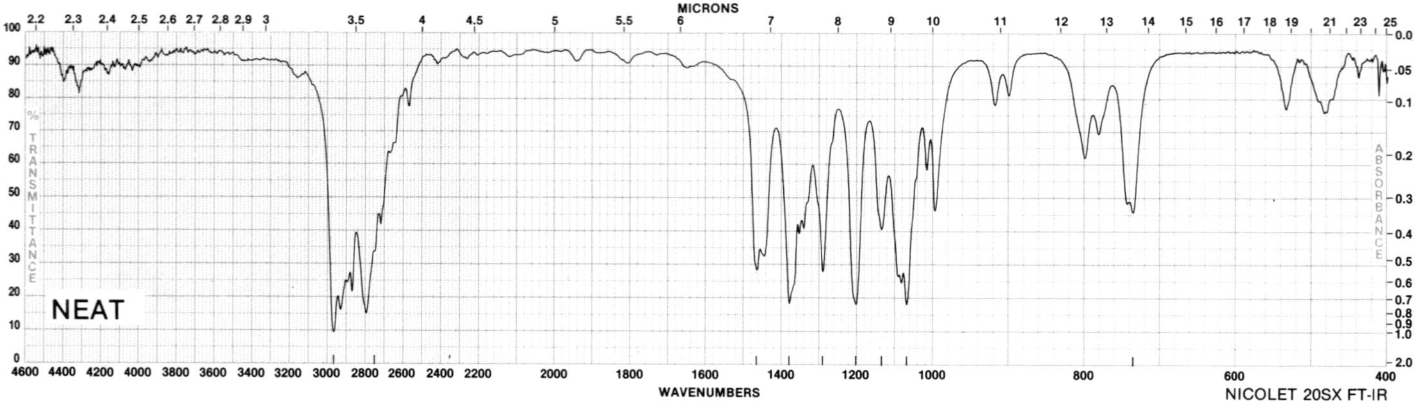

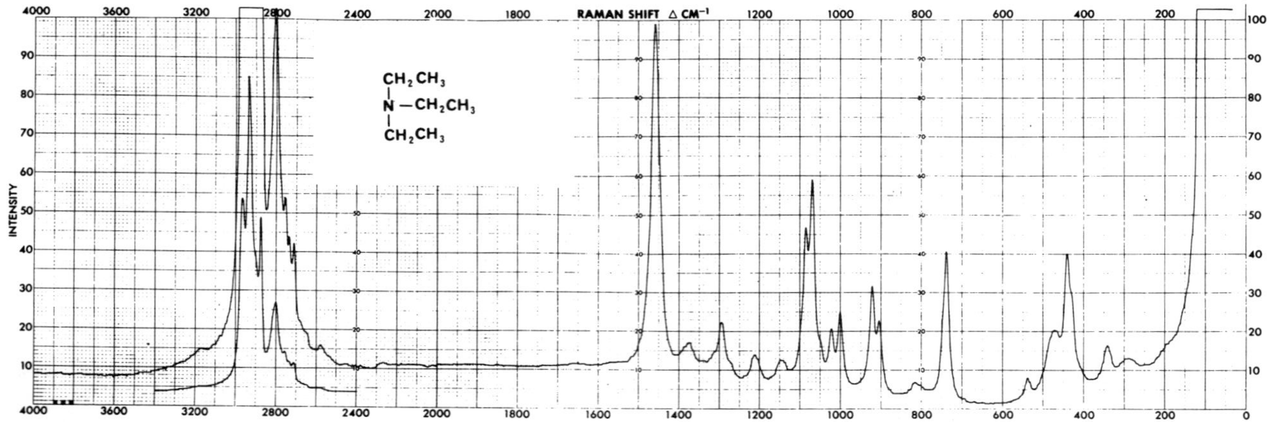

APPENDIX 2

Infrared and Raman Spectra of Selected Carbohydrate, Steroid, Organometallic, and Polymer Spectra from the Schrader Raman/Infrared Atlas of Organic Compounds*

Although there is little discussion of carbohydrates, steroids, or polymers in Chapters 2–17, these are important classes of compounds and they contain many of the characteristic groups that have been discussed. A selection of IR and Raman spectra of such substances are presented here. They are from the Raman/IR Atlas of Organic Compounds by B. Schrader. The instruments used and the sample preparations are noted on the spectra. These compounds are also included in the composite name and molecular formula indices at the end of this book.

INDEX TO SPECTRA

* B. Schrader, *Raman/Infrared Atlas of Organic Compounds*, 2nd Ed., VCH-Verl.-Ges., Weinkeim (1989), with permission.

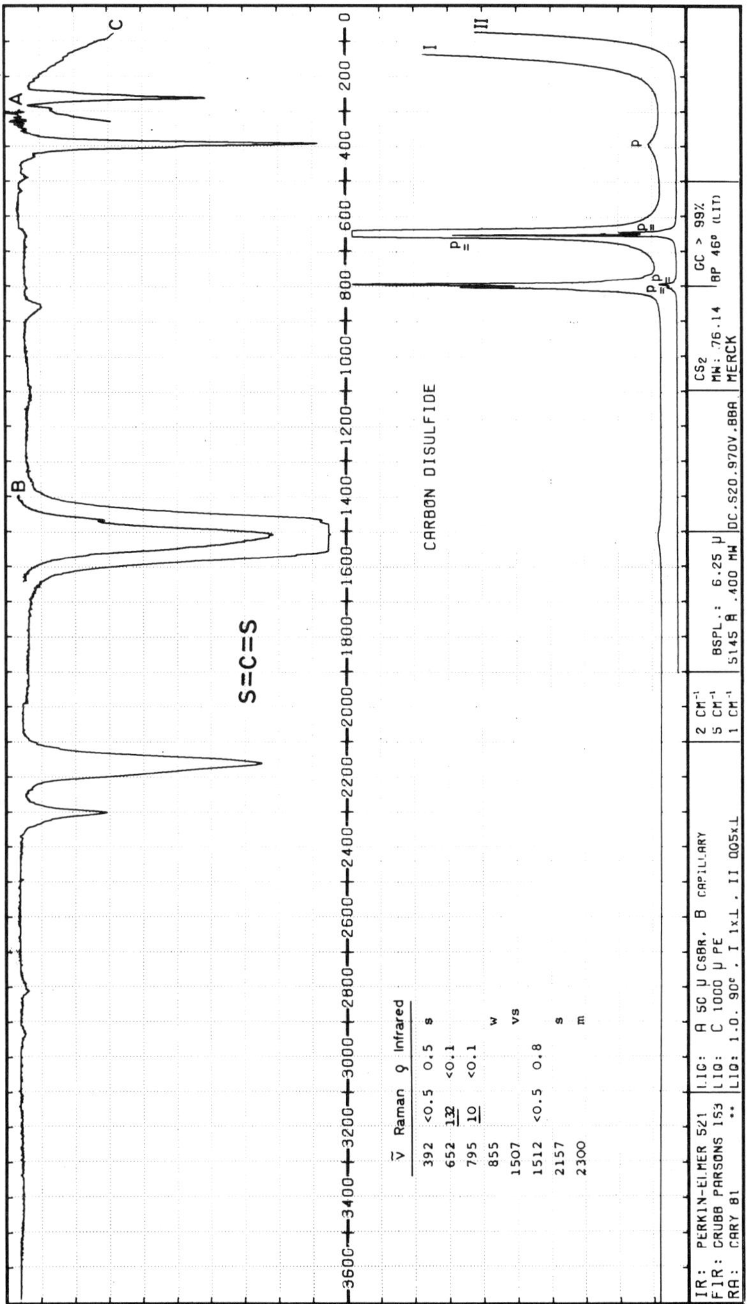
S=C=S
CARBON DISULFIDE
ν̃ Raman ρ Infrared
392 <0.5 0.5 s
652 132 <0.1
795 10 <0.1
855 w
1507 vs
1512 <0.5 0.8
2157 s
2300 m
IR: PERKIN-ELMER 521
FIR: GRUBB PARSONS 153
RA: CARY 81
2 CM⁻¹
5 CM⁻¹
1 CM⁻¹
BSPL.: 6.25 μ
DC.S20.970V.BBA
CS2
MW: 76.14
MERCK
GC > 99%
BP 46° (LIT)

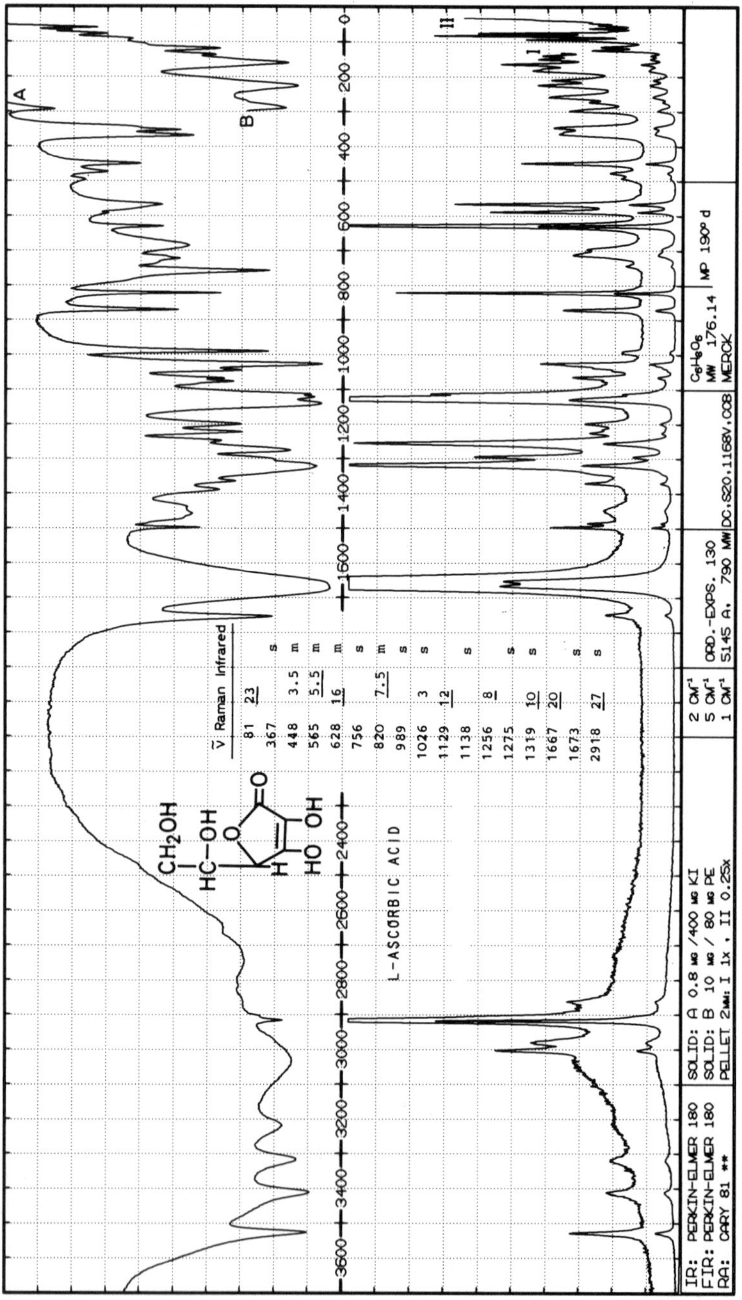
L-ASCORBIC ACID
CH2OH
HC—OH
O
O
H
HO OH
A
B
I
II
3600 3400 3200 3000 2800 2600 2400 1600 1400 1200 1000 800 600 400 200 0
ν̃ Raman Infrared
81 23
367 s
448 3.5 m
565 5.5 m
628 16 m
756 s
820 7.5 m
989 s
1026 3 s
1129 12
1138 s
1256 8
1275 s
1319 10 s
1667 20
1673 s
2918 27 s
IR: PERKIN-ELMER 180
FIR: PERKIN-ELMER 180
RA: CARY 81 **
SOLID: A 0.8 MG /400 MG KI
SOLID: B 10 MG / 80 MG PE
PELLET 2MM: I 1x , II 0.25x
2 CM-1
5 CM-1
1 CM-1
ORD.-EXPS. 130
5145 A. 790 MW
DC:S20.1168V.C08
C6H8O6
MW 176.14
MERCK
MP 190° d

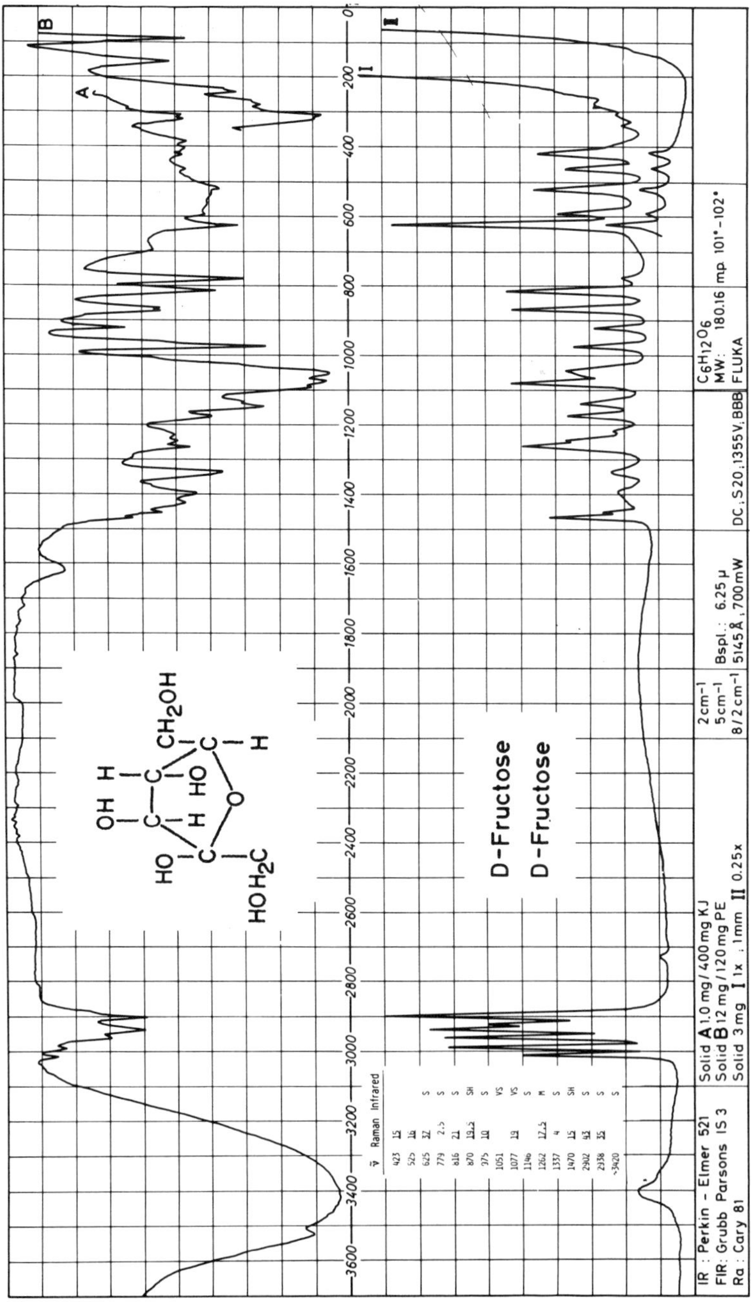

ṽ	Raman	Infrared
423	15	
525	16	
625	32	S
779	2.5	S
816	21	S
870	19.5	SH
975	10	S
1051		VS
1077	19	VS
1146		S
1262	17.5	M
1337	4	S
1470	15	SH
2902	43	S
2938	35	S
~3420		S

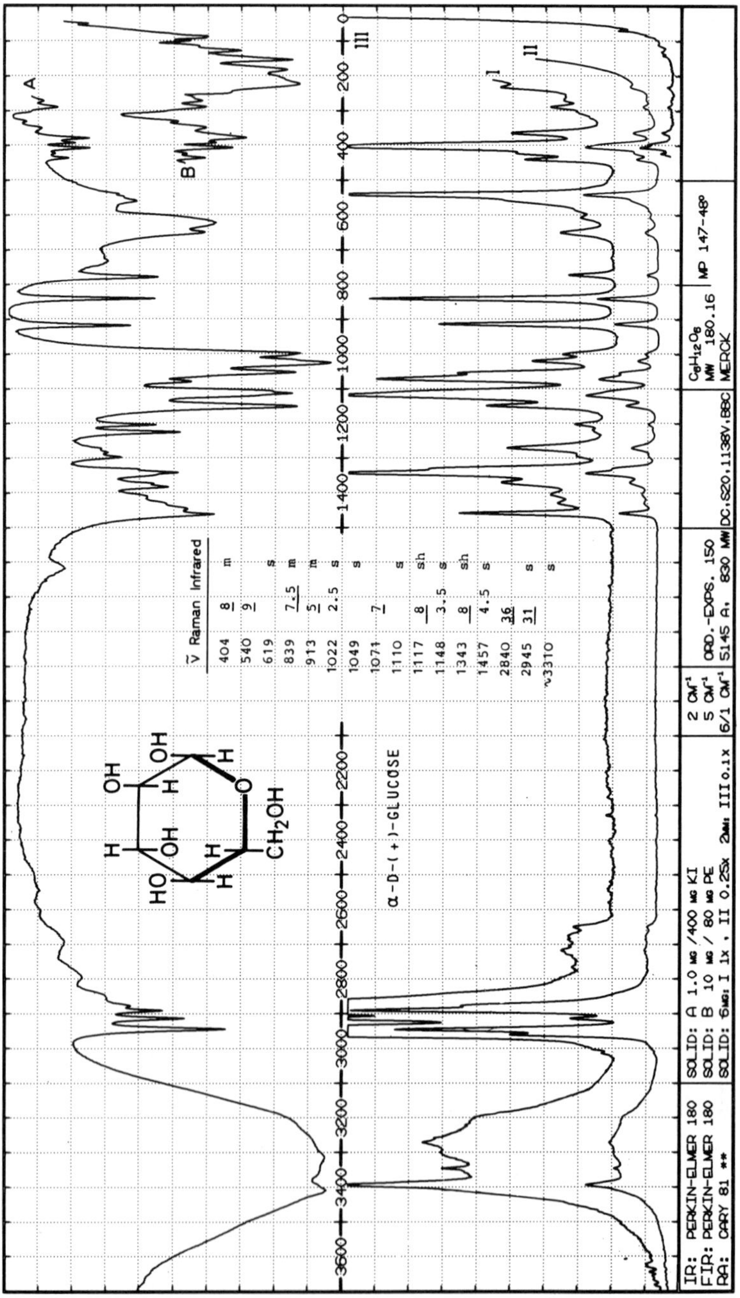
α-D-(+)-GLUCOSE
A
B
I
II
III
3600 3400 3200 3000 2800 2600 2400 2200
1400 1200 1000 800 600 400 200 0
ṽ Raman Infrared
404 8 m
540 9
619 s
839 7.5 m
913 5 m
1022 2.5 s
1049 s
1071 7
1110 s
1117 8 sh
1148 3.5 s
1343 8 sh
1457 4.5 s
2840 36
2945 31 s
~3310 s
IR: PERKIN-ELMER 180
FIR: PERKIN-ELMER 180
RA: CARY 81 **
SOLID: A 1.0 MG /400 MG KI
SOLID: B 10 MG / 80 MG PE
SOLID: 6MG: I 1x , II 0.25x 2MM: III 0.1x
2 CM⁻¹
5 CM⁻¹
6/1 CM⁻¹
ORD.-EXPS. 150
5145 A, 830 MW
DC,S20,1138V,BBC
C6H12O6
MW 180.16
MERCK
MP 147-48°

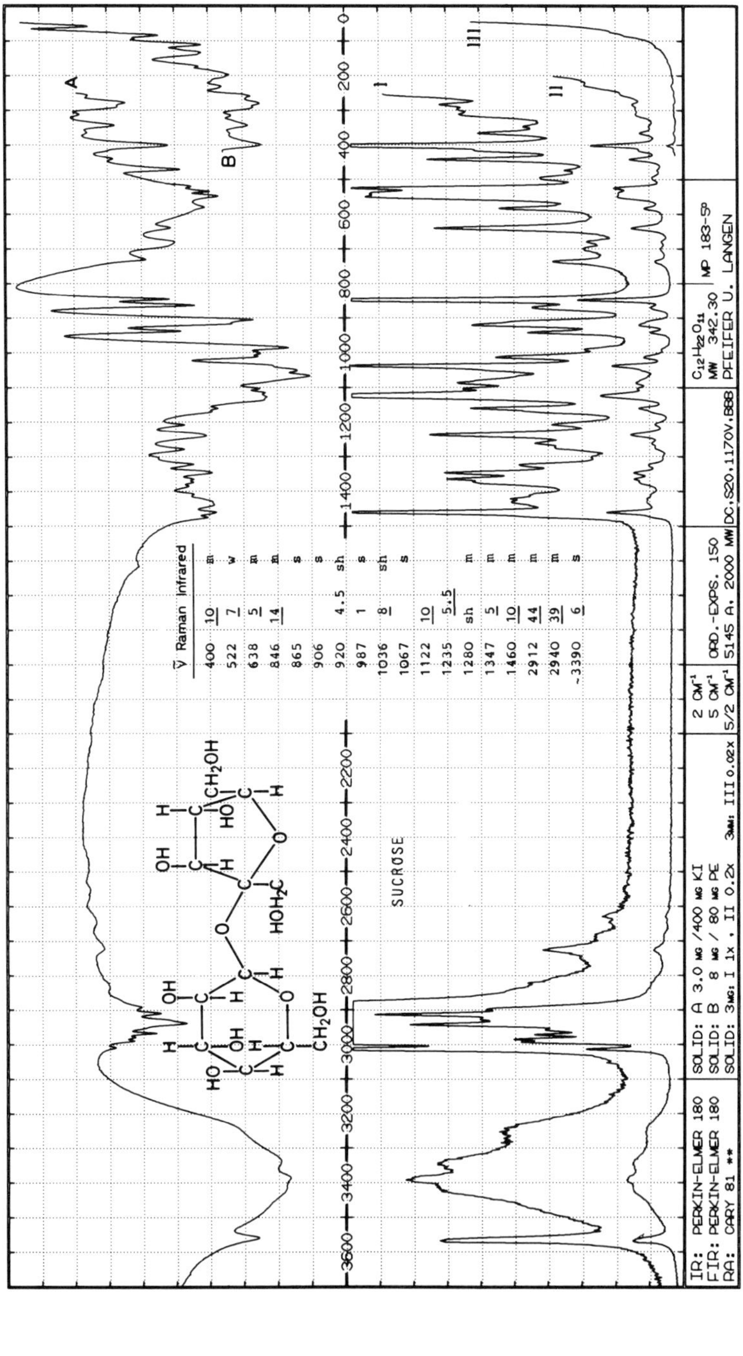
A
B
I
II
III
SUCROSE
ν̄ Raman Infrared
400 10 m
522 7 w
638 5 m
846 14 m
865 s
906 s
920 4.5 sh
987 1 s
1036 8 sh
1067 s
1122 10
1235 5.5
1280 sh m
1347 5 m
1460 10 m
2912 44 m
2940 39 m
~3390 6 s
3600 3400 3200 3000 2800 2600 2400 2200 1400 1200 1000 800 600 400 200 0
IR: PERKIN-ELMER 180
FIR: PERKIN-ELMER 180
RA: CARY 81 **
SOLID: A 3.0 MG /400 MG KI
SOLID: B 8 MG / 80 MG PE
SOLID: 3MG: I 1x , II 0.2x 3MM: III 0.02x
2 CM-1
5 CM-1
5/2 CM-1
ORD.-EXPS. 150
5145 A. 2000 MW
DC:S20.1170V.888
C12H22O11
MW 342.30
MP 183-5°
PFEIFER U. LANGEN

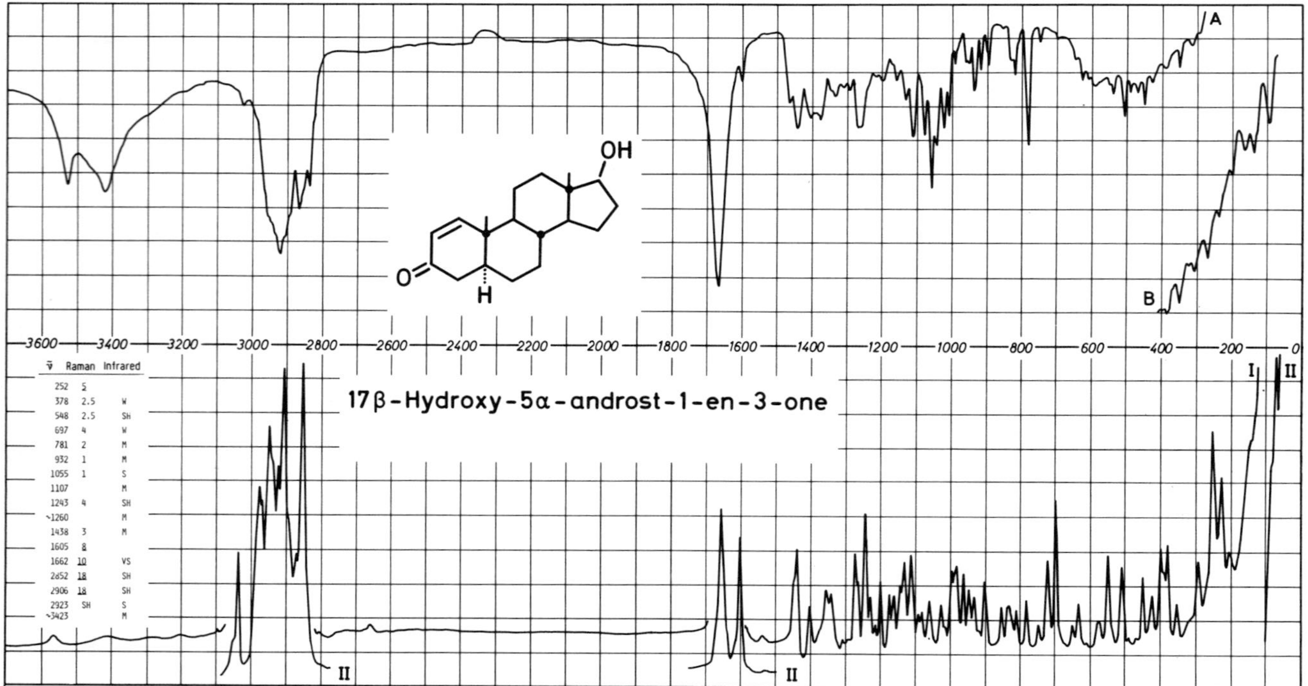

17β-Hydroxy-5α-androst-1-en-3-one

$\tilde{\nu}$	Raman	Infrared
252	5	
378	2.5	W
548	2.5	SH
697	4	W
781	2	M
932	1	M
1055	1	S
1107		M
1243	4	SH
~1260		M
1438	3	M
1605	8	
1662	10	VS
2852	18	SH
2906	18	SH
2923	SH	S
~3423		M

IR: PERKIN-ELMER 521 FIR: GRUBB PARSONS IS3 RA: CARY 81	SOLID: A 1.5 MG /400 MG KJ SOLID: B 30 MG /150 MG PE PELLET 2 MM: I 1x, II 0.5x	2 CM^{-1} 5 CM^{-1} 4 CM^{-1}	BSPL: 6.25 μ 5145 Å .300 MW	DC.S20.1395V.CCC	$C_{19}H_{28}O_2$ MW: 288.44	MP 158°

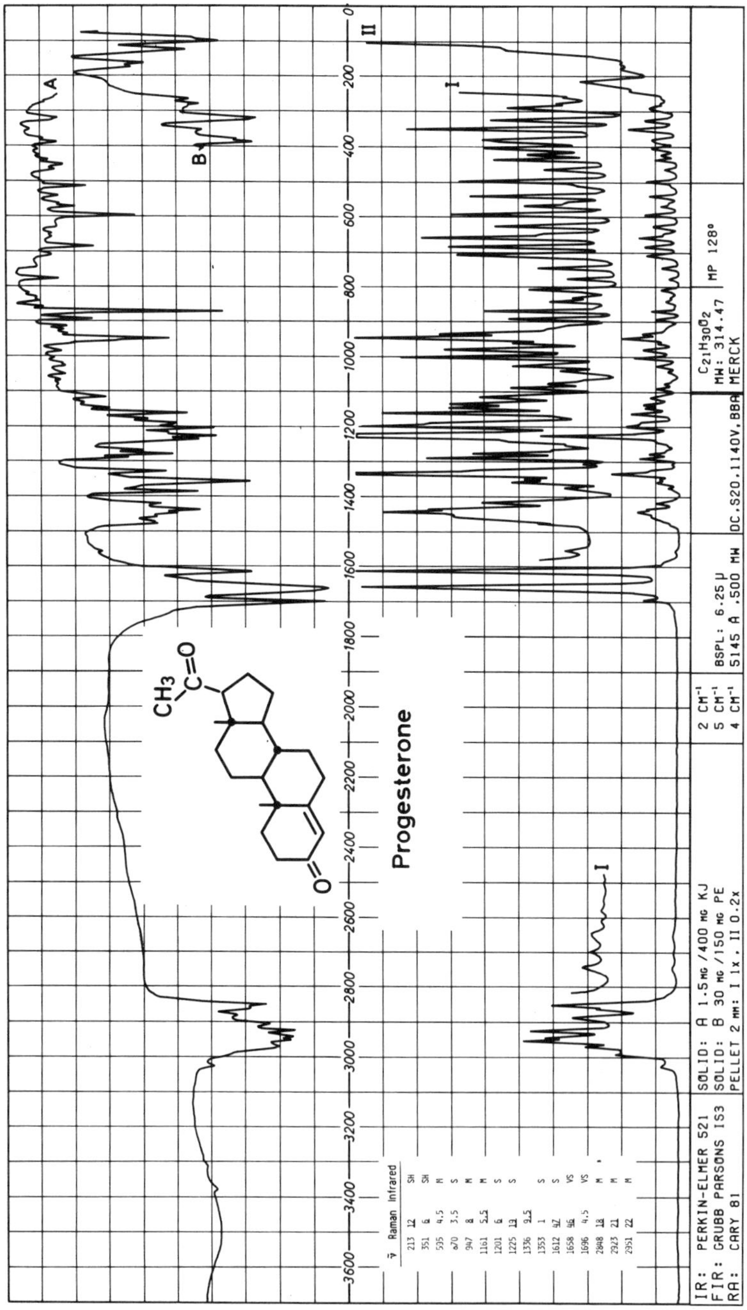

$\tilde{\nu}$	Raman	Infrared
213	12	SH
351	6	SH
535	4.5	M
670	3.5	S
947	8	M
1161	5.5	M
1201	6	S
1225	13	S
1336	9.5	
1353	1	S
1612	47	S
1658	46	VS
1696	4.5	VS
2848	18	M
2923	21	M
2951	22	M

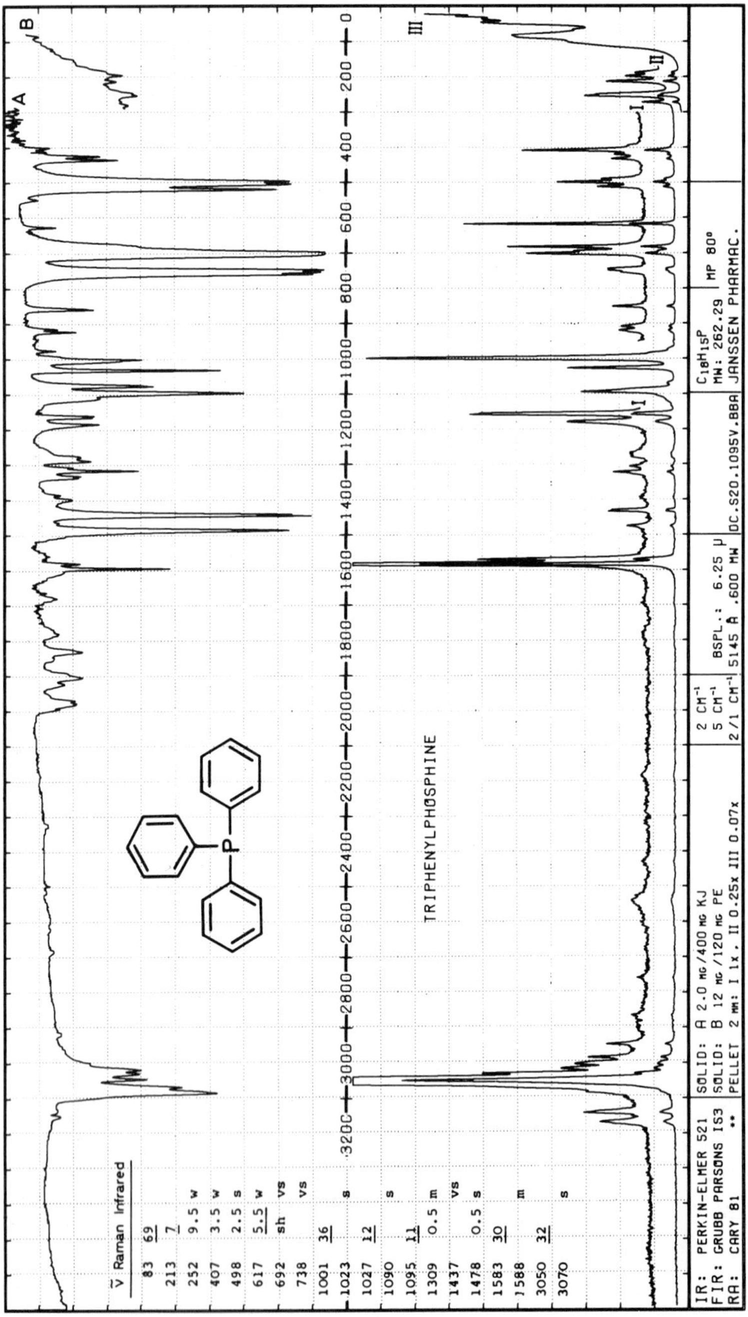
ṽ Raman Infrared
83 69
213 7
252 9.5 w
407 3.5 w
498 2.5 s
617 5.5 w
692 sh vs
738 vs
1001 36
1023 s
1027 12
1090 s
1095 11
1309 0.5 m
1437 vs
1478 0.5 s
1583 30
1588 m
3050 32
3070 s
3200 3000 2800 2600 2400 2200 2000 1800 1600 1400 1200 1000 800 600 400 200 0
A
B
I
II
III
P
TRIPHENYLPHOSPHINE
IR: PERKIN-ELMER 521
FIR: GRUBB PARSONS IS3
RA: CARY 81 **
SOLID: A 2.0 MG/400 MG KJ
SOLID: B 12 MG/120 MG PE
PELLET 2 MM: I 1x, II 0.25x III 0.07x
2 CM-1
5 CM-1
2/1 CM-1
BSPL.: 6.25 µ
5145 Å .600 MW
DC.S20.1095V.BBA
C18H15P
MW: 262.29
MP 80°
JANSSEN PHARMAC.

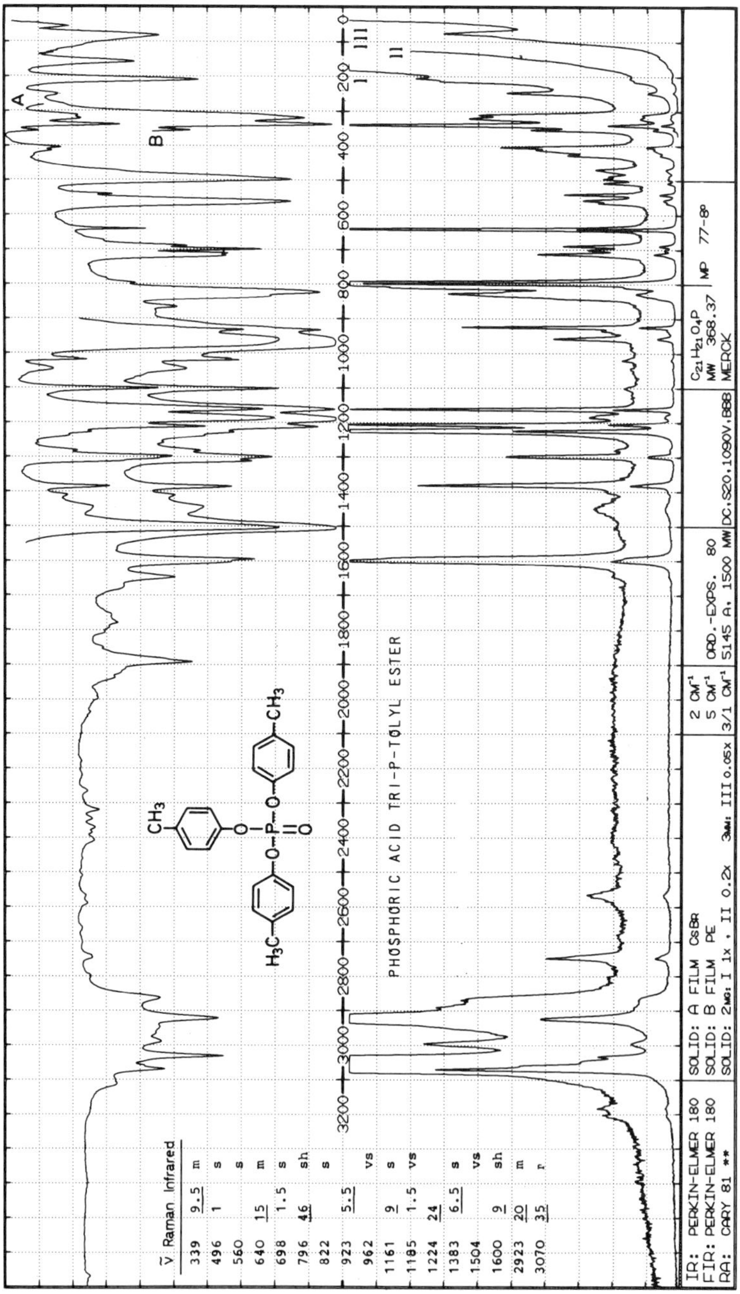
PHOSPHORIC ACID TRI-P-TOLYL ESTER
A
B
I
II
III
3200 3000 2800 2600 2400 2200 2000 1800 1600 1400 1200 1000 800 600 400 200 0
ṽ Raman Infrared
339 9.5 m
496 1 s
560 s
640 15 m
698 1.5 s
796 46 sh
822 s
923 5.5
962 vs
1161 9 s
1185 1.5 vs
1224 24
1383 6.5 s
1504 vs
1600 9 sh
2923 20 m
3070 35 r
IR: PERKIN-ELMER 180
FIR: PERKIN-ELMER 180
RA: CARY 81 **
SOLID: A FILM CsBR
SOLID: B FILM PE
SOLID: 2MG: I 1x , II 0.2x 3MG: III 0.05x
2 CM⁻¹
5 CM⁻¹
3/1 CM⁻¹
ORD.-EXPS. 80
5145 A, 1500 MW
DC.S20.1090V.BBB
$C_{21}H_{21}O_4P$
MW 368.37
MERCK
MP 77-8°

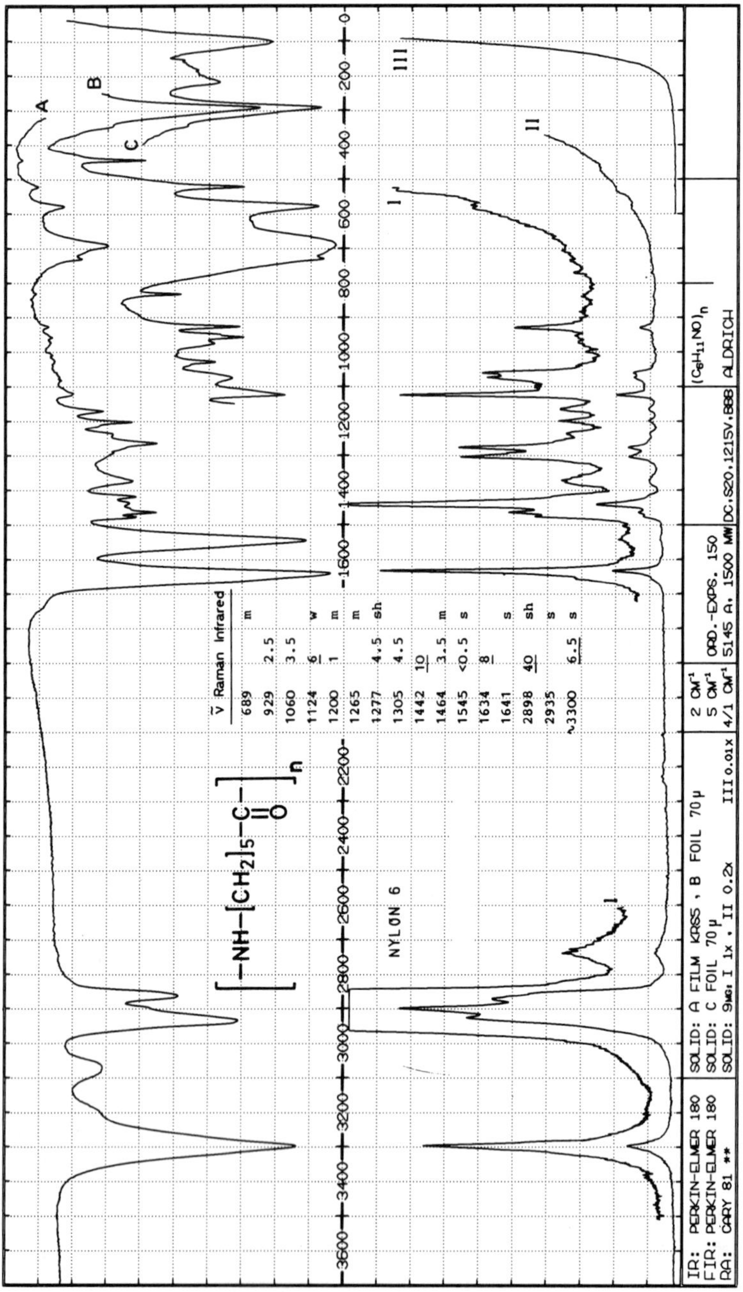

A
B
C
I
II
III
3600 3400 3200 3000 2800 2600 2400 2200 1600 1400 1200 1000 800 600 400 200 0
[—NH—[CH2]5—C(=O)—]n
NYLON 6
ν̃ Raman Infrared
689 m
929 2.5
1060 3.5
1124 6 w
1200 1 m
1265 m
1277 4.5 sh
1305 4.5
1442 10
1464 3.5 m
1545 <0.5 s
1634 8
1641 s
2898 40 sh
2935 s
~3300 6.5 s
IR: PERKIN-ELMER 180
FIR: PERKIN-ELMER 180
RA: CARY 81 **
SOLID: A FILM KRS5 , B FOIL 70 µ
SOLID: C FOIL 70 µ
SOLID: 9mg: I 1x , II 0.2x III 0.01x
2 CM-1
5 CM-1
4/1 CM-1
ORD.-EXPS. 150
5145 A. 1500 MW
DC:S20:1215V.BBB
(C6H11NO)n
ALDRICH

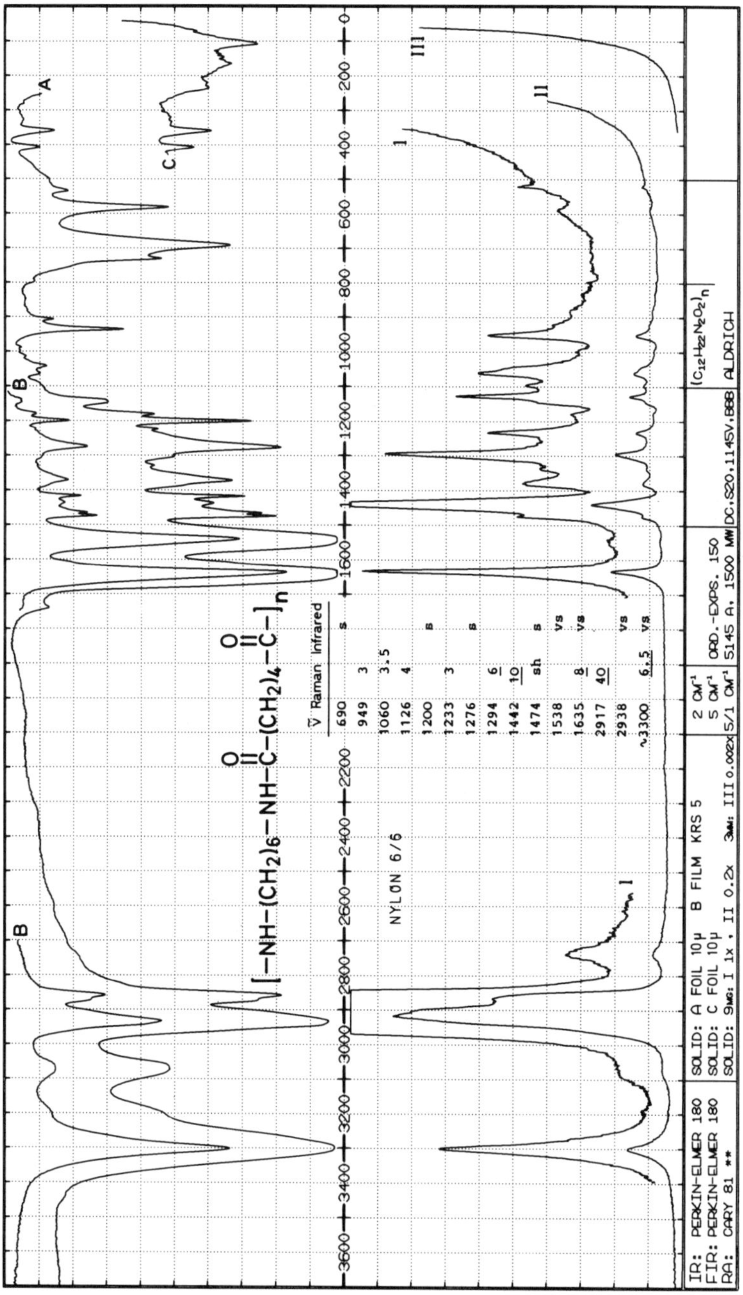

NYLON 6/6
[-NH-(CH2)6-NH-C(=O)-(CH2)4-C(=O)-]n
ṽ Raman Infrared
690 s
949 3
1060 3.5
1126 4
1200 s
1233 3
1276 s
1294 6
1442 10
1474 sh s
1538 vs
1635 8 vs
2917 40
2938 vs
~3300 6.5 vs
A
B
C
I
II
III
IR: PERKIN-ELMER 180 SOLID: A FOIL 10µ B FILM KRS 5 2 CM-1
FIR: PERKIN-ELMER 180 SOLID: C FOIL 10µ 5 CM-1 ORD.-EXPS. 150
RA: CARY 81 ** SOLID: 9MG: I 1x, II 0.2x 3MM: III 0.002X 5/1 CM-1 5145 A. 1500 MW DC.S20.1145V.BBB
(C12H22N2O2)n
ALDRICH

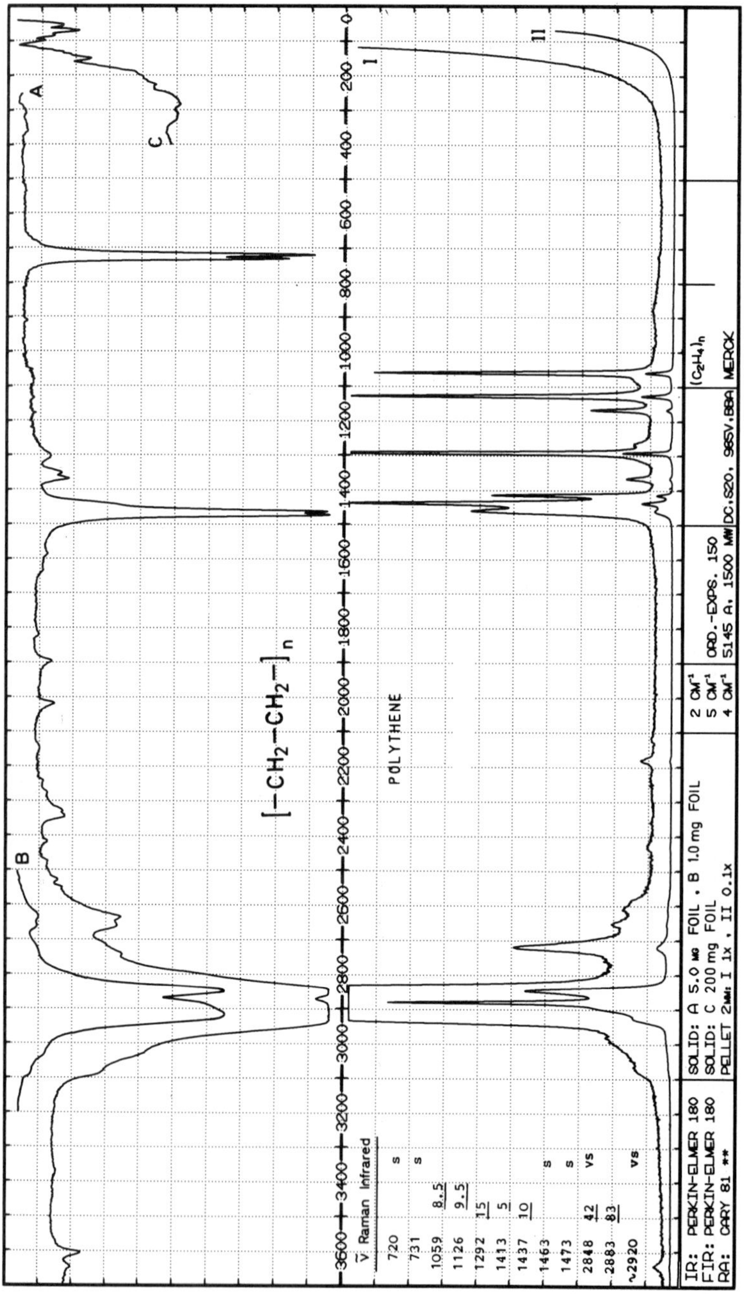

A
B
C
I
II
[—CH2—CH2—]n
POLYTHENE
3600 3400 3200 3000 2800 2600 2400 2200 2000 1800 1600 1400 1200 1000 800 600 400 200 0
ṽ Raman Infrared
720 s
731 s
1059 8.5
1126 9.5
1292 15
1413 5
1437 10
1463 s
1473 s
2848 42 vs
2883 83
~2920 vs
IR: PERKIN-ELMER 180
FIR: PERKIN-ELMER 180
RA: CARY 81 **
SOLID: A 5.0 MG FOIL . B 1.0 mg FOIL
SOLID: C 200 mg FOIL
PELLET 2MM: I 1x . II 0.1x
2 CM-1
5 CM-1
4 CM-1
ORD.-EXPS. 150
5145 A, 1500 MW
DC.S20, 985V.BBA
(C2H4)n
MERCK

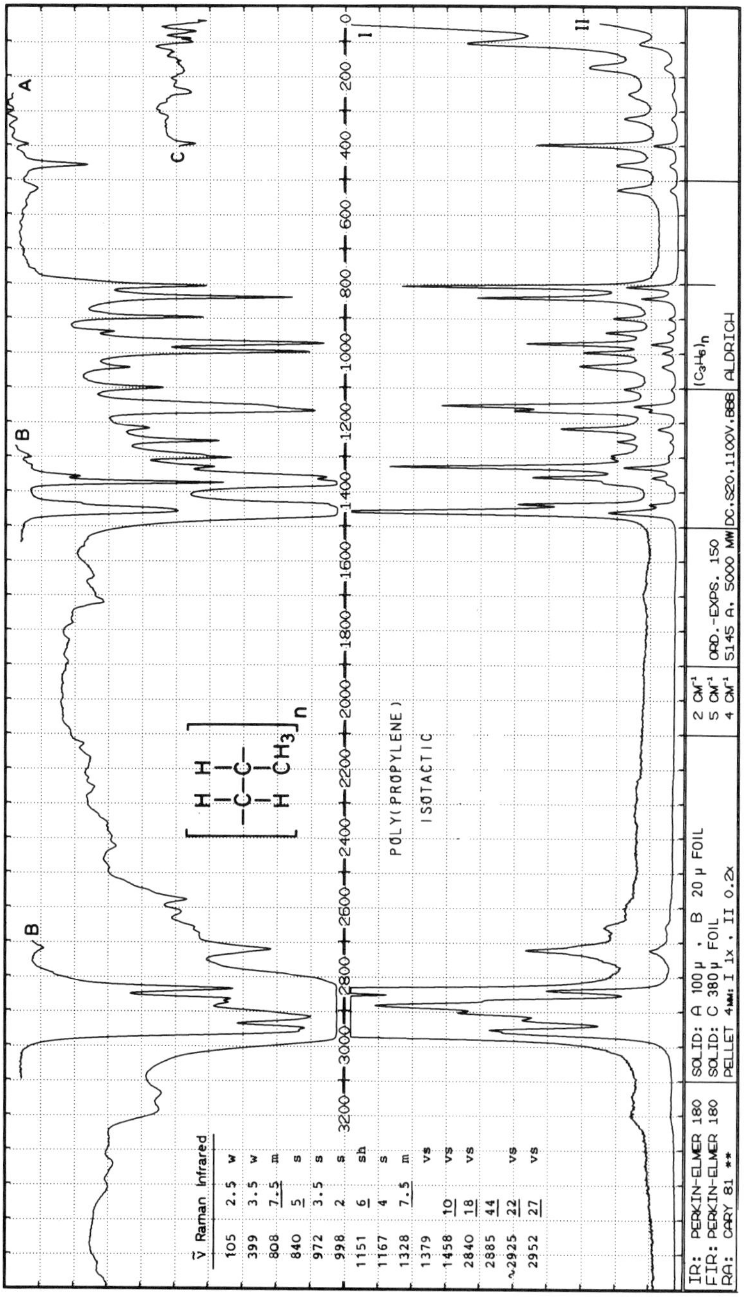
ṽ Raman Infrared
105 2.5 w
399 3.5 w
808 7.5 m
840 5 s
972 3.5 s
998 2 s
1151 6 sh
1167 4 s
1328 7.5 m
1379 vs
1458 10 vs
2840 18 vs
2885 44
~2925 22 vs
2952 27 vs
POLY(PROPYLENE)
ISOTACTIC
A
B
C
I
II
3200 3000 2800 2600 2400 2200 2000 1800 1600 1400 1200 1000 800 600 400 200 0
IR: PERKIN-ELMER 180
FIR: PERKIN-ELMER 180
RA: CARY 81 **
SOLID: A 100 μ , B 20 μ FOIL
SOLID: C 380 μ FOIL
PELLET 4MM: I 1x , II 0.2x
2 CM⁻¹
5 CM⁻¹
4 CM⁻¹
ORD.-EXPS. 150
5145 A, 5000 MW
DC,S20,1100V,BBB
(C3H6)n
ALDRICH

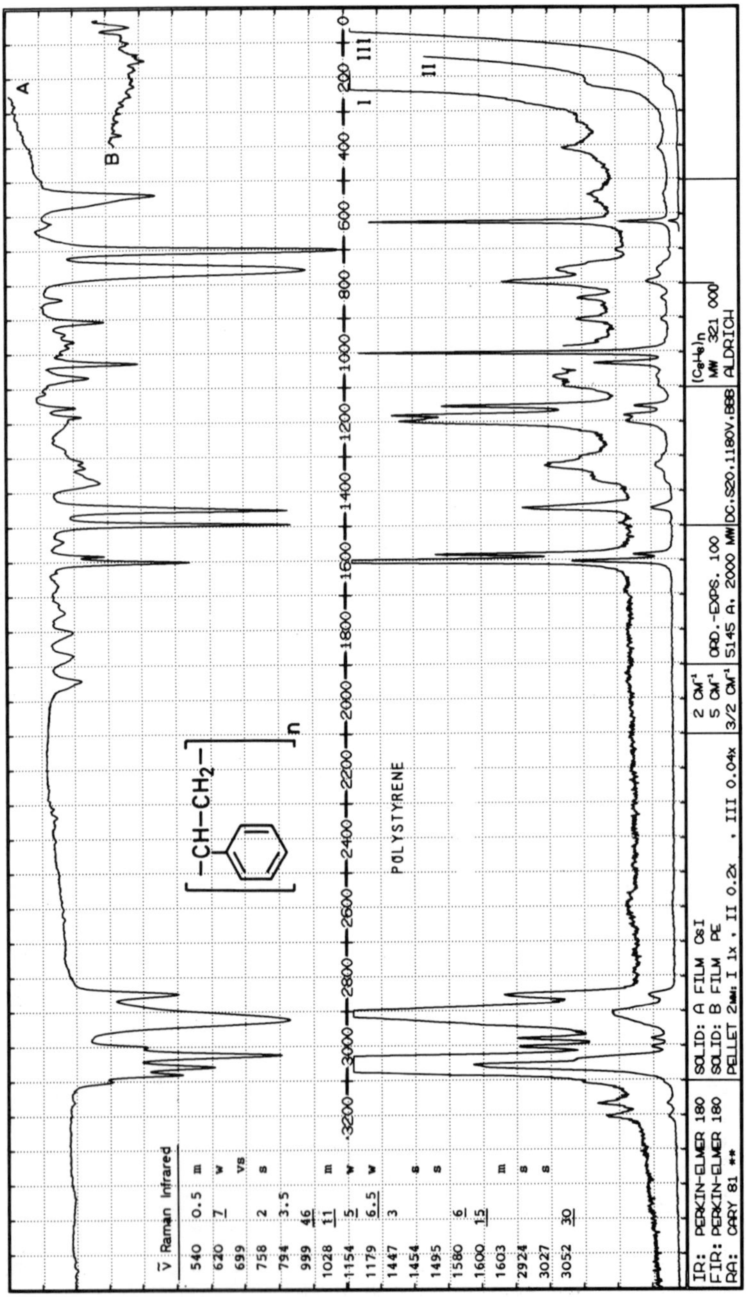

ṽ Raman Infrared
540 0.5 m
620 7 w
699 vs
758 2 s
794 3.5
999 46
1028 11 m
1154 5 w
1179 6.5 w
1447 3
1454 s
1495 s
1580 6
1600 15
1603 m
2924 s
3027 s
3052 30
POLYSTYRENE
A
B
I
II
III
IR: PERKIN-ELMER 180 SOLID: A FILM CsI 2 CM⁻¹ ORD.-EXPS. 100
FIR: PERKIN-ELMER 180 SOLID: B FILM PE 5 CM⁻¹
RA: CARY 81 ** PELLET 2MM: I 1x, II 0.2x, III 0.04x 3/2 CM⁻¹ 5145 A, 2000 MW DC:S20,1180V,BBB
(C8H8)n MW 321 000 ALDRICH

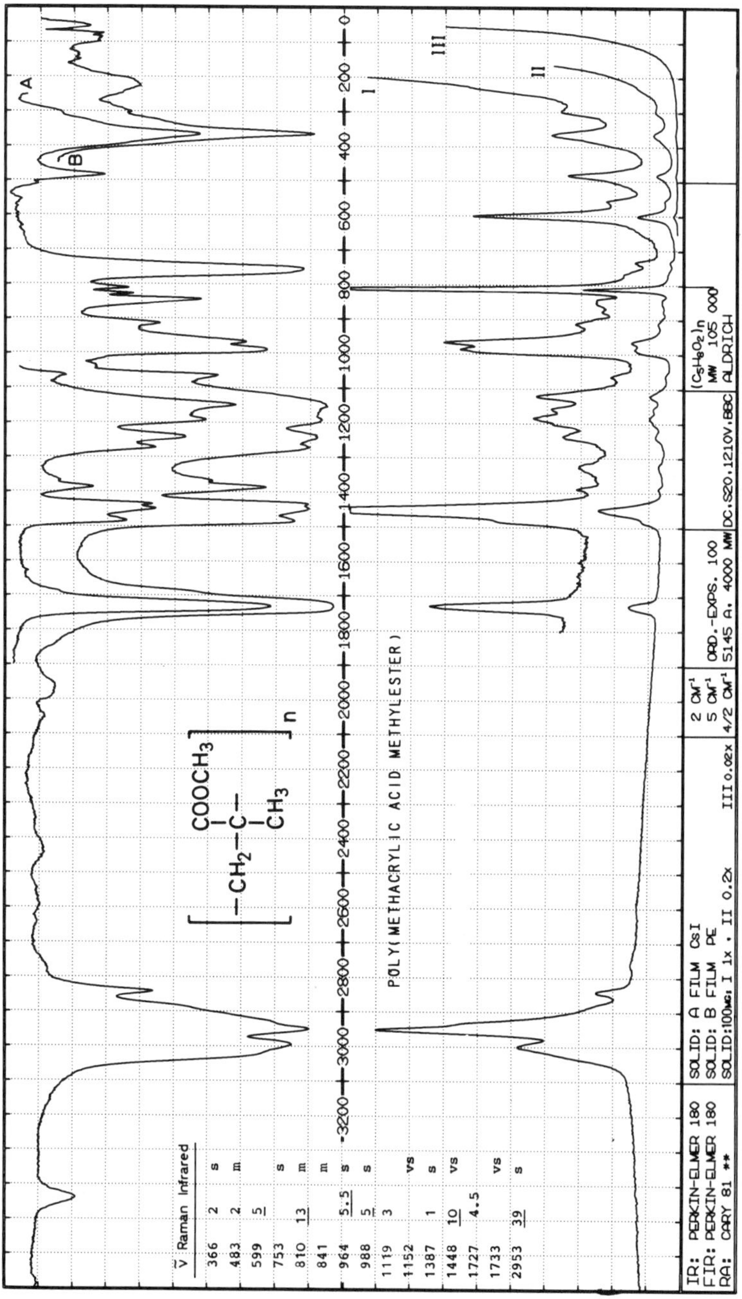

ṽ Raman Infrared
366 2 s
483 2 m
599 5
753 s
810 13 m
841 m
964 5.5 s
988 5 s
1119 3
1152 vs
1387 1 s
1448 10 vs
1727 4.5
1733 vs
2953 39 s
[—CH2—C(COOCH3)(CH3)—]n
POLY(METHACRYLIC ACID METHYLESTER)
3200 3000 2800 2600 2400 2200 2000 1800 1600 1400 1200 1000 800 600 400 200 0
A
B
I
II
III
IR: PERKIN-ELMER 180
FIR: PERKIN-ELMER 180
RA: CARY 81 **
SOLID: A FILM CsI
SOLID: B FILM PE
SOLID:100MG: I 1x . II 0.2x III 0.02x
2 CM⁻¹
5 CM⁻¹
4/2 CM⁻¹
ORD.-EXPS. 100
5145 A. 4000 MW
DC.S20.1210V.BBC
(C5H8O2)n
MW 105 000
ALDRICH

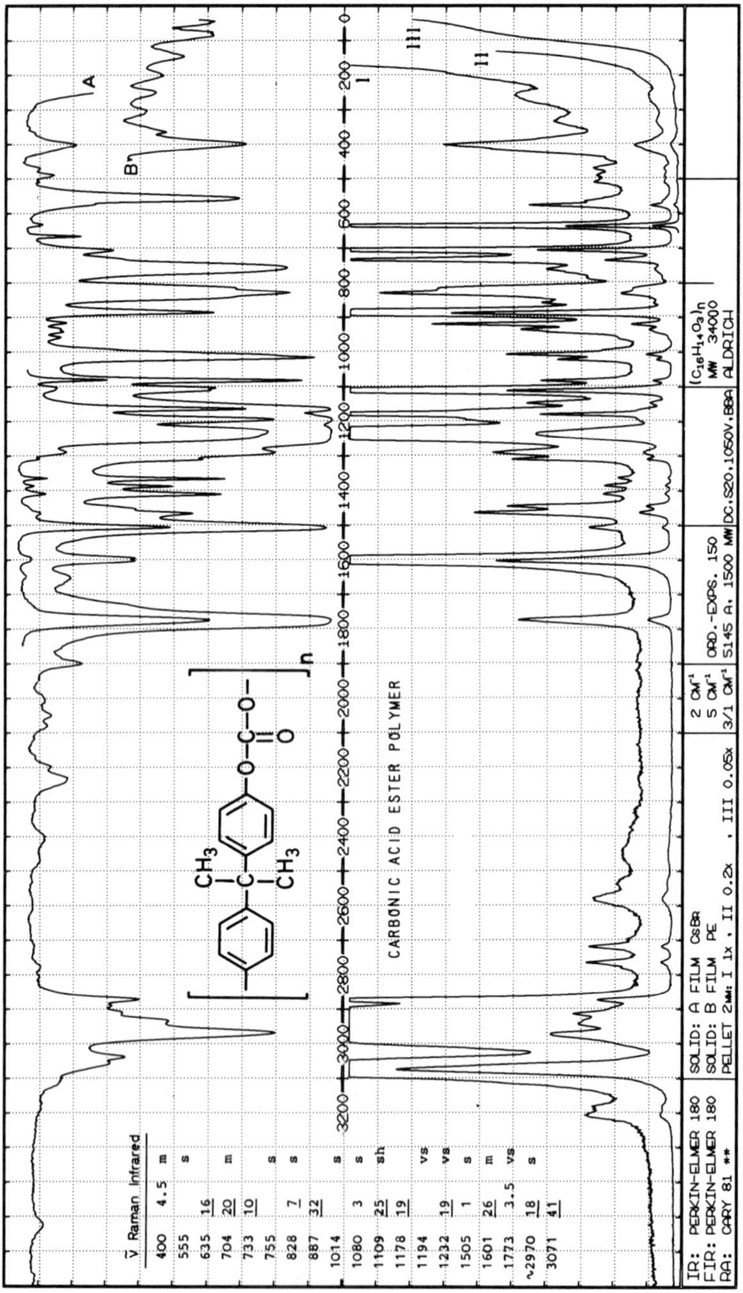

ṽ Raman Infrared
400 4.5 m
555 s
635 16
704 20 m
733 10
755 s
828 7 s
887 32
1014 s
1080 3 s
1109 25 sh
1178 19
1194 vs
1232 19 vs
1505 1 s
1601 26 m
1773 3.5 vs
~2970 18 s
3071 41
CARBONIC ACID ESTER POLYMER
A
B
I
II
III
3200 3000 2800 2600 2400 2200 2000 1800 1600 1400 1200 1000 800 600 400 200 0
IR: PERKIN-ELMER 180
FIR: PERKIN-ELMER 180
RA: CARY 81 **
SOLID: A FILM CsBr
SOLID: B FILM PE
PELLET 2MM: I 1x , II 0.2x , III 0.05x
2 CM⁻¹
5 CM⁻¹
3/1 CM⁻¹
ORD.-EXPS. 150
5145 A. 1500 MW
DC:S20,1050V,BBA
(C16H14O3)n
MW 34000
ALDRICH

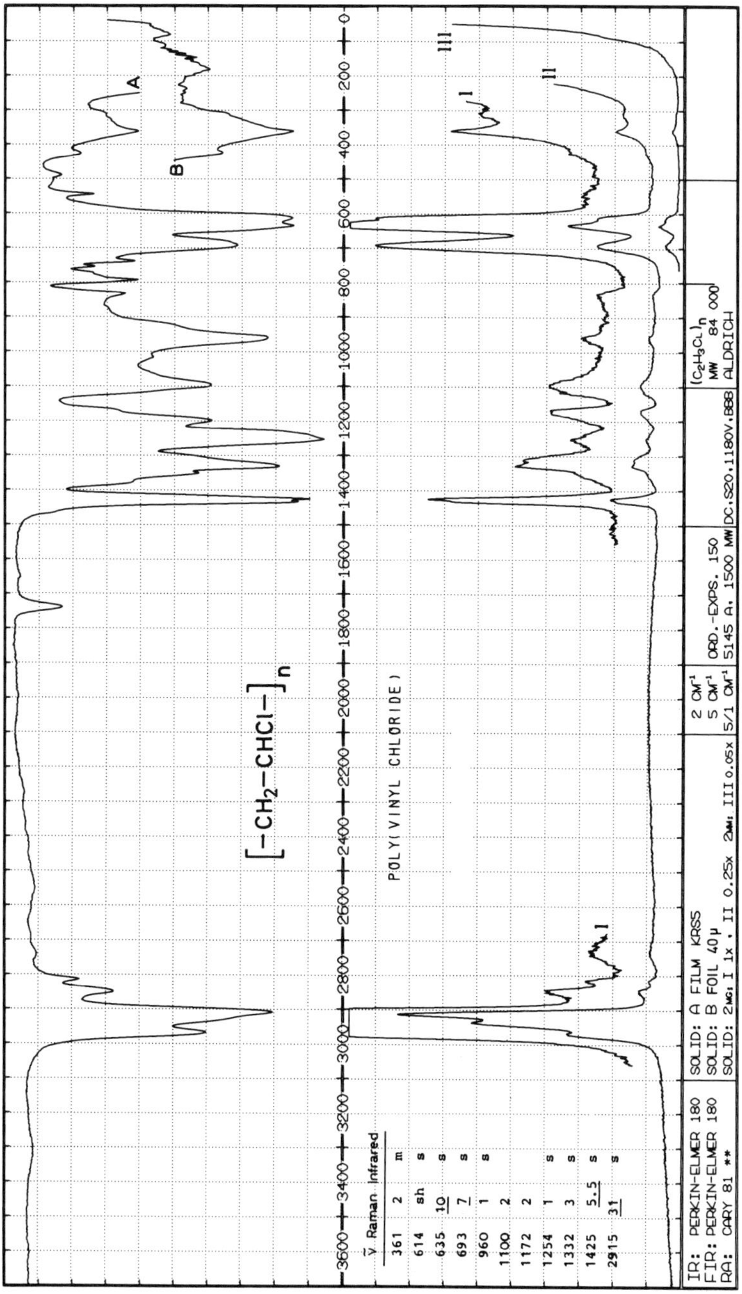
$[-CH_2-CHCl-]_n$
POLY(VINYL CHLORIDE)
ṽ Raman Infrared
361 2 m
614 sh s
635 10 s
693 7 s
960 1 s
1100 2
1172 2
1254 1 s
1332 3 s
1425 5.5 s
2915 31 s
A
B
I
II
III
IR: PERKIN-ELMER 180 SOLID: A FILM KRSS 2 CM-1
FIR: PERKIN-ELMER 180 SOLID: B FOIL 40µ 5 CM-1 ORD.-EXPS. 150
RA: CARY 81 ** SOLID: 2MG: I 1x, II 0.25x 2MM: III 0.05x 5/1 CM-1 5145 A. 1500 MW DC.S20.1180V.BBB
(C2H3CL)n MW 84 000 ALDRICH

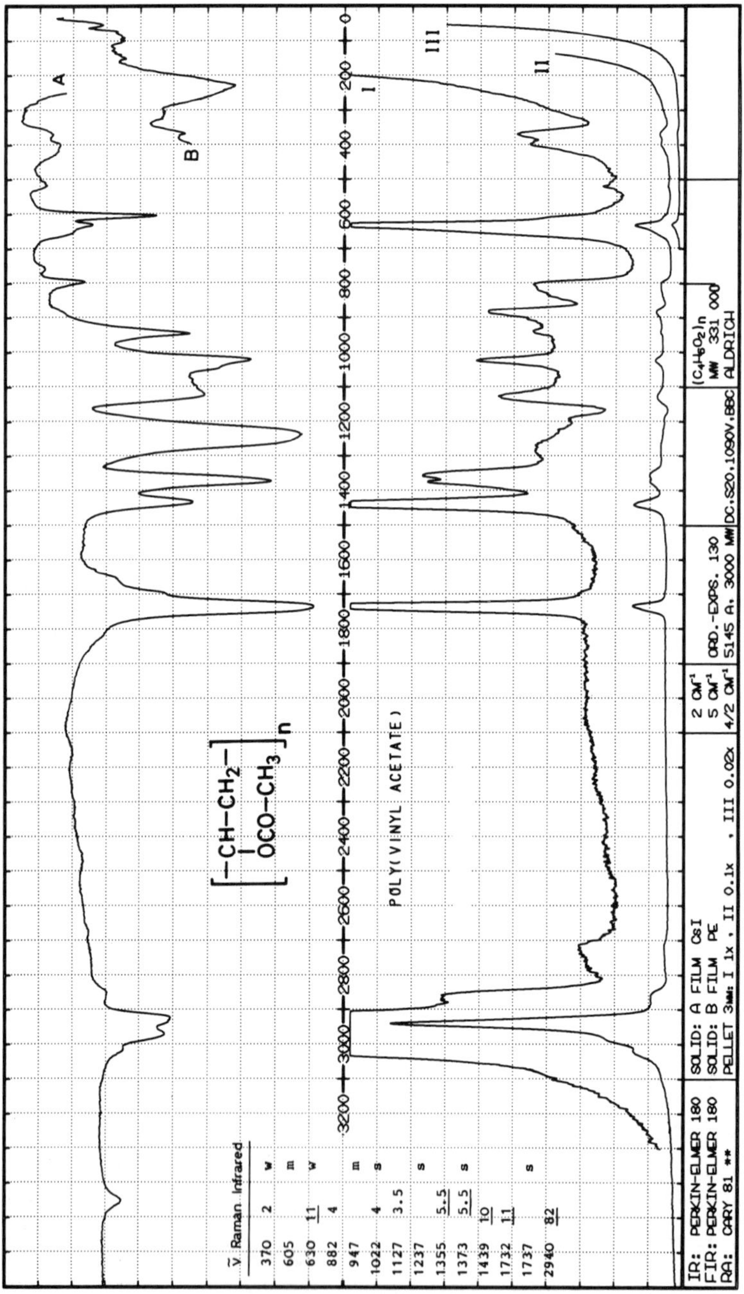
A
B
I
II
III
POLY(VINYL ACETATE)
ṽ Raman Infrared
370 2 w
605 m
630 11 w
882 4
947 m
1022 4 s
1127 3.5
1237 s
1355 5.5
1373 5.5 s
1439 10
1732 11
1737 s
2940 82
IR: PERKIN-ELMER 180
FIR: PERKIN-ELMER 180
RA: CARY 81 **
SOLID: A FILM CsI
SOLID: B FILM PE
PELLET 3mm: I 1x , II 0.1x , III 0.02x
2 CM⁻¹
5 CM⁻¹
4/2 CM⁻¹
ORD.-EXPS. 130
5145 A. 3000 MW
DC.S20.1090V.BBC
(C4H6O2)n
MW 331 000
ALDRICH

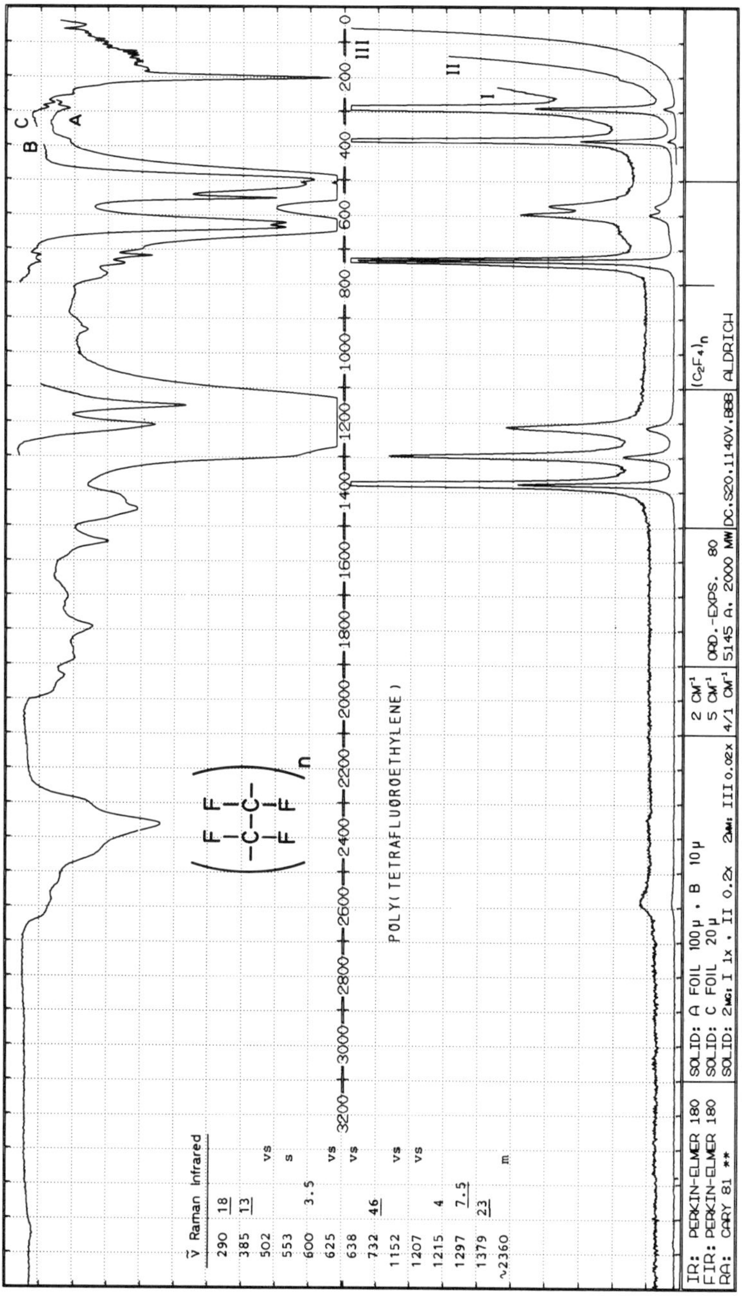

ṽ Raman Infrared
290 18
385 13
502 vs
553 s
600 3.5
625 vs
638 vs
732 46
1152 vs
1207 vs
1215 4
1297 7.5
1379 23
~2360 m
POLY(TETRAFLUOROETHYLENE)
IR: PERKIN-ELMER 180 SOLID: A FOIL 100 μ . B 10 μ
FIR: PERKIN-ELMER 180 SOLID: C FOIL 20 μ
RA: CARY 81 ** SOLID: 2MG; I 1x . II 0.2x 2MM: III 0.02x
2 CM⁻¹
5 CM⁻¹
4/1 CM⁻¹
ORD.-EXPS. 80
5145 A. 2000 MW
DC:S20.1140V.BBB
(C2F4)n
ALDRICH

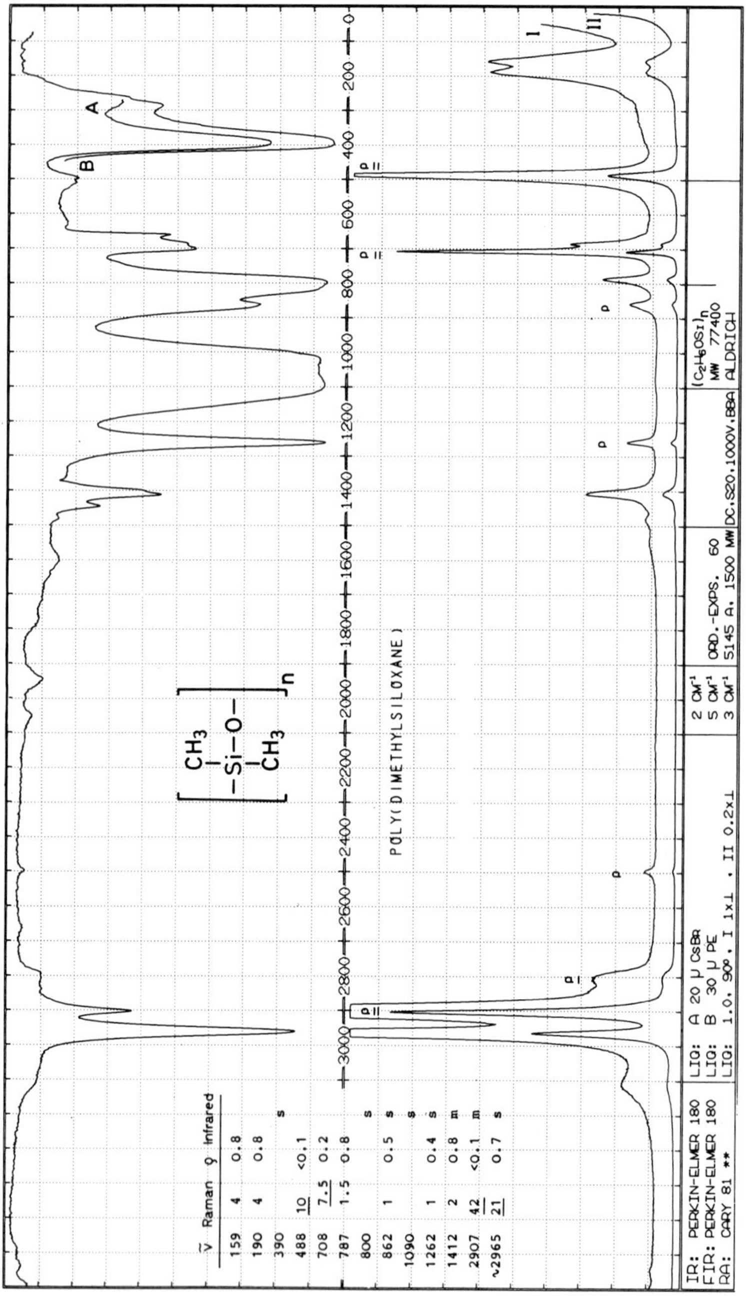
ν̃ Raman ρ Infrared
159 4 0.8
190 4 0.8
390 s
488 10 <0.1
708 7.5 0.2
787 1.5 0.8
800 s
862 1 0.5 s
1090 s
1262 1 0.4 s
1412 2 0.8 m
2907 42 <0.1 m
~2965 21 0.7 s
POLY(DIMETHYLSILOXANE)
A
B
I
II
IR: PERKIN-ELMER 180
FIR: PERKIN-ELMER 180
RA: CARY 81 **
LIQ: A 20 μ CsBr
LIQ: B 30 μ PE
LIQ: 1.0, 90°, I 1x⊥, II 0.2x⊥
2 CM-1
5 CM-1
3 CM-1
ORD.-EXPS. 60
5145 A. 1500 MW
DC.S20.1000V.BBA
(C2H6OSi)n
MW 77400
ALDRICH

APPENDIX 3

A Summary of Characteristic Raman and Infrared Frequencies

A. Raman*

B. Infrared†

A.

A SUMMARY OF CHARACTERISTIC RAMAN FREQUENCIES

Frequency (cm^{-1})	Vibration	Compound
3400–3330	Bonded antisymmetric NH_2 stretch	Primary amines
3380–3340	Bonded OH stretch	Aliphatic alcohols
3374	CH stretch	Acetylene (gas)
3355–3325	Bonded antisymmetric NH_2 stretch	Primary amides
3350–3300	Bonded NH stretch	Secondary amines
3335–3300	≡CH stretch	Alkyl acetylenes
3300–3250	Bonded symmetric NH_2 stretch	Primary amines
3310–3290	Bonded NH stretch	Secondary amides
3190–3145	Bonded symmetric NH_2 stretch	Primary amides
3175–3154	Bonded NH stretch	Pyrazoles

*F. R. Dollish, W. G. Fateley, and F. F. Bentley, *Characteristic Raman Frequencies of Organic Compounds*, John Wiley & Sons Inc., New York, 1974, with permission.

†N. B. Colthup, L. H. Daly, and S. E. Wiberly, *Introduction to Infrared and Raman Spectroscopy*, Third Edition, Academic Press, Inc., 1990.

Frequency (cm^{-1})	Vibration	Compound
3103	Antisymmetric $=CH_2$ stretch	Ethylene (gas)
3100–3020	CH_2 stretches	Cyclopropane
3100–3000	Aromatic CH stretch	Benzene derivatives
3095–3070	Antisymmetric $=CH_2$ stretch	$C=CH_2$ derivatives
3062	CH stretch	Benzene
3057	Aromatic CH stretch	Alkyl benzenes
3040–3000	CH stretch	C=CHR derivatives
3026	Symmetric $=CH_2$ stretch	Ethylene (gas)
2990–2980	Symmetric $=CH_2$ stretch	$C=CH_2$ derivatives
2986–2974	Symmetric NH_3^+ stretch	Alkyl ammonium chlorides (aqueous solution)
2969–2965	Antisymmetric CH_3 stretch	*n*-Alkanes
2929–2912	Antisymmetric CH_2 stretch	*n*-Alkanes
2884–2883	Symmetric CH_3 stretch	*n*-Alkanes
2861–2849	Symmetric CH_2 stretch	*n*-Alkanes
2850–2700	CHO group (2 bands)	Aliphatic aldehydes
2590–2560	SH stretch	Thiols
2316–2233	C≡C stretch (2 bands)	$R-C\equiv C-CH_3$
2301–2231	C≡C stretch (2 bands)	$R-C\equiv C-R'$
2300–2250	Pseudoantisymmetric N=C=O stretch	Isocyanates
2264–2251	Symmetric C≡C—C≡C stretch	Alkyl diacetylenes
2259	C≡N stretch	Cyanamide
2251–2232	C≡N stretch	Aliphatic nitriles
2220–2100	Pseudoantisymmetric N=C=S stretch (2 bands)	Alkyl isothiocyanates
2220–2000	C≡N stretch	Diaklyl cyanamides
2172	Symmetric C≡C—C≡C stretch	Diacetylene
2161–2134	$\overset{+}{N}\equiv\overset{-}{C}$ stretch	Aliphatic isonitriles
2160–2100	C≡C stretch	Alkyl acetylenes
2156–2140	C≡N stretch	Alkyl thiocyanates
2104	Antisymmetric N=N=N stretch	CH_3N_3
2094	C≡N stretch	HCN
2049	Pseudoantisymmetric C=C=O stretch	Ketene
1974	C≡C stretch	Acetylene (gas)
1964–1958	Antisymmetric C=C=C stretch	Allenes
1870–1840	Symmetric C=O stretch	Saturated 5-membered ring cyclic anhydrides
1820	Symmetric C=O stretch	Acetic anhydride
1810–1788	C=O stretch	Acid halides

Frequency (cm^{-1})	Vibration	Compound
1807	C=O stretch	Phosgene
1805–1799	Symmetric C=O stretch	Noncyclic anhydrides
1800	C=C stretch	$F_2C{=}CF_2$ (gas)
1795	C=O stretch	Ethylene carbonate
1792	C=C stretch	$F_2C{=}CFCH_3$
1782	C=O stretch	Cyclobutanone
1770–1730	C=O stretch	Halogenated aldehydes
1744	C=O stretch	Cyclopentanone
1743–1729	C=O stretch	Cationic α-amino acids (aqueous solution)
1741–1734	C=O stretch	*O*-Alkyl acetates
1740–1720	C=O stretch	Aliphatic aldehydes
1739–1714	C=C stretch	$C{=}CF_2$ derivatives
1736	C=C stretch	Methylene cyclopropane
1734–1727	C=O stretch	*O*-Alkyl propionates
1725–1700	C=O stretch	Aliphatic ketones
1720–1715	C=O stretch	*O*-Alkyl formates
1712–1694	C=C stretch	RCF=CFR
1695	Nonconjugated C=O stretch	Uracil derivatives (aqueous solution)
1689–1644	C=C stretch	Monofluoroalkenes
1687–1651	C=C stretch	Alkylidene cyclopentanes
1686–1636	Amide I band	Primary amides (solids)
1680–1665	C=C stretch	Tetraalkyl ethylenes
1679	C=C stretch	Methylene cyclobutane
1678–1664	C=C stretch	Trialkyl ethylenes
1676–1665	C=C stretch	*trans*-Dialkyl ethylenes
1675	Symmetric C=O stretch (cyclic dimer)	Acetic acid
1673–1666	C=N stretch	Aldimines
1672	Symmetric C=O stretch (cyclic dimer)	Formic acid (aqueous solution)
1670–1655	Conjugated C=O stretch	Uracil, cytosine, and guanine derivatives (aqueous solution)
1670–1630	Amide I band	Tertiary amides
1666–1652	C=N stretch	Ketoximes
1665–1650	C=N stretch	Semicarbazones (solid)
1663–1636	Symmetric C=N stretch	Aldazines, ketazines
1660–1654	C=C stretch	*cis*-Dialkyl ethylenes
1660–1650	Amide I band	Secondary amides
1660–1649	C=N stretch	Aldoximes

Frequency (cm^{-1})	Vibration	Compound
1660–1610	C=N stretch	Hydrazones (solid)
1658–1644	C=C stretch	$R_2C{=}CH_2$
1656	C=C stretch	Cyclohexene, cycloheptene
1654–1649	Symmetric C=O stretch (cyclic dimer)	Carboxylic acids
1652–1642	C=N stretch	Thiosemicarbazones (solid)
1650–1590	NH_2 scissors	Primary amines
1649–1625	C=C stretch	Allyl derivatives
1648–1640	N=O stretch	Alkyl nitrites
1648–1638	C=C stretch	$H_2C{=}CHR$
1647	C=C stretch	Cyclopropene
1638	C=O stretch	Ethylene dithiocarbonate
1637	Symmetric C=C stretch	Isoprene
1634–1622	Antisymmetric NO_2 stretch	Alkyl nitrates
1630–1550	Ring stretches (doublet)	Benzene derivatives
1623	C=C stretch	Ethylene (gas)
1620–1540	Three or more coupled C=C stretches	Polyenes
1616–1571	C=C stretch	Chloroalkenes
1614	C=C stretch	Cyclopentene
1596–1547	C=C stretch	Bromoalkenes
1581–1465	C=C stretch	Iodoalkenes
1575	Symmetric C=C stretch	1,3-Cyclohexadiene
1573	N=N stretch	Azomethane (in solution)
1566	C=C stretch	Cyclobutene
1560–1550	Antisymmetric NO_2 stretch	Primary nitroalkanes
1555–1550	Antisymmetric NO_2 stretch	Secondary nitroalkanes
1548	N=N stretch	1-Pyrazoline
1545–1535	Antisymmetric NO_2 stretch	Tertiary nitroalkanes
1515–1490	Ring stretch	2-Furfuryl group
1500	Symmetric C=C stretch	Cyclopentadiene
1480–1470	OCH_3, OCH_2 deformations	Aliphatic ethers
1480–1460	Ring stretch	2 Furfurylidene or 2-furoyl group
1473–1446	CH_3, CH_2 deformations	*n*-Alkanes
1466–1465	CH_3 deformation	*n*-Alkanes
1450–1400	Pseudosymmetric N=C=O stretch	Isocyanates
1443–1398	Ring stretch	2-Substituted thiophenes
1442	N=N stretch	Azobenzene
1440–1340	Symmetric CO_2^- stretch	Carboxylate ions (aqueous solution)

Frequency (cm^{-1})	Vibration	Compound
1415–1400	Symmetric CO_2^- stretch	Dipolar and anionic α-amino acids (aqueous solution)
1415–1385	Ring stretch	Anthracenes
1395–1380	Symmetric NO_2 stretch	Primary nitroalkanes
1390–1370	Ring stretch	Naphthalenes
1385–1368	CH_3 symmetric deformation	*n*-Alkanes
1375–1360	Symmetric NO_2 stretch	Secondary nitroalkanes
1355–1345	Symmetric NO_2 stretch	Tertiary nitroalkanes
1350–1330	CH deformation	Isopropyl group
1320	Ring vibration	1,1-Dialkyl cyclopropanes
1314–1290	In-plane CH deformation	*trans*-Dialkyl ethylenes
1310–1250	Amide III band	Secondary amides
1310–1175	CH_2 twist and rock	*n*-Alkanes
1305–1295	CH_2 in-phase twist	*n*-Alkanes
1300–1280	CC bridge bond stretch	Biphenyls
1282–1275	Symmetric NO_2 stretch	Alkyl nitrates
1280–1240	Ring stretch	Epoxy derivatives
1276	Symmetric N=N=N stretch	CH_3N_3
1270–1251	In-plane CH deformation	*cis*-Dialkyl ethylenes
1266	Ring "breathing"	Ethylene oxide (oxirane)
1230–1200	Ring vibration	Para-disubstituted benzenes
1220–1200	Ring vibration	Mono- and 1,2-dialkyl cyclopropanes
1212	Ring "breathing"	Ethylene imine (aziridine)
1205	C_6H_5—C vibration	Alkyl benzenes
1196–1188	Symmetric SO_2 stretch	Alkyl sulfates
1188	Ring "breathing"	Cyclopropane
1172–1165	Symmetric SO_2 stretch	Alkyl sulfonates
1150–950	CC stretches	*n*-Alkanes
1145–1125	Symmetric SO_2 stretch	Dialkyl sulfones
1144	Ring "breathing"	Pyrrole
1140	Ring "breathing"	Furan
1130–1100	Symmetric C=C=C stretch (2 bands)	Allenes
1130	Pseudosymmetric C=C=O stretch	Ketene
1112	Ring "breathing"	Ethylene sulfide
1111	NN stretch	Hydrazine
1070–1040	S=O stretch (1 or 2 bands)	Aliphatic sulfoxides

Frequency (cm^{-1})	Vibration	Compound
1065	C=S stretch	Ethylene trithiocarbonate
1060–1020	Ring vibration	Ortho-disubstituted benzenes
1040–990	Ring vibration	Pyrazoles
1030–1015	In-plane CH deformation	Monosubstituted benzenes
1030–1010	Trigonal ring "breathing"	3-Substituted pyridines
1030	Trigonal ring "breathing"	Pyridine
1029	Ring "breathing"	Trimethylene oxide (oxetane)
1026	Ring "breathing"	Trimethylene imine (azetidine)
1010–990	Trigonal ring "breathing"	Mono-, meta-, and 1,3,5-substituted benzenes
1001	Ring "breathing"	Cyclobutane
1000–985	Trigonal ring "breathing"	2- and 4-Substituted pyridines
992	Ring "breathing"	Benzene
992	Ring "breathing"	Pyridine
939	Ring "breathing"	1,3-Dioxolane
933	Ring vibration	Alkyl cyclobutanes
930–830	Symmetric COC stretch	Aliphatic ethers
914	Ring "breathing"	Tetrahydrofuran
906	ON stretch	Hydroxylamine
905–837	CC skeletal stretch	*n*-Alkanes
900–890	Ring vibration	Alkyl cyclopentanes
900–850	Symmetric CNC stretch	Secondary amines
899	Ring "breathing"	Pyrrolidine
866	Ring "breathing"	Cyclopentane
877	OO stretch	Hydrogen peroxide
851–840	Pseudosymmetric CON stretch	*O*-Alkyl hydroxylamines
836	Ring "breathing"	Piperazine
835–749	Skeletal stretch	Isopropyl group
834	Ring "breathing"	1,4-Dioxane
832	Ring "breathing"	Thiophene
832	Ring "breathing"	Morpholine
830–720	Ring vibration	Para-disubstuted benzenes
825–820	C_3O skeletal stretch	Secondary alcohols
818	Ring "breathing"	Tetrahydropyran
815	Ring "breathing"	Piperidine
802	Ring "breathing"	Cyclohexane (chair form)
785–700	Ring vibration	Alkyl cyclohexanes
760–730	C_4O skeletal stretch	Tertiary alcohols

Frequency (cm^{-1})	Vibration	Compound
760–650	Symmetric skeletal stretch	*tert*-Butyl group
740–585	CS stretch (1 or more bands)	Alkyl sulfides
735–690	"C=S stretch"	Thioamides, thioureas (solid)
733	Ring "breathing"	Cycloheptane
730–720	CCl stretch, P_C conformation	Primary chloroalkanes
715–620	CS stretch (1 or more bands)	Dialkyl disulfides
709	CCl stretch	CH_3Cl
703	Ring "breathing"	Cyclooctane
703	Symmetric CCl_2 stretch	CH_2Cl_2
690–650	Pseudosymmetric N=C=S stretch	Alkyl isothiocyanates
688	Ring "breathing"	Tetrahydrothiophene
668	Symmetric CCl_3 stretch	$CHCl_3$
660–650	CCl stretch, P_H conformation	Primary chloroalkanes
659	Symmetric CSC stretch	Pentamethylene sulfide
655–640	CBr stretch, P_C conformation	Primary bromoalkanes
630–615	Ring deformation	Monosubstituted benzenes
615–605	CCl stretch, S_{HH} conformation	Secondary chloroalkanes
610–590	CI stretch, P_C conformation	Primary iodoalkanes
609	CBr stretch	CH_3Br
577	Symmetric CBr_2 stretch	CH_2Br_2
570–560	CCl stretch, T_{HHH} conformation	Tertiary chloroalkanes
565–560	CBr stretch, P_H conformation	Primary bromoalkanes
540–535	CBr stretch, S_{HH} conformation	Secondary bromoalkanes
539	Symmetric CBr_3 stretch	$CHBr_3$
525–510	SS stretch	Dialkyl disulfides
523	CI stretch	CH_3I
520–510	CBr stretch, T_{HHH} conformation	Tertiary bromoalkanes
510–500	CI stretch, P_H conformation	Primary iodoalkanes
510–480	SS stretch	Dialkyl trisulfides
495–485	CI stretch, S_{HH} conformation	Secondary iodoalkanes
495–485	CI stretch, T_{HHH} conformation	Tertiary iodoalkanes
484–475	Skeletal deformation	Dialkyl diacetylenes
483	Symmetric CI_2 stretch	CH_2I_2
459	Symmetric CCl_4 stretch	CCl_4
437	Symmetric CI_3 stretch	CHI_3 (in solution)
425–150	"Chain expansion"	*n*-Alkanes
355–335	Skeletal deformation	Monoalkyl acetylenes
267	Symmetric CBr_4 stretch	CBr_4 (in solution)
200–160	Skeletal deformation	Aliphatic nitriles
178	Symmetric CI_4 stretch	CI_4 (solid)

B.

COLTHUP SPECTRA-STRUCTURE CORRELATION CHARTS FOR INFRARED FREQUENCIES IN THE 4000–600 cm^{-1} REGION

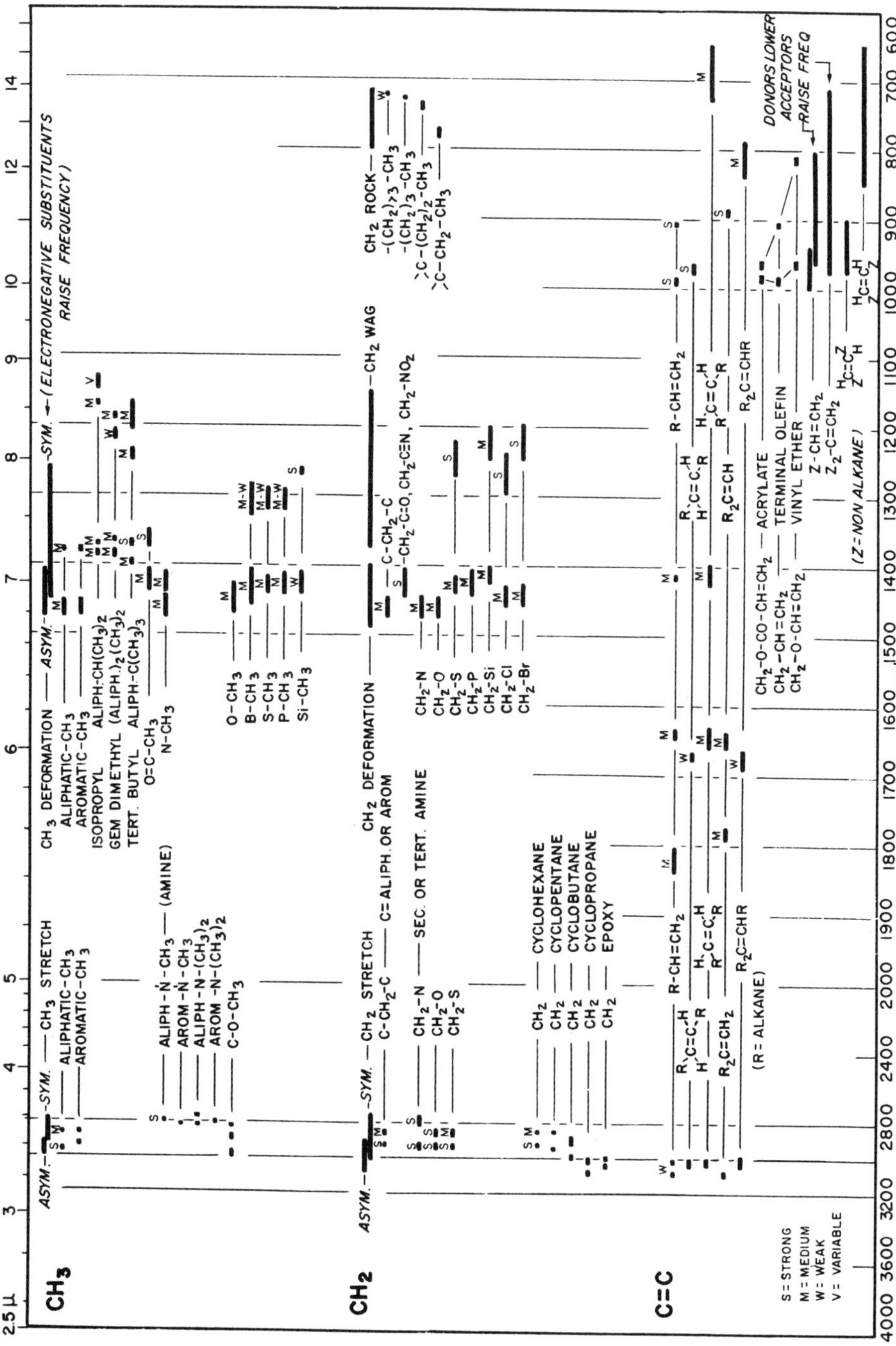
CH3
CH2
C=C
CH3 STRETCH
ALIPHATIC-CH3
AROMATIC-CH3
CH3 DEFORMATION
ISOPROPYL
GEM DIMETHYL
TERT. BUTYL
(ELECTRONEGATIVE SUBSTITUENTS RAISE FREQUENCY)
(AMINE)
CH2 STRETCH
C=ALIPH. OR AROM
SEC. OR TERT. AMINE
CH2 DEFORMATION
CH2 WAG
CH2 ROCK
CYCLOHEXANE
CYCLOPENTANE
CYCLOBUTANE
CYCLOPROPANE
EPOXY
(R = ALKANE)
ACRYLATE
TERMINAL OLEFIN
VINYL ETHER
(Z = NON ALKANE)
DONORS LOWER
ACCEPTORS RAISE FREQ
S = STRONG
M = MEDIUM
W = WEAK
V = VARIABLE
CM⁻¹
2.5 μ

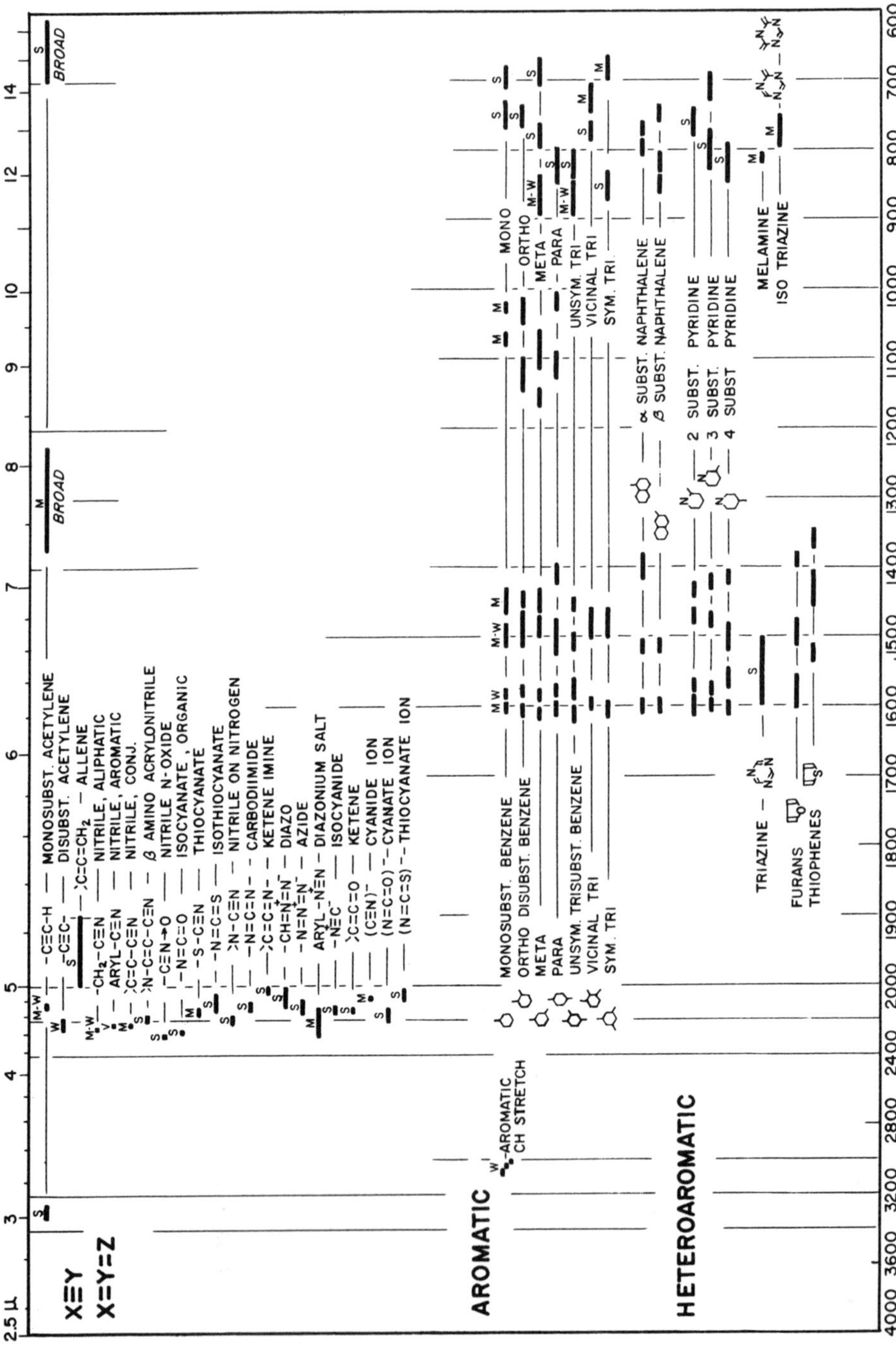
X≡Y
X=Y=Z
AROMATIC
HETEROAROMATIC
-C≡C-H — MONOSUBST. ACETYLENE
-C≡C- — DISUBST. ACETYLENE
>C=C=CH2 — ALLENE
-CH2-C≡N — NITRILE, ALIPHATIC
ARYL-C≡N — NITRILE, AROMATIC
>C=C-C≡N — NITRILE, CONJ.
>N-C=C-C≡N — β AMINO ACRYLONITRILE
-C≡N→O — NITRILE N-OXIDE
-N=C=O — ISOCYANATE, ORGANIC
-S-C≡N — THIOCYANATE
-N=C=S — ISOTHIOCYANATE
>N-C≡N — NITRILE ON NITROGEN
-N=C=N- — CARBODIIMIDE
>C=C=N- — KETENE IMINE
-CH=N=N — DIAZO
-N=N=N — AZIDE
ARYL-N≡N — DIAZONIUM SALT
-N≡C — ISOCYANIDE
>C=C=O — KETENE
(C≡N)⁻ — CYANIDE ION
(N=C=O)⁻ — CYANATE ION
(N=C=S)⁻ — THIOCYANATE ION
BROAD
AROMATIC CH STRETCH
MONOSUBST. BENZENE
ORTHO DISUBST. BENZENE
META
PARA
UNSYM. TRISUBST. BENZENE
VICINAL TRI
SYM. TRI
MONO
ORTHO
META
PARA
UNSYM. TRI
VICINAL TRI
SYM. TRI
α SUBST. NAPHTHALENE
β SUBST. NAPHTHALENE
2 SUBST. PYRIDINE
3 SUBST. PYRIDINE
4 SUBST. PYRIDINE
MELAMINE
ISO TRIAZINE
TRIAZINE
FURANS
THIOPHENES
S M W M-W V
2.5 μ 3 4 5 6 7 8 9 10 12 14
4000 3600 3200 2800 2400 2000 1900 1800 1700 1600 1500 1400 1300 1200 1100 1000 900 800 700 600
CM⁻¹

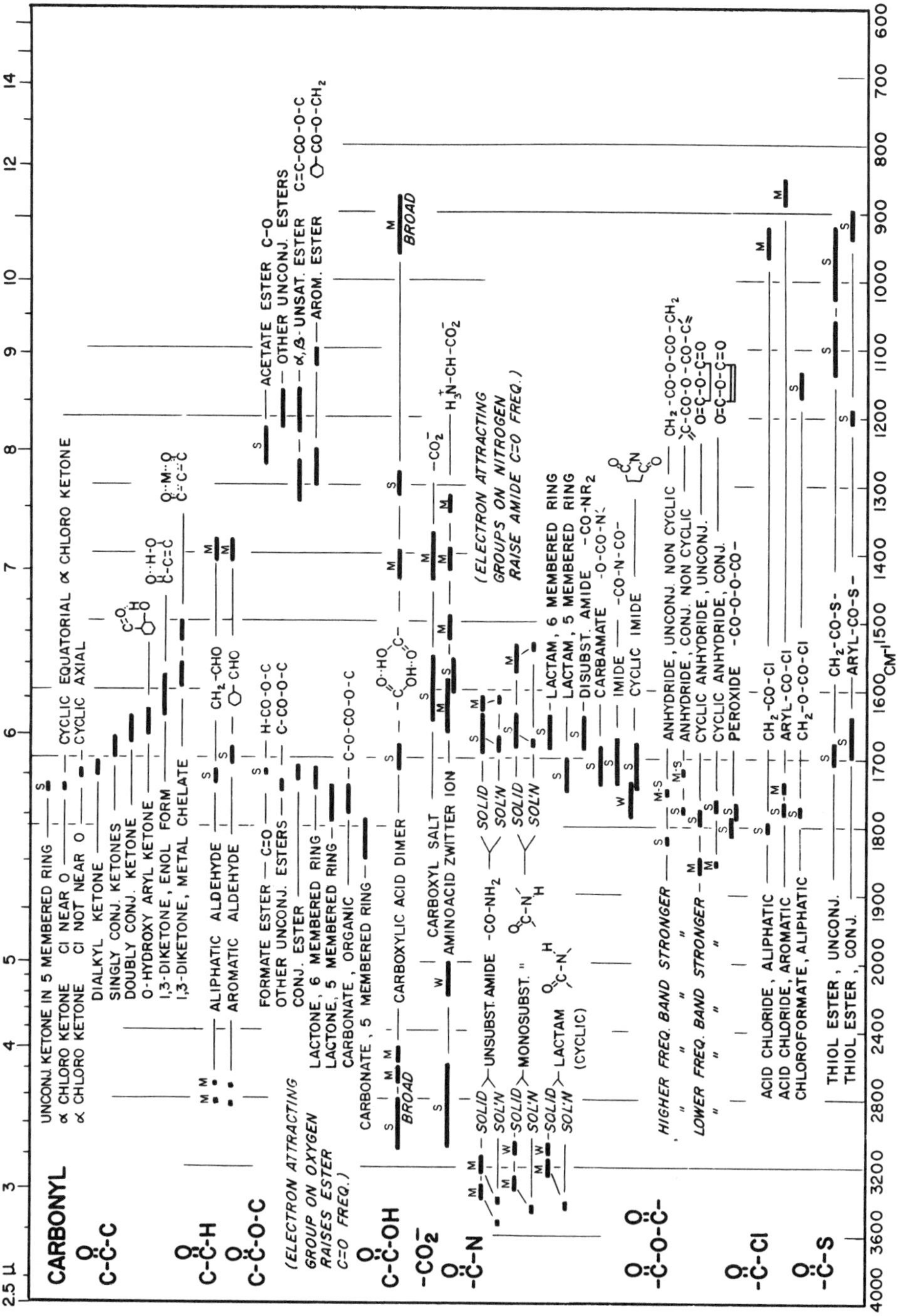
CARBONYL
UNCONJ. KETONE IN 5 MEMBERED RING
α CHLORO KETONE Cl NEAR O
α CHLORO KETONE Cl NOT NEAR O
CYCLIC EQUATORIAL α CHLORO KETONE
CYCLIC AXIAL
DIALKYL KETONE
SINGLY CONJ. KETONES
DOUBLY CONJ. KETONE
O-HYDROXY ARYL KETONE
1,3-DIKETONE, ENOL FORM
1,3-DIKETONE, METAL CHELATE
ALIPHATIC ALDEHYDE
AROMATIC ALDEHYDE
FORMATE ESTER
OTHER UNCONJ. ESTERS
CONJ. ESTER
LACTONE, 6 MEMBERED RING
LACTONE, 5 MEMBERED RING
CARBONATE, ORGANIC
CARBONATE, 5 MEMBERED RING
(ELECTRON ATTRACTING GROUP ON OXYGEN RAISES ESTER C=O FREQ.)
ACETATE ESTER C-O
OTHER UNCONJ. ESTERS
α,β-UNSAT. ESTER
AROM. ESTER
CARBOXYLIC ACID DIMER
BROAD
CARBOXYL SALT
AMINOACID ZWITTER ION
UNSUBST. AMIDE -CO-NH₂
MONOSUBST. "
LACTAM (CYCLIC)
SOLID
SOL'N
(ELECTRON ATTRACTING GROUPS ON NITROGEN RAISE AMIDE C=O FREQ.)
LACTAM, 6 MEMBERED RING
LACTAM, 5 MEMBERED RING
DISUBST. AMIDE -CO-NR₂
CARBAMATE -O-CO-N<
IMIDE -CO-N-CO-
CYCLIC IMIDE
HIGHER FREQ. BAND STRONGER
LOWER FREQ. BAND STRONGER
ANHYDRIDE, UNCONJ. NON CYCLIC
ANHYDRIDE, CONJ. NON CYCLIC
CYCLIC ANHYDRIDE, UNCONJ.
CYCLIC ANHYDRIDE, CONJ.
PEROXIDE -CO-O-O-CO-
ACID CHLORIDE, ALIPHATIC
ACID CHLORIDE, AROMATIC
CHLOROFORMATE, ALIPHATIC
THIOL ESTER, UNCONJ.
THIOL ESTER, CONJ.
CH₂-CO-S-
ARYL-CO-S-
CM⁻¹
4000 3600 3200 2800 2400 2000 1900 1800 1700 1600 1500 1400 1300 1200 1100 1000 900 800 700 600
2.5 μ 3 4 5 6 7 8 9 10 12 14

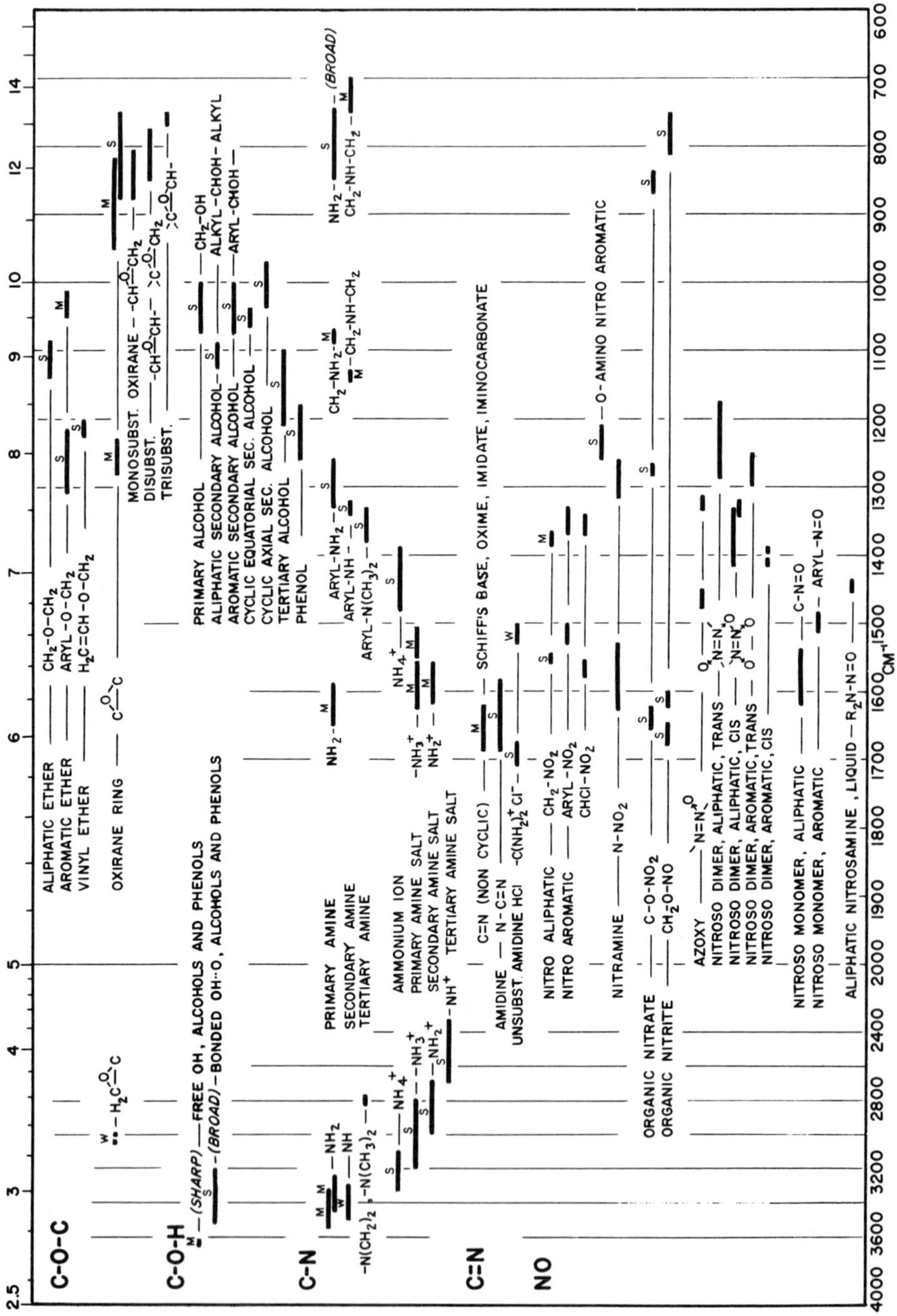
C-O-C
ALIPHATIC ETHER — CH2-O-CH2
AROMATIC ETHER — ARYL-O-CH2
VINYL ETHER — H2C=CH-O-CH2
OXIRANE RING
MONOSUBST. OXIRANE
DISUBST.
TRISUBST.
C-O-H
(SHARP) — FREE OH, ALCOHOLS AND PHENOLS
(BROAD) — BONDED OH··O, ALCOHOLS AND PHENOLS
PRIMARY ALCOHOL — CH2-OH
ALIPHATIC SECONDARY ALCOHOL — ALKYL-CHOH-ALKYL
AROMATIC SECONDARY ALCOHOL — ARYL-CHOH
CYCLIC EQUATORIAL SEC. ALCOHOL
CYCLIC AXIAL SEC. ALCOHOL
TERTIARY ALCOHOL
PHENOL
C-N
PRIMARY AMINE
SECONDARY AMINE
TERTIARY AMINE
AMMONIUM ION
PRIMARY AMINE SALT
SECONDARY AMINE SALT
TERTIARY AMINE SALT
ARYL-NH2
ARYL-NH
ARYL-N(CH3)2
CH2-NH2
CH2-NH-CH2
(BROAD)
C=N
C=N (NON CYCLIC) — SCHIFF'S BASE, OXIME, IMIDATE, IMINOCARBONATE
AMIDINE — N-C=N
UNSUBST. AMIDINE HCl
NO
NITRO ALIPHATIC — CH2-NO2
NITRO AROMATIC — ARYL-NO2
CHCl-NO2
O-AMINO NITRO AROMATIC
NITRAMINE — N-NO2
ORGANIC NITRATE — C-O-NO2
ORGANIC NITRITE — CH2-O-NO
AZOXY
NITROSO DIMER, ALIPHATIC, TRANS
NITROSO DIMER, ALIPHATIC, CIS
NITROSO DIMER, AROMATIC, TRANS
NITROSO DIMER, AROMATIC, CIS
NITROSO MONOMER, ALIPHATIC — C-N=O
NITROSO MONOMER, AROMATIC — ARYL-N=O
ALIPHATIC NITROSAMINE, LIQUID — R2N-N=O
2.5 3 4 5 6 7 8 9 10 12 14
4000 3600 3200 2800 2400 2000 1900 1800 1700 1600 1500 1400 1300 1200 1100 1000 900 800 700 600
CM-1

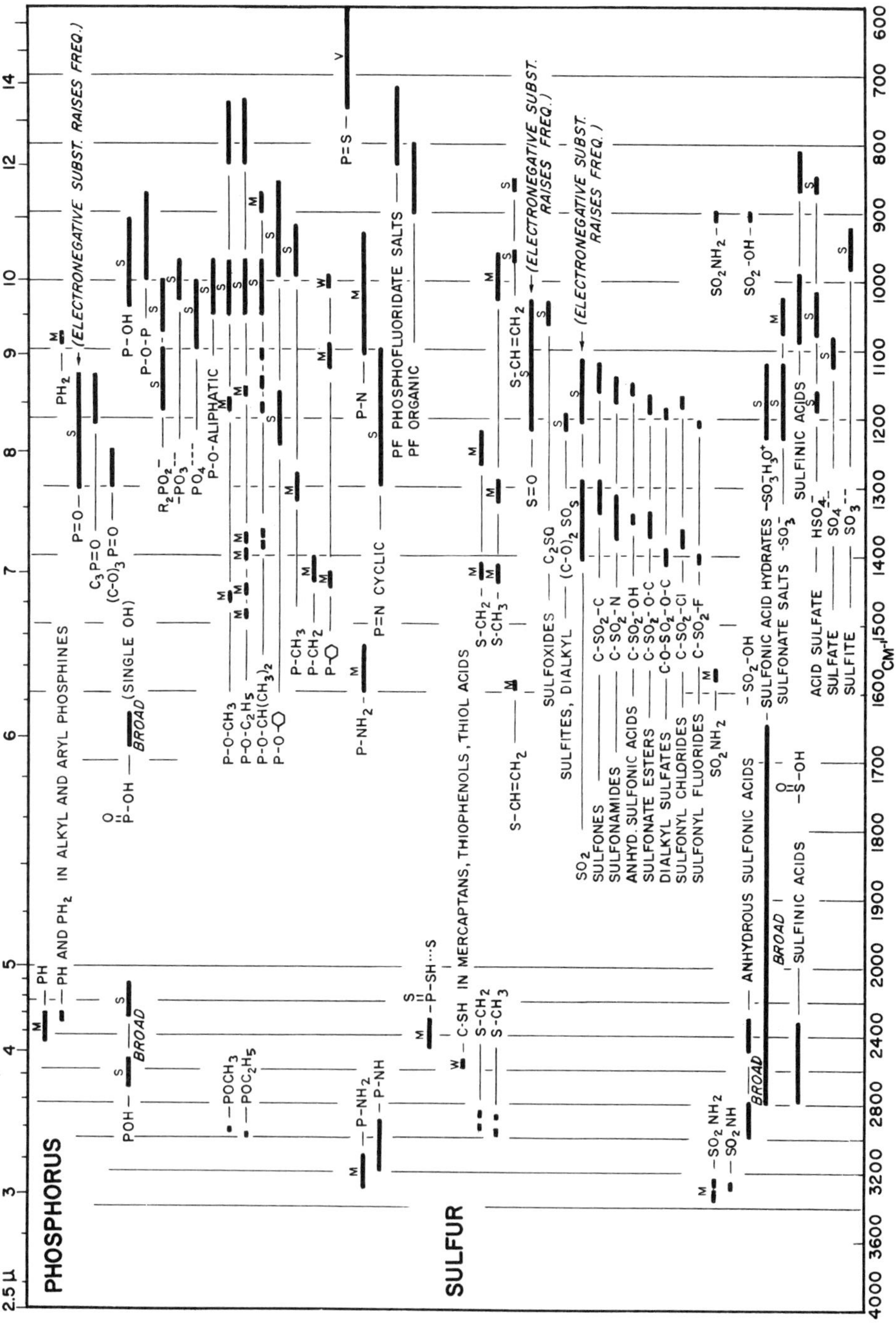
PHOSPHORUS
SULFUR
PH AND PH_2 IN ALKYL AND ARYL PHOSPHINES
(ELECTRONEGATIVE SUBST. RAISES FREQ.)
P-O-ALIPHATIC
PF PHOSPHOFLUORIDATE SALTS
PF ORGANIC
P=N CYCLIC
C-SH IN MERCAPTANS, THIOPHENOLS, THIOL ACIDS
SULFOXIDES
SULFITES, DIALKYL
SULFONES
SULFONAMIDES
ANHYD. SULFONIC ACIDS
SULFONATE ESTERS
DIALKYL SULFATES
SULFONYL CHLORIDES
SULFONYL FLUORIDES
ANHYDROUS SULFONIC ACIDS
SULFONIC ACID HYDRATES
SULFONATE SALTS
SULFINIC ACIDS
ACID SULFATE
SULFATE
SULFITE
CM⁻¹

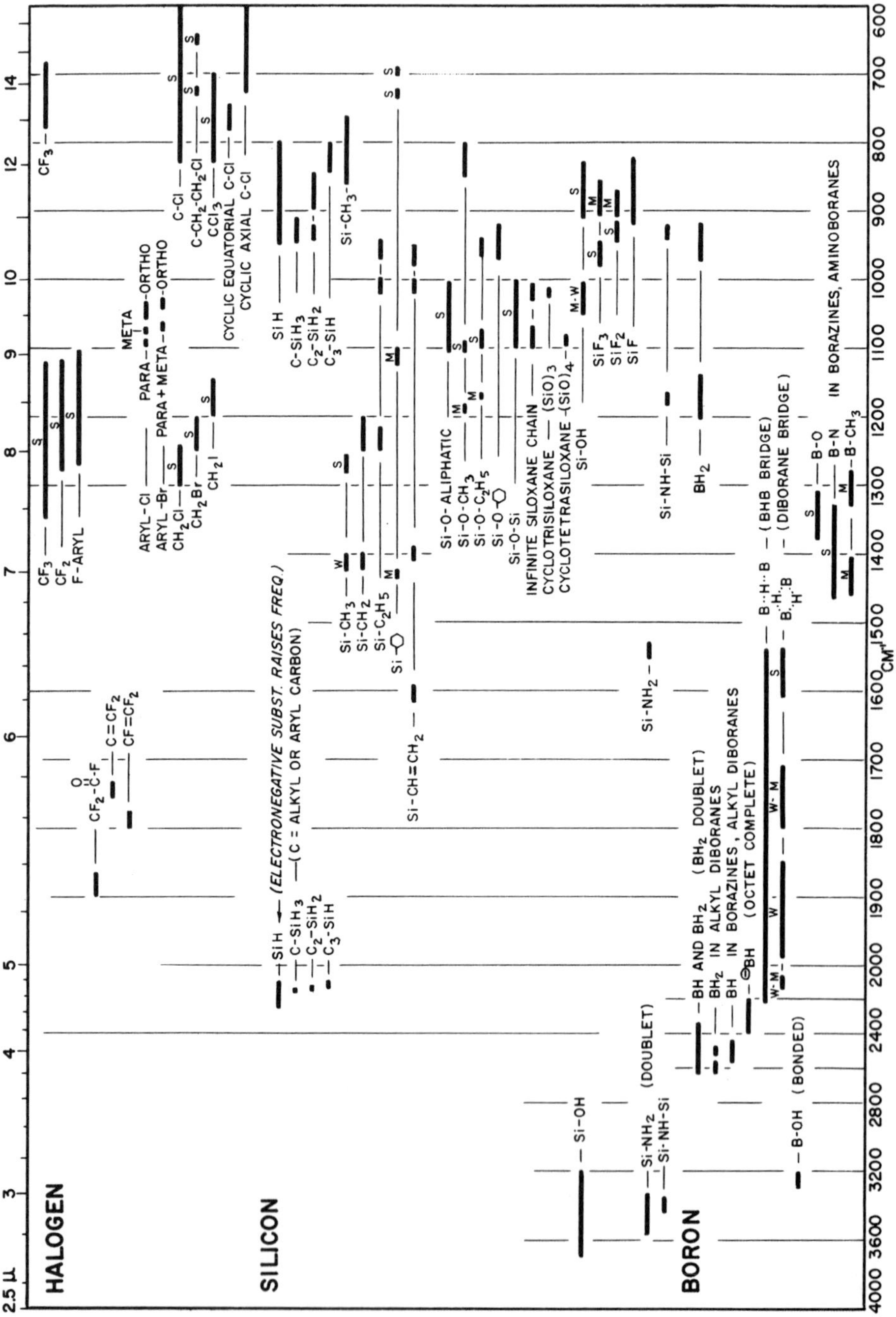
HALOGEN
SILICON
BORON
CF3
CF2
F-ARYL
ARYL-Cl
ARYL-Br
CH2Cl
CH2Br
CH2I
PARA
META
ORTHO
PARA + META
C-Cl
C-CH2-CH2-Cl
CCl3
CYCLIC EQUATORIAL C-Cl
CYCLIC AXIAL C-Cl
C=CF2
CF=CF2
SiH
C-SiH3
C2-SiH2
C3-SiH
(ELECTRONEGATIVE SUBST. RAISES FREQ.)
(C = ALKYL OR ARYL CARBON)
Si-CH3
Si-CH2
Si-C2H5
Si-CH=CH2
Si-O-ALIPHATIC
Si-O-CH3
Si-O-C2H5
Si-O-Si
INFINITE SILOXANE CHAIN
CYCLOTRISILOXANE
CYCLOTETRASILOXANE
Si-OH
SiF3
SiF2
SiF
Si-NH-Si
Si-NH2
(DOUBLET)
BH2
BH AND BH2 (BH2 DOUBLET)
BH2 IN ALKYL DIBORANES
BH IN BORAZINES, ALKYL DIBORANES
(OCTET COMPLETE)
(BHB BRIDGE)
(DIBORANE BRIDGE)
B-OH (BONDED)
B-O
B-N
B-CH3
IN BORAZINES, AMINOBORANES
2.5 μ
3
4
5
6
7
8
9
10
12
14
4000 3600 3200 2800 2400 2000 1900 1800 1700 1600 1500 1400 1300 1200 1100 1000 900 800 700 600
CM⁻¹

Alphabetical Name Index of Spectra

Molecular Formula Index of Spectra

Molecular Formula	Name	Page No.
A. Compounds		
CS_2	carbon disulfide	457
CH_3NO	formamide	439
C_2Cl_6	ethane, hexachloro-	318
$C_2H_3O_2NaO_2$	acetic acid, sodium salt (sodium acetate)	359
C_2H_4OS	acetic acid, thio-	358
$C_2H_4O_2$	acetic acid	355
$C_2H_4O_2$	formic acid, methyl ester (methyl format)	440
C_2H_5NO	formamide, *N*-methyl-	440
C_2H_6OS	methyl sulfoxide	396
$C_2H_6O_3S$	dimethyl sulfite	397
C_3H_3N	acrylonitrile	338
$C_3H_4N_2$	pyrazole	422
C_3H_5NS	thiocyanic acid, ethyl ester	342
C_3H_6O	propane, 1,2-epoxy-	327
$C_3H_6O_2$	acetic acid, methyl ester	348
C_3H_7NO	formamide, *N*,*N*-dimethyl-	441
$C_3H_7NO_2$	propane, 1-nitro-	382
$C_3H_7NO_3$	propyl nitrate	389
C_3H_9N	propylamine	372
$C_4H_2N_2$	fumaronitrile (trans)	340
$C_4H_2O_3$	maleic anhydride	361
$C_4H_4N_2O_2$	pyrazine, 1,4-dioxide	386
C_4H_4O	furan	421
C_4H_4S	thiophene	453
$C_4H_6O_2$	acetic acid, vinyl ester	350
$C_4H_6O_2$	butyrolactone	353
C_4H_7N	butyronitrile	339
C_4H_8O	2-butanone	345
C_4H_8O	butyraldehyde	343
$C_4H_8O_2$	*p*-dioxane	326
C_4H_9NO	acetamide, *N*-ethyl-	365
C_4H_9NO	butyraldehyde, oxime	387
C_4H_9NO	butyramide	364
C_4H_9NO	propane, 2-methyl-2-nitroso-	384
$C_4H_9O_3SNa$	1-butanesulfonic acid, sodium salt	402

B. Polymers

Index

A

B

C

D

E

F

G

H

I

K

L

M

N

O

P

U

V

W

X

ISBN 0-12-451160-0

90018

9 780124 511606